To Ed,

With my ver ░░░░░ her,
admiration, and love,

Mitt

Physics of Charged Macromolecules

Charged molecules percolate all aspects of our modern lives including food, health care, and water-based technologies. As a concise introduction to the physics of charged macromolecules, this book covers the basics of electrostatics as well as cutting-edge modern research developments. This accessible book provides a clear and intuitive view of concepts and theory, and features appendices detailing mathematical methodology. Supported by results from real-world experiments and simulations, this book equips the reader with a vital foundation for performing experimental research. Topics include living matter and synthetic materials including polyelectrolytes, polyzwitterions, polyampholytes, proteins, intrinsically disordered proteins, and DNA/RNA. Serving as a gateway to the growing field of charged macromolecules and their applications, this concept-driven book is a perfect guide for students beginning their studies in charged macromolecules, providing new opportunities for research and discovery.

Murugappan Muthukumar is the Wilmer D. Barrett Professor of Polymer Science and Engineering at the University of Massachusetts Amherst. He is a fellow of the American Physical Society and a recipient of the Dillon Medal and the Polymer Physics Prize of the American Physical Society, as well as the American Chemical Society Polymer Chemistry Award.

Physics of Charged Macromolecules

Synthetic and Biological Systems

MURUGAPPAN MUTHUKUMAR
University of Massachusetts Amherst

CAMBRIDGE
UNIVERSITY PRESS

Shaftesbury Road, Cambridge CB2 8EA, United Kingdom

One Liberty Plaza, 20th Floor, New York, NY 10006, USA

477 Williamstown Road, Port Melbourne, VIC 3207, Australia

314–321, 3rd Floor, Plot 3, Splendor Forum, Jasola District Centre, New Delhi – 110025, India

103 Penang Road, #05–06/07, Visioncrest Commercial, Singapore 238467

Cambridge University Press is part of Cambridge University Press & Assessment, a department of the University of Cambridge.

We share the University's mission to contribute to society through the pursuit of education, learning and research at the highest international levels of excellence.

www.cambridge.org
Information on this title: www.cambridge.org/9780521864879

DOI: 10.1017/9781139046749

First published 2023

A catalogue record for this publication is available from the British Library.

Library of Congress Cataloging-in-Publication Data
Names: Muthukumar, Murugappan, author.
Title: Physics of charged macromolecules : synthetic and biological systems / Murugappan Muthukumar.
Description: Cambridge ; New York, NY : Cambridge University Press, 2023. | Includes bibliographical references.
Identifiers: LCCN 2022025102 | ISBN 9780521864879 (hardback) | ISBN 9781139046749 (ebook)
Subjects: LCSH: Macromolecules – Electric properties.
Classification: LCC QD381.9.E38 M88 2023 | DDC 547/.70457–dc23/eng20221123
LC record available at https://lccn.loc.gov/2022025102

ISBN 978-0-521-86487-9 Hardback

Contents

Preface

Charged macromolecules constitute the vocabulary used by Mother Nature to express life as we know it. These molecules are also abundant in the synthetic world. Yet, understanding the behavior of charged macromolecules is one of the grand challenges of the biological and physical sciences. The difficulty of this challenge lies in several long-range forces operating simultaneously that endow assemblies of charged macromolecules with amazing functions. The electrostatic forces, topological correlations emanating from macromolecular connectivity, and hydrodynamics are all long-ranged. These aspects collectively guide the structure, dynamics, and movement of individual macromolecules and their assemblies. In view of the prevalence of charged macromolecules in nature, extensive phenomenology has been cultivated during the past many decades. The rich collective behaviors of charged macromolecules are unique to the presence of charges and deviate significantly from those of uncharged macromolecular systems. In efforts to understand these behaviors, there has been extensive theoretical effort during the past seven decades with variable success. More and more theories are being vigorously pursued, with different levels of assumptions and approximations, in order to obtain a comprehensive description of the behavior of charged macromolecules.

This complicated field of charged macromolecules inevitably demands highly sophisticated mathematical techniques accompanied by clear physical pictures. The goal of this book is to focus on the most important concepts pertinent to electrolyte solutions, charged macromolecules, and their assemblies. Heavy mathematics and fine details of theories are relegated to the original literature. I am painfully aware that I have left out many interesting developments particularly in theoretical aspects. Most of the material in this book deals with fundamentals and general concepts that are yet to enter textbooks.

This book is an introduction to the vast field of charged macromolecules practiced by biologists, physicists, chemists, and chemical engineers. It is written at the level of an entering graduate student who is interested in the burgeoning fields of living matter, soft matter physics, polyelectrolyte physics, and biotechnology. The book is structured in a manner that it provides a gateway to a large number of topics at the center of the physics of charged macromolecules. The goal is to provide a common language of fundamental concepts to a broader audience, independent of the expertise of the reader. Creating a common conceptual framework in such a difficult subject for use by a diverse readership is not an easy task. However, I have attempted to start from the basics of charged and uncharged polymers, and then I have combined these concepts to describe the collective behavior of charged systems. Effort is made to make the discussion qualitative and concept based. Technical details of a few important concepts are provided in appendices. Highly sophisticated mathematical details are referred to original publications.

The outline is as follows: After an exposure to the scope of the topics in Chapter 1, the second chapter provides a synopsis of models of uncharged polymer chains in isolation and experimental results in dilute solutions. Chapter 2 is a convenient introduction to readers who might be interested solely in basic polymer physics. The next two chapters, Chapters 3 and 4, deal with the basics of electrostatics, dielectric media, interactions among electrolyte ions, and the nature of electrolyte solutions with physical boundaries. Chapter 5 is devoted to a survey of experimental results in dilute solutions of charged macromolecules and various theoretical approaches to comprehend these facts. Chapter 6 describes how the fundamental principles developed in dilute solutions are modified by crowding of charged macromolecules in homogeneous nondilute solutions. This situation has been a long-standing challenge to understand requiring advanced theoretical apparatus. Readers interested in biological systems in dilute conditions can skip this chapter without losing the thread of concepts. The dynamics of solutions of charged macromolecules is dealt with in Chapter 7. This chapter, along with Appendix 6, constitutes a separate mini-book of their own and deals with the dynamics and mobility of charged rigid particles and flexible macromolecules. Chapter 8 deals with the fundamentals of liquid–liquid phase separation in solutions of charged macromolecules after addressing the situation with uncharged macromolecules. Micellization, fibrillization, and microphase separation are also dealt with in Chapter 8. Applications of the various concepts developed in the above chapters to the phenomena of adsorption, virus packing, and coacervation are presented in Chapter 9. The final chapter is on charged gels, which are ubiquitous in health care and other industries. The gap between what we understand today and what needs to be accomplished in the future is briefly mentioned in the Epilogue.

I hope the readers will benefit from this introduction to the field and implement the main concepts given here in their own journeys with charged macromolecules.

Acknowledgments

I want to thank my students and postdocs for the intellectual stimulation they have challenged me with over the past two decades and more. Without their collaboration, this book would not be possible. In particular, I am grateful to Khatcher Margossian for reading almost every page and contributing valuable suggestions to improve the clarity of presentation. Most significantly, his encouragement and conviction for the need for such a book catalyzed the completion of this work. I am also grateful to Siao-Fong Li who has gone through most of the derived equations. My special thanks are to Alexis Batakis for drawing most of the figures in the book.

I also gratefully acknowledge support from the National Science Foundation for my research on charged macromolecules. Without such support, this book would not have been a possibility. I am also indebted to many collaborators outside my laboratory. In particular, it is a pleasure to thank Professor Manfred Schmidt for his friendship and numerous stimulating discussions on the intricacies of polyelectrolyte behavior.

Finally, it is my greatest pleasure to thank my wife Lalitha for all support she has provided throughout my career. During the difficult time of writing this book, I was fortunate to spend time with my grandson Kanna who inspired boundless energy, imagination, and purity of thought. I am immensely grateful to him for support and play.

1 Introduction to Charged Macromolecules

1.1 General Premise

Charged macromolecules are everywhere in nature. Examples in biological contexts are deoxyribonucleic acid (DNA), ribonucleic acid (RNA), polysaccharides, and proteins, which are dispersed in aqueous solutions usually under crowded conditions. These macromolecules have existed since the dawn of life on earth some billions of years ago, as they constitute the makeup of various organisms and their functions. There is also a plethora of synthetic charged macromolecules such as polyacrylic acid, poly(styrene sulfonate), and poly(vinyl pyridinium) salts, which are of significance in the materials world. These charged macromolecules have led to water-based materials with amazing attributes in the healthcare industry and biotechnology. A proper understanding of the structure and dynamics of charged macromolecules and their collections is a difficult challenge due to multiple forces acting simultaneously over long distances. In this overview chapter, we shall present the key conceptual features responsible for the behavior of charged macromolecules and some tantalizing experimental data exhibited by charged macromolecules in water.

Charged macromolecules are made up of many electrically charged monomers that are connected contiguously into long polymer chains. The examples of single-stranded DNA (ssDNA), a polysaccharide, a polypeptide, and poly(styrene sulfonate) are illustrated in Fig. 1.1. Each repeat unit bears a charge as indicated by the circles in the figure.

We begin with the nomenclature of these macromolecules. We shall refer to the charged macromolecules made up of identical repeat units as **homo polyelectrolytes**. Typically, the skeleton of such polyelectrolytes is flexible under physiologically relevant conditions. Therefore, these are called **flexible polyelectrolytes**. In contrast to homo polyelectrolytes, **hetero polyelectrolytes** are composed of more than one kind of charged repeat unit. Classic examples are DNA and RNA. The incipient characteristic of these information-carrying charged macromolecules is their ability to form hydrogen bonds among the bases of the various repeat units, as illustrated in Fig. 1.2a for double-stranded DNA (dsDNA) and in Fig. 1.2b for single-stranded RNA (ssRNA). The base pairing in dsDNA results in the formation of double helices where the continuous backbones are stiff, in contrast to flexible ssDNA. Such molecules are called **semiflexible polyelectrolytes**, and they can even be rod-like if their contour

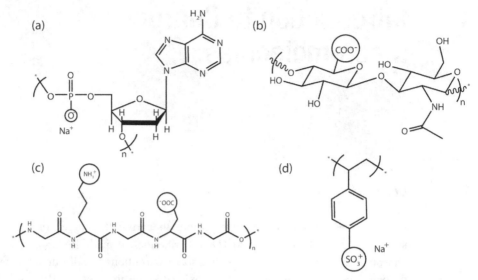

Figure 1.1 Chemical structures of (a) poly(deoxyadenylic acid), (b) the polysaccharide hyaluronate, (c) a polypeptide, and (d) poly(styrene sulfonate), showing the contiguous chain connectivity of charged repeat units. The charged groups are indicated by circles.

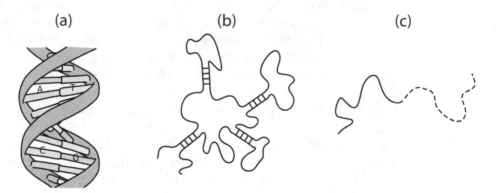

Figure 1.2 (a) Base pairing due to hydrogen bonding in dsDNA results in a stiff polyelectrolyte. (b) Base pairing in ssRNA results in a branched polyelectrolyte. (c) Diblock copolymer of poly(styrene sulfonate) (solid curve), which is charged, and uncharged poly(ethylene oxide) (broken curve).

length is short. On the other hand, base pairing within a single chain, as in ssRNA depicted in Fig. 1.2b, results in highly branched structures. In such structures, the hydrogen-bonded stems function as branch points. In addition, the branch points in a branched architecture can arise from either chemical cross-linking or physical association between several chemical groups. Such branched molecules either with physical association or chemical cross-linking are labeled as **branched polyelectrolytes**. In terms of synthetic polyelectrolytes, the simplest example of a hetero polyelectrolyte is a diblock copolymer with one block being a polyelectrolyte and the other block

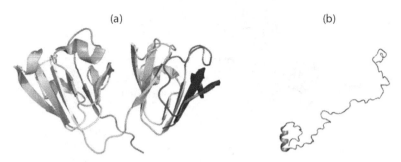

Figure 1.3 (a) Folded state of the human γ-D crystallin protein. (b) An example of intrinsically disordered proteins.

being an uncharged polymer, as illustrated in Fig. 1.2c for the poly(styrene sulfonate-co-ethylene oxide) diblock copolymer. Such synthetic systems offer a broad range of opportunities for formulating novel self-assembling structures of technological relevance.

In the examples of Figs. 1.1a, 1.1b, and 1.1d, all repeat units bear charges of the same sign (as in the phosphate group in Fig. 1.1a). If a polymer is made up of positively charged repeat units, it is called a polycation, and if it is made up of negatively charged repeat units, it is called a polyanion. There is another class of charged macromolecules, as shown in Fig. 1.1c, where each molecule has both positive and negative charges in a certain sequencing order, as exemplified by proteins. Macromolecules containing both positive and negative charges are known as **polyampholytes**. Depending on the extent and sequence of positive and negative charges and their tunability by the ionic environment through pH, salt concentration, and hydrophobicity, such molecules can either appropriately fold into stable tertiary structures (Fig. 1.3a) or lack this ability by forming intrinsically disordered proteins (Fig. 1.3b).

For polyelectrolytes and polyampholytes, ionization equilibria of the repeat units in the polymer sequences dictate where the charges ought to be present at a given pH and salt concentration. A prescription of specific charge locations along the chain contour is referred to as **quenched** charge distribution. In reality, however, the charge on a particular repeat unit can fluctuate between on and off over a period of time depending on the local electrostatic environment. As a result, the charges on the polymer can be imagined to be smeared over the chain backbone. Such a description is called **annealed** charge distribution. Almost all theoretical formulations on charged macromolecules treat only this class of charged macromolecules.

The above descriptions of polyelectrolytes are only for the backbone of the chains, which cannot exist on their own, since the accumulation of similar charges on one molecule would be energetically prohibitive. Because the whole polyelectrolyte solution must be electrically neutral in equilibrium, there is an equal number of oppositely charged ions, called **counterions**, to balance the number of ionized groups of the polyelectrolyte chains. Depending on the temperature, the amount and identity of added salt, and the polarizability of the dipolar aqueous background medium, small ions hover around the polyelectrolyte chains to variable degrees. Furthermore, hydrogen

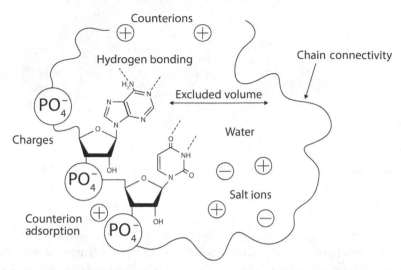

Figure 1.4 Sketch of a portion of ssRNA in salty water emphasizing the simultaneous presence of various factors: chain connectivity, charges on the repeat units, counterions, salt ions, ion-pair formation between charged repeat units and the counterions, polarizability of water, hydrogen bonding, and van der Waals type excluded volume interaction.

bonding, structural reorganization of water, and short-ranged van der Waals interactions (called **excluded volume interactions**) are also present. As an example, a small portion of an isolated ssRNA molecule in salty water is sketched in Fig. 1.4. The system is endowed with the omnipresent electrostatic interaction between charges (phosphate groups, counterions, and salt ions), hydrogen bonding, dipolar interactions, and excluded volume interactions. The confluence of all of these attributes in addition to the topological character of chain connectivity is responsible for the structure and dynamics of the system. This description of the various forces is not unique to only ssRNA portrayed in Fig. 1.4 but is ubiquitous to all charged macromolecules in aqueous solutions.

In general, charged macromolecules dispersed in aqueous solutions are strongly correlated. Every molecule interacts with every other molecule in the entire system. The various structural and functional properties of solutions of macromolecules are collective behaviors involving essentially all molecules in the system mediated by its electrolyte background. The origin of the various distinct properties of charged macromolecules, which are qualitatively different from those of uncharged polymers, lies in the long-ranged electrostatic interactions among the various molecules and ionic constituents of the system. Modulations of these Coulomb forces by van der Waals type interactions are responsible for the collective properties of charged macromolecules. In crowded environments, as are common in many biological situations, the system appears as a Coulomb soup. The structure of charged macromolecules, their self-assembly into large-scale mesoscopic structures, and the movement of the various macromolecules in such a Coulomb soup form the subject of this book.

Table 1.1 Both polymeric and ionic properties control polyelectrolyte precipitation

Solute	$BaCl_2$	$AlCl_3$
NaPSS	Yes	Yes
$ClSO_3H$	Yes	No
p-Toluenesulfonate	Yes	No

The behavior of charged macromolecules in solutions cannot be separately attributed to either solely polymeric properties or solely ionic properties. As an example, consider separate aqueous solutions of sodium poly(styrene sulfonate) (NaPSS), chlorosulfonic acid ($ClSO_3H$), and p-toluenesulfonate, each containing either $BaCl_2$ or $AlCl_3$ at pH 3.2 and room temperature [Narh & Keller (1993)]. Except for the solutions of chlorosulfonic acid and p-toluenesulfonate containing $AlCl_3$, all other solutions yield solid precipitates under these experimental conditions, as given in Table 1.1 (where "Yes" and "No" indicate the presence and absence of a precipitate, respectively). The ionic groups of NaPSS, $ClSO_3H$, and p-toluenesulfonate are essentially the same, while NaPSS is a polymer and the other two are monomers. As shown in Table 1.1 under the column of $BaCl_2$, precipitation in these three solutions indicates that the precipitation is an ionic property and that the polymer nature of NaPSS is not critically relevant. On the other hand, NaPSS precipitates in $AlCl_3$ solutions, whereas $ClSO_3H$ and p-toluenesulfonate do not, as shown in Table 1.1 under the column of $AlCl_3$. This indicates that precipitation in $AlCl_3$ solutions is essentially a polymer property. Thus, polyelectrolyte behavior must be a combination of both polymeric and ionic properties. We briefly introduce the basic features of polymeric and ionic properties in the following two sections, before deeply dwelling on them in the following chapters.

1.2   Chain Connectivity and Long-Ranged Topological Correlation

Consider an uncharged large flexible macromolecule of N repeat units in isolation. On average, it adopts a coil-like conformation, as depicted in Fig. 1.5a, due to chain connectivity. It appears as a rough sphere. A measure of its physical size is its radius of gyration R_g. In general, the dependence of R_g on N (which is proportional to the molecular weight of the polymer) is given by the proportionality relation [de Gennes (1979)],

$$R_g \sim N^\nu, \qquad (1.2.1)$$

where ν is called the size exponent. The value of ν depends on the chemical details of the repeat units of the polymer and the solvent, and temperature, whose roles are collectively called **excluded volume interactions**. The range of values of ν is between $1/3$ and 1, corresponding to the extreme limits of globule-like and rod-like conformations for the chain. The local monomer density $\rho(r)$ at radial distance r from

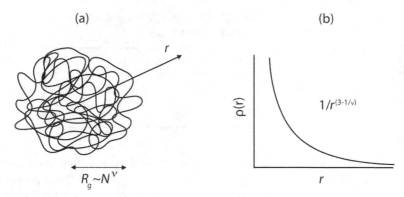

Figure 1.5 (a) Random coil conformation of an uncharged flexible chain with the radius of gyration R_g, which depends on the number of repeat units N as $R_g \sim N^\nu$, where $(1/3 < \nu < 1)$. (b) Long-ranged topological correlation due to chain connectivity, as shown by the dependence of local monomer density $\rho(r)$ on the radial distance r.

the center of mass of the chain decays with r as an inverse power law [de Gennes (1979)],

$$\rho(r) \sim \frac{1}{r^{3-1/\nu}}, \tag{1.2.2}$$

as sketched in Fig. 1.5b. The value of the exponent in the power-law decay depends on the value of ν. For example, for a flexible chain without consideration of excluded volume interactions, $\nu = 1/2$, and hence r dependence of $\rho(r)$ is long-ranged,

$$\rho(r) \sim \frac{1}{r}. \tag{1.2.3}$$

This long-ranged correlation of monomer density is entirely due to chain connectivity. Hence, it is called the topological correlation of the chain. The quantitative nature of the topological correlation of the chain given in Eq. (1.2.2) is modified by both the excluded volume interactions and electrostatic interactions as represented through the size exponent ν.

1.3   Scales of Energy, Length, and Time in Charged Systems

Energy: Consider the electrostatic interaction energy U_{ij} between a pair of ions with charges $z_i e$ and $z_j e$ separated by a distance r_{ij} (Fig. 1.6a) in a medium with a uniform dielectric constant ϵ (in the absence of any other charges). This is given by Coulomb's law,

$$\frac{U_{ij}}{k_B T} = z_i z_j \frac{e^2}{4\pi\epsilon_0\epsilon k_B T} \frac{1}{r_{ij}}, \tag{1.3.1}$$

where $k_B T$ is the Boltzmann constant times the absolute temperature, z_i is the valency of the ith ion, e is the electronic charge, and ϵ_0 is the permittivity of vacuum. The

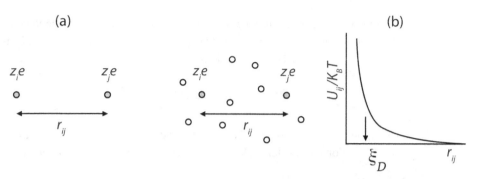

Figure 1.6 (a) The interaction energy between two ions of charges $z_i e$ and $z_j e$ in the absence of any other charges in their neighborhood is the long-ranged Coulomb interaction energy. (b) In the presence of other ions, shown as open circles, the electrostatic interaction energy between two ions is a screened Coulomb energy, also known as the Debye–Hückel electrostatic energy. The range ξ_D over which the electrostatic interaction decays by $1/e$ is the Debye length.

middle factor in Eq. (1.3.1) has the dimension of length and is called the **Bjerrum length** ℓ_B,

$$\ell_B \equiv \frac{e^2}{4\pi\epsilon_0\epsilon k_B T}. \qquad (1.3.2)$$

If the separation distance r_{ij} is comparable to ℓ_B, then the electrostatic interaction energy of two monovalent ions ($z_i = 1 = z_j$) is comparable to the thermal energy $k_B T$, as is evident from Eq. (1.3.1). For oppositely charged ions, if the distance between them is shorter than ℓ_B, then the attraction is stronger than $k_B T$, and the ions are more likely to be ion pairs instead of completely dissociated ions. The Bjerrum length sets the scale for energy. It is inversely proportional to ϵT, where ϵ itself is temperature dependent. For aqueous solutions at 25°C, the Bjerrum length (with ϵ =80) is

$$\ell_B \simeq 0.7 \text{ nm}. \qquad (1.3.3)$$

Length: When other ions are present in the solution, the electrostatic interaction energy between a pair of ions is no longer the Coulomb energy given by Eq. (1.3.1). Instead, it is a screened Coulomb energy. According to the approximate theory of Debye and Hückel for electrolyte solutions, the pairwise interaction energy between two ions of charges $z_i e$ and $z_j e$ in an electrolyte solution is given by the screened Coulomb energy,

$$\frac{U_{ij}}{k_B T} = z_i z_j \ell_B \frac{e^{-\kappa r_{ij}}}{r_{ij}}, \qquad (1.3.4)$$

where

$$\kappa^2 = \frac{e^2}{\epsilon_0\epsilon k_B T} \sum_i z_i^2 n_{i0}, \qquad (1.3.5)$$

where n_{i0} is the average number concentration of the ith ion. While the strength of the electrostatic interaction between ions is given by the Bjerrum length, its range is given by κ^{-1}, known as the **Debye length** $\xi_D \equiv \kappa^{-1}$,

$$\xi_D = \left(\frac{e^2}{\epsilon_0 \epsilon k_B T} \sum_i z_i^2 n_{i0} \right)^{-1/2}. \tag{1.3.6}$$

The range decreases if ϵT is reduced and/or if the electrolyte concentration is increased. For monovalent salts in water ($\epsilon = 80$ at $25\,°C$), the Debye length is

$$\xi_D \simeq \frac{0.3}{\sqrt{c_s}} \text{ nm}, \tag{1.3.7}$$

where c_s is the salt concentration in units of moles per liter. The screened potential (Debye–Hückel potential) given by Eq. (1.3.4) is only approximate. Nevertheless, the Debye length is a fundamental scale of length in the treatment of polyelectrolyte solutions. For interionic distances larger than ξ_D, the electrostatic interaction energy between any pair of ions is weak; for distances shorter than ξ_D, the electrostatic interaction energy is strong. Thus, the Debye length denotes the range of distance over which a pair of ions interact significantly between them, as sketched in Fig. 1.6b.

The above two fundamental parameters (ℓ_B and ξ_D) have very important consequences in the physics of solutions of all kinds of charged macromolecules. Consider a polyelectrolyte solution at a finite concentration. Although the polyelectrolyte molecule is fully ionizable, the chance that a counterion would be near a monomer and be bound to the polymer is finite, if the distance between the monomer and the counterion is less than ℓ_B. In particular, the effective dielectric constant in the neighborhood of the chain backbone can be quite low and hence the binding energy is very high, as can be seen from Eq. (1.3.1).

As a result, a certain number of counterions are bound to the polymer as depicted in Fig. 1.7, and the naked charge of the molecule is never realizable in equilibrium at finite polyelectrolyte concentrations and ambient temperatures. Furthermore, when a counterion is bound to the monomer of the polymer, a dipole is formed. Such dipoles are only temporary but can be sufficiently long-lived to interact with other dipoles formed along the chain contour. These dipole–dipole interactions can be quite strong compared to $k_B T$. For example, for two freely rotating dipoles $\mathbf{p}_1$ and $\mathbf{p}_2$ of unit charge separated by the distance r in an electrolyte solution, the angularly averaged interaction energy is attractive (Appendix 3),

$$\frac{U_{\text{dipole–dipole}}(r)}{k_B T} = -\frac{p_1^2 p_2^2 \ell_B^2}{3 r^6} e^{-2\kappa r} \left[1 + 2\kappa r + \frac{5}{3}(\kappa r)^2 + \frac{2}{3}(\kappa r)^3 + \frac{1}{6}(\kappa r)^4 \right]. \tag{1.3.8}$$

For typical physiological values of monovalent salt concentrations of about 150 mM and a typical length of the ion pairs, the effective pairwise interaction energy is about $10\,k_B T$. The dipole–dipole interaction energy can therefore be much higher than the thermal energy. In view of such strong attraction energies associated with dipole–dipole pairs, we readily anticipate that some segments of the chains would cling together due to the formation of quadrupoles and the rest of the chains would repel

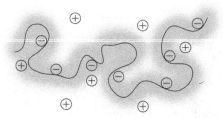

Figure 1.7 Counterions can adsorb on chain backbone forming temporary dipoles. The shaded area denotes the local environment around the oil-like chain backbone where the local dielectric constant can be substantially lower than the bulk value away from the chain.

each other. Naturally, such a scenario can result in nonuniform structures and even attraction between similarly charged polymers.

Time: Another message conveyed by Fig. 1.7 is that there is a hierarchy of time scales: diffusion time for a free counterion, lifetime of an adsorbed counterion, segmental relaxation time, characteristic relaxation time for the whole polymer, characteristic time for the relaxation of the collection of dipole pairs, entanglement time, etc. As a result, the dynamics of polyelectrolyte chains and the transport properties of polyelectrolyte solutions exhibit very rich phenomenology. In qualitative terms, sometimes the large macromolecules follow the motion of small free counterions, and some other times, the small ions hover over the large macromolecules, both by maintaining electroneutrality over a reasonably small volume.

1.4 Confluence of Electrostatic and Topological Correlations

The cooperative features of chain connectivity and electrostatic interactions among all charged species in the system lead to amazing properties as evident in various biological processes in life itself. Many challenges arise in a fundamental understanding of the behavior of charged macromolecules. We now briefly mention a few representative examples of some of the enigmatic properties displayed by charged macromolecules.

1.4.1 Similarly Charged Macromolecules Can Attract Each Other

As the polymer concentration of dsDNA of molar mass 10^5 Da in a 0.1 M NaCl solution is increased, the macromolecules can aggregate into large clusters of micrometer size and settle down at the bottom of the solution over a period of time [Wissenburg *et al.* (1995)]. This phenomenon is seen through cryoelectron microscopy (Fig. 1.8a) and light scattering (Fig. 1.8b). In Fig. 1.8b, the ordinate is proportional to the inverse of scattered intensity from the solution and the abscissa is the concentration of dsDNA. At a DNA concentration of about 40 g/L or above, thermodynamic instability sets

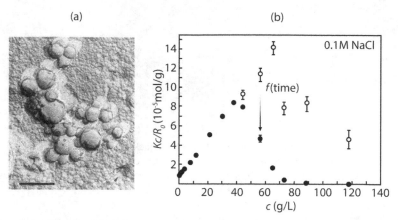

(a)   (b)

Figure 1.8 dsDNA molecules can aggregate although each molecule bears the same negative charge. (a) Electron micrograph for a 73 g/L DNA solution at 0.01 M NaCl. The bar corresponds to 300 nm. (b) Inverse scattering intensity versus DNA concentration c at 0.1 M NaCl. Thermodynamic instability and precipitation occur above a threshold value of c, and depend on time history, f(time). Open symbols denote the condition after filtration of the precipitate and the filled symbols denote the condition before filtration [Wissenburg *et al.* (1995)].

in and the solution is not homogeneous anymore. The precipitation depends on time history, f(time), as indicated in Fig. 1.8b. This clumping phenomenon is surprising because dsDNA molecules are all similarly charged and are expected to repel each other to remain fully dispersed in the solution. Even more remarkably, the instability occurs at lower DNA concentrations if the salt concentration is lower where the electrostatic repulsion among the polymer chains is expected to be higher.

1.4.2   Ordinary–Extraordinary Transition

In dilute solutions of uncharged macromolecules, the measured diffusion coefficient D using the dynamic light-scattering technique follows the Stokes–Einstein law, enabling accurate characterization of the macromolecule in terms of its hydrodynamic radius R_h (which is proportional to the radius of gyration R_g),

$$D = \frac{k_B T}{6\pi \eta_0 R_h}, \qquad (1.4.1)$$

where η_0 is the viscosity of the solvent. This methodology fails in the case of charged macromolecules when the concentration c_s of the added low molar mass salt in the solution is not high. As c_s is reduced, the value of D increases [Lin *et al.* (1978)] (Fig. 1.9a), contrary to the expectation that the chain would be bigger due to electrostatic repulsion among the charged repeat units at lower salt concentrations, and hence from Eq. (1.4.1) D would be smaller. In spite of this counterintuitive result, the increase in D as c_s is decreased is known as the "ordinary" or "fast" diffusive behavior. In addition, upon further decrease in c_s, a new "slow" diffusive mode of relaxation

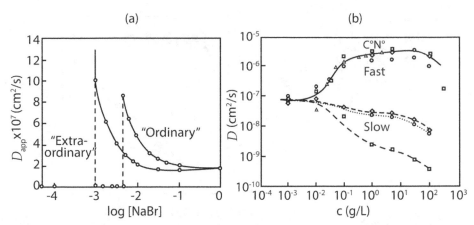

Figure 1.9 Ordinary–extraordinary transition. (a) Apparent diffusion coefficient D_{app} of poly(L-lysine) versus NaBr concentration. Circles and squares denote 1.0 mg/mL and 2.0 mg/mL of the polymer, respectively. At higher salt concentrations, ordinary behavior is observed. At lower salt concentrations, both the extraordinary and ordinary (not shown) behaviors are simultaneously present [Lin *et al.* (1978)]. (b) The emergence of the fast and slow modes in salt-free solutions of poly(vinyl pyridinium) salt as the polymer concentration is increased. Different symbols denote different molecular weights [Förster *et al.* (1990)].

appears in conjunction with the fast mode. The value of c_s at which the slow mode appears depends on the polymer concentration. Analysis of the diffusion coefficient of this slow mode indicates the presence of very large aggregates amidst unaggregated macromolecules. In view of the unexpected emergence of large aggregates formed by multiple similarly charged polymers, especially at lower salt concentrations, the slow diffusion is known as the "extraordinary" behavior [Lin *et al.* (1978)]. The simultaneous appearance of the ordinary and extraordinary behaviors for a given set of salt concentration and polyelectrolyte concentration is known as the "ordinary-extraordinary transition." Illustrative data [Lin *et al.* (1978), Förster *et al.* (1990)] for poly(L-lysine) and poly(vinyl pyridinium) salt solutions are shown in Fig. 1.9. It must be noted that the fast diffusion coefficient is independent of polymer concentration and molecular weight over several orders of magnitude. The numerical value of the fast D is only a factor of about 4 smaller than that of a small metallic ion such as K^+, although the molecular weight of the polymer is in the millions of Daltons.

1.4.3 Reentrant Precipitation

When a trivalent salt, such as $LaCl_3$ or spermidine chloride, is added to a solution of negatively charged polyelectrolytes such as NaPSS, a universal reentrant precipitation phenomenon is often observed. At low salt concentrations, a homogeneous single phase is observed. As the salt concentration c_s is increased, precipitation occurs as a system of two coexisting phases. Upon further increase in c_s, the solution again becomes a single phase (Fig. 1.10). This reentrant phenomenon is universal in the

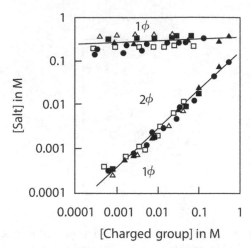

Figure 1.10 Universality of phase diagrams for trivalent cations and different anionic polyelectrolytes. Different symbols denote different polymers and the trivalent cations are La^{3+} and spermidine [Sabbagh & Delsanti (2000)].

sense that the boundaries of the two-phase region at room temperature are insensitive to the chemical nature of the polymer or the identity of the trivalent cation. The data for NaPSS, sodium poly(vinyl sulfonate), potassium poly(vinyl sulfate), sodium poly(acrylate), and sodium dextran sulfate in solutions with either La^{3+} or spermidine all collapse on the same phase diagram [Sabbagh & Delsanti (2000)].

1.4.4   Encapsulation of DNA via Complexation

Encapsulation of DNA and RNA (which are negatively charged) and other such cargo molecules using an assembly of other charged macromolecules (which are typically positively charged) is a ubiquitous phenomenon in biological contexts. An example is the assembly of viruses [Knipe & Howley (2001), Montiel-Garcia *et al.* (2021)] (Fig. 1.11a). A fundamental understanding of the electrostatic principles behind polyelectrolyte encapsulation and formulations of synthetic platforms for storage and controlled delivery of macromolecular cargo continues to attract considerable attention. In general, the topological correlation in the large macromolecule arising from chain connectivity plays a major role in the process of encapsulation. As an example, DNA (pUC19-supercoiled DNA of 2686 base pairs) is complexed with a cylindrical brush polymer made up of quaternized 2-vinyl pyridinium side chains (with the degree of polymerization of about 26), and the resultant structure of the complex is shown in Fig. 1.11b [Störkle *et al.* (2007)]. Methods to control the size of the complex, length of DNA loops, and structural correlations inside the complex are ongoing research.

1.4.5   Gel Swelling

When many polyelectrolyte chains are cross-linked by permanent chemical junctions to form a large polyelectrolyte network in water, the resulting hydrogel exhibits many

(a)   (b)

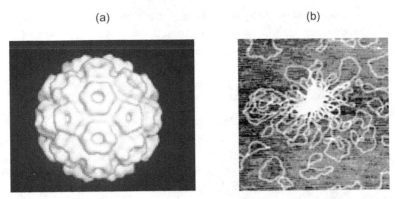

Figure 1.11 (a) Assembly of proteins carrying net positive charge and negatively charged ssRNA to form the Cowpea chlorotic mottle virus. The figure is from VIPERdb. (http://viperdb.scripps.edu) [Montiel-Garcia *et al.* (2021)]. (b) AFM picture of a complex of DNA and a cationically charged polymer brush [Störkle *et al.* (2007)].

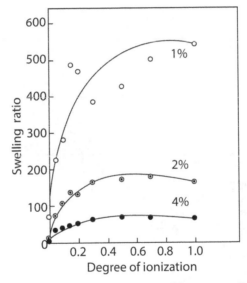

Figure 1.12 Swelling ratio versus degree of ionization at different cross-link concentrations (1%, 2%, and 4% from top to bottom) [Kuhn *et al.* (1950)].

fantastic properties, such as superabsorbancy of water and simultaneous exhibition of solid- and liquid-like behaviors. Such charged hydrogels are also of significant importance in the contexts of polymeric engines of life. As an example of the extent to which a cross-linked network can swell upon exposure to water, the dependence of the swelling ratio of the equilibrated swollen gel to the dry network is given in Fig. 1.12 in terms of the degree of ionization and cross-link density of the polymer network [Kuhn *et al.* (1950)]. As shown in Fig. 1.12, the swelling ratio can be as high as 500 for weakly cross-linked gels. The electrostatic interactions and the elasticity of the network control the hydrogel behavior. An understanding of the key concepts of thermodynamic stability and deformation upon stress in hydrogels in terms of the role of electrostatics is vital for optimal designs of hydrogel-based engines.

1.4.6   Outlook

The scope of the following chapters is to address the basic conceptual framework behind the rich phenomena exhibited by charged macromolecules as illustrated in the aforementioned sections. The primary goal is to provide a description of the current understanding of the fundamental forces behind the structural organization and dynamics of charged macromolecules. Specifically, we shall address the charge and size of charged macromolecules, structures in homogeneous solutions, onset of thermodynamic instability and phase transitions, structural evolution with oppositely charged macromolecules, coacervation, membraneless organelles, charged brushes, dynamics in polyelectrolyte solutions, kinetics of phase separation, mobility of charged macromolecules between compartments, and implications to biological systems.

2 Models of Uncharged Macromolecules

Macromolecules are naturally very large in size in comparison with the dimensions of the atoms and monomers which make up such gigantic molecules. The manner in which macromolecules assume their structure, interact among themselves, and move around, involves large scales of space and time. Therefore we require a conceptual framework that can cope with such large scales of space and time, without getting bogged down by very minute details at atomistic levels. Also, an atomistic description of very large collections of huge molecules spanning over macroscopic spatial lengths and time durations is a formidable challenge. In view of this challenge, the field of macromolecules has successfully evolved into building coarse-grained models of individual macromolecules by absorbing atomistic details into a few parameters. These models are generic to all macromolecules, independent of whether they carry ionic groups or not. In this chapter we describe the basic models of macromolecules in their uncharged state, with the role of ionic groups and the surrounding aqueous electrolyte background relegated to a later chapter. First, we outline the various models of chain connectivity for flexible, semiflexible, branched, and circular macromolecules. Apart from the entropic forces arising from chain connectivity, an important aspect of macromolecules is the intermonomer potential interactions among all nonneighbor monomers of the molecules. These interactions can be short-ranged and/or long-ranged depending on their chemical and ionic nature of the monomers. The intersegment interactions among nonnearest neighbors are collectively called excluded volume interactions. As the next step to a description of chain connectivity, we address the role of excluded volume interactions on the properties of macromolecules, such as chain swelling and coil–globule transitions. One of the key quantities of fundamental interest in the physics of macromolecules is the size exponent of a macromolecule, which defines how its size depends on its mass. We conclude this chapter with a discussion of how the size exponent is determined experimentally.

2.1 Flexible Chains

Consider a flexible chain with a large number N_0 of repeat units $-(-C(XY)-)-_{N_0}$ connected in a linear manner. A typical conformation of the chain is portrayed in Fig. 2.1a. X and Y denote the chemical identities of the side groups attached to each of the carbon (C) atoms forming the skeleton of the chain, and the C–C bond distance is a.

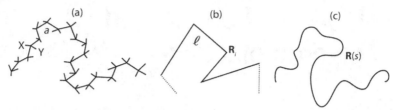

Figure 2.1 (a) Cartoon of a conformation of a portion of a real linear chain, $(–C(XY)–)_{N_0}$, with bond length a. (b) Sketch of the Kuhn chain model of N Kuhn segments, each of average length ℓ and average number of repeat units m. $\mathbf{R}_i$ denotes the position of the center of the ith segment. $N \sim N_0$ and $\ell \sim a$. (c) Continuous curve description of the Kuhn chain model with contour length $L = N\ell$. $\mathbf{R}(s)$ is equivalent to $\mathbf{R}_i$, with the arc length variable s in the range $0 \leq s \leq L$.

Depending on the chemical nature of these side groups, there are potential energy barriers for free rotation around the various bonds along the chain contour. For flexible chains, these rotational barriers are low and, as a result, the direction of orientation of a specified bond persists only for a very short distance along the chain contour. Beyond a certain length ℓ along the skeleton of the chain, the bond orientations are uncorrelated. Therefore the real chain can be imagined as an freely jointed chain (FJC) made of N segments, each of average length ℓ and an average number of monomers m, as sketched in Fig. 2.1b. This model of the chain is known as the **Kuhn chain model**, and N and ℓ are the number of Kuhn segments per chain and the Kuhn length, respectively. The local chemical details are parameterized by the Kuhn length and its value is different for different polymers. The two Kuhn model parameters, N and ℓ, are directly proportional, respectively, to the number of repeat units N_0 and the bond length a of the real chain,

$$N \sim N_0 \qquad \text{and} \qquad \ell \sim a. \tag{2.1.1}$$

The numerical prefactors in these proportional relations depend on the chemical identity of the repeat units of the polymer in terms of relative orientations of adjacent bonds, rotational barriers around bonds, and temperature. At lower temperatures, the effective rotational barriers are higher and as a result the Kuhn length increases representing the enhanced local chain stiffness along the chain backbone [Flory (1953a), Flory (1969), Yamakawa (1971)].

By definition of the Kuhn model, the orientations of the adjacent Kuhn segments are uncorrelated. Thus each chain conformation is equivalent to a trajectory of a particle undergoing random walk. Consequently, for sufficiently large values of N, the Kuhn model chain is also referred to as the **random walk chain** of N steps, each of step length ℓ. Each Kuhn step is connected by centers of Kuhn segments. These centers of segments are commonly referred to as "beads." Sometimes each Kuhn step is taken as connecting two beads so that there are $N + 1$ beads for a Kuhn chain of N steps. The distinction between these two definitions of "beads" is immaterial for $N \to \infty$. Another equivalent representation of the Kuhn chain is the continuous curve model. By taking the limit of $N \to \infty$ and $\ell \to 0$, but their product fixed by the contour length $N\ell$, the random walk chain is represented as a continuous curve. The product $N\ell$ defines the connection between N and the contour length L as

$$L = N\ell. \tag{2.1.2}$$

In this continuous representation of the Kuhn chain, each conformation can be treated as a random path starting from one end to the other end of the chain with a contour length L. The position $\mathbf{R}_i$ of a particular segment i along the polymer backbone is denoted equivalently by the position $\mathbf{R}(s)$ of the arc length variable $s = i\ell$ along the continuous path, where $0 \leq s \leq L$ (Fig. 2.1c). Following are some of the key properties of the Kuhn chain model.

2.1.1 Mean Square End-to-End Distance

Averaging over all possible conformations of a flexible chain of N Kuhn steps, without any regard to segment–segment interactions among nonadjacent segments, the mean square end-to-end distance of the chain follows from the well-known results of random walk statistics and Einstein's law on Brownian motion as

$$\langle R^2 \rangle_0 = N\ell^2. \tag{2.1.3}$$

Here, the angular brackets denote averaging over chain conformations. The subscript 0 in Eq. (2.1.3) is a reminder that the segment–segment interactions resulting in excluded volume effects are ignored in the Kuhn model chain. The above result can be easily derived for the chain statistics as follows. The end-to-end distance vector $\mathbf{R}$ for any chain conformation is $\sum_i^N \mathbf{r}_i$, where $\mathbf{r_i}$ is the "bond vector" of the ith Kuhn step. The square of the end-to-end distance has N diagonal terms $\mathbf{r}_i^2$ and off-diagonal terms $\mathbf{r}_i \cdot \mathbf{r}_j$. Since the vectors $\mathbf{r}_i$ and $\mathbf{r}_j$ are uncorrelated in the Kuhn model for $i \neq j$, the off-diagonal terms vanish upon averaging over all chain conformations. Only the diagonal terms survive upon averaging. Since each of the diagonal terms $\mathbf{r}_i^2$ is ℓ^2, Eq. (2.1.3) is obtained. The root mean square end-to-end distance, $R = \langle R^2 \rangle_0^{1/2}$ is a measure of the three-dimensional size of the polymer chain (coil) for a given contour length.

2.1.2 Radius of Gyration

Another measure of polymer size, which is easily measurable in experiments, is the radius of gyration R_{g0} (proportional to $\langle R^2 \rangle_0^{1/2}$),

$$R_{g0} = \frac{\sqrt{N}\ell}{\sqrt{6}} \sim N^{1/2}. \tag{2.1.4}$$

2.1.3 Segment Density Profile

For the Kuhn model chain, the manner in which the segments are spatially distributed at a radial distance r from the center of mass of the chain is given by the experimentally measurable monomer density distribution function $\rho(r)$,

$$\rho(r) \sim \frac{1}{r}. \tag{2.1.5}$$

The topological correlation arising from the chain connectivity leads to a long-ranged correlation among the various segments as mentioned in Section 1.2 [de Gennes (1979)]. The above law is valid for radial distances larger than the Kuhn length ℓ but shorter than the radius of gyration R_{g0}.

2.1.4 Probability of End-to-End Distance

The probability distribution function $G_0(\mathbf{R}, \mathbf{0}; N)$ of a Kuhn chain of N steps with a fixed end-to-end distance $\mathbf{R}$ is derived from the theory of random walks. For long chains, satisfying the condition $R \ll N\ell$, the probability distribution function is given by

$$G_0(\mathbf{R}, \mathbf{0}; N) = \left(\frac{3}{2\pi N \ell^2}\right)^{3/2} e^{-\left(\frac{3R^2}{2N\ell^2}\right)}. \tag{2.1.6}$$

In view of the Gaussian nature of the distribution function $G_0(\mathbf{R}, \mathbf{0}; N)$, the Kuhn model chain for large N values such that $R \ll N\ell$ is called the **Gaussian chain**.

It is often convenient to write the above Gaussian distribution as a diffusion equation or as a Feynman path integral [Yamakawa (1971), Freed (1972, 1987), Doi & Edwards (1986), Fredrickson (2006)]. Such alternative representations facilitate a more transparent description of interactions among a collection of macromolecules, allowing easier conceptualization of various quantities of experimental interest. Eq. (2.1.6) is exactly equivalent to the diffusion equation

$$\left(\frac{\partial}{\partial N} - \frac{\ell^2}{6}\nabla_{\mathbf{R}}^2\right) G_0(\mathbf{R}, \mathbf{0}; N) = \delta(\mathbf{R})\delta(N), \tag{2.1.7}$$

where δ is the Dirac delta function, and ∇^2 is the Laplacian. Eq. (2.1.6) is also equivalent to the Feynman path integral

$$G_0(\mathbf{R}, \mathbf{0}; N) = \int_{\mathbf{0}}^{\mathbf{R}} \mathcal{D}[\mathbf{R}(s)] \exp\left[-\left(\frac{3}{2\ell^2}\right)\int_0^N ds \left(\frac{\partial \mathbf{R}(s)}{\partial s}\right)^2\right]. \tag{2.1.8}$$

Here, each chain conformation is treated as a continuous path starting from one end of the chain at $\mathbf{0}$ to the other end of the chain at $\mathbf{R}$ by taking the label of the Kuhn steps of the chain as a continuous variable s. The spatial location along the path is $\mathbf{R}(s)$, where s is the arc length variable ($0 \le s \le N$). The symbol $\int \mathcal{D}[\mathbf{R}(s)]$ denotes the sum over all paths and the limits of the integral sign denote the fixed positions of the two chain ends.

2.1.5 Form Factor

The form factor of a single macromolecule, which is measured in scattering experiments, is generally defined as

$$P(\mathbf{k}) = \frac{1}{N^2} \sum_i \sum_j \left\langle \exp\left[i\mathbf{k} \cdot \left(\mathbf{R}_i - \mathbf{R}_j\right)\right]\right\rangle, \tag{2.1.9}$$

where $\mathbf{k}$ is the scattering wave vector with magnitude $k = (4\pi/\lambda)\sin(\theta/2)$ (with λ being the wavelength of the incident radiation and θ the scattering angle). $\mathbf{k}$ is also the Fourier variable conjugate to the spatial position variable $\mathbf{r}$. The angular brackets in Eq. (2.1.9) denote the average over polymer conformations. For a Gaussian chain, $P(\mathbf{k})$ can be evaluated using Eq. (2.1.6), and the Debye structure factor $S_D(\mathbf{k})$, defined as $NP(\mathbf{k})$, is obtained as

$$S_D(k) = \frac{2N}{k^4 R_{g0}^4}\left[k^2 R_{g0}^2 - 1 + \exp\left(-k^2 R_{g0}^2\right)\right], \qquad (2.1.10)$$

where $R_{g0}^2 = N\ell^2/6$ as given in Eq. (2.1.4). For large k such that $k R_{g0} > 1$, $S_D(k) \sim 1/k^2$, which is the Fourier transform of segmental distribution function given in Eq. (2.1.5).

2.1.6   Free Energy

Thermodynamic properties such as the conformational entropy and free energy of a chain, where one chain end is fixed at the origin of the coordinate system $\mathbf{0}$ and the other end is fixed at $\mathbf{R}$, are derived from the probability distribution function $G_0(\mathbf{R},\mathbf{0}; N)$ of a Kuhn chain of N steps with the fixed end-to-end distance $\mathbf{R}$. The conformational entropy of the Kuhn chain, $S(R) = k_B \ln G_0(\mathbf{R})$ is obtained from Eq. (2.1.6) as

$$S(R) = \text{constant} - \frac{3k_B}{2}\left(\frac{R^2}{N\ell^2}\right), \qquad (2.1.11)$$

where k_B is the Boltzmann constant, and the constant term is independent of R.

The free energy $F(R)$ of the chain at the absolute temperature T is

$$F(R) = E(R) - TS(R), \qquad (2.1.12)$$

where $E(R)$ is the internal energy of the chain. For the Kuhn model chain in the absence of any excluded volume interaction, the internal energy is the number of Kuhn steps times the "bond energy" of the Kuhn step which is independent of R. Using Eq. (2.1.11), the free energy of a Kuhn chain with a fixed end-to-end distance $\mathbf{R}$ is given by

$$F(R) = \text{constant} + \frac{3k_B T}{2}\left(\frac{R^2}{N\ell^2}\right). \qquad (2.1.13)$$

The dependence of $F(R)$ on R is equivalent to the stretching energy of a Hookean spring, with the spring constant $3k_B T/N\ell^2$. As the chain length is increased, the spring constant becomes smaller representing the highly elastic nature of flexible macromolecules. The expression given by Eq. (2.1.13) for the free energy of a Gaussian chain is valid only for chain expansions from its equilibrium size, $\langle R^2 \rangle_0^{1/2} = \sqrt{N}\ell$, but not for chain compressions. Accounting for fluctuations in each Kuhn segment

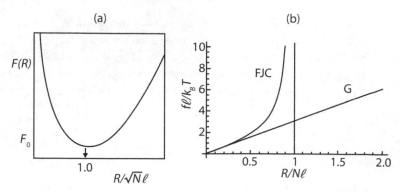

Figure 2.2 (a) Dependence of free energy of a Gaussian chain on its end-to-end distance R, based only on conformational entropy. For $R > \sqrt{N}\ell$, the chain is equivalent to a Hookean spring with spring constant proportional to T/N. (b) Dependence of tensile force $f\ell/k_B T$ on chain extension $R/N\ell$ according to the Gaussian chain model (G) and the freely jointed chain (FJC) model.

length, the correct free energy of the Gaussian chain is [Flory (1953a), des Cloiuzeaux & Jannink (1990)]

$$F(R) = \text{constant} + \frac{3k_B T}{2}\left(\frac{R^2}{N\ell^2} - 1 - \ln\frac{R^2}{N\ell^2}\right). \tag{2.1.14}$$

This result is sketched in Fig. 2.2a, where the constant term is denoted as F_0. The minimum free energy occurs at the equilibrium value $R = \langle R^2 \rangle_0^{1/2} = \sqrt{N}\ell$. For chain expansion around the equilibrium value, the free energy is quadratic in R, whereas for chain compression, the dependence of free energy on R is steeper.

2.1.7   Tensile Force

The tensile force **f** required to maintain a Gaussian chain at a certain end-to-end distance R is obtained from Eq. (2.1.14) as $\mathbf{f} = -\partial F(\mathbf{R})/\partial \mathbf{R}$,

$$\frac{f\ell}{k_B T} = 3\left(\frac{R}{N\ell} - \frac{\ell}{R}\right). \tag{2.1.15}$$

The tensile force is 0 when the end-to-end distance is equal to its equilibrium value $\sqrt{N}\ell$. For large values of the end-to-end distance, such that $R \gg \sqrt{N}\ell$, the tensile force is linear in R, as shown in Fig. 2.2b. For very large values of R approaching the contour length of the chain $N\ell$, the assumption $R \ll N\ell$ used in obtaining the Gaussian distribution function G_0 of Eq. (2.1.6) is not valid. Implementing that the chain ends can be extended only up to the finite chain length, the relation between the tensile force and the end-to-end distance is [Flory (1953a), Yamakawa (1971)]

$$\frac{R}{N\ell} = \coth\left(\frac{f\ell}{k_B T}\right) - \frac{k_B T}{f\ell} \equiv \mathcal{L}\left(\frac{f\ell}{k_B T}\right), \tag{2.1.16}$$

where $\mathcal{L}$ is the Langevin function as defined in this equation. The Kuhn chain model with finite extensibility is known as the **freely jointed chain (FJC) model**. The result of Eq. (2.1.16) for FJC is included in Fig. 2.2b as a comparison with that of a Gaussian chain.

2.2 Semiflexible Chains

Many polymer molecules, such as dsDNA, are stiff along their backbone. In addressing chain stiffness due only to chain connectivity, one approach is to modify the Kuhn chain model (which is a freely rotating chain) by introducing a small angle θ between two consecutive Kuhn steps, as shown in Fig. 2.3a. This model is called the **Kratky–Porod** or **worm-like** chain model. The average orientational correlation function between the directions of ith and jth Kuhn steps can be shown [Yamakawa (1971)] to depend on the distance along the chain contour, $|i - j|\ell$, as

$$\langle \ell_i \cdot \ell_j \rangle_0 = \ell^2 e^{-\ell|i-j|/\ell_p}, \tag{2.2.1}$$

where ℓ_p, called the **persistence length**, is a measure of the distance along the chain contour over which segmental orientations are correlated. The angular brackets denote the averaging over all possible conformations of the chain, and the subscript 0 denotes the absence of excluded volume interactions. For large N and small θ values, the model chain of Fig. 2.3a can be treated as a continuous stiff wire (Fig. 2.3b) with a bending force constant ϵ_b at every arc length variable s ($0 \le s \le L = N\ell$) along the chain contour, where the tangent vector is $\mathbf{u}(s)$. Using the theory of elasticity of rods [Landau & Lifshitz (1980)], the orientational correlation function between the tangent vectors at two locations along the chain backbone is given by the same equation, Eq. (2.2.1), with the persistence length ℓ_p given as the bending force constant in units of thermal energy,

$$\ell_p = \frac{\epsilon_b}{k_B T}. \tag{2.2.2}$$

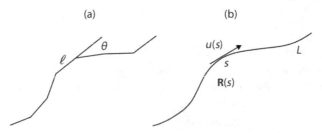

Figure 2.3 Kratky–Porod or worm-like chain model. (a) Discretized version as a modification of Kuhn chain model. (b) Continuous representation as a stiff worm-like wire with bending elasticity.

The mean square end-to-end distance for the Kratky–Porod (worm-like) chain model is [Yamakawa (1971)]

$$\langle R^2 \rangle_0 = 2\ell_p L \left[1 - \frac{\ell_p}{L} \left(1 - e^{-L/\ell_p} \right) \right], \tag{2.2.3}$$

where $L = N\ell$ is the contour length of the chain. If the contour length is shorter than the persistence length, as in a short fragment of dsDNA, Eq. (2.2.3) leads to the rodlike limit of the end-to-end distance being the contour length so that $\langle R^2 \rangle_0 = L^2$. In the other limit of contour length being much larger than the persistence length, as in a very long dsDNA or in flexible chains, Eq. (2.2.3) reduces to the Gaussian chain result of Eq. (2.1.3), $\langle R^2 \rangle_0 = L\ell$,

$$\langle R^2 \rangle_0 = \begin{cases} 2\ell_p L \equiv \ell L, & \ell_p \ll L \\ L^2 & \ell_p \gg L. \end{cases} \tag{2.2.4}$$

As defined in the above equation, the Kuhn length ℓ is equivalent to twice the persistence length ℓ_p,

$$\ell = 2\ell_p, \tag{2.2.5}$$

and thus the Kuhn length is a measure of chain stiffness as evident from the construction of the Kuhn chain model. The corresponding two limits for the radius of gyration of the Kratky–Porod (worm-like) chain, in the absence of any intersegment excluded volume interactions, are [Yamakawa (1971)]

$$R_{g0}^2 = \begin{cases} \frac{L\ell}{6} & \ell_p \ll L \\ \frac{L^2}{12} & \ell_p \gg L. \end{cases} \tag{2.2.6}$$

2.3   Other Architectures

2.3.1   Branched Macromolecules

The radius of gyration of randomly branched macromolecules, both synthetic and biological, as depicted in Figs. 2.4a and 2.4b, is given by [Zimm & Stockmayer (1949), de Gennes (1968)]

$$R_{g0} = A_b N^{1/4}, \tag{2.3.1}$$

where N is the total number of segments in the whole branched architecture and intersegment excluded volume interactions are absent. The numerical prefactor A_b depends on the number of branches from a branch point and the frequency of branching inside the macromolecule. The reduced value of the size exponent (1/4) for a randomly branched polymer, compared to the value (1/2) for a random walk chain, is a measure of the compactness of the structure arising from multiple branching within the macromolecule. The above result is valid only for randomly branched architectures in the asymptotic limit of large N. Again, the excluded volume interactions are not

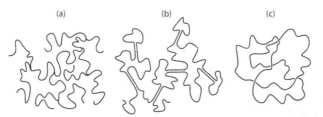

Figure 2.4 Sketches of randomly branched and circular macromolecules. (a) A synthetic polymer with branch points at random locations along a chosen skeletal path, and random branches emanating from these branch points. In this example, three branches emanate from every branch point. (b) Cartoon of a tertiary structure from a random sequence of ssRNA or ssDNA where the stems denote the hydrogen-bonded double helical domains and the rest denote the unhybridized segments of the molecule. (c) Cartoon of a circular polymer.

included in the exact result of Eq. (2.3.1) which represents only the chain connectivity and random branching.

2.3.2   Circular Macromolecules

When the two ends of a linear chain are fused together to obtain a circular polymer or a ring polymer (Fig. 2.4c), the three-dimensional size is expected to be smaller. The exact result for the radius of gyration of a circular macromolecule of N Kuhn segments each of segment length ℓ is [Yamakawa (1971)]

$$R_{g0,\text{ring}} = \frac{1}{\sqrt{2}} R_{g0,\text{linear}} = \sqrt{\frac{N}{12}} \ell. \qquad (2.3.2)$$

This result is obtained without any explicit consideration of knots in the conformations of the circular macromolecule.

The above models of flexible and semiflexible chains account for only the chain connectivity and the various excluded volume interactions among the nonadjacent segments are ignored, which we address in the next section.

2.4   Excluded Volume Interactions

Polymer chains cannot intersect themselves. In addition to the chain connectivity described in the preceding sections, each monomer will exclude a certain volume for the approach of another monomer resulting in the excluded volume interaction between a pair of monomers. Any two monomers i and j separated by a distance $\mathbf{r}_{ij}$ interact between themselves with the interaction energy $u(\mathbf{r}_{ij})$, as depicted in Fig. 2.5a. Their interaction is mediated by the surrounding solvent molecules as well. We expect the net effective interaction to be infinitely repulsive at very close distances, 0 at far distances, and attractive at intermediate distances due to van der Waals type forces, as sketched in Fig. 2.5b. The depth and range of the attraction depend on the specificities of the monomers and the solvent, and temperature.

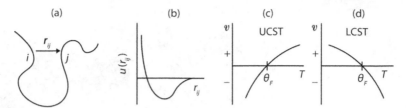

Figure 2.5 (a) Excluded volume interaction between ith and jth monomers with separation distance $\mathbf{r}_{ij}$. (b) Sketch of the dependence of the pairwise interaction energy between two monomers on their separation distance. (c) Temperature dependence of the excluded volume parameter v specific to the interaction energy for a particular combination of the polymer and solvent, for an upper critical solution temperature (UCST) system. (d) Same as (c), but now for a lower critical solution temperature (LCST) system. For both (c) and (d), v is 0 at the Flory ideal temperature θ_F, due to a cancellation of repulsive and attractive interactions.

A quantitative description of the precise details of the pairwise potential energy between monomers mediated by solvent is presently an impossible task, as it requires computation of quantum mechanical forces among all atoms in such large collections of atoms. In view of this, we define a parameter, called the **excluded volume parameter** v, that is specific to a particular combination of the polymer and solvent at a prescribed temperature. This parameter is the second viral coefficient for the monomer in the solvent. Since the chain connectivity is already parametrized using Kuhn segments, the excluded volume parameter is also defined in terms of the second viral coefficient for a pair of Kuhn segments,

$$v = \frac{1}{\ell^3} \int d\mathbf{r}_{ij} \left[1 - e^{-\frac{1}{k_B T} u(\mathbf{r}_{ij})} \right], \qquad (2.4.1)$$

where i and j denote the Kuhn segments, and $u(\mathbf{r}_{ij})$ is the effective interaction energy between the two segments. Since the integral on the right-hand side of the above equation has the dimension of volume, the dimensionless excluded volume parameter v is expressed in units of the cube of Kuhn length. Since the intersegment pairwise potential $u(\mathbf{r}_{ij})$ is short-ranged in the absence of any charges on the polymer, it can be written as a pseudo potential,

$$\frac{u(\mathbf{r}_{ij})}{k_B T} \equiv v\ell^3 \delta(\mathbf{r}_{ij}), \qquad (2.4.2)$$

where $\delta(\mathbf{r}_{ij})$ is the Dirac delta function. This definition of the pseudo potential is chosen to be consistent with the high temperature limit of Eq. (2.4.1). The physical meaning of Eq. (2.4.2) is that whenever two segments are in contact, the cost of energy is v in units of $k_B T$.

The excluded volume parameter v is equivalently written as

$$v \equiv (1 - 2\chi). \qquad (2.4.3)$$

The parameter χ appearing in this equation is the Flory–Huggins parameter [Flory (1953a)] representing the chemical mismatch between the polymer and solvent in polymer solutions. According to the Flory–Huggins theory [Flory (1953a)] of thermodynamics of polymer solutions, which will be described in a later chapter, the various polymer segments and solvent molecules are randomly cast into a regular lattice of certain coordination number z and the average nearest neighbor interaction energy in the lattice in units of $k_B T$ is given by

$$\chi = \frac{z}{k_B T} \left[w_{ps} - \frac{1}{2}(w_{pp} + w_{ss}) \right],$$ (2.4.4)

where w_{pp}, w_{ss}, and w_{ps} are the nearest neighbor interaction energies for polymer–polymer segments, solvent–solvent, and solvent–polymer segment pairs, which themselves are parameters. Eq. (2.4.3) gives the connection between the excluded volume parameter v for polymer chains in dilute solutions and the Flory–Huggins chemical mismatch parameter χ employed in describing the thermodynamics of concentrated polymer solutions.

At lower temperatures, the attractive part of the intersegment interaction energy $u(\mathbf{r}_{ij})$ dominates so that the second term inside the square brackets of Eq. (2.4.1) is larger than the first term and hence v is negative. At higher temperatures, the segments repel each other ($u(\mathbf{r}_{ij})$ is positive), and now the first term dominates over the second term so that v is positive. This behavior of the temperature dependence of the excluded volume parameter v is sketched in Fig. 2.5c.

At the special temperature marked θ_F in Fig. 2.5c, called the ideal temperature, or the **Flory temperature**, or the theta temperature, the excluded volume parameter $v = 0$ due to a compensation between attractive and repulsive interactions, although the chain is still nonintersecting. In view of the inverse proportionality between χ and temperature, given by Eq. (2.4.4), and since $v = 0$ at θ_F, χ and v can be written as

$$\chi = \frac{\theta_F}{2T}, \qquad v = \left(1 - \frac{\theta_F}{T} \right).$$ (2.4.5)

For temperatures above the ideal temperature ($T > \theta_F$) for a particular polymer–solvent system, the excluded volume parameter is positive ($v > 0$). Now, the chain swells due to net repulsive excluded volume interactions among its segments relative to polymer–solvent interactions. In other words, $\chi < 1/2$, with the solvent-polymer interaction being more attractive than the polymer-polymer and solvent–solvent interactions. As a result, the solvent is fully miscible with the polymer and consequently the chain swells. When this occurs, the solvent is called a **good solvent**. On the other hand, for $T < \theta_F$, the chain shrinks due to net attractive intersegment interactions and tends to form globule-like structures. Under such conditions the solvent is called a **poor solvent** for the polymer.

The temperature dependence of the excluded volume parameter v can be described using Eqs. (2.4.3) and (2.4.4). At the Flory temperature θ_F, $\chi = 1/2$ so that $v = 0$. If the term inside the square brackets in Eq. (2.4.4) are insensitive to temperature, then an increase in temperature beyond θ_F will reduce the value of χ so that v becomes positive and the solvent is a good solvent. On the other hand, if the temperature is reduced below θ_F, $v < 0$ and $\chi > 1/2$ making the solvent a poor solvent. If the temperature is low enough, the polymer solution undergoes a phase separation into a polymer-rich phase and a polymer-poor phase which coexist. Such a situation where a polymer solution becomes homogeneous without phase separation upon heating is known as the **upper critical solution temperature (UCST)** system. Fig. 2.5c corresponds to the UCST behavior.

On the other hand, if the temperature dependence of the term inside the square brackets of Eq. (2.4.4) is such that the parameter χ increases with temperature, then v becomes negative at higher temperatures. For this situation, $v > 0$ at lower temperatures and $v < 0$ at higher temperatures, as sketched in Fig. 2.5d, and phase separation occurs at higher temperatures. Polymer solutions following this behavior are called **lower critical solution temperature (LCST)** systems. LCST behavior can easily arise from reorganization of water around charged macromolecules. It is a common practice to interpret experiments in terms of temperature dependent χ or equivalently v instead of the molecular origins of the temperature dependence of various properties of macromolecular systems. Simply put, if χ varies inversely with temperature, we expect UCST behavior, and if it varies proportionally with temperature then we expect LCST behavior.

In the presence of excluded volume interactions $u(\mathbf{r}_{ij})$ between the two segments i and j parametrized by Eqs. (2.4.3) and (2.4.4), the probability distribution function for the end-to-end distance $\mathbf{R}$ of a Kuhn chain with N segments is given by the generalization of Eq. (2.1.8), known as the Edwards path integral [Freed (1972), Doi & Edwards (1986)],

$$G(\mathbf{R}, \mathbf{0}; N) = \int_{\mathbf{0}}^{\mathbf{R}} \mathcal{D}[\mathbf{R}(s)] \exp\left[-\left(\frac{3}{2\ell^2}\right) \int_0^N ds \left(\frac{\partial \mathbf{R}(s)}{\partial s}\right)^2\right.$$
$$\left. -\frac{1}{2} \int_0^N ds \int_0^N ds' u[\mathbf{R}(s) - \mathbf{R}(s')]\right], \quad (2.4.6)$$

where continuous notation for the chain connectivity is used. The term inside the square brackets is called the Edwards Hamiltonian. The above integral representation of a Gaussian chain model with excluded volume interactions can be equivalently written as a diffusion equation,

$$\left(\frac{\partial}{\partial N} - \frac{\ell^2}{6}\nabla^2_{\mathbf{R}(s)} + u[\mathbf{R}(s)]\right) G(\mathbf{R}(s), \mathbf{R}(s'); s-s') = \delta(\mathbf{R}(s)-\mathbf{r}(s'))\delta(s-s'). \quad (2.4.7)$$

This diffusion equation is nonlinear, since the term $u[\mathbf{R}(s)]$ depends on G; hence, a self-consistent procedure is required to determine the probability distribution function G. This protocol is often called in polymer literature as the self-consistent field theory. Eq. (2.4.7) looks like a "time-dependent Schrödinger equation" and several techniques developed in quantum mechanics for solving such equations can be implemented to address many polymer problems such as adsorption of charged macromolecules to charged interfaces and virus assembly [Wiegel (1986), Muthukumar (1987), Belyi & Muthukumar (2006), Muthukumar (2012b)].

Treatment of the omnipresent excluded volume interactions in addressing the various structural and thermodynamic properties of macromolecules involves solving Eqs. (2.4.6) and (2.4.7). Using reasonable approximations, significant progress has been achieved in solving these equations pertinent to various macromolecular systems even when long-ranged electrostatic interactions are present. For now, we avoid mathematical details and provide some of the classical arguments which are conceptually transparent.

2.5 Chain Swelling

Let us consider the consequences of excluded volume interactions on a Gaussian chain. First, let the excluded volume interaction be repulsive, that is, $v > 0$, corresponding to temperatures higher than the Flory temperature (in the UCST system). We expect the chain to swell due to excluded volume interactions under these conditions. The optimal size adopted by a flexible chain with repulsive excluded volume interactions ($v > 0$) is addressed by considering the free energy contribution from the conformational entropy arising from chain connectivity $F_{\text{connectivity}}$ and energy from excluded volume interactions F_{exc},

$$F = F_{\text{connectivity}} + F_{\text{exc}}. \tag{2.5.1}$$

Let us take the root mean square end-to-end distance R as a measure of chain size. The coil volume is proportional to R^3, and the average segment density inside the chain is proportional to N/R^3. For a Gaussian chain, $F_{\text{connectivity}}$ is given by Eq. (2.1.13). F_{exc} is the product of the energy cost $v\ell^3$ per contact within the excluded volume element ℓ^3, the probability of finding two segments to form the contact at any spatial location inside the chain, and the volume of the coil ($\sim R^3$). Since the probability of finding two segments at a location is proportional to the square of the average segment density ($\sim (N/R^3)^2$), we get

$$\frac{F_{\text{exc}}}{k_B T} \sim \frac{v\ell^3}{2} \left(\frac{N}{R^3} \right)^2 R^3, \tag{2.5.2}$$

where the factor 2 in the denominator is to avoid double counting of the segment–segment interactions. All other prefactors such as $4\pi/3$, etc., are left out here. Combining Eqs. (2.1.13), (2.5.1), and (2.5.2), the free energy of the chain is given by

$$\frac{F}{k_B T} = \frac{3R^2}{2N\ell^2} + \frac{v\ell^3}{2} \frac{N^2}{R^3}, \tag{2.5.3}$$

where terms independent of R are left out. The repulsive excluded volume interaction favors larger values of R by lowering F_{exc}, and the entropic part due to chain connectivity favors smaller values of R by lowering $F_{\text{connectivity}}$. As a result, an optimum is reached between these two opposite driving forces. The optimal chain radius is obtained by minimizing the free energy of the chain given in Eq. (2.5.3) with respect to R. Leaving out the coefficients, the result is

$$\frac{\partial F}{\partial R} = \frac{R}{N\ell^2} - \frac{v\ell^3 N^2}{R^4} = 0, \tag{2.5.4}$$

so that the optimum chain radius, called the Flory radius R_F, follows as

$$\frac{R_F}{\ell} \sim v^{1/5} N^{3/5}. \qquad (2.5.5)$$

This general result is universally valid for all flexible chains in good solvents, in the asymptotic limit of very large excluded volume parameter. The dimensionless variable that determines the strength of the excluded volume effect is $v\sqrt{N}$, which is known as the Fixman parameter. The result of Eq. (2.5.5) is valid for the condition $v\sqrt{N} \gg 1$. In this limit, the chain is referred to as the self-avoiding walk (SAW) chain. The exponent 3/5 is only approximate, since the above derivation is based on mean field arguments without consideration of conformational fluctuations inside the polymer. The value obtained from proper field theory formulation and renormalization group calculations [Freed (1987), Muthukumar & Nickel (1987), des Cloiuzeaux & Jannink (1990)] is 0.5886. Because the mean field value of 0.6 for the Flory exponent is close enough to the more accurate value and because the difference is usually within experimental error, we adopt the value of 3/5 in the rest of the book.

Experimental validation of Eq. (2.5.5) is shown in Fig. 2.6 for diversely different polymer systems. According to Eq. (2.5.5), because the radius of gyration R_g is proportional to R_F, R_g^2/N should scale linearly with $N^{1/5}$ for all polymer systems in good solvents. Additionally, the $v^{2/5}$ dependence of the slope differs for different polymer–solvent combinations, since v depends on the specificity of the polymer–solvent pair. These conclusions are precisely borne out in Fig. 2.6a where experimental data for poly(α-methylstyrene) (PαMS) in toluene [Kato *et al.* (1970)], poly(isobutylene) (PIB) in cyclohexane [Matsumoto *et al.* (1972)], poly(D-β-hydroxybutyrate (PHB) in trifluoroethanol [Miyaki *et al.* (1977)], and polystyrene (PS) in benzene [Miyaki *et al.* (1978)] are presented. Furthermore, many denatured proteins, such as ubiquitin, lysozyme, RNase A, cytochrome C, and carbonic anhydrase, also follow the same size exponent of 3/5 given by Eq. (2.5.5), as shown in Fig. 2.6b [Kohn *et al.* (2004), Fitzkee & Rose (2004)].

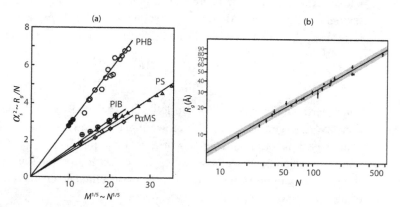

Figure 2.6 (a) Chain swelling above the ideal temperature, with repulsive excluded volume interactions ($v > 0$). The radius of gyration R_g is proportional to $N^{3/5}$ for $v\sqrt{N} \gg 1$. The symbols are explained in the text [Fujita (1990)]. (b) Unfolded proteins follow the law $R_g \sim N^{0.598}$ [Kohn *et al.* (2004)].

For weaker excluded volume effects ($v\sqrt{N} \ll 1$), due to either shorter chains or lower pairwise interaction energy (but still $v > 0$), the free energy of Eq. (2.1.14) due to chain entropy needs to be used instead of Eq. (2.1.13) in deriving the optimal coil radius. Including the correct prefactor for F_{exc} (obtained using an exact perturbation calculation from Eq. (2.4.6) [Muthukumar & Nickel (1984), Muthukumar (1984)], or equivalently a variational calculation [Muthukumar (1987)], the free energy of a Gaussian chain of radius R with excluded volume interaction is (as derived in Appendix 2)

$$\frac{F(R)}{k_BT} = \frac{3}{2}\left(\frac{R^2}{N\ell^2} - 1 - \ln\frac{R^2}{N\ell^2}\right) + \frac{4}{3}\left(\frac{3}{2\pi}\right)^{3/2}\frac{v\ell^3N^2}{R^3}.$$ (2.5.6)

The unnecessary R-independent constant terms are ignored, except for the $(-3/2)$ term. Minimization of $F(R)$ of Eq. (2.5.6) with respect to R yields

$$\left(\frac{R^2}{N\ell^2}\right)^{5/2} - \left(\frac{R^2}{N\ell^2}\right)^{3/2} = \frac{4}{3}\left(\frac{3}{2\pi}\right)^{3/2}v\sqrt{N}.$$ (2.5.7)

The asymptotic limits of this equation are

$$\frac{R}{\ell} = \begin{cases} \sqrt{N} & v = 0 \\ (\frac{4}{3})^{1/5}(\frac{3}{2\pi})^{3/10}v^{1/5}N^{3/5}, & v\sqrt{N} \gg 1 \end{cases}$$ (2.5.8)

in agreement with Eqs. (2.1.4) and (2.5.5). The crossover between these two limits is given by Eq. (2.5.7) for $v > 0$ as sketched in Fig. 2.7a where $R/\sqrt{N}\ell$ is plotted against $v\sqrt{N}$.

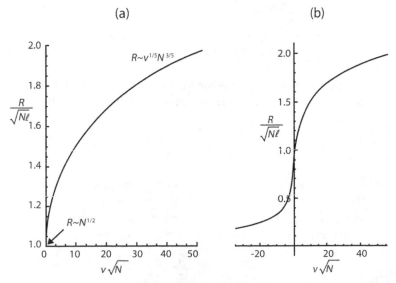

Figure 2.7 (a) Crossover curve for chain expansion from the ideal condition to a swollen condition as $v\sqrt{N}$ is increased. (b) Coil–globule transition given by Eq. (2.6.2) with $w = 0.05$.

The general chain swelling behavior due to excluded volume effects, as portrayed in Fig. 2.7a, is generic to all flexible macromolecules in good solvents. The dependence on N and v for the coil size in the asymptotic regimes is the same for all macromolecular solutions in good solvents, independent of chemical specificities. In the crossover region, an effective size exponent v_{eff} can be identified bounded between 1/2 and 3/5. Also, the crossover formula for the radius of gyration is given by the same Eq. (2.5.7) derived for the root mean square end-to-end distance, by replacing the factor (4/3) on the right-hand side of the equation with the factor (134/105), without any substantial change in the conclusions from Fig. 2.7a.

2.6   Coil–Globule Transition

When the excluded volume parameter v is negative, either by choosing a poor solvent for the polymer, or changing the temperature such that $\chi > 1/2$, the macromolecule tends to shrink its size. For $v < 0$, the minimum of the free energy of the chain as given by Eqs. (2.5.3) and (2.5.6) is at $-\infty$, corresponding to $R \to 0$. This unphysical collapse is prevented by including many-body interactions among the segments in addition to the two-body interactions included in Eqs. (2.5.3) and (2.5.7). It is sufficient to include only three-body interactions which turn out to be repulsive [Grosberg & Khokhlov (1994)]. These repulsive three-body excluded volume interactions prevent the chain collapse to an unphysical point-like state arising from only two-body attractive interactions, and stabilize a globule-like structure of the macromolecule.

The energy per chain due to three-body excluded volume interactions is proportional to the product of the triple-contact energy w (in units of $k_B T$), probability of finding three segments in a volume ℓ^3 (proportional to the cube of segment density, $(N/R^3)^3$), and the volume of the coil (proportional to R^3). Adding this three-body term to Eq. (2.5.6), we get

$$\frac{F(R)}{k_B T} = \frac{3}{2}\left(\frac{R^2}{N\ell^2} - 1 - \ln\frac{R^2}{N\ell^2}\right) + \frac{4}{3}\left(\frac{3}{2\pi}\right)^{3/2}\frac{v\ell^3 N^2}{R^3} + \frac{w\ell^6 N^3}{R^6}, \qquad (2.6.1)$$

where the numerical prefactor for the repulsive three-body term is absorbed in the parameter $w(> 0)$. Minimization of $F(R)$ with respect to R yields

$$\left(\frac{R^2}{N\ell^2}\right)^{5/2} - \left(\frac{R^2}{N\ell^2}\right)^{3/2} = \frac{4}{3}\left(\frac{3}{2\pi}\right)^{3/2} v\sqrt{N} + 2w\left(\frac{N\ell^2}{R^2}\right)^{3/2}. \qquad (2.6.2)$$

When the coil size shrinks for $v < 0$, becoming smaller than the value $\sqrt{N}\ell$ at $v = 0$, the terms on the right-hand side dominate over the entropic term on the left-hand side of Eq. (2.6.2). Therefore, for sufficiently large attractive v values, the two terms on the right hand-side with opposite signs give the optimum coil radius as

$$\frac{R}{\ell} \sim \left(\frac{w}{|v|}\right)^{1/3} N^{1/3}. \qquad (2.6.3)$$

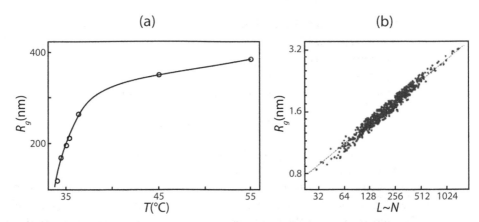

Figure 2.8 (a) Chain collapses at temperatures below the Flory temperature [Slagowski et al. (1976)]. (b) Many proteins in their native states follow the universal law for globules, $R_g \sim N^{1/3}$ [Banavar *et al.* (2005)].

Therefore, for negative v corresponding to poor solvents or $\chi > 1/2$, the chain attains a state where its volume is proportional to its mass ($\sim N$). This state is the globular state. The crossover between the ideal coil size and the globular size is given by Eq. (2.6.2), as presented in Fig. 2.7b for the particular choice of the parameter $w = 0.05$. It can be shown that inclusion of higher order contacts beyond the three body contacts does not alter the above results for the globule state [des Cloizeaux & Jannink (1990)].

An example of experimental data for the collapse of a coil as temperature decreases is shown in Fig. 2.8a for polystyrene in cyclohexane [Slagowski *et al.* (1976)]. For the molar mass of 44×10^6 Da in this system, the Flory temperature is 35.4°C at which the radius of gyration R_g is 207 nm. As the temperature is reduced from 55°C to 34°C, the radius of gyration decreases from 348 nm to 121 nm, exhibiting a substantial chain collapse below the Flory temperature. Occurrence of globules is quite common among folded proteins as well. In their native states, hundreds of proteins follow the same size exponent 1/3 given by Eq. (2.6.3), as seen from the slope of the double logarithmic plot of R_g versus N in Fig. 2.8b [Banavar *et al.* (2005)].

2.7 Swelling of Randomly Branched Macromolecules

In the absence of excluded volume interactions, the radius of gyration of randomly branched macromolecules depends on the total number of segments N, according to $R_g \sim N^{1/4}\ell$ (Eq. (2.3.1)). Inclusion of repulsive two-body interactions among segments, as in Eq. (2.5.3), results in the free energy of a randomly branched macromolecule as

$$\frac{F(R)}{k_B T} = \frac{3}{2}\frac{R^2}{\sqrt{N}\ell^2} + \frac{v\ell^3 N^2}{2R^3},$$

(2.7.1)

where $N\ell^2 (= \langle R^2 \rangle_0)$ for a linear Gaussian chain is replaced by its equivalent result for the branched chain, $\sqrt{N}\ell^2$, as R_{g0} is proportional to R in Eq. (2.3.1) and the prefactors are taken as unspecified constants. The equilibrium radius of a branched macromolecule with strong excluded volume interaction in a good solvent is obtained by minimizing $F(R)$ of Eq. (2.7.1) with respect to R to get

$$\frac{R}{\ell} \sim v^{1/5} \sqrt{N}. \tag{2.7.2}$$

Although the N-exponent for the dependence of the coil size is the same (1/2) for both a Gaussian coil at the Flory temperature and a randomly branched macromolecule in a good solvent, the distinction between these systems is in the dependence on the excluded volume parameter. Predictions of the Flory theory for randomly branched polymers are in qualitative agreement with results from simulations [Everaers *et al.* (2017)].

2.8   Summary of Size Exponents

The above results on the role of excluded volume effects on the size of macromolecules are summarized as

$$R_g \sim N^\nu, \tag{2.8.1}$$

where ν is the size exponent. Analogous to a compact sphere of radius R, where $R^3 \sim N$, Eq. (2.8.1) can be rewritten as

$$R^{d_f} \sim N, \qquad d_f \equiv \frac{1}{\nu}, \tag{2.8.2}$$

where d_f is the fractal dimension of the macromolecule. The value of ν depends on the quality of the solvent, that is the temperature relative to the ideal temperature θ_F for a particular combination of the macromolecule and solvent. The three reference states are the Gaussian chain at the ideal temperature with $\nu = 1/2$, fully swollen state due to repulsive excluded volume interactions with $\nu = 3/5$, and the globular state with $\nu = 1/3$,

$$\nu = \begin{cases} \frac{3}{5} & T \gg \theta_F & \text{(Good solvent, SAW)} \\ \frac{1}{2} & T = \theta_F & \text{(Ideal solvent, Gaussian)} \\ \frac{1}{3} & T \ll \theta_F & \text{(Poor solvent, Globule).} \end{cases} \tag{2.8.3}$$

These states and the crossover behaviors between them are sketched in Fig. 2.9 for the UCST system. The temperature dependence of the size of a semiflexible polymer is also included in Fig. 2.9. As the temperature is lowered, the persistence length of a semiflexible chain becomes larger and hence the average coil size increases. Asymptotically at lower temperatures, the chain becomes rodlike so that the chain radius is proportional to its contour length, that is $\nu = 1$. For a randomly branched macromolecule in a good solvent, $\nu = 1/2$. Although this result is the same as that for a linear flexible chain in ideal solvents, the radius of gyration of a randomly branched

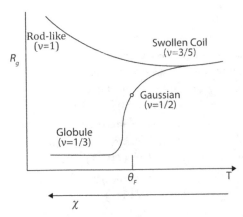

Figure 2.9 A summary of excluded volume effect on the size of a flexible polymer chain. The size exponent is 1/3, 1/2, and 3/5, for poor solvents, ideal conditions, and good solvents, respectively. An effective size exponent $1/3 < \nu_{eff} < 3/5$ may be identified in the crossover regions. The temperature dependence of a semiflexible chain is also included. For UCST systems, the dependence of polymer size on temperature follows opposite trends for flexible and semiflexible chains. In terms of χ, decreasing along the abscissa, behavior of R_g is the same for both UCST and LCST.

macromolecule scales as $v^{1/5}\sqrt{N}$, whereas there is no dependence on v for a Gaussian chain at the Flory temperature.

For the general situations of UCST or LCST, the same sketch in Fig. 2.9 for flexible chain is valid if the Flory–Huggins parameter χ decreases along the abscissa.

2.9   Experimental Determination of Size Exponent

Many structural and thermodynamic properties of a macromolecule in a solvent are dictated by its size exponent ν. For example, the average monomer density profile $\rho(r)$ of a chain as a function of the radial distance r from the center-of-mass of the chain is given as [de Gennes (1979)]

$$\rho(r) \sim \frac{1}{r^{3-\frac{1}{\nu}}}, \tag{2.9.1}$$

for $\ell < r < R_g$. For example, $\rho \sim 1/r$ for ideal solutions ($\nu = 1/2$), $\rho \sim 1/r^{4/3}$ for good solutions ($\nu = 3/5$), and so on. The Fourier transform of $\rho(r)$ in three dimensions is the scattering form factor $P(k)$ determined in various scattering experiments,

$$P(k) \sim \frac{1}{k^{1/\nu}}, \qquad \frac{2\pi}{\ell} > k > \frac{2\pi}{R_g}, \tag{2.9.2}$$

where k is the scattering wave number $k = \frac{4\pi}{\lambda}\sin(\frac{\theta}{2})$, where λ is the wavelength of the incident radiation, and θ is the scattering angle. In view of Eq. (2.8.2), the form factor in the above equation is written as

$$P(k) \sim \frac{1}{k^{d_f}}, \tag{2.9.3}$$

where d_f is the fractal dimension of the chain. This result is generally valid for all statistical fractal objects when their internal self-similar structure is explored using appropriate scattering wave vectors. According to Eq. (2.9.2), a double logarithmic plot of the form factor versus scattering wave number should be linear with a slope of $-1/\nu$. Therefore, the size exponent of a polymer chain in dilute solutions under various conditions can be experimentally determined by measuring the form factor from the scattering intensity at many scattering wave vectors.

In addition, in the limit of $kR_g \ll 1$, the structure factor, $S(k) = NP(k)$, of any scattering object is [Yamakawa (1971), Higgins & Benoit (1994)]

$$S(k) = N\left(1 - \frac{k^2 R_g^2}{3} + \cdots\right), \qquad kR_g \ll 1. \tag{2.9.4}$$

This important result is exemplified by the Debye structure factor given in Eq. (2.1.10) for Gaussian chains. Therefore, from a plot (known as the Zimm plot) of inverse scattering intensity, which is proportional to the inverse of the structure factor, against k^2 is linear, with the intercept and the slope giving, respectively, the reciprocal of the molar mass and one-third of the square of the radius of gyration divided by the molar mass of the macromolecule. From a table of R_g and the corresponding molecular weight of the macromolecule, the size exponent is determined using Eq. (2.8.1).

The determination of ν in dilute solutions is extremely important. The knowledge of the value of ν for a particular polymer–solvent combination allows an adequate description of the system at higher polymer concentrations, as we shall see numerous times in later chapters.

3   Water, Oil, and Salt

Aqueous solutions of charged macromolecules are inherently inhomogeneous at molecular length scales in terms of charge distribution. As the various ionic species move around due to thermal energy, the charge distribution and the accompanying force fields undergo perpetual changes. In addition, entropic changes associated with reorganization of water molecules around the various charged species play a significant role in the properties of the system. The reorganization of solvent molecules occurs simultaneously with the conformational changes of the macromolecules and rearrangement of various charged species in the system. As a result, a self-consistent description of the charge distribution, force, electric field, electric potential, potential energy, solvent entropy, and free energy is required to adequately describe the properties of solutions of charged macromolecules. In order to understand the various rich phenomena emerging from such collective behavior, it is first necessary to have a good grasp on the properties of small molecules relevant to solutions of charged macromolecules. These include the solvent water, oil constituting the backbone of the macromolecule, and low molecular weight salt, which is used to tune the properties of the various charged macromolecular systems. An overview of these small molecular systems is the focus of the present chapter. We first remind ourselves of the basics of electrostatics in vacuum. Next, we discuss dielectric media and electric polarization with a particular emphasis on water, which is the most commonly used solvent in the study of charged macromolecules. We shall then discuss the energetics of isolated ions, ion pairs, dipoles, and the collective behavior of strong electrolytes in dilute solutions. We conclude by discussing the fundamental scales of energy and length in charged systems. Several key concepts in the field of charged macromolecules, such as the Born free energy, Bjerrum length, entropy of released bound water, Poisson–Boltzmann formalism, Debye–Hückel theory, and electrostatic screening length, are presented in this chapter.

3.1   Basics of Electrostatics in Vacuum

Consider a collection of charges in vacuum, as in Fig. 3.1, where q_j is the charge of the jth ionic species at the spatial location $\mathbf{r}'$, and Q is the charge of a specific ionic species at $\mathbf{r}$, which is at a vectorial distance $\mathcal{R} = \mathbf{r} - \mathbf{r}'$ from j.

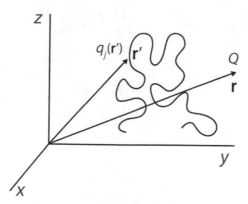

Figure 3.1 An arrangement of a collection of charges in three-dimensional space, where j, q_j, and $\mathbf{r}'$ characterize the charge distribution. The charge Q at $\mathbf{r}$ is at a distance $\mathcal{R}$ from q_j. $(x, y,$ and z are the Cartesian coordinates.)

3.1.1   Force and Electric Field

The force $\mathbf{f}_j(\mathbf{r})$ exerted by the charge q_j located at $\mathbf{r}'$ on a charge Q at $\mathbf{r}$ is given by the **Coulomb law**,

$$\mathbf{f}_j(\mathbf{r}) = \frac{1}{4\pi\epsilon_0} \frac{q_j Q}{\mathcal{R}^2} \hat{\mathcal{R}}, \tag{3.1.1}$$

where $\mathcal{R} = \mathbf{r} - \mathbf{r}'$ and $\hat{\mathcal{R}}$ is the unit vector along $\mathcal{R}$. The unit of charge is the electronic charge e and ϵ_0 is the permittivity of the vacuum, and their values in SI units are [Young & Freedman (2000)]

$$e = 1.602 \times 10^{-19} \text{C} \qquad \text{and} \qquad \frac{1}{4\pi\epsilon_0} = 8.988 \times 10^9 \frac{\text{N.m}^2}{\text{C}^2}, \tag{3.1.2}$$

where force is in Newtons (N), distance in meters (m), and charge in Coulombs (C). The force is attractive or repulsive, if q_1 and Q have opposite signs or the same sign, respectively. The electric field $\mathbf{E}_j(\mathbf{r})$ at $\mathbf{r}$ due to q_j at $\mathbf{r}'$ is the force per unit charge at $\mathbf{r}$,

$$\mathbf{E}_j(\mathbf{r}) = \frac{1}{Q}\mathbf{f}_j(\mathbf{r}) = \frac{1}{4\pi\epsilon_0} \frac{q_j}{\mathcal{R}^2} \hat{\mathcal{R}}. \tag{3.1.3}$$

When the source of the electric field is a distribution of charges given by $\rho(\mathbf{r}')$, the net electric field $\mathbf{E}(\mathbf{r})$ at $\mathbf{r}$ is the superposition of electric fields from all individual source charges at various locations,

$$\mathbf{E}(\mathbf{r}) = \frac{1}{4\pi\epsilon_0} \int d\mathbf{r}' \rho(\mathbf{r}') \frac{\hat{\mathcal{R}}}{\mathcal{R}^2}. \tag{3.1.4}$$

This integral expression for $\mathbf{E}(\mathbf{r})$ can be equivalently written [Griffiths (1999)] in its differential form as the **Gauss law**,

$$\nabla \cdot \mathbf{E}(\mathbf{r}) = \frac{1}{\epsilon_0}\rho(\mathbf{r}), \tag{3.1.5}$$

where the left-hand side is the divergence of the electric field.

3.1.2   Electric Potential

The electric field can be expressed [Griffiths (1999)] as a gradient of a scalar function ψ, called electric potential, defined by

$$\mathbf{E}(\mathbf{r}) = -\nabla \psi(\mathbf{r}). \tag{3.1.6}$$

The electric potential $\psi(\mathbf{r})$ is defined within a constant that can be determined by choosing convenient reference points in space [Griffiths (1999)]. Substitution of Eq. (3.1.6) into Eq. (3.1.5) gives the **Poisson equation**,

$$\nabla^2 \psi(\mathbf{r}) = -\frac{\rho(\mathbf{r})}{\epsilon_0}. \tag{3.1.7}$$

The electric potential for a collection of charges prescribed by the charge distribution $\rho(\mathbf{r}')$ is given by the integral form of the Poisson equation as [Griffiths (1999)]

$$\psi(\mathbf{r}) = \frac{1}{4\pi\epsilon_0} \int d\mathbf{r}' \frac{\rho(\mathbf{r}')}{|\mathbf{r} - \mathbf{r}'|}. \tag{3.1.8}$$

If there is a point charge q at the origin of the coordinate system, then the electric potential at $\mathbf{r}$ follows from this equation as

$$\psi(\mathbf{r}) = \frac{q}{4\pi\epsilon_0 r}, \tag{3.1.9}$$

where r is the magnitude of the distance vector $\mathbf{r}$. For a collection of charges, if the location $\mathbf{r}$ is far away from the localized charge distribution, the electric potential at $\mathbf{r}$ of Eq. (3.1.8) can be expressed [Jackson (1999), Griffiths (1999)] as a **multipole expansion** by writing $1/|\mathbf{r} - \mathbf{r}'|$ as a series in $1/r$, resulting in

$$\psi(\mathbf{r}) = \frac{q}{4\pi\epsilon_0 r} + \frac{1}{4\pi\epsilon_0} \frac{\hat{\mathbf{r}} \cdot \hat{\mathbf{p}}}{r^2} + \cdots, \tag{3.1.10}$$

where the hats denote unit vectors, q is the total charge, and $\mathbf{p}$ is the dipole moment of the charge distribution,

$$q = \int d\mathbf{r}' \rho(\mathbf{r}') \tag{3.1.11}$$

$$\mathbf{p} = \int d\mathbf{r}' \mathbf{r}' \rho(\mathbf{r}'). \tag{3.1.12}$$

The terms on the right-hand side of Eq. (3.1.10) are the monopole, dipole, etc., contributions to the electric potential at $\mathbf{r}$. Even when the total charge q of the charge distribution is 0, the electric potential can be nonzero with the dipolar contribution as the leading term.

3.1.3   Electrostatic Potential Energy

The electrostatic potential energy W of a charge Q located at $\mathbf{r}$, where the electric potential due to other charges in the assembly is $\psi(\mathbf{r})$, is given by

$$W(\mathbf{r}) = Q\psi(\mathbf{r}). \tag{3.1.13}$$

As an example, the electrostatic potential energy W_{ij} of a pair of charges q_i at $\mathbf{r}_i$ and q_j at $\mathbf{r}_j$ is the product of q_i and the electric potential at $\mathbf{r}_i$ given by Eq. (3.1.9) due to q_j at $\mathbf{r}_j$,

$$W_{ij} = \frac{q_i q_j}{4\pi\epsilon_0 r_{ij}}, \qquad (3.1.14)$$

where $r_{ij} = |\mathbf{r}_i - \mathbf{r}_j|$ is the separation distance between the two charges. As a generalization of Eq. (3.1.14), the electrostatic potential energy of an assembly of charges with charge distribution $\rho(\mathbf{r})$ is given by

$$W = \frac{1}{2} \int d\mathbf{r} \rho(\mathbf{r}) \psi(\mathbf{r}), \qquad (3.1.15)$$

where the factor 1/2 avoids double counting of interactions among charges and the integral is over the whole space of charge distribution.

Using the Gauss law (Eq. (3.1.5)) and (3.1.6) in Eq. (3.1.15), the electrostatic potential energy of an assembly of charges can be written [Griffiths (1999)] in terms of only the electric field as

$$W = \frac{\epsilon_0}{2} \int d\mathbf{r} E^2(\mathbf{r}), \qquad (3.1.16)$$

where the integral is over the whole space.

In summary, the electric force $\mathbf{f}$, electric field $\mathbf{E}$, electric potential ψ, and electrostatic potential energy W are given, respectively, by Eqs. (3.1.1), (3.1.4), (3.1.8), and (3.1.15) for any given charge distribution $\rho(\mathbf{r})$.

3.2   Dielectric Media

The strength of an electric field is reduced inside a dielectric medium, such as water, in comparison with its value in vacuum. In general, for an isotropic and homogeneous dielectric, the electric field inside the dielectric is given by

$$\mathbf{E} = \frac{1}{\epsilon} \mathbf{E}_{\text{vacuum}}, \qquad (3.2.1)$$

where ϵ is the static dielectric constant, which is taken as a representation of the dielectric medium, and $\mathbf{E}_{\text{vacuum}}$ is the electric field if the dielectric medium was absent. The reduction in the electric field strength is due to orientation of molecules constituting the medium in response to the electric field [Debye (1928), Frohlich (1958)]. As a result, the medium is polarized with polarization $\mathbf{P}$, which is the dipole moment per unit volume, or equivalently, the medium acquires bound charges of density ρ_b in the bulk of the medium [Griffiths (1999)],

$$\rho_b = -\nabla \cdot \mathbf{P}. \qquad (3.2.2)$$

Therefore, the Gauss law, Eq. (3.1.5), for the electric field inside a dielectric, is

$$\nabla \cdot \mathbf{E}(\mathbf{r}) = \frac{1}{\epsilon_0}[\rho(\mathbf{r}) + \rho_b(\mathbf{r})], \qquad (3.2.3)$$

where ρ and ρ_b denote the densities of free charges and bound charges, respectively.

Defining the electric displacement $\mathbf{D}$ as

$$\mathbf{D}(\mathbf{r}) = \epsilon_0 \mathbf{E}(\mathbf{r}) + \mathbf{P}(\mathbf{r}), \tag{3.2.4}$$

we obtain from Eqs. (3.2.2) to (3.2.4) the Gauss law for a dielectric medium,

$$\nabla \cdot \mathbf{D}(\mathbf{r}) = \rho(\mathbf{r}). \tag{3.2.5}$$

Since the extent of polarization in the medium depends on the external electric field, we write $\mathbf{P}$ as

$$\mathbf{P}(\mathbf{r}) \equiv \epsilon_0 \chi_e \cdot \mathbf{E}(\mathbf{r}), \tag{3.2.6}$$

which defines the electric susceptibility tensor χ_e. Substitution of Eq. (3.2.6) into Eq. (3.2.4) gives

$$\mathbf{D}(\mathbf{r}) = \epsilon_0 \epsilon \cdot \mathbf{E}(\mathbf{r}), \tag{3.2.7}$$

where ϵ is the dielectric tensor,

$$\epsilon = \mathbf{1} + \chi_e, \tag{3.2.8}$$

with $\mathbf{1}$ being the unit tensor. In general, the dielectric tensor depends on the spatial location and the electric field as explicitly expressed as $\epsilon(\mathbf{r}, \mathbf{E}(\mathbf{r}))$. Using Eq. (3.1.6) for the electric field, $\mathbf{E}(\mathbf{r}) = -\nabla\psi(\mathbf{r})$, and combining with Eqs. (3.2.5) and (3.2.7), the Poisson equation for a dielectric medium is

$$\nabla \cdot [\epsilon_0 \epsilon(\mathbf{r}, \mathbf{E}(\mathbf{r})) \cdot \nabla\psi(\mathbf{r})] = -\rho(\mathbf{r}). \tag{3.2.9}$$

This is a general result embodying the spatial heterogeneity of the electrical properties of the dielectric medium and the nonlinear response of the dielectric tensor to the applied electric field.

The electrostatic potential energy W, which is the work done in assembling a collection of charges in a dielectric medium is [Griffiths (1999)]

$$W = \frac{1}{2} \int d\mathbf{r} \mathbf{D}(\mathbf{r}) \cdot \mathbf{E}(\mathbf{r}), \tag{3.2.10}$$

where the electric displacement is defined through Eqs. (3.2.4) and (3.2.7).

3.2.1   Homogeneous Isotropic Linear Dielectrics

For a homogeneous isotropic linear dielectric,

$$\epsilon = \epsilon \mathbf{1}, \tag{3.2.11}$$

where ϵ is the uniform static dielectric constant appearing in Eq. (3.2.1). Under these special conditions, the Poisson equation, Eq. (3.2.9), becomes

$$\nabla^2 \psi(\mathbf{r}) = -\frac{\rho(\mathbf{r})}{\epsilon_0 \epsilon}. \tag{3.2.12}$$

Therefore, for homogeneous isotropic linear dielectric media, the permittivity ϵ_0 of vacuum appearing in all formulas of Section 3.1 can simply be replaced by $\epsilon_0 \epsilon$, which is now the permittivity of the dielectric medium. As an example, the electric potential and the electric field at a distance $\mathbf{r}$ inside the dielectric from a charge q at the origin follow from Eqs. (3.1.3) and (3.1.9) as

$$\psi(\mathbf{r}) = \frac{q}{4\pi\epsilon_0\epsilon r}, \qquad \mathbf{E}(\mathbf{r}) = \frac{q}{4\pi\epsilon_0\epsilon r^2}\hat{\mathbf{r}}. \tag{3.2.13}$$

The electrostatic potential energy W_{ij} for two charges q_i and q_j separated by the distance $\mathbf{r}_{ij}$ follows from Eq. (3.1.14) as

$$W_{ij} = \frac{q_i q_j}{4\pi\epsilon_0\epsilon r_{ij}}. \tag{3.2.14}$$

Analogously, the work done in assembling a collection of charges in a dielectric medium follows from Eq. (3.2.10) as

$$W = \frac{\epsilon_0\epsilon}{2} \int d\mathbf{r} E^2(\mathbf{r}). \tag{3.2.15}$$

In fact, the electrostatic potential energy, which is the amount of work during a reversible process of assembly of charges at constant temperature and volume, is equal to the change in the Helmholtz free energy ΔF_{el} associated with the assembly of the charges,

$$\Delta F_{el} = W. \tag{3.2.16}$$

The Helmholtz free energy F of a dielectric is written as

$$F = F_0 + \Delta F_{el}, \tag{3.2.17}$$

where F_0 is the Helmholtz free energy in the absence of any electric field (both internal and external), and ΔF_{el} is given by Eqs. (3.2.15) and (3.2.16), so that F is given as

$$F = F_0 + \epsilon\left(\frac{\epsilon_0}{2} \int d\mathbf{r} E^2(\mathbf{r})\right). \tag{3.2.18}$$

3.2.2   Entropy of Dielectrics

If the uniform dielectric constant ϵ is independent of temperature, ΔF_{el} given in Eq. (3.2.18) is also a temperature-independent quantity and is simply the change in the internal energy of the system due to the external electric field. However, in general, the dielectric constant of a typical dielectric medium such as water is dependent on temperature. Therefore, ΔF_{el} represents both the internal energy change and entropic change associated with reorientation of molecules constituting the dielectric medium in response to the external electric field. Using the thermodynamic relation $F = U - TS$ and $S = -(\partial F/\partial T)_{V,D}$, and Eq. (3.2.10), the energetic and entropic parts of ΔF_{el} can be shown as [Frohlich (1958)]

$$\Delta F_{el} = \Delta U_{el} - T\Delta S_{el} = \epsilon\left(\frac{\epsilon_0}{2} \int d\mathbf{r} E^2(\mathbf{r})\right) \tag{3.2.19}$$

$$\Delta U_{el} = \left(\epsilon + T\frac{\partial \epsilon}{\partial T}\right)\left(\frac{\epsilon_0}{2}\int d\mathbf{r}E^2(\mathbf{r})\right) \tag{3.2.20}$$

$$\Delta S_{el} = \frac{\partial \epsilon}{\partial T}\left(\frac{\epsilon_0}{2}\int d\mathbf{r}E^2(\mathbf{r})\right). \tag{3.2.21}$$

When $\partial \epsilon/\partial T < 0$, which is the case in aqueous solutions, $\Delta S_{el} < 0$, as seen from Eq. (3.2.21). This demonstrates that the entropy of the system decreases due to the presence of the external electric field as it creates more ordering of the constituent molecules. Therefore, ΔF_{el} given in Eq. (3.2.19) for dielectrics is not simply energy, but Helmholtz free energy. In fact, the entropic contribution can dominate over the energy contribution as illustrated in the next section for the most commonly used solvent in assembling charged macromolecules, namely water.

3.3   Water

Water is the matrix of life, since it is the solvent in all known biological systems [Franks (2000)]. It is also the most commonly used solvent in *in vitro* experiments on charged macromolecules. In view of the preponderance of water as the solvent, we describe its key properties pertinent to the discussion of solutions of charged macromolecules. Each water molecule is electrically neutral, but has a dipole moment $p = 1.85$ D, where 1 D = 3.3×10^{-30} C m (Fig. 3.2a). The various water molecules interact among themselves through dipole–dipole interactions and hydrogen bonding. The average hydrogen-bond energy is about 23 kJ/mol and is about an order of magnitude stronger than the thermal energy (2.5 kJ/mol at 25°C). Due to the relatively high energy of hydrogen bonds, which is still smaller than the covalent O–H bond energy of about 470 kJ/mol, the water molecules in the liquid state arrange themselves as a network made from hydrogen bonds (Fig. 3.2b). Yet, each molecule is able to rotate and diffuse. At 25°C, the rotational relaxation time is 8.3 ps and the diffusion coefficient is 2.5×10^{-9} m^2/s.

For length and time scales at which macroscopic samples of water are investigated, we adopt a continuum description where a uniform dielectric constant ϵ is ascribed

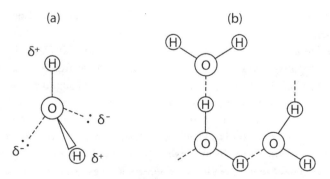

Figure 3.2 (a) Water molecule is a dipole. (b) Hydrogen bonding network of water.

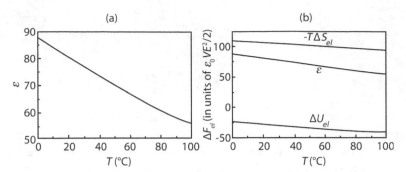

Figure 3.3 (a) Temperature dependence of static dielectric constant of water. (b) Entropic contribution $-T\Delta S_{el}$ to the Helmholtz free energy ΔF_{el} dominates over the energetic contribution ΔU_{el} in the presence of an electric field. The ordinate is in units of $\epsilon_0 V E^2/2$.

to water as an empirical parameter. Phenomenologically, ϵ is dependent on temperature [Malmberg & Maryott (1956)] as given in Fig. 3.3a. Based on the discussion in Section 3.2.2, the temperature dependence of ϵ implies that ϵ contains both energetic and entropic contributions.

For a uniform electric field $\mathbf{E}$ and a homogeneous sample of volume V, the electrostatic contributions to the Helmholtz free energy, internal energy, and entropy due to the presence of the electric field follow from Eqs. (3.2.19) to (3.2.21), in units of $\epsilon_0 V E^2/2$, as

$$\Delta F_{el} = \epsilon \tag{3.3.1}$$

$$\Delta U_{el} = \epsilon + T\frac{\partial \epsilon}{\partial T} \tag{3.3.2}$$

$$\Delta S_{el} = \frac{\partial \epsilon}{\partial T}. \tag{3.3.3}$$

As seen in Fig. 3.3a, $\partial \epsilon/\partial T$ is nonzero, so entropy plays a significant role in the response of an aqueous solution to an electric field. In fact, at the room temperature of 25°C,

$$\frac{\partial \ln \epsilon}{\partial \ln T} = -1.36. \tag{3.3.4}$$

The quantities $\Delta F_{el}, \Delta U_{el}$, and $-T\Delta S_{el}$, obtained from Eqs. (3.3.1) to (3.3.3) and the temperature dependence of ϵ given in Fig. 3.3a, are plotted in Fig. 3.3b as functions of temperature in the range between 0°C and 100°C. As seen from the figure, ΔU_{el} is negative and the entropic contribution to the free energy, $-T\Delta S_{el}$, dominates over ΔU_{el} resulting in a positive value of ΔF_{el} throughout the temperature range pertinent to water in its liquid state. Thus the entropic loss associated with ordering of water molecules due to the external electric field is the dominant feature of water. This phenomenon is generic to water independent of the source of the electric field. The electric fields from ions and charged macromolecules dispersed in water create significant reorganization of water molecules around themselves, and the free energy of the solution is often dominated by entropic changes in water.

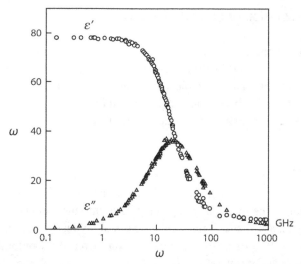

Figure 3.4 Frequency dependence of ϵ' and ϵ'' for water at 25°C in the microwave region of the electromagnetic radiation [Ellison *et al.* (1996)].

In general, the dielectric response of a medium to a time-dependent electric field is also time dependent. This response is measured as a frequency-dependent dielectric constant using experimental techniques such as impedance spectroscopy. It is convenient to represent ϵ as a complex quantity,

$$\epsilon(\omega) = \epsilon'(\omega) - i\epsilon''(\omega), \tag{3.3.5}$$

where ϵ' and ϵ'' are the real and imaginary parts, which depend on the frequency ω. While ϵ' is a measure of the influx of energy into the dielectric due to the electric field, ϵ'' is a measure of loss of energy in the medium, and naturally ϵ' and ϵ'' are inter-related [Frohlich (1958)]. At low frequencies, the complex dielectric constant approaches the static value. The frequency dependence of the complex dielectric constant of water at 25°C in the microwave region is given in Fig. 3.4 [Ellison *et al.* (1996)]. At low frequencies, $\epsilon(\omega)$ approaches the static value of 80. At higher frequencies, ϵ' and ϵ'' reflect the dynamic processes in the system. From the rate of decay of ϵ' with ω or the position of the maximum in $\epsilon''(\omega)$, the characteristic relaxation times associated with various molecular dynamics, such as rotational diffusion, can be determined. As seen in Fig. 3.4, care must be exercised in associating the value of 80 for the dielectric constant of water in considerations of dynamics of charged macromolecules at shorter time scales (that is, higher frequencies).

As ϵ' and ϵ'' are related to the refractive index n_r and absorption coefficient of the medium [Jackson (1999)], the refractive index of a medium is also frequency dependent. At the optical wave length of 589 nm and at 15°C, the refractive index of water is 1.333. The corresponding value for proteins is in the range between 1.51 and 1.54, providing excellent contrast between proteins and water in the optical range of the electromagnetic spectrum. As a result, structures of charged macromolecules in aqueous solutions can be readily investigated using optical techniques.

3.4   Dissolution of Salt and Oil in Water

Let us consider the stabilities of a pinch of salt and a drop of an oil in terms of dissolution in water. While the salt can completely dissociate into ions, the oil molecules resist dissolution. The oil molecules induce significant reorientation of water molecules accompanied by entropic changes, and are surrounded by cage-like structures, called clathrates, constituted by water molecules.

3.4.1   Salt

As an example, consider a sample of NaCl crystals consisting of N_s cations and N_s anions. The ions are arranged regularly into the salt crystal lattice structure shown in Fig. 3.5a with the nearest neighbor cation–anion distance $R_0 = 0.282$ nm. The potential energy U_{salt} of this ionic crystal arising from the Coulombic interactions among all ions is approximately given as [Kittel (1996), Berry $et\ al.$ (1980)]

$$U_{salt} = -\frac{N_s e^2 M_s}{4\pi \epsilon_0 R_0},\qquad(3.4.1)$$

where M_s is the Madelung constant with $M_s = 1.7476$ for NaCl lattice. For a pair of adjacent ions inside the lattice, the binding energy of the ion pair is

$$\Delta U_{pair} = \frac{e^2}{4\pi \epsilon_0 R_0} \simeq 200 k_B T.\qquad(3.4.2)$$

Therefore, Eqs. (3.4.1) and (3.4.2) show that even a nanoscopic NaCl crystal of linear dimension 30 nm is incredibly stable with binding energy of about $4\times10^8\ k_B T$. The dissociation constant K_{salt} of one NaCl molecule in the vapor phase is

$$K_{salt,\ vapor} = \frac{C_{Na^+} C_{Cl^-}}{C_{NaCl}} = k_{coll}\, e^{-\Delta U_{pair}/k_B T},\qquad(3.4.3)$$

where the concentrations are in molarity and k_{coll}, arising from the collision frequency of the ions, is about 62 mol/L [Fowler (1966)]. Using Eq. (3.4.2), the dissociation constant is estimated as

$$K_{salt,\ vapor} \simeq 62 \times e^{-200} \quad \sim 10^{-85}\ \text{mol/L.}\qquad(3.4.4)$$

This is a very small number showing that NaCl does not dissociate easily in the vapor phase.

On the other hand, in the aqueous phase, the permittivity is reduced by a factor of 80 (dielectric constant of water) so that the dissociation constant is

$$K_{salt,\ aq} \simeq 62 \times e^{-200/80} \quad \sim 5\ \text{mol/L.}\qquad(3.4.5)$$

Thus, NaCl and other such metallic salts are spontaneously soluble in water.

Table 3.1 Free energy of solvation of nonpolar molecules

Solute	ΔG_s (kJ/mol)
Methane	14.5
n-Butane	24.5
Cyclohexane	28.2
Methyl cyclohexane	31.9
Benzene	19.3
Toluene	22.7
Ethyl benzene	26.1

(a)   (b)

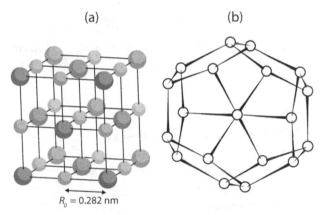

$R_0 = 0.282$ nm

Figure 3.5 (a) Face-centered cubic structure of NaCl. (b) An example of a clathrate cage formed by water molecules around a nonpolar solute molecule. The cage has pentagonal dodecahedral symmetry, and circles denote oxygen atoms [Franks (2000)].

3.4.2   Oil and Hydrophobic Effect

Oil-like substances are made of nonpolar molecules such as alkanes. The Gibbs free energy of solvation ΔG_s associated with a transfer of various nonpolar molecules from a nonpolar solvent into water at 25°C is given in Table 3.1. Generally speaking, the backbones of charged macromolecules are made up of such oil-like nonpolar moieties.

As evident from Table 3.1, the free energy to dissolve a nonpolar molecule in water is prohibitive with a cost of about $10k_BT$ (the thermal energy k_BT is 2.5 kJ/mol at 25°C). The equilibrium constant for dissolution of a nonpolar molecule is

$$K_{\text{nonpolar, aq}} \simeq 62 \times e^{-10} \quad \sim 10^{-4} \text{ mol/L,} \tag{3.4.6}$$

where the equivalent of Eq. (3.4.3) is used. Thus oil and water do not mix. Therefore, the backbones of most of the synthetic charged macromolecules are insoluble in water and the charges endowed on such molecules make them soluble.

The origin of the insolubility of nonpolar molecules in water lies primarily on the entropy associated with reorganization of the molecules. The nonpolar molecules tend to minimize contact with polar molecules and form small clusters. This disrupts the native hydrogen bonding network of water, and water molecules need to rearrange themselves in order to accommodate the presence of the solute molecules. This is reflected in the higher contribution of entropy to ΔG_s than the enthalpic part. For example, the division of ΔG_s for n-butane into the enthalpic part (ΔH_s) and the entropic part $-T\Delta S_s$ is [Israelachvilli (2011)]

$$\Delta G_s = \Delta H_s - T\Delta S_s = -4.3 + 28.8 = 24.5\text{kJ/mol}. \qquad (3.4.7)$$

While the enthalpic part (-4.3 kJ/mol) is favorable for dissolution, the entropic part (28.8 kJ/mol) is highly unfavorable.

The induced reordering of water molecules around nonpolar molecules is revealed by **clathrate cages** formed by water molecules surrounding a guest molecule such as a lower hydrocarbon or ethylene oxide [Franks (2000)]. These cages can have highly symmetric organization as illustrated in Fig. 3.5b. The phenomenon of immiscibility of guest nonpolar molecules with the host water molecules, dominated by entropy associated with solvent reorganization, is generically called the **hydrophobic effect** [Tanford (1980), Israelachvilli (2011)].

3.5   Ion Solvation and Born Free Energy

When an ion is introduced into a dielectric solvent, the electric charge on the ion modifies the solvent structure in its neighborhood. As a result of change in entropy due to solvent reorganization and energetic interactions between the ion and solvent molecules, the free energy of the system changes. The free energy associated with placing an ion in the interior of the solvent, called the "self-energy" of the ion, can be calculated based on electrostatics using the **Born model**, without resorting to molecular details of the solvent. The Born model considers the transfer of a spherical ion of radius r_i and charge $z_i e$ (z_i is the valency of the ion and e is the electronic charge) from vacuum with permittivity ϵ_0 to a solvent of permittivity $\epsilon_0 \epsilon$, as depicted in Fig. 3.6a.

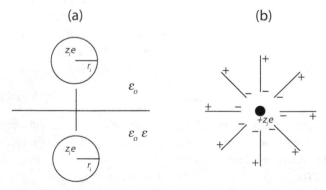

Figure 3.6 (a) Born model for free energy of ion solvation. A spherical ion of radius r_i and charge $z_i e$ is transferred from vacuum to the interior of a solvent with dielectric constant ϵ. (b) Polarization of solvent due to a positively charged ion of charge $z_i e$.

The free energy W_s of the ion inside the solvent follows from Eqs. (3.2.13) and (3.2.15) as

$$W_s = \frac{\epsilon_0 \epsilon}{2} \int d\mathbf{r} E^2(\mathbf{r}) = \frac{\epsilon_0 \epsilon}{2} \int d\mathbf{r} \left(\frac{z_i e}{4\pi \epsilon_0 \epsilon r^2} \right)^2. \tag{3.5.1}$$

The electric field is outside the ion and the radial distance variable in Eq. (3.5.1) cannot be less than the ionic radius r_i. With this cutoff in r, the angular integration of Eq. (3.5.1) gives

$$W_s = \frac{\epsilon_0 \epsilon}{2} 4\pi \int_{r_i}^{\infty} dr r^2 \left(\frac{z_i e}{4\pi \epsilon_0 \epsilon r^2} \right)^2 = \frac{(z_i e)^2}{8\pi \epsilon_0 \epsilon} \int_{r_i}^{\infty} dr \frac{1}{r^2}. \tag{3.5.2}$$

Performing the integral yields

$$W_s = \frac{(z_i e)^2}{8\pi \epsilon_0 \epsilon r_i}. \tag{3.5.3}$$

The self-energy of an ion is proportional to the square of its charge, independent of whether it is positively charged or negatively charged, and inversely proportional to its radius. We shall call this self-energy of an ion as the **Born free energy** of the ion.

The free energy W_0 associated with placing the ion in vacuum (where $\epsilon = 1$) follows from Eq. (3.5.3) as

$$W_0 = \frac{(z_i e)^2}{8\pi \epsilon_0 r_i}. \tag{3.5.4}$$

Therefore, the free energy of ion solvation, defined as the change in free energy due to transfer of an ion from vacuum to the interior of the solvent, $W_s - W_0$, is

$$\Delta F_{\text{ion}} = -\frac{(z_i e)^2}{8\pi \epsilon_0 r_i} \left(1 - \frac{1}{\epsilon} \right). \tag{3.5.5}$$

We shall call this as the **Born free energy of solvation** of the ion. Since $\epsilon > 1$, ΔF_{ion} must be negative and hence immersion of an ion in any dielectric solvent is a spontaneous process.

The changes in the internal energy ΔU_{ion} and entropy ΔS_{ion} accompanying ion solvation are obtained from thermodynamic relations by addressing the temperature dependence of the dielectric constant. The change in entropy accompanying ion solvation follows from Eq. (3.5.5) as

$$\Delta S_{\text{ion}} = -\frac{\partial (\Delta F_{\text{ion}})}{\partial T} = \frac{(z_i e)^2}{8\pi \epsilon_0 r_i} \frac{1}{\epsilon^2} \frac{\partial \epsilon}{\partial T}. \tag{3.5.6}$$

The change in energy ΔU_{ion} follows from Eqs. (3.5.5) and (3.5.6) as

$$\Delta U_{\text{ion}} = \Delta F_{\text{ion}} + T \Delta S_{\text{ion}} = -\frac{(z_i e)^2}{8\pi \epsilon_0 r_i} \left(1 - \frac{1}{\epsilon} - \frac{1}{\epsilon} \frac{\partial \ln \epsilon}{\partial \ln T} \right). \tag{3.5.7}$$

For water as the solvent, the dielectric constant decreases with temperature, as shown in Fig. 3.3a. Also, around room temperatures, $\epsilon \simeq 80$ and $\partial \ln \epsilon / \partial \ln T \simeq -1.36$

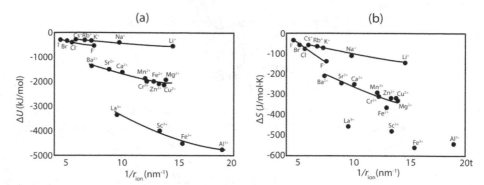

Figure 3.7 Inverse relation between solvation free energy of an ion and its radius. (a) ΔU_{ion} and (b) ΔS_{ion} versus the reciprocal of the crystallographic radius for various ions. ΔU_{ion} dominates over ΔS_{ion} and the behavior of ΔF_{ion} is similar to that of ΔU_{ion}.

(Eq. (3.3.4)). Therefore, both ΔU_{ion} and ΔS_{ion} are negative for ion solvation in water, according to the Born model. The reduction in entropy associated with ion solvation is a representation of reorganization of dipoles of the solvent around the ion, as depicted in Fig. 3.6b. The electric field from the immersed ion polarizes the solvent molecules and the partial negative charges of the dipoles point toward a positively charged ion and vice versa. As a result of such constraints, the entropy of the solvent medium decreases. In terms of magnitudes, ΔU_{ion} is much more negative than ΔS_{ion}, as suggested by Eqs. (3.5.6) and (3.5.7), resulting in overall negative value for ΔF_{ion}. Both ΔU_{ion} and ΔS_{ion} are proportional to $(z_i e)^2 / r_i$,

$$\Delta U_{ion}, \Delta S_{ion} \sim -\frac{(z_i e)^2}{r_i}. \tag{3.5.8}$$

Therefore, the Born model predicts universal dependence of ΔU_{ion} and ΔS_{ion} on the ionic radius and charge of an ion. A plot of these quantities against $1/r_i$ should be linear with slope proportional to the square of the ionic charge independent of its sign. Experimental values of ΔU_{ion} and ΔS_{ion} for a few representative ions in water are plotted in Fig. 3.7, when the ionic radius r_i is taken from crystallographic data [Berry *et al.* (2000), Marcus (2015)]. It is evident from Fig. 3.7, where the data are from Marcus (2015), that ΔU_{ion} is highly negative in comparison with ΔS_{ion}, and both ΔU_{ion} and ΔS_{ion} decrease with $1/r_i$, as predicted by the Born model. However, the dependence on $1/r_i$ is not linear, and the sign of the ion charge leads to significant differences between cations and anions, in contrast to the predictions of the Born model.

The deviations of the experimental values from the Born prediction emphasizes the importance of the specificity of the ion. The charge and the crystallographic radius of the ion are insufficient to fully describe ion solvation. This well-known effect is referred to as the **Hofmeister series** and depicts the fact that various ions of the same sign and charge have profoundly different capacities to be solvated or to influence the solubility of macromolecules such as proteins [Marcus (2015)]. The discrepancy between the Born free energy and experimental results is attributed to the fact that the

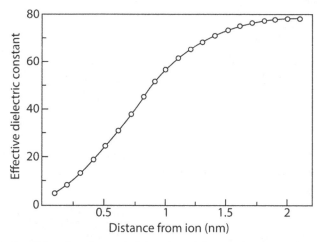

Figure 3.8 Effective dielectric constant as a function of distance from a monovalent ion (adapted from [Mehler & Eichele (1984)]).

hydrated ionic radius is larger than the crystallographic radius. As an empirical procedure, adding 0.085 nm to the radii of cations and 0.01 nm to the radii of anions has been shown to exhibit a universal law for ion solvation free energy of monovalent ions [Bockris & Reddy (1970)]. Another significant source of the discrepancy between the Born free energy and experimental results is the breakdown of the assumption of uniform dielectric constant near an ion. The dependence of the average dielectric constant of water on the distance r from a univalent ion, based on experimental data [Debye (1925), Webb (1926), Schwarzenbach (1936), Conway $et\,al.$ (1951), Mehler & Eichele (1984)] is presented in Fig. 3.8. The effective dielectric constant is significantly lower in the proximity of the ion in comparison with the bulk value. Even at distances of about 1 nm from the ion, the dielectric constant is only about 50 instead of the bulk value of about 80. It is therefore necessary to take into account the nonuniform nature of ϵ as a function of the distance from the location of an ion in treatments of interactions among charged species in water. The inhomogeneity of the dielectric constant will be recognized only phenomenologically in the later chapters on electrostatic interactions in solutions of macromolecules. A fundamental theory for the specific effects of ions is lacking at present.

3.6   Ion-Pair Formation and Bjerrum Length

3.6.1   Release of Bound Water Drives Attraction

The work in bringing two charges q_1 and q_2 to a distance r from infinity in a medium of dielectric constant ϵ is the free energy change given by Eq. (3.2.14) as

$$\Delta F = \frac{q_1 q_2}{4\pi\epsilon_0\epsilon r}. \qquad (3.6.1)$$

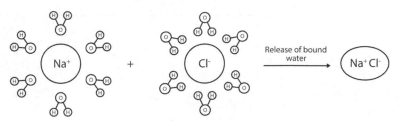

Figure 3.9 Cartoon of release of bound water molecules during ion-pair formation.

This can be resolved into the energetic and entropic parts as in the preceding sections. The entropic contribution to the free energy of formation of an ion pair follows from Eq. (3.6.1) as

$$\Delta S = -\left(\frac{\partial(\Delta F)}{\partial T}\right)_V = (\Delta F)\frac{1}{\epsilon}\frac{\partial \epsilon}{\partial T}, \qquad (3.6.2)$$

where the constant volume condition is indicated as a subscript. For oppositely charged ions, ΔF is negative and $\partial\epsilon/\partial T$ is negative for water and so ΔS is positive,

$$\Delta S > 0. \qquad \text{(oppositely charged ions)} \qquad (3.6.3)$$

The gain in entropy favors ion-pair formation. The energy part is given by

$$\Delta U = \Delta F + T\Delta S = \Delta F\left(1 + \frac{\partial \ln \epsilon}{\partial \ln T}\right), \qquad (3.6.4)$$

where Eqs. (3.6.1) and (3.6.2) are used. Since $\partial \ln \epsilon/\partial \ln T$ is -1.36 around room temperature for water, and $\Delta F < 0$ for oppositely charged ions, the energy contribution is positive

$$\Delta U > 0, \qquad \text{(oppositely charged ions)} \qquad (3.6.5)$$

and is unfavorable for the formation of an ion pair in water.

The above result is counterintuitive. However, for the specific example of formation of an ion pair from Na^+ and Cl^- ions, the values of ΔU and $-T\Delta S$ are, respectively, $0.92\ k_BT$ and $-3.56\ k_BT$, and so the free energy of formation is [Evans & Wennerström (1999)]

$$\Delta F = \Delta U - T\Delta S = 0.92k_BT - 3.56k_BT = -2.64k_BT, \qquad (3.6.6)$$

emphasizing that the dominant driving force for ion-pair formation is entropic in origin, as sketched in Fig. 3.9. Isolated ions are surrounded by water molecules in a coordinated manner such that the oxygen atoms are closer to cations and are facing away from anions. Upon encounter between the cation and anion to form an ion pair, the bound water molecules on the individual ions are released leading to an increase in entropy of the system. The entropy change accompanying the weak reorganization of water molecules around the newly formed ion pair is insufficient to overcome the entropy gain due to release of bound water, resulting in a net gain in entropy. This feature of entropically driven ion-pair formation, operative even for Na^+ and Cl^- ions, is massively amplified when a polycation and a polyanion complex together in aqueous media.

Table 3.2 Bjerrum length at 25°C

Solvent	ϵ	ℓ_B (nm)
Cyclohexane (CH)	2.02	28.0
Toluene (T)	2.38	23.3
Chloroform (C)	4.72	11.86
Tetra hydro furan (THF)	7.4	7.57
iso-Propanol (iP)	19.4	2.89
Ethanol (E)	24.0	2.3
Urea (U)	24.0	2.3
Dimethyl formamide (DMF)	36.7	1.53
Water	80.0	0.7
N-methyl formamide (NFM)	186.9	0.3

3.6.2   Bjerrum Length

In general, the distance at which the electrostatic free energy for a pair of ions in a dielectric medium equals the thermal energy is the Bjerrum length, as introduced in Section 1.3,

$$\ell_B = \frac{e^2}{4\pi\epsilon_0\epsilon k_B T}. \tag{3.6.7}$$

The Bjerrum length depends on T and ϵ (which on its own is temperature dependent). As the dielectric constant varies from one solvent to another, the Bjerrum length changes as shown in Table 3.2. The dependence of ℓ_B on ϵ for different solvents at 25°C is given by the solid curve in Fig. 3.10. For inter-ion distances shorter than ℓ_B, the ions interact more strongly than the thermal energy. For this condition, corresponding to the area below the solid curve in Fig. 3.10, oppositely charged ions can form ion pairs as marked in the figure. On the other hand, for inter-ion distances longer than ℓ_B, the interaction energy is weaker than the thermal energy, and no ion-pair formation is expected, as marked in the area above the solid curve in Fig. 3.10.

In general, the formation of an ion pair at a separation distance r when one ion is brought from an initial distance far away (Fig. 3.11a) depends both on the electrostatic interaction and translational entropy of the ions. The free energy of ion-pair formation at a finite distance r between the ions is

$$\frac{\Delta F}{k_B T} = -\frac{\ell_B}{r} + \ln V, \tag{3.6.8}$$

where the $\ln V$ term is due to the loss of translational entropy in the formation of the ion pair. The qualitative feature of Eq. (3.6.8) is sketched in Fig. 3.11b. If the volume of the system is very large, then the ions would not form the ion pair, because $\Delta F > 0$ for such very dilute conditions. The formation of ion pair thus depends not only on the Bjerrum length but also on the concentration of ions in the solution. Furthermore, the dielectric constant near an ion, particularly near charged groups in a macromolecule, is not the bulk value used in constructing the solid curve (ℓ_B versus ϵ)

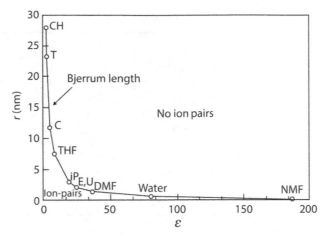

Figure 3.10 Plot of Bjerrum length as a function of uniform dielectric constant. The symbols denote the names of the solvent as in Table 3.2. Below the solid curve where the inter-ion distance r is less than ℓ_B, the free energy of ion pair is stronger than thermal energy facilitating ion pair formation. Above the solid curve, formation of ion pair is unlikely.

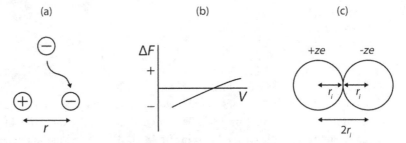

Figure 3.11 (a) Approach of an ion from a large distance to r from another oppositely charged ion. (b) Translational entropy contributes to ion pair formation. (c) Two ions of charges $+ze$ and $-ze$, each of finite size r_i forming an ion pair.

in Fig. 3.10. The local dielectric constant near the backbone, which is essentially oil-like, can be quite low and hence ion-pair formation can be significantly strong at short inter-ion distances leading to a considerable modification on the net charge of charged macromolecules as discussed in Section 1.3.

3.6.3   Effect of Ion Size

Instead of point-like charges, consider a pair of oppositely charged ions, each of radius r_i and charge $\pm ze$ (Fig. 3.11c). Let us choose a reference state for the ions where they are initially in vacuum with their separation distance being infinitely large. These ions are then bought into a dielectric medium to form an ion pair. The free energy of this ion pair, which is equivalent to a dipole of charge ze and length $2r_i$ is the sum of the Born free energy of the ions and the interaction free energy between the ions,

$$\Delta F = \frac{(+ze)^2}{8\pi\epsilon_0\epsilon r_i} + \frac{(-ze)^2}{8\pi\epsilon_0\epsilon r_i} + \frac{(+ze)(-ze)}{4\pi\epsilon_0\epsilon(2r_i)} = \frac{(ze)^2}{8\pi\epsilon_0\epsilon r_i}, \tag{3.6.9}$$

where Eqs. (3.5.3) and (3.2.14) are used for the Born free energy and the interaction energy, respectively. Thus the self-energy of an ion pair (dipole) is the same as that of an individual ion, based on the Born model. Therefore, we expect the solubility of an ion pair to be comparable to that of an isolated ion.

3.7 Dipoles

When a counterion adsorbs on a polyelectrolyte chain at a particular ionic group, as in Fig. 1.7, a temporary dipole is formed. There is also a class of polymers, called polyzwitterions, where every repeat unit is a dipole. Examples of polyzwitterions are the polymers from the dipolar monomers, such as sulfobetaine and phosphoryl choline (Fig. 3.12). In view of the prevalence of dipoles in charged macromolecules, even in the context of homo polyelectrolytes, we summarize below some of the key aspects of dipoles.

For a collection of charges in any medium, such as a distribution of both positive and negative electrolyte ions (Fig. 3.13a), or a segment of chromatin (Fig. 3.13b) [Alberts *et al.* (2016)], the dipole moment **p** is given by Eq. (3.1.12) as

$$\mathbf{p} = \int d\mathbf{r} \, \mathbf{r} \, \rho(\mathbf{r}),\tag{3.7.1}$$

where $\rho(\mathbf{r})$ is the charge density. For a simple electric dipole as in Fig. 3.13c, where the charges $+q$ and $-q$ are separated by a distance d, the dipole moment is a vector pointing toward the positive charge with a magnitude

$$p = |\mathbf{p}| = qd.\tag{3.7.2}$$

Depending on the nature of the assembly of charges, the dipole moment can be very high and hence interactions among dipoles can be quite significant. The dipole

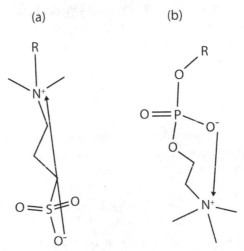

Figure 3.12 (a) Sulfobetaine. (b) Phosphorylcholine. R is a chemical group, and the arrows indicate the dipoles.

Table 3.3 Biological examples of dipole moments

System	Dipole moment p (D)
Water	1.85
Peptide bond	3.5
α-Helix	63.1
Proteins	100–1000
DNA (calf thymus)	32,000

(a)   (b)   (c)

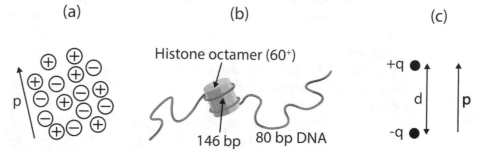

Figure 3.13 (a) Collection of charges. (b) Cartoon of a nucleosome assembled from dsDNA and histone octamer. About 146 bp of DNA wind around the histone octamer which carries about 60 positive charges. This assembly is repeated through 80-bp-long linker dsDNA. (c) Simple electric dipole of moment qd.

moments of water and some charged assemblies are given in Table 3.3 in units of Debye ($1D = 3.3 \times 10^{-30}$C m) [Daune (1999)].

3.7.1   Properties of Simple Dipoles

a. Self-energy

Consider a dipole made from two spherical ions of radius r_i and charges $+ze$ and $-ze$, separated by a distance d, as in Fig. 3.14a. Following the same arguments as in Section 3.6.3 and using the same reference state, the Helmholtz free energy ΔF of the dipole is given by

$$\Delta F = \frac{(+ze)^2}{8\pi\epsilon_0\epsilon r_i} + \frac{(-ze)^2}{8\pi\epsilon_0\epsilon r_i} + \frac{(+ze)(-ze)}{4\pi\epsilon_0\epsilon d}, \tag{3.7.3}$$

where the first, second, and the third terms on the right-hand side are the Born free energy of the ion with charge $+ze$, the Born free energy of the ion with charge $-ze$, and the Coulomb energy for the interaction between the two ions. Eq. (3.7.3) simplifies to

$$\Delta F = \frac{(ze)^2}{4\pi\epsilon_0\epsilon r_i}\left(1 - \frac{r_i}{d}\right). \tag{3.7.4}$$

This result is a generalization of Eq. (3.6.9) for the self-energy of a dipole with separation distance $2r_i$ to arbitrary separation distance d.

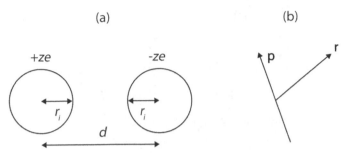

Figure 3.14 (a) Dipole made of two charges separated by the distance d. (b) Arbitrary orientation of the dipole moment **p** with respect to the direction of a spatial location.

b. Electric potential

The electric potential $\psi(\mathbf{r})$ at a distance $\mathbf{r}$ from the center of the dipole **p** (Fig. 3.14b), given by [Griffiths (1999)]

$$\psi(\mathbf{r}) = \frac{\mathbf{p} \cdot \mathbf{r}}{4\pi\epsilon_0\epsilon r^3}, \tag{3.7.5}$$

falls off as the inverse square of the distance.

c. Electric field

The electric field at **r** from the center of the dipole follows from Eqs. (3.1.6) and (3.7.5) as [Griffiths (1999)]

$$\mathbf{E}(\mathbf{r}) = -\frac{1}{4\pi\epsilon_0\epsilon} \frac{[\mathbf{p} - 3\hat{\mathbf{r}}(\hat{\mathbf{r}} \cdot \mathbf{p})]}{r^3}, \tag{3.7.6}$$

where $\hat{\mathbf{r}}$ is the unit vector along **r**.

d. Force on a dipole in an electric field

The force **f** on a dipole of moment **p** in an electric field **E** is

$$\mathbf{f} = (\mathbf{p} \cdot \nabla)\mathbf{E}. \tag{3.7.7}$$

e. Free energy of a dipole in an electric field

The change in Helmholtz free energy ΔF due to an electric field is given as

$$\Delta F = -\mathbf{p} \cdot \mathbf{E}. \tag{3.7.8}$$

f. Orientation of a dipole in an electric field

The average orientation of a dipole in an electric field is obtained by calculating the projection of the dipole moment along the direction of the electric field (Fig. 3.15a) as [Debye (1928)]

$$< p\cos\theta > = \int_0^\pi d\theta \, \sin\theta \, (p\cos\theta) \, e^{pE\cos\theta/k_BT}, \tag{3.7.9}$$

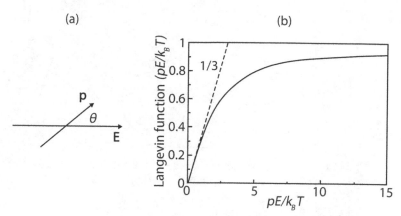

Figure 3.15 (a) Electric field tends to orient the dipole along its direction, which is opposed by thermal energy. θ is the angle between **p** and **E**. (b) Plot of the average orientation $< \cos \theta >$ given as the Langevin function $\mathcal{L}(pE/k_BT)$ defined in Eq. (3.7.10). For weak electric fields, the dependence is linear with a slope of 1/3. The dipole is fully aligned with the electric field at high field strengths.

where the angular brackets denote the averaging over all possible orientations of the dipole and the exponential factor is the Gibbs distribution function $\exp(-\Delta F/k_BT)$ given in terms of Eq. (3.7.8). Performing the integration yields

$$< \cos \theta >= \coth\left(\frac{pE}{k_BT}\right) - \frac{k_BT}{pE} \equiv \mathcal{L}\left(\frac{pE}{k_BT}\right), \tag{3.7.10}$$

where $\mathcal{L}$ is the Langevin equation introduced in Section 2.1.7. A plot of the Langevin function is given in Fig. 3.15b. The limits of the average orientation in the weak and strong electric fields are

$$< \cos \theta >= \begin{cases} \frac{1}{3}\frac{pE}{k_BT} & \frac{pE}{k_BT} << 1 \\ 1 & \frac{pE}{k_BT} >> 1. \end{cases} \tag{3.7.11}$$

For weak electric fields, $< \cos \theta >$ is linear with pE/k_BT with a slope of 1/3 as shown in Fig. 3.15b. At high electric field strengths, the dipole is in alignment with the electric field.

The corresponding limits for the average free energy change follow from Eq. (3.7.8) as

$$< \Delta F >= \begin{cases} -\frac{1}{3}\frac{p^2E^2}{k_BT} & \frac{pE}{k_BT} << 1 \\ -pE & \frac{pE}{k_BT} >> 1. \end{cases} \tag{3.7.12}$$

The change in free energy of the dipole due to the electric field depends quadratically on pE and inversely on the temperature for weak fields or at higher temperatures. On the other hand, $< \Delta F >$ depends linearly on pE and is independent of temperature for strong fields or at lower temperatures.

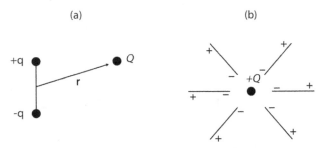

(a)   (b)

Figure 3.16 (a) A charge Q is at a distance $\mathbf{r}$ from the center of a dipole. (b) Dipoles of solvent molecules orient toward the charge Q.

g. Ion–dipole interaction

Consider a point charge Q at a distance $\mathbf{r}$ from the center of a simple dipole as in Fig. 3.16a. The free energy of the ion–dipole interaction follows from Eq. (3.2.13) as

$$\Delta F = -\mathbf{p} \cdot \mathbf{E} = -\mathbf{p} \cdot \frac{Q}{4\pi\epsilon_0\epsilon} \frac{\hat{r}}{r^2} \sim \frac{1}{r^2}, \tag{3.7.13}$$

where the electric field is from the charge Q and the orientation of the dipole is fixed. The interaction free energy is attractive and is inversely proportional to the square of the separation distance between the ion and the dipole.

On the other hand, if the dipole is allowed thermal motion, the free energy at higher temperatures $(pE/k_BT \ll 1)$ is $-p^2E^2/3k_BT$ as given in Eq. (3.7.12). Substituting the value of E from Eq. (3.2.13), the free energy change associated with ion–dipole interaction becomes

$$\Delta F = -\frac{1}{3}\frac{p^2}{k_BT}\frac{Q^2}{(4\pi\epsilon_0\epsilon)^2}\frac{1}{r^4} \sim -\frac{1}{r^4}, \tag{3.7.14}$$

which is attractive and falls off with the separation distance as $-1/r^4$. As a result, the permanent dipole moments of water orient in the radial direction of the field from the ion (Fig. 3.16b).

h. Dipole–dipole interaction

Combining Eqs. (3.7.6) and (3.7.8), the free energy of interaction between two dipoles of moments $\mathbf{p}_1$ and $\mathbf{p}_2$ separated by a distance $\mathbf{r}$ (Fig. 3.17a) is given by

$$\Delta F = \frac{1}{4\pi\epsilon_0\epsilon}\frac{1}{r^3}[\mathbf{p}_1 \cdot \mathbf{p}_2 - 3(\hat{\mathbf{r}} \cdot \mathbf{p}_1)(\hat{\mathbf{r}} \cdot \mathbf{p}_2)], \tag{3.7.15}$$

where r and $\hat{\mathbf{r}}$ are, respectively, the magnitude and unit vector of $\mathbf{r}$.

The free energy of a dipole–dipole interaction depends on the orientations of the dipoles and the separation distance vector. Consider the four relative alignments of two dipoles of dipole moments $\mathbf{p}_1$ and $\mathbf{p}_2$ with respect to the separation distance vector as given in Fig. 3.17b. Let ℓ be the fixed distance between the centers of the two dipoles. The corresponding values of the free energy given by Eq. (3.7.15), from left to right, are $J, -J, -2J$, and $2J$, where $J = p_1p_2/(4\pi\epsilon_0\epsilon\ell^3)$. The most stable arrangement occurs when the two dipoles are aligned with respect to $\mathbf{r}$. The next most stable configuration is when the dipoles are antiparallel and orthogonal to $\mathbf{r}$. The other two

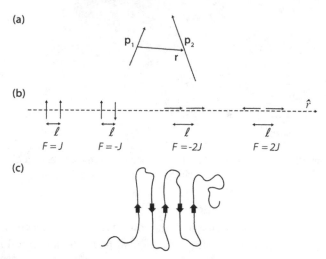

Figure 3.17 (a) Dipole–dipole interaction. (b) Free energy of dipole–dipole interaction for four configurations in units of $J = p^2/(4\pi\epsilon_0\epsilon\ell^3)$, where ℓ is the separation distance between the two dipoles of equal dipole moment p. (c) Folded conformation of a macromolecule due to antiparallel dipolar orientations in different portions of the molecule, in conjunction with parallel dipolar orientation along the chain axis.

configurations (parallel dipoles orthogonal to **r** and antiparallel dipoles along **r**) are unstable, since the free energy change is positive.

In view of the above discussion, a polymer chain carrying dipoles can adopt folded conformations where dipoles align along the chain's skeletal direction to form "stems": these stems arrange in an antiparallel manner as sketched in Fig. 3.17c. Such folded conformations can be further stabilized, at nonzero temperatures, by the conformational entropy associated with the hairpins connecting the stems.

Finally, we note that the free energy of a pair of dipoles $\mathbf{p}_1$ and $\mathbf{p}_2$ which are randomly oriented due to thermal motion is

$$< \Delta F > = -\frac{2}{3} \frac{p_1^2 p_2^2}{k_B T (4\pi\epsilon_0\epsilon)^2} \frac{1}{r^6}. \qquad (3.7.16)$$

This result is similar to the induction and dispersion energies derived in theories of intermolecular interactions [Israelachvilli (2011)]. We shall call all of these r^{-6} interaction energies collectively as the van der Waals interaction energy,

$$\Delta F = \frac{C}{r^6}, \qquad (3.7.17)$$

where C is a molecular parameter. The van der Waals interaction is short-ranged.

3.8   Dilute Solutions of Strong Electrolytes

Consider a dilute solution of a completely dissociated simple electrolyte. The charges of the ions are denoted by $z_a e$, where the subscript a refers to the different kinds (cations and anions) of ions. e is the absolute value of the unit electronic charge. z_a

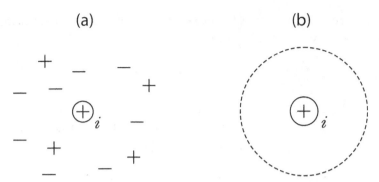

Figure 3.18 Sketches of (a) distribution of ions and (b) the ion cloud around the reference ion i.

is a positive or a negative integer. Let n_{a0} be the number concentration of ions of the a-type, namely the number of ions of the a-type per unit volume of the solution. Since the solution as a whole is electrically neutral, we have

$$\sum_a n_{a0} z_a = 0. \tag{3.8.1}$$

Due to the electrostatic interaction being long-ranged, the ions are not randomly distributed in the solution. Although the distance between any two nearby ions is about the same (which is set by the electrolyte concentration), the distribution of the ion charge is not uniform. The immediate neighborhood of a positively charged ion is likely to have more negative ions than the average number that would be at distances sufficiently far away from the reference ion. As a result, we can imagine that every ion in the solution is surrounded by a cloud of ions such that the net charge of the cloud is exactly the same as that of the reference ion, but of opposite sign (Fig. 3.18). We call this the "counterion cloud."

We shall address the essential features of the counterion cloud around a spherical ion, variations in ion distribution and electric potential, and thermodynamic properties of dilute electrolyte solutions using two models: (i) point-charge model and the Debye–Hückel theory and (ii) restricted primitive model (RPM) and the extended Debye–Hückel theory. The starting point of these models is the Poisson–Boltzmann formalism as described below.

3.8.1   Poisson–Boltzmann Formalism

Consider a large volume of an electrolyte solution consisting of cations and anions, which is overall electrically neutral. Let us choose any one of these ions as the reference ion labeled as the ith ion of charge $z_i e$ located at the origin of the coordinate system. The electric potential around a reference ion in an electrolyte solution is determined by the correlated distribution of other ions in the system. The electric potential $\psi(\mathbf{r})$ at any spatial location $\mathbf{r}$ from the reference ion at the origin of the coordinate system is given by the Poisson equation, Eq. (3.2.12), in dielectric media, as

$$\nabla^2 \psi(\mathbf{r}) = -\frac{\rho(\mathbf{r})}{\epsilon_0 \epsilon}, \tag{3.8.2}$$

where $\rho(\mathbf{r})$ is the local charge density of ions at $\mathbf{r}$. The total charge density due to all ions (except the reference ion) at $\mathbf{r}$ is given by

$$\rho(\mathbf{r}) = \sum_a e z_a n_a(\mathbf{r}), \qquad (3.8.3)$$

where $n_a(\mathbf{r})$ is the local number concentration of a-type ions (cations or anions). According to statistical mechanics [Hill (1986), McQuarrie (1976)], the local number concentration of a-type ions is given by

$$n_a(\mathbf{r}) = n_{a0} g_{ia}(\mathbf{r}), \qquad (3.8.4)$$

where n_{a0} is the average number concentration of a-type ions and $g_{ia}(\mathbf{r})$ is the pair correlation function between the ath ion and the reference ion i. The pair correlation function is given by [Hill (1986), McQuarrie (1976)]

$$g_{ia}(\mathbf{r}) = \exp\left[-\frac{e z_a w_{ia}(\mathbf{r})}{k_B T}\right], \qquad (3.8.5)$$

where w_{ia} is the potential of mean force between the reference ion and the ath ion. Substitution of Eqs. (3.8.3)–(3.8.5) into Eq. (3.8.2) gives the Poisson equation as

$$\nabla^2 \psi(\mathbf{r}) = -\frac{1}{\epsilon_0 \epsilon} \sum_a e z_a n_{a0} \exp\left[-\frac{e z_a w_{ia}(\mathbf{r})}{k_B T}\right]. \qquad (3.8.6)$$

This general expression for the electric potential $\psi(\mathbf{r})$ includes correlations among all ions involving two-body, three-body, etc., forces, making the solution of Eq. (3.8.6) impractical. In order to simplify the analysis, we assume that $n_a(\mathbf{r})$ is given by the Boltzmann distribution,

$$n_a(\mathbf{r}) = n_{a0} \exp\left[-\frac{e z_a \psi(\mathbf{r})}{k_B T}\right], \qquad (3.8.7)$$

where the potential of mean force in Eq. (3.8.6) is replaced by the average electric potential $\psi(\mathbf{r})$ at the location of the particular ath ion, so that $e z_a \psi(\mathbf{r})$ is the potential energy of an ion of a-type in the average electric potential $\psi(\mathbf{r})$ at its location.

Combining the Poisson equation (Eq. (3.8.2)) and the Boltzmann equation (Eq. (3.8.7)), and using Eq. (3.8.3), we get the **Poisson–Boltzmann equation**,

$$\nabla^2 \psi(\mathbf{r}) = -\frac{e}{\epsilon_0 \epsilon} \sum_a z_a n_{a0} \exp\left[-\frac{e z_a}{k_B T} \psi(\mathbf{r})\right]. \qquad (3.8.8)$$

We must be aware that the Boltzmann formula for the distribution of ions is not exact, and the potential of mean force from all ions should be used in the argument of the exponential instead of the average potential. As a result, the Poisson–Boltzmann equation is known to lead to some incorrect results [McQuarrie (1976), Barthel *et al.* (1998)]. Nevertheless, this approximation is adequate in explaining many experimental situations pertinent to solutions of charged macromolecules.

3.8.2 Debye–Hückel Theory

The Poisson–Boltzmann equation for the electric potential, or equivalently the distribution of ions, is nonlinear and the solution of the equation requires numerical methods, except for a few special situations. However, considerable simplification arises, without losing the major aspects of ion correlations, by linearizing the above nonlinear equation, Eq. (3.8.8).

Let us consider experimental conditions where the electrical potential energy of an ion interacting with its ionic environment, $ez_a\psi$, is small in comparison with the thermal energy k_BT. Under these conditions, the exponential of Eq. (3.8.8) can be expanded as a series,

$$\nabla^2\psi(\mathbf{r}) = -\frac{e}{\epsilon_0\epsilon}\sum_a z_a n_{a0}\left[1 - \frac{ez_a}{k_BT}\psi(\mathbf{r}) + \cdots\right]. \tag{3.8.9}$$

Due to the electroneutrality condition of Eq. (3.8.1), the first term on the right-hand side of the above equation vanishes. By ignoring all the higher order terms inside the square brackets except the linear term in ψ, we get the **linearized Poisson–Boltzmann equation**,

$$\nabla^2\psi(\mathbf{r}) = \kappa^2\psi(\mathbf{r}), \tag{3.8.10}$$

where

$$\kappa^2 \equiv \frac{e^2}{\epsilon_0\epsilon k_BT}\sum_a z_a^2 n_{a0}. \tag{3.8.11}$$

The linearized Poisson–Boltzmann equation is referred to as the **Debye–Hückel equation**. The coefficient κ^2 appearing in the above equation is one of the important parameters in the discussion of electrolyte solutions and charged macromolecules. We shall describe its physical interpretation in Section 3.8.3.

Consider the electric potential around the reference ion i with charge ez_i. Since the ion cloud is spherically symmetric on average, Eq. (3.8.10) can be rewritten as

$$\frac{1}{r^2}\frac{d}{dr}\left(r^2\frac{d\psi}{dr}\right) = \kappa^2\psi, \tag{3.8.12}$$

where r is the radial distance from the reference ion i. The general solution of Eq. (3.8.12) is

$$\psi(r) = \frac{A_0}{r}e^{-\kappa r} + \frac{B_0}{r}e^{\kappa r}, \tag{3.8.13}$$

where the constants A_0 and B_0 are to be determined using the boundary conditions at $r \to 0$ and $r \to \infty$. As $r \to \infty$, the electric potential must vanish, and therefore the constant B_0 must be 0. For $r \to 0$, the potential must be that of a point charge ez_i as given by Eq. (3.2.13), so that

$$A_0 = \frac{ez_i}{4\pi\epsilon_0\epsilon}. \tag{3.8.14}$$

Therefore, the solution of Eq. (3.8.12) is

$$\psi(r) = \frac{ez_i}{4\pi\epsilon_0\epsilon}\frac{e^{-\kappa r}}{r}. \tag{3.8.15}$$

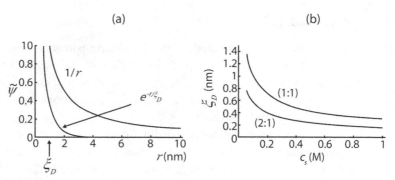

Figure 3.19 (a) Distinction between the Coulomb potential and the screened Coulomb (Debye–Hückel) potential from a reference ion for $\xi_D = \kappa^{-1} = 1$ nm. The screening length ξ_D (Debye length) is a measure of the range of electrostatic interaction. The ordinate is $\tilde{\psi} = 4\pi\epsilon_0\epsilon\psi/ez_i$. (b) Dependence of the Debye length on salt concentration for (1:1) and (2:1) type electrolytes.

Thus the potential is Coulomb-like at distances shorter than κ^{-1}, and it becomes small at distances longer than κ^{-1}. The net effect of correlations among all ions is to screen the potential from an ion and make it less long-ranged than the Coulomb potential. The potential given by Eq. (3.8.15) is called the Debye–Hückel potential, screened Coulomb potential, or the Yukawa potential. κ^{-1} is called the **Debye length** or the electrostatic screening length.

The electrostatic energy between two charges ez_i and ez_j separated by a distance r_{ij} is the product of ez_j and the potential due to the charge ez_i,

$$u_{ij}(r_{ij}) = \frac{e^2 z_i z_j}{4\pi\epsilon_0\epsilon} \frac{e^{-\kappa r_{ij}}}{r_{ij}}. \qquad (3.8.16)$$

The electrostatic potential energy between two ions given by Eq. (3.8.16) is called the Debye–Hückel potential energy. The collective effect of the ions in the solution is to screen the Coulomb interaction between a pair of ions given by Eq. (3.2.14) resulting in the screened electrostatic interaction given by Eq. (3.8.16). The distinction between the Coulomb potential and the screened Coulomb potential (Debye–Hückel potential) is shown in Fig. 3.19a, where $4\pi\epsilon_0\epsilon\psi(r)/ez_i$ is plotted against the distance r from the reference ion.

3.8.3 Electrostatic Screening Length (Debye Length)

The electrostatic screening length, also called the Debye length ξ_D, is the key measure of the range of electrostatic interaction among charges in a medium. The electric potential given by Eq. (3.8.15) is reduced by the factor $1/e$ when the distance from the reference ion is κ^{-1}. Taking this length as a measure of the range of electrostatic interaction, we define κ^{-1} as the screening length or the Debye length,

$$\xi_D \equiv \kappa^{-1}. \qquad (3.8.17)$$

Substitution of Eq. (3.8.11) in the above definition gives

$$\xi_D = \left(\frac{e^2}{\epsilon_0 \epsilon k_B T} \sum_a z_a^2 n_{a0} \right)^{-1/2} . \tag{3.8.18}$$

This result can be expressed in terms of the electrostatic strength parameter (Bjerrum length ℓ_B), by using Eq. (3.6.7) in the above equation, as

$$\xi_D = (4\pi \ell_B \sum_a z_a^2 n_{a0})^{-1/2} . \tag{3.8.19}$$

Thus ξ_D is inversely proportional to the square root of the Bjerrum length for a fixed concentration of the electrolyte,

$$\xi_D \sim \frac{1}{\sqrt{\ell_B}} . \tag{3.8.20}$$

In spite of the relation between ξ_D and ℓ_B, it must be recognized that changes in temperature and dielectric constant affect both ξ_D and ℓ_B, whereas changes in electrolyte concentration affect only ξ_D. While the Bjerrum length is a property of the solvent, the Debye length is a property of the solution. The dependence of ξ_D on the electrolyte concentration can be written in terms of the experimentally convenient concentration unit (molarity). The number concentration n_{a0} of ions of a-type is

$$n_{a0} = 10^3 N_A c_{a0}, \tag{3.8.21}$$

where c_{a0} is the concentration of ions of a-type in moles per liter. N_A is the Avogadro number 6.023×10^{23}, and the factor of 10^3 arises from the relation $1000 \ dm^3 = 1 \ m^3$, with a liter being $1 \ dm^3$. Utilizing the above conversion factor, κ^2 becomes

$$\kappa^2 = 4000\pi \ell_B N_A \sum_a z_a^2 c_{a0}, \tag{3.8.22}$$

where c_{a0} is in moles per liter and κ^{-1} and ℓ_B are in the SI unit of m. It is also convenient to group the valencies of ions and their concentrations together by defining the ionic strength I of the electrolyte solution as

$$I \equiv \frac{1}{2} \sum_a z_a^2 c_{a0} . \tag{3.8.23}$$

In terms of the ionic strength, κ^2 becomes

$$\kappa^2 = 8000\pi \ell_B N_A I, \tag{3.8.24}$$

or more explicitly

$$\kappa^2 = \frac{2000 e^2 N_A}{\epsilon_0 \epsilon k_B T} I . \tag{3.8.25}$$

For an aqueous electrolyte solution at 25°C (with $\epsilon = 78.54$),

$$\kappa = 2.32 \times 10^9 \sqrt{2I} \ m^{-1}, \tag{3.8.26}$$

Table 3.4 Debye length in nm for different electrolyte concentrations and electrolyte type (aqueous solutions at 25°C).

c_s (mol/L)	(1:1)-type	(2:1)-type
10^{-6} (salt-free)	304	176
10^{-4}	30.4	17.6
10^{-3}	9.6	5.55
10^{-2}	3.04	1.76
10^{-1}	0.96	0.555
0.150	0.78	0.453
1.0	0.304	0.176

and the Debye length is

$$\xi_D = \frac{0.43}{\sqrt{2I}} \text{ nm.} \tag{3.8.27}$$

For monovalent salts, ξ_D is conveniently given by

$$\xi_D \simeq \frac{0.3}{\sqrt{c_s}} \text{ nm,} \tag{3.8.28}$$

where c_s is the salt concentration in units of moles per liter.

The typical values of the Debye length are given in Table 3.4 for (1:1) (sodium chloride type) and (2:1) (calcium chloride type) salts at concentration c_s. The salt-free condition in experiments is usually associated with a background salt concentration of 10^{-6} mol/L [Beer *et al.* (1997)]. The dependence of ξ_D on c_s, for (1:1) and (2:1) type salts, is presented in Fig. 3.19b. In general, ξ_D decreases with salt concentration, and more sharply for ions with higher valencies. Specifically, the Debye length is 0.78 nm (which is comparable to the value of the Bjerrum length) at 150 mM monovalent salt in the solution. It progressively decreases with c_s and becomes merely 0.3 nm (which is less or comparable to monomer sizes) at 1 M of monovalent salt. These values for the electrostatic range become even smaller if multivalent ions such as magnesium and calcium are present in the solution. Thus, for higher salt concentrations, the electrostatic correlation is not long-ranged, and it is possible to combine the electrostatic interaction among monomers with the short-ranged excluded volume interaction described in Chapter 2 in addressing equilibrium properties of charged macromolecules.

3.8.4   Ion Size and Extended Debye–Hückel Theory

Instead of taking the ions as point charges, as done above, the Debye–Hückel theory can be extended to ions with finite size. Let each of the ions (cations and anions) be modeled as a hard sphere of diameter a with its charge located at the origin of the hard sphere (Fig. 3.20a). Further, let us assume that each of the ions is made of a material with the same dielectric constant as the solvent. This model is known as

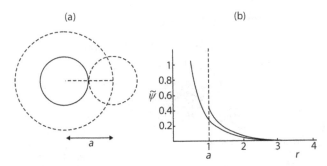

Figure 3.20 Extended Debye–Hückel theory. (a) Ions with diameter a. (b) Electric potential around a charge ez_i with diameter a for $a = 1$ nm and $\kappa^{-1} = 0.78$ nm. The lower curve is the Debye–Hückel potential for point charges. The ordinate is $4\pi\epsilon_0\epsilon\psi/ez_i$.

the **restricted primitive model** (RPM) of electrolytes [McQuarrie (1976)]. Now, the linearized Poisson–Boltzmann equation for the electric potential around a reference ion of diameter a and charge ez_i at its center is given by

$$\nabla^2\psi = \begin{cases} \kappa^2\psi & r > a \\ 0 & r \le a. \end{cases} \tag{3.8.29}$$

Using the boundary conditions, (i) $\psi(r) \to 0$ as $r \to \infty$, and (ii) ψ and $\partial\psi/\partial r$ are continuous at $r = a$, the solution of Eq. (3.8.29) is the extension of the Debye–Hückel theory to finite ion sizes, given as

$$\psi(r) = \frac{ez_i}{4\pi\epsilon_0\epsilon} \frac{1}{(1 + \kappa a)} \frac{e^{-\kappa(r-a)}}{r}, \quad r > a, \tag{3.8.30}$$

and

$$\psi(r) = \frac{ez_i}{4\pi\epsilon_0\epsilon r} - \frac{ez_i\kappa}{4\pi\epsilon_0\epsilon(1 + \kappa a)}, \quad 0 < r \le a. \tag{3.8.31}$$

The electric potential given by Eq. (3.8.30) is called the extended Debye–Hückel potential and is plotted in Fig. 3.20b. For comparison, the result of Eq. (3.8.15) for point charges is included in the figure. The finite size of the ion modifies the prefactor by a factor of $\exp(\kappa a)/(1 + \kappa a)$, and the beginning of the decay of the potential is shifted to the ion diameter. The distance dependence is the same as the screened Coulomb potential for $r > a$. For $a \to 0$, Eq. (3.8.30) reduces to the Debye–Hückel potential.

The first term on the right-hand side of Eq. (3.8.31) is obviously the contribution from the reference ion itself (according to the Coulomb law, Eq. (3.2.13)). As a result, the second term must be the contribution of all ions outside the reference ion. Hence, the second term is identified as the electric potential ψ_{cloud} from the ion cloud acting on the reference ion,

$$\psi_{\text{cloud}} = -\frac{ez_i\kappa}{4\pi\epsilon_0\epsilon(1 + \kappa a)}. \tag{3.8.32}$$

We now consider the charge distribution constituting the cloud that leads to the potential given by Eq. (3.8.32).

3.8.5　Size and Shape of Ion Cloud

The charge density distribution of the ion cloud around a reference ion i of charge ez_i follows from the linearized Poisson–Boltzmann equations (Eqs. (3.8.2) and (3.8.29)) as

$$\rho_{\text{cloud}}(\mathbf{r}) = -\epsilon_0 \epsilon \kappa^2 \psi(\mathbf{r}), \qquad r > a. \tag{3.8.33}$$

Substituting Eq. (3.8.30) for the electric potential, we get

$$\rho_{\text{cloud}}(\mathbf{r}) = -\frac{ez_i \kappa^2}{4\pi(1 + \kappa a)} \frac{e^{-\kappa(r-a)}}{r}. \tag{3.8.34}$$

The charge density around an ion has a sign opposite to that of the reference ion and is proportional to the potential in the Debye–Hückel approximation. The total charge surrounding the reference ion is

$$ez_{\text{cloud}} = \int_a^\infty \rho_{\text{cloud}}(r) 4\pi r^2 dr, \tag{3.8.35}$$

where radial symmetry is used and the lower limit of the integral reflects that the centers of ions in the cloud cannot approach the reference ion within the diameter a of the ion (Fig. 3.20a). The evaluation of the integral of Eq. (3.8.35) leads to the expected result that the total charge around the reference ion i is equal and opposite to the charge of the reference ion,

$$ez_{\text{cloud}} = -ez_i. \tag{3.8.36}$$

The fraction of net charge of the cloud, $d(ez_{\text{cloud}})$, between r and $r + dr$ in the cloud is $4\pi r^2 \rho_{\text{cloud}}(r)$,

$$d(ez_{\text{cloud}}) = -\frac{ez_i \kappa^2}{(1 + \kappa a)} r e^{-\kappa(r-a)}, \tag{3.8.37}$$

where Eq. (3.8.34) is used. Rewriting this result,

$$\rho_{\text{cloud}}(r) \equiv \left| -\frac{(1 + \kappa a) d(ez_{\text{cloud}}) \exp(-\kappa a)}{ez_i \kappa} \right| = \kappa r e^{-\kappa r}, \qquad r > a. \tag{3.8.38}$$

The shape and the extent of the cloud around an ion are illustrated in Fig. 3.21, where $\rho_{\text{cloud}}(r)$ is plotted against κr. The maximum occurs at the Debye length $\xi_D \equiv \kappa^{-1}$, and the most important region of the ion cloud is in the neighborhood of $r \sim \kappa^{-1}$. As seen in Table 3.4 and Fig. 3.19b, the location of this region moves to larger values as the solution becomes more dilute in salt concentration.

As we shall see later in dealing with charged macromolecules, the clouds of counterions around each of the charged monomers dominate the behavior of these macromolecules. In fact, the physics of polyelectrolytes should really be called counterion physics. It is thus essential to have a good grasp of the results presented in this section to understand the structure and mobility of charged macromolecules.

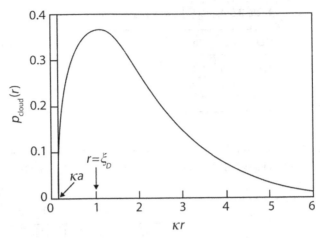

Figure 3.21 Shape and size of ion cloud: fraction of net charge of the cloud versus the radial distance in unit of Debye length $\xi_D = \kappa^{-1}$.

3.8.6 Free Energy

The electrostatic correlations of the ions in the electrolyte solution contribute to the Helmholtz free energy of the system, depending on the Debye length and ion size. Based on the Debye–Hückel theory for the RPM (Fig. 3.20a), the result [McQuarrie (1976)] is

$$F = F_0 + F_{el}, \qquad (3.8.39)$$

where F_{el} is the electrostatic contribution and F_0 is the contribution from translational entropy of ions under ideal mixing conditions,

$$F_0 = -n_+ \ln V - n_- \ln V, \qquad (3.8.40)$$

where n_+ and n_- are the number of cations and anions, respectively, in volume V. The electrostatic correlations lower the free energy of the solution, and F_{el} is given by [McQuarrie (1976)]

$$\frac{F_{el}}{V k_B T} = -\frac{1}{4\pi a^3}\left[\ln(1 + \kappa a) - \kappa a + \frac{1}{2}\kappa^2 a^2\right]. \qquad (3.8.41)$$

For low salt concentrations (that is $\xi_D \equiv \kappa^{-1}$ is large), κa may become so small that the above equation simplifies to

$$\frac{F_{el}}{V k_B T} = -\frac{\kappa^3}{12\pi}, \qquad \kappa a \to 0. \qquad (3.8.42)$$

In this limit, the electrostatic part of free energy density is inversely proportional to the cube of the Debye length, namely the electrostatic correlation volume.

The reduction in the free energy due to electrostatic correlations among all ions in aqueous solutions arises mainly from changes in the entropy of water. This seen by resolving F_{el} of Eq. (3.8.42) into energy and entropy contributions,

$$F_{el} = U_{el} - T S_{el}. \qquad (3.8.43)$$

Table 3.5 Entropy dominates electrostatic stabilization in water

Quantity	Temperature-independent ϵ	Temperature-dependent ϵ
F_{el}	< 0	< 0
U_{el}	< 0	> 0
$-TS_{el}$	> 0	< 0

Using the thermodynamic relation $S_{el} = -\partial F_{el}/\partial T$, and accounting for the temperature dependence of the dielectric constant (which appears in the definition of κ), we get

$$-\frac{TS_{el}}{Vk_BT} = \frac{1}{24\pi}\kappa^3 + \frac{\kappa^3}{8\pi}\frac{\partial \ln \epsilon}{\partial \ln T}. \tag{3.8.44}$$

Substituting this result into Eq. (3.8.43), the energy part follows as

$$\frac{U_{el}}{Vk_BT} = -\frac{\kappa^3}{8\pi}\left(1 + \frac{\partial \ln \epsilon}{\partial \ln T}\right). \tag{3.8.45}$$

If the temperature dependence of ϵ is suppressed, then it follows from Eqs. (3.8.42) to (3.8.45) that $F_{el} < 0, U_{el} < 0$, and $-TS_{el} > 0$, and thus the stabilization of the system from electrostatic correlations is driven by energy. On the other hand, if the temperature dependence of ϵ is taken into account for water ($\partial \ln \epsilon/\partial \ln T = -1.36$ at room temperature), then $F_{el} < 0, U_{el} > 0$, and $-TS_{el} < 0$ demonstrating that entropy dominates the free energy of aqueous electrolyte solutions. These contrasting scenarios are summarized in Table 3.5, which emphasizes the role of entropy of water in electrolyte solutions and the importance of accounting for the temperature dependence of ϵ in any attempt at describing the experimental results on charged macromolecular systems.

3.8.7   Osmotic Pressure

Using the standard procedure [Hill (1986), McQuarrie (1976)] of calculating the osmotic pressure of a solution from its Helmholtz free energy, the osmotic pressure Π of a dilute electrolyte solution is given by

$$\Pi = \Pi_0 + \Pi_{el}. \tag{3.8.46}$$

Here Π_0 is the ideal part (based on the van't Hoff law),

$$\frac{\Pi_0}{k_BT} = c_+ + c_-, \tag{3.8.47}$$

with c_+ and c_- being the number concentrations of the cations and anions, and Π_{el} is the electrostatic part,

$$\frac{\Pi_{el}}{k_BT} = -\frac{\kappa^3}{24\pi}, \tag{3.8.48}$$

in the limit of $\kappa a \rightarrow 0$. Since κ^2 is proportional to salt concentration c_s, the contribution to the osmotic pressure from electrostatic correlation depends on salt concentration as $c_s^{3/2}$, strongly deviating from the viral series observed for uncharged systems. Deviations in the osmotic pressure from the ideal result is expressed as the **osmotic coefficient** ϕ_{os} defined as

$$\phi_{os} = \frac{\Pi}{\Pi_0}. \tag{3.8.49}$$

According to the Debye–Hückel theory, the osmotic coefficient follows from Eqs. (3.8.46) to (3.8.48) as

$$\phi_{os} = \left[1 - \frac{\kappa^3}{24\pi(c_+ + c_-)} \right], \tag{3.8.50}$$

exhibiting significant deviation from ideality in electrolyte solutions. Such deviations from ideality are even more pronounced in solutions of charged macromolecules.

4 Charged Interfaces and Geometrical Objects

Large charged macromolecules and self-assembled structures, such as globular proteins, lipid bilayers, collagen bundles, viruses, and colloidal particles, expose large interfaces with the background electrolyte solution. These interfaces bear charges due to either ionic equilibria, chemical reactions, or adsorption of specific ions at the interface. Even the walls of the container of the solution can carry charges. Also, charges can be controllably injected onto the interfaces by using electrodes with an externally applied voltage. The charge of an interface and how the electrolyte ions hover around the interface lead to spatial variations in the electric potential near the interface. These features control the manner in which other charged macromolecules organize at these interfaces and how they move around in the solution.

In general, the behavior of electrolyte solutions near charged interfaces can be quite complex. However, several fundamental concepts associated with charged interfaces can be identified by considering idealized model systems with simplifying assumptions. As examples, we shall consider in this chapter planar surfaces, large spherical and cylindrical macroions, and an infinitely thin line of charges, as sketched in Fig. 4.1. In these systems, the charged objects are rigid without any entropic contributions associated with conformational fluctuations of the molecules making up the systems.

The universal feature that is common to all these interfaces is the emergence of ionic clouds, reminiscent of those around point charges described in Chapter 3. The surface charge influences the distribution of ions in the medium. Ions of opposite charge (counterions) are attracted toward the surface, and ions of like charge (co-ions) are repelled away from the surface. In addition, thermal motion of ions in the solution influences the distribution of counterions and co-ions. The net result is the formation of a thin adsorbed layer, where a certain number of counterions pile up on the surface, followed by a thick diffuse layer where the counterions are in excess over the co-ions. Beyond the diffuse layer, the distributions of counterions and co-ions are uniform as in the bulk of the solution. These three regions are sketched in Fig. 4.2. The charged surface and the neutralizing excess of counterions over co-ions distributed in a diffuse manner are collectively called an electric double layer.

To gain further insight into the constitution and extent of the electric double layer, and the variations in the charge distribution and electric potential, we shall also consider thermodynamic properties such as the free energy of the system. In this chapter, we shall introduce important concepts such as the Gouy–Chapman length, double layer thickness, Manning condensation, and charge regularization of charged objects

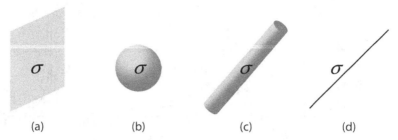

Figure 4.1 Charged interfaces of (a) a planar surface, (b) a spherical particle, (c) a cylindrical particle, and (d) an infinitely thin line charge. σ is the charge density.

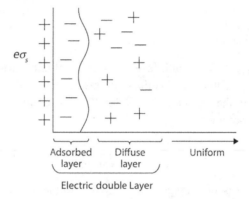

Figure 4.2 Generic scenario of ion distribution near a charged surface as an electric double layer. The abscissa is the direction orthogonal to the interface of charge density $e\sigma_s$.

in electrolyte solutions. We shall also mention the consequences of different dielectric constants across an interface. The inevitable ionic clouds around rigid macroions, or around flexible macromolecules, dominantly control the effective charge and mobility of these objects in electrolyte solutions. Therefore, it is invaluable to first acquire a good understanding of counterion distribution around charged interfaces based on ideal models, which is the primary focus of this chapter.

4.1   Planar Interfaces

Consider an electrolyte solution at an infinitely wide planar interface with uniform charge density $\sigma = e\sigma_s$ (Fig. 4.3a). σ_s is the uniform number charge density. The solution has a uniform dielectric constant ϵ and the ions inside are point-charges. The image charges due to the dielectric mismatch at the interface are ignored. The spatial variation of the electric potential away from the interface and the density distributions of the counterions and co-ions can be obtained exactly from the Poisson–Boltzmann equation for this model. This theory is known as the **Gouy–Chapman theory**. In fact, this situation is one of the few cases where the nonlinear Poisson–Boltzmann equation can be solved exactly, and helps to establish confidence in using the Debye–Hückel theory under appropriate experimental conditions.

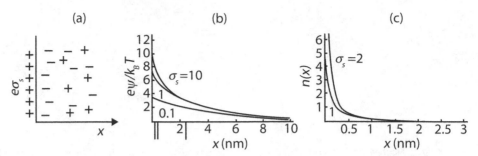

Figure 4.3 (a) An electrolyte solution near an interface with surface charge density $e\sigma_s$. (b) Decay of electric potential with distance from the interface in salt-free solutions. The vertical marks denote the Gouy–Chapman length for $\sigma_s = 10, 1.0$, and 0.1 in nm^{-2} (from left to right). (c) Density profile of counterions showing the crowding of counterions near a charged interface [Muthukumar (2011)].

Since the planar interface is taken as infinitely wide, the spatial variations of the electric potential and ion distributions occur in only one dimension along the x-axis normal to the interface (Fig. 4.3a). Therefore, the Poisson–Boltzmann equation, Eq. (3.8.8), becomes

$$\frac{d^2\psi(x)}{dx^2} = -\frac{e}{\epsilon_0\epsilon} \sum_a z_a n_{a0} \exp\left[-\frac{ez_a}{k_BT}\psi(x)\right], \tag{4.1.1}$$

where n_{a0} is the average number density of the a-type ions in the solution. The electric field due to the interface vanishes at distances far away from the interface. Let ψ_s be the surface potential. Using these as boundary conditions,

$$\psi(x = 0) = \psi_s \tag{4.1.2}$$

and

$$\frac{d\psi}{dx}\Big|_{x\to\infty} = 0. \tag{4.1.3}$$

The surface potential ψ_s is related to the surface charge density σ_s, as will be shown in the following text. We will treat the system either with a fixed surface charge density or equivalently with a fixed surface charge potential. Eq. (4.1.1) is solved to obtain the electric potential and distribution of ions near the charged surface. We now give the results of this calculation for the two cases: (a) salt-free solutions and (b) solutions with salt.

4.1.1   Salt-Free Solutions

a. Gouy–Chapman Theory

In salt-free solutions, the only ions in the solution are counterions to the charges on the surface in order to maintain the overall electroneutrality of the system. Let the charge of the counterion be ze. By solving Eq. (4.1.1), the electric potential depends logarithmically on the distance from the interface [Israelachvilli (2011)],

$$\frac{ze\psi(x)}{k_BT} = 2\ln\left(1 + \frac{x}{\lambda}\right) + \frac{ze\psi_s}{k_BT}, \tag{4.1.4}$$

where λ is

$$\lambda \equiv \frac{1}{2\pi\ell_B \mid z\sigma_s \mid}, \qquad (4.1.5)$$

with the Bjerrum length ℓ_B defined in Eq. (1.3.2). Since the distance dependence of the electric potential appears only as x/λ, λ defines a distance characteristic of the system. This distance λ is called the **Gouy–Chapman length**, which is a measure of the range of attraction for the counterion cloud from the interface. The fall of the electric potential with distance from the interface is illustrated in Fig. 4.3b for mono-valent counterions and three surface charge densities ($\sigma_s = 0.1, 1$, and 10 per nm^2) and $\ell_B = 0.7$ nm. The values of the Gouy–Chapman length are also included in these figures to indicate the characteristic length for the decay of the potential.

b. Electric Field and Contact Theorem

The electric field follows from Eqs. (3.1.6) and (4.1.4) as

$$E(x) = -\frac{k_B T}{ze} \frac{2}{(x + \lambda)}. \qquad (4.1.6)$$

For positively charged interfaces, the counterions are negatively charged, and the electric field is normal to the surface pointing into the electrolyte solution. The opposite is true for negatively charged interfaces. The electric field at the interface E_s in salt-free solutions follows from the above equation and Eq. (4.1.5) as

$$E_s = \frac{e\sigma_s}{\epsilon_0\epsilon}, \qquad (4.1.7)$$

a general result known as the contact theorem [Israelachvili (2011)].

c. Counterion Crowding at Interfaces

The density profile $n(x)$ of the counterions can be derived by substituting Eq. (4.1.4) in the Boltzmann law (Eq. (3.8.7)) to get

$$n(x) = \frac{2\pi\ell_B\sigma_s^2}{(1 + \frac{x}{\lambda})^2}. \qquad (4.1.8)$$

This expression for monovalent counterions is given in Fig. 4.3c for two surface charge densities ($\sigma_s = 1$, and 2 per nm^2) and $\ell_B = 0.7$ nm. The number density of counterions at the interface ($x = 0$) is

$$n(x = 0) = 2\pi\ell_B\sigma_s^2. \qquad (4.1.9)$$

The counterion density, due only to the electrostatic attraction and thermal motion, at the interface is proportional to the square of the surface charge density and to the Bjerrum length. Thus, the counterions pile up at the interface in salt-free solutions.

4.1.2   Solutions with Salt

a. Electric Potential

When a strongly dissociating salt is present in the solution in addition to the counte-rions, the general behavior of the electric potential is the same as in salt-free solutions. In terms of the density distributions of ions, we need to consider the co-ions as well. For simplicity, let us assume that the added salt is of the symmetric type (z:z) with the valencies of the cations and anions of the salt and the counterions being the same as z. Only the signs are different depending on whether they are positively charged or negatively charged. According to the Gouy–Chapman theory [Evans & Wenner-ström (1999)], the exact result for the electric potential $\psi(x)$ is given by [Hiemenz & Rajagopalan (1997), Israelachvilli (2011)]

$$\tanh\left(\frac{ze\psi(x)}{4k_BT}\right) = e^{-\kappa x}\tanh\left(\frac{ze\psi_s}{4k_BT}\right), \tag{4.1.10}$$

where ψ_s is the surface potential and κ is the inverse Debye length. According to the Gouy–Chapman theory, $\tanh(ze\psi/k_BT)$ decreases exponentially with x.

Since k_BT/e sets the scale for the electric potential in terms of the temperature, we give a special symbol to it,

$$\psi_\theta \equiv \frac{k_BT}{e}. \tag{4.1.11}$$

At 25°C, $\psi_\theta = 25.7$ mV. For weak potentials such that $z\psi \ll \psi_\theta$, the Poisson–Boltzmann equation (Eq. (4.1.1)) can be linearized to give the linearized Poisson–Boltzmann equation,

$$\frac{d^2\psi(x)}{dx^2} = \kappa^2\psi(x), \tag{4.1.12}$$

where κ is the inverse Debye length, as defined in Section 3.8. The solution of Eq. (4.1.12) with the same boundary conditions as in Eqs. (4.1.2) and (4.1.3) is

$$\psi(x) = \psi_s e^{-\kappa x}. \tag{4.1.13}$$

This is the Debye–Hückel law for the one-dimensional variation along the x-axis, normal to the two-dimensional interface. Eq. (4.1.10) naturally reduces to Eq. (4.1.13) for $z\psi \ll \psi_\theta$ and $z\psi_s \ll \psi_\theta$.

b. Difference between Debye–Hückel and Poisson–Boltzmann Predictions

The extent of error arising from the Debye–Hückel approximation to the full Gouy–Chapman theory is presented in Fig. 4.4a, by comparing Eqs. (4.1.10) and (4.1.13). Here, the electric potential ψ is plotted against x for a 0.01 M solution of a (1:1) electrolyte and at the constant surface potential of 77.1 mV (i.e., three times the room temperature value, $3\psi_\theta$). For this electrolyte solution, the Debye length is 3 nm and the potential is significant even at 10 nm. The discrepancy between the Gouy–Chapman (nonlinear Poisson–Boltzmann) and Debye–Hückel (linearized Poisson–Boltzmann) is rather mild for surface potentials that are even three times k_BT/e. This offers

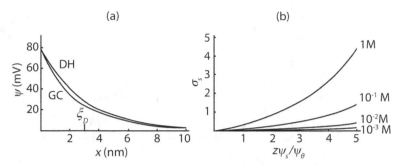

Figure 4.4 (a) Comparison of Gouy–Chapman (GC) and Debye–Hückel (DH) results for the electric potential. $\xi_D = 3$ nm (0.01 M monovalent salt) and the surface potential is 77.1 mV corresponding to three times the thermal energy k_BT. (b) Relation between the number of charges on the surface per nm^2, σ_s, and the surface potential ψ_s, for different concentrations of a monovalent salt [Muthukumar (2011)].

confidence in utilizing the Debye–Hückel approximation to extract the main qualitative features of more complex situations of charged macromolecules near charged interfaces.

c. Surface Charge and Surface Potential

It can be shown that the surface charge density σ_s, namely the number of unit charges per nm^2, is uniquely related to the surface potential ψ_s as a function of the salt concentration. The result is [Evans & Wennerström (1999)]

$$\sigma_s = \frac{\kappa}{2\pi z \ell_B} \sinh\left(\frac{z\psi_s}{2\psi_\theta}\right). \tag{4.1.14}$$

Substituting the values of the Bjerrum length and the Debye length (in terms of monovalent salt concentration c_s in moles per liter) for aqueous solutions at room temperatures, a more convenient form is

$$\sigma_s = 0.758 \sqrt{c_s} \sinh\left(\frac{\psi_s}{2\psi_\theta}\right), \tag{4.1.15}$$

with $\psi_\theta = 25.7 mV$. The dimension of σ_s is nm^{-2}. The dependence of this relation between the surface charge density and the surface potential on c_s is given in Fig. 4.4b. The dependence is progressively steeper as c_s increases. For a fixed surface potential, the surface charge density increases as the electrolyte concentration is increased. For example, the number of surface charges per nm^2 doubles as the electrolyte concentration is increased from 0.1 M to 1 M, at $\psi_s = 4\psi_\theta$. Equivalently, the surface potential decreases with an increase in c_s for fixed surface charge density.

As seen in Fig. 4.4a, the electric potential falls off roughly exponentially with the distance from the interface for a constant surface potential. A more realistic experimental scenario occurs when the surface maintains a constant surface charge density. By combining Eqs. (4.1.10) and (4.1.14), the electric potential can be obtained in terms of σ_s. The typical result for $\psi(x)$ is illustrated in Fig. 4.5a where $c_s = 0.1$ M and 1 M monovalent salt in water at T = 25°C, by keeping σ_s at a constant value of 1 per nm^2.

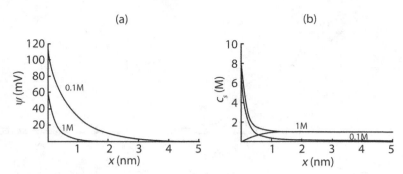

Figure 4.5 (a) Effect of electrolyte concentration on the dependence of electric potential on the distance from the interface, for fixed surface charge density $\sigma_s = 1$ nm^{-2}. (b) Concentration profiles of counterions and co-ions in molarity for bulk concentrations of monovalent electrolyte at 0.1 M and 1.0 M. Surface charge density is fixed at $\sigma_s = 1$ nm^{-2}. [Muthukumar (2011)].

The electric potential decreases roughly exponentially with x with the characteristic distance being the Debye length. Also, the electric field at the interface is independent of the salt concentration when the surface charge is fixed (as seen from the contact theorem, Eq. (4.1.7)).

d. Enhancement of Counterions and Depletion of Co-ions

The distribution of ions in solution from an interface can be readily obtained from the Boltzmann law, Eq. (3.8.7), by using the result for the potential. The distributions of counterions and co-ions are given in Fig. 4.5b for 0.1 M and 1 M monovalent electrolyte solutions in water at T = 25°C, by keeping the surface charge density σ_s at the constant value of 1 per nm^2. For each electrolyte concentration, the upper curve represents the distribution of excess counterions near the surface. There is a depletion of co-ions near the surface given by the lower curve. There exists a threshold distance beyond which the interface has little influence on the counterion excess. The value of this threshold distance increases as the electrolyte concentration decreases. The area enveloped by the curves of counterions and co-ions for each electrolyte concentration is the net charge of the solution and must be exactly the opposite of the surface charge.

The key result is that the counterions prefer to be closer to highly charged interfaces, although they are not permanently adsorbed. It can be shown [Evans & Wennerström (1999), Israelachvili (2011)] that the total number density of all ions at the interface is given by

$$n(x = 0) = n_\infty + \frac{\epsilon_0 \epsilon}{2 k_B T} E_s^2,$$

(4.1.16)

where n_∞ is the number density of all ions in the bulk solution far away from the interface and E_s is the electric field at the interface, given by Eq. (4.1.7). The above equation is a general result, known as the Grahame equation, and can be used to assess the accumulation of various ions at a charged interface essentially as a monolayer.

e. Electric Double Layer

Since the excess of counterions and depletion of co-ions occur only within a distance comparable to the Debye length, it is useful to construct a geometrical representation of the ion cloud near the interface. The capacitance of the interface, $C_{\text{interface}}$, is given by $d\sigma_s e/d\psi_s$ and it follows from Eq. (4.1.14) that

$$C_{\text{interface}} \equiv \frac{d\sigma_s e}{d\psi_s} = \frac{\epsilon_0 \epsilon}{\kappa^{-1}} \cosh\left(\frac{z\psi_s}{2\psi_\theta}\right). \qquad (4.1.17)$$

For $z\psi_s \ll \psi_\theta$, the capacitance of the interface assumes the simple form,

$$C_{\text{interface}} = \frac{\epsilon_0 \epsilon}{\kappa^{-1}}. \qquad (4.1.18)$$

This expression is exactly the capacitance of a parallel-plate capacitor [Young & Freedman (2000)] with two parallel plates with a separation distance of the Debye length κ^{-1} confining a dielectric medium of permittivity $\epsilon_0 \epsilon$. In view of this equivalence between the capacitance of a charged interface (for sufficiently weak surface potentials) and that of a parallel-plate capacitor, a charged interface is called an "electric double layer."

As described above, the counterions pile up very close to the charged interface forming an adsorbed layer, referred to as the Stern layer [Hiemenz & Rajagopalan (1997), Evans & Wennerström (1999), Israelachvilli (2011)]. The effective electric potential at the interface is the value after accounting for the presence of the adsorbed layer. This potential is the Stern potential ψ_δ. The exponential decay of the electric potential with distance from the interface is practically only from the Stern layer. Furthermore, when a fluid moves tangential to the interface, a rough boundary separating the immobilized ions and solvent molecules (due to the no-slip boundary condition at the interface) and the mobile fluid may be identified as the surface of shear at which the electric potential is called the zeta potential ψ_ζ. The surface of shear is inside the diffuse electric double layer. When the distance from the Stern layer is the Debye length ξ_D, the electric potential is reduced to the value ψ_δ/e. Therefore ξ_D is taken as a measure of the thickness of the electric double layer. Combining the above described concepts of strong accumulation (adsorption) of counterions near the interface, surface of shear, finite range for the excess of counterions and depletion of co-ions, and the electric double layer, a cartoon of a charged interface is depicted in Fig. 4.6.

f. Entropy Dominance in Charging an Interface

The work ΔU required to charge a double layer follows from Eq. (3.1.15) as

$$\Delta U = \frac{1}{2} \int d\mathbf{r} \rho(\mathbf{r}) \psi(\mathbf{r}). \qquad (4.1.19)$$

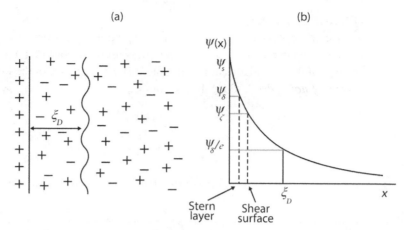

Figure 4.6 (a) Cartoon of electric double layer of thickness ξ_D. (b) Typical profile of the electric potential $\psi(x)$ as a function of the distance x from the interface (ξ_D is much larger than the distances of the Stern surface and surface of shear from the interface). At $x = \xi_D$, the Stern potential is reduced by the Euler number e.

Substituting Eq. (4.1.10) for $\psi(x)$ and Eq. (3.2.12) for $\rho(\mathbf{r})$, the change in internal energy due to charging the interface per unit area is given by

$$\frac{\Delta U}{k_B T} = \frac{\sigma}{ze} \tanh\left(\frac{ze\psi_s}{4k_B T}\right). \tag{4.1.20}$$

The Helmholtz free energy change ΔF for the interface is obtained from ΔU by using the thermodynamic relation,

$$U = -T^2 \frac{\partial(F/T)}{\partial T}. \tag{4.1.21}$$

Integrating $(-\Delta U/T^2)$ with respect to temperature in accordance with Eqs. (4.1.20) and (4.1.21), ΔF due to charging is obtained as

$$\frac{\Delta F}{k_B T} = \frac{2\sigma}{ze}\left[\frac{ze\psi_s}{2k_B T} - \tanh\left(\frac{ze\psi_s}{4k_B T}\right)\right]. \tag{4.1.22}$$

The change in entropy ΔS accompanying the charging of the interface, given by $\Delta S = -(\Delta F - \Delta U)/T$, is

$$\frac{\Delta S}{k_B} = -\frac{2\sigma}{ze}\left[\frac{ze\psi_s}{2k_B T} - \frac{3}{2}\tanh\left(\frac{ze\psi_s}{4k_B T}\right)\right]. \tag{4.1.23}$$

The contributions from internal energy (ΔU) from Eq. (4.1.20), and entropy (ΔS) from Eq. (4.1.23) to the free energy ΔF are presented in Fig. 4.7, where these contributions, in units of $\sigma k_B T/ze$ per unit area, are plotted against $ze\psi_s/k_B T$. It is evident from this figure that entropy is the major component to the free energy in assembling a charged interface. The entropic dominance over energy is a key underlying theme behind essentially all assemblies of charged structures in aqueous solutions.

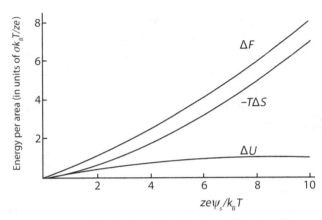

Figure 4.7 Dominance of entropy in the formation of a charged interface. Free energy ΔF, energy ΔU, and entropic contribution $-T\Delta S$ per unit area in units of $\sigma k_B T/ze$ as a function of the dimensionless surface potential $ze\psi_s/k_B T$.

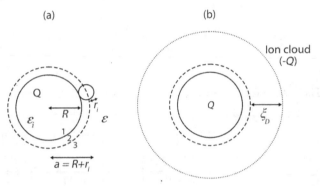

Figure 4.8 (a) Effective radius of a spherical macroion, $a = R + r_i$. (b) The counterion cloud is at the Debye length $\xi_D = \kappa^{-1}$ from the effective radius a. For electrolyte ions much smaller than the macroion, $a \simeq R$.

4.2 Charged Spherical Particles

The essential features of the electric potential and distribution of electrolyte ions around proteins, icosahedral viruses, sphere-shaped colloidal particles, macroions, etc., in electrolyte solutions can be obtained by studying spherical charged particles. As a generic example (Fig. 4.8), consider a spherical macroion of radius R with its total charge Q distributed uniformly on its surface, present in an electrolyte solution. The surface charge density of the particle is $\sigma = Q/(4\pi R^2)$. Let the static dielectric constant inside the particle be ϵ_i and that of the electrolyte solution be ϵ. Since the total system is electrically neutral, the solution contains the counterions of the macroion and the dissociated salt ions. Let r_i be the mean radius of small ions. The effective radius of the spherical macroion that excludes the small ions (Fig. 4.8) is

$$a = R + r_i. \tag{4.2.1}$$

As shown in Fig. 4.8, there are three regions: (1) interior of the macroion, (2) excluded region for small ions, and (3) electrolyte solution.

The electrical potential and distributions of counterions and co-ions around the macroion are obtained by solving the Poisson–Boltzmann equation (Eq. (3.8.8)). No analytical solution is possible for this nonlinear equation in this geometry. Therefore, we use the linearized Poisson–Boltzmann equation as in the extended Debye–Hückel theory (Section 3.8.4),

$$\nabla^2\psi_1 = 0, \quad \nabla^2\psi_2 = 0, \quad \nabla^2\psi_3 = \kappa^2\psi_3, \tag{4.2.2}$$

where the subscripts 1, 2, and 3 denote the three regions of space (Fig. 4.8). κ^2 is defined as in Eq. (3.8.11) and Section 3.8.3. The above Debye–Hückel equations are solved by using radial symmetry with the radial distance r from the center of the macroion, as done in Eq. (3.8.12), and the boundary conditions of potential and electric field at the two interfaces at $r = R$ and $r = a$,

$$\psi_1(R) = \psi_2(R), \quad \epsilon_i\frac{d\psi_1}{dr}\bigg|_R - \epsilon\frac{d\psi_2}{dr}\bigg|_R = \sigma, \quad \psi_2(a) = \psi_3(a), \quad \frac{d\psi_2}{dr}\bigg|_a = \frac{d\psi_3}{dr}\bigg|_a. \tag{4.2.3}$$

The results for the electric potential $\psi(r)$ in the three regions are

$$\psi(r) = \frac{Q}{4\pi\epsilon_0\epsilon}\frac{1}{(1+\kappa a)}\frac{e^{-\kappa(r-a)}}{r}, \quad r > a. \tag{4.2.4}$$

$$\psi(r) = \frac{Q}{4\pi\epsilon_0\epsilon}\left[\frac{1}{r} - \frac{1}{(a+\kappa^{-1})}\right], \quad R < r < a. \tag{4.2.5}$$

$$\psi(r) = \frac{Q}{4\pi\epsilon_0\epsilon}\left[\frac{1}{R} - \frac{1}{(a+\kappa^{-1})}\right], \quad 0 < r \leq R. \tag{4.2.6}$$

The electric potential due to the macroion in the electrolyte solution, given by Eq. (4.2.4) for $r > a$, is exactly the same as the extended Debye–Hückel result given by Eq. (3.8.30) and described in Fig. 3.20. The potential is screened with the screening length $\xi_D = \kappa^{-1}$. As in the extended Debye–Hückel theory, the ion cloud surrounding the macroion is represented as a spherical shell with the opposite charge of the macroion, located at the Debye length ξ_D away from the effective excluded surface, with radius $a = R + r_i$, of the macroion. This result is sketched in Fig. 4.8b.

For distances between the macroion radius and the sum of macroion radius and small ion radius, the first term on the right-hand side of Eq. (4.2.5) is due to the macroion and the second term is due to the counterion cloud situated at a distance $a + \kappa^{-1}$. Similarly, the surface potential of the macroion ψ_s at $r = R$ follows from Eq. (4.2.6) as

$$\psi_s = \frac{Q}{4\pi\epsilon_0\epsilon}\left[\frac{1}{R} - \frac{1}{(a+\kappa^{-1})}\right], \quad r = R. \tag{4.2.7}$$

The first term on the right-hand side is due to the macroion itself and the second term is due to the counterion cloud at $a + \xi_D$. If the radius of the small ion is negligible in comparison with the radius R of the macroion or a spherical colloidal particle, $a \simeq R$. In this case,

$$\psi(r) = \frac{Q}{4\pi\epsilon_0\epsilon} \frac{1}{(1 + \kappa R)} \frac{e^{-\kappa(r-R)}}{r}, \qquad r > R, \tag{4.2.8}$$

$$\psi_s = \frac{Q}{4\pi\epsilon_0\epsilon} [\frac{1}{R} - \frac{1}{(R + \kappa^{-1})}], \qquad r = R. \tag{4.2.9}$$

It is clear from the above equations that the macroion immersed in an electrolyte solution cannot be treated as an isolated particle. It is mandatory to treat any charged particle as a collective quasiparticle consisting of the particle and its counterion cloud. This ubiquitous result is essential in understanding the mobility of macroions and other charged macromolecules in electrolyte solutions.

4.3   Charged Cylindrical Assemblies

Short dsDNA, collagen bundles, cylindrical viruses such as the tobacco mosaic virus, anisotropic nanoparticles, etc., adopt cylindrical shapes with their charges exposed to the background electrolyte solution. Insight into the electrical potential emanating from these objects and the accompanying ion cloud is essential to understand how such charged assemblies interact among themselves, self-organize, and move in electrolyte solutions. In view of this general relevance, we now present two scenarios: (i) salt-free solutions and (ii) solutions with salt. Although the first case is of limited experimental relevance, it allows an exact solution of the Poisson–Boltzmann equation, forming the fundamental basis for the concept of counterion condensation [Alfrey et al. (1951), Fuoss et al. (1951)]. The second case is directly pertinent to many experimental situations and the approximate linearized Poisson–Boltzmann description portrays the features of the omnipresent counterion cloud around cylinder-shaped macromolecules and their assemblies.

4.3.1   Counterion Distribution in Salt-Free Solutions

Consider a collection of long charged cylinders in a salt-free solution. Let each cylinder be of contour length L, made of N segments of segment length ℓ with $L = N\ell$. The cylinder is taken to be of uniform radius a. Let identical charges in the cylinder be uniformly distributed along the cylinder axis with ℓ being the charge separation, as shown in Fig. 4.9. Let ez_p be the charge of the individual charges. Equivalently, the magnitude of z_p can be identified as the degree of ionization α if we were to export the present calculations to experimental situations. We further assume that the cylinders are regularly organized in space and orientation, as illustrated in Fig. 4.9. As seen from this figure, a cylindrical cell of diameter $2R$ can be defined for each of the cylinders

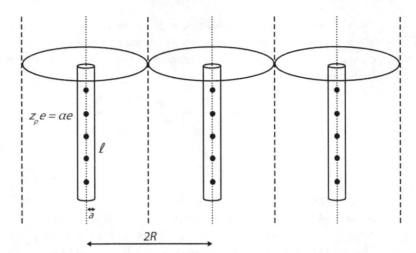

Figure 4.9 Cell model for a solution of charged cylinders. The charge of each repeat unit is ez_p or equivalently αe with α being the degree of ionization. The charge separation along the cylinder axis is ℓ, and the radius of the cylinder is a. The radius of the cylindrical cell is R, which is determined by the concentration of cylinders.

[Alfrey *et al.* (1951), Fuoss *et al.* (1951)]. As we shall see below, it is sufficient to treat one cylindrical cell, instead of the whole collection of rods.

Since the concentration of repeat units in each cell is the same by construction, it is given by

$$c = \frac{N}{\pi R^2 L}. \tag{4.3.1}$$

Using $L = N\ell$, the cell radius is

$$R = \frac{1}{\sqrt{\pi c \ell}} \tag{4.3.2}$$

and is dictated by the concentration. Since each cell is electrically neutral, the concentration of counterions in the present salt-free solution is αc, where the counterions are assumed to be monovalent.

The electric potential around the cylinder is given by the Poisson–Boltzmann equation (Eq. (3.8.8)), as

$$\nabla^2 \psi = -\frac{\rho(\mathbf{r})}{\epsilon_0 \epsilon}, \tag{4.3.3}$$

where $\rho(\mathbf{r})$ is the charge distribution given by the Boltzmann distribution. Let us take the cylinder to be positively charged and the monovalent counterions to be negatively charged. Therefore, the charge distribution is given as

$$\rho(\mathbf{r}) = -e\alpha c \, \exp\left(\frac{e\psi}{k_B T}\right). \tag{4.3.4}$$

Substituting Eq. (4.3.4) into Eq. (4.3.3), the Poisson–Boltzmann equation becomes

$$\nabla^2 \psi = \frac{e\alpha c}{\epsilon_0 \epsilon} \exp\left(\frac{e\psi}{k_B T}\right). \tag{4.3.5}$$

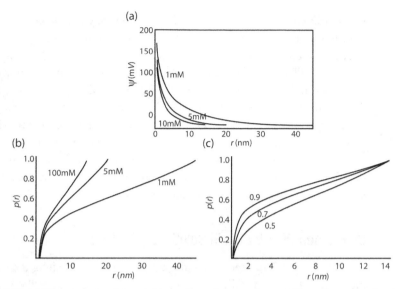

Figure 4.10 (a) Electric potential and (b) fraction of counterions within a distance r from the cylinder axis for $c = 10^{-3}$, 0.005, and 0.01 M, $\ell_B = 0.7$ nm, $\alpha = 0.5$, and $a = 0.6$ nm. (c) Fraction of counterions for different degrees of ionization $\alpha = 0.5, 0.7$, and 0.9 (from bottom to top) at $c = 0.01$ M, $\ell_B = 0.7$ nm, and $a = 0.6$ nm.

In the limit of infinitely long cylinders, the cylindrical symmetry allows us to write the Laplacian ∇^2 in the cylindrical coordinate system as

$$\frac{1}{r}\frac{d}{dr}\left(r\frac{d\psi(r)}{dr}\right) = \frac{e\alpha c}{\epsilon_0\epsilon}\exp\left(\frac{e\psi}{k_B T}\right). \qquad (4.3.6)$$

This equation can be solved exactly by using the boundary conditions that the electric field at the interface between two adjacent cells vanishes and that each cell is electrically neutral,

$$\frac{d\psi}{dr}\bigg|_{R} = 0, \qquad (4.3.7)$$

$$\frac{e\alpha}{\ell} + \int_{r=a}^{R} dr\, 2\pi r\, \rho(r) = 0. \qquad (4.3.8)$$

The first and second terms in Eq. (4.3.8) correspond to the charge on the cylinder per charge separation length ℓ and the total charge from the ion cloud per ℓ, where $\rho(r)$ is given by Eq. (4.3.4). In addition to calculating the electric potential $\psi(r)$, the fraction $p(r)$ of counterions within r from the cylinder axis can be obtained as

$$p(r) = \frac{1}{(\alpha/\ell)}\int_{a}^{r} dr\, 2\pi r\, \alpha c \exp\left(\frac{e\psi}{k_B T}\right). \qquad (4.3.9)$$

The exact results for $\psi(r)$ and $p(r)$ from Eqs. (4.3.6) to (4.3.9) are given in Figs. 4.10a and 4.10b at different repeat unit concentrations ($c = 10^{-3}$, 0.005, and 0.01 M) for $\ell_B = 0.7$ nm, $\alpha = 0.5$, and $a = 0.6$ nm. The value of the Bjerrum length is chosen to be pertinent to aqueous solutions at ambient temperatures. The cylinder radius is typical of the backbone of simple polyelectrolyte molecules. At lower concentrations, the electric potential extends far beyond the cylinder diameter. As the concentration is lowered, the potential becomes more long-ranged. As a direct consequence, the counterion cloud extends further and further as the cylinder concentration is lowered. The dependence of the fraction of counterions within r on the degree of ionization α is given in Fig. 4.10c for $c = 0.01$ M, $\ell_B = 0.7$ nm, and $a = 0.6$ nm. As intuitively expected, the fraction of counterions near the cylinder is higher if the concentration of counterions inside each cell is higher due to ionization.

It must be emphasized that the fraction of counterions around the cylinder increases continuously with the distance from the cylinder. There are no discontinuities in the distribution of ions around the charged cylinders. Indeed, there are no phase transitions such as the condensation of a gas into its liquid state.

4.3.2   Counterion Distribution in Solutions with Salt

Consider a cylinder of radius a and length L with its axis along the z-direction and the radial distance r perpendicular to the z-axis (Fig. 4.11a). Let the identical charges in the cylinder be distributed uniformly along the cylinder axis with ℓ being the charge separation. The charge density is ez_p/ℓ and the total charge is $Q = ez_pL/\ell$, where z_p represents the valency and sign of the individual charges. Outside the cylinder, the solution has the symmetric $(z_c{:}z_c)$ type electrolyte, and the counterions from the salt and the cylinder are taken to be identical. Analogous to the restricted primitive model of simple electrolytes discussed in Section 3.8.4, the volume of the cylinder of radius a excludes counterions and co-ions. We again make the simplifying assumption that there is no dielectric mismatch between the inside and outside of the cylinder.

The electric potential $\psi(r)$ at the radial distance from the axis of the cylinder is given by the Poisson–Boltzmann equation (Eq. (3.8.8)). For symmetric electrolytes,

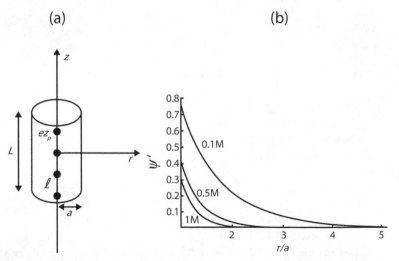

Figure 4.11 (a) A cylinder with charges along the central axis. (b) Dependence of electric potential on r/a for monovalent salt concentrations at 0.1 M ($\xi_D \simeq 0.98$ nm), 0.5 M ($\xi_D \simeq$ 0.424 nm), and 1 M ($\xi_D \simeq 0.3$ nm) ($\psi' = 2\pi\epsilon_0\epsilon\ell\psi/ez_p$) [Muthukumar (2011)].

the valency z_c is the same for both the anions and cations, except for the sign. Also, it turns out that it is a good approximation to take n_{a0} in Eq. (3.8.8) as the average electrolyte number concentration n_0 for both types of ions. As a result, Eq. (3.8.8) becomes,

$$\nabla^2 \psi(\mathbf{r}) = -\frac{ezn_0}{\epsilon_0 \epsilon} \left[\exp\left(\frac{-ez\psi(\mathbf{r})}{k_B T} \right) - \exp\left(\frac{ez\psi(\mathbf{r})}{k_B T} \right) \right]. \tag{4.3.10}$$

Using the relation, $\sinh(x) = (e^x - e^{-x})/2$, we get

$$\nabla^2 \psi(\mathbf{r}) = \frac{2ezn_0}{\epsilon_0 \epsilon} \sinh\left(\frac{ez\psi(\mathbf{r})}{k_B T} \right). \tag{4.3.11}$$

If we consider long cylinders to ignore end effects, as in Section 4.3.1, there is cylindrical symmetry about the cylinder axis. For this two-dimensional problem, the above Poisson–Boltzmann equation is solved in the cylindrical coordinate system of Fig. 4.11. Rewriting the Laplacian ∇^2 in the cylindrical coordinate system, we get

$$\frac{1}{r} \frac{d}{dr} \left(r \frac{d\psi(r)}{dr} \right) = \frac{2ezn_0}{\epsilon_0 \epsilon} \sinh\left(\frac{ez\psi(r)}{k_B T} \right). \tag{4.3.12}$$

In general, this equation needs to be solved numerically to get the spatial variations of the electric potential and charges [Gross & Osterle (1968)]. Therefore, as usual, we make the Debye–Hückel approximation by linearizing the above equation,

$$\frac{1}{r} \frac{d}{dr} \left(r \frac{d\psi(r)}{dr} \right) = \kappa^2 \psi(r), \tag{4.3.13}$$

where

$$\kappa^2 = \frac{2e^2 z^2 n_0}{\epsilon_0 \epsilon k_B T}. \tag{4.3.14}$$

κ^{-1} is the Debye length for the present situation. In order to solve Eq. (4.3.13), we need two boundary conditions. One is that the electric field vanishes at distances far away from the cylinder and the other is that the solution satisfies the electric field at its surface in the absence of the electrolyte.

We first consider the situation without the electrolyte. The electric field and the potential near the surface of the cylinder are obtained as follows. Using the Gauss law [Young & Freedman (2000)], the electric field $E(r)$ at a radial distance r from the cylinder axis, without any regard to the presence of ions in the electrolyte solution, is given by

$$E(r) = \frac{ez_p}{2\pi\epsilon_0 \epsilon \ell r}, \tag{4.3.15}$$

so that the electric field at the surface of the cylinder is

$$E_s = \frac{ez_p}{2\pi\epsilon_0 \epsilon \ell a}. \tag{4.3.16}$$

The electric potential due to the charge distribution inside the cylinder follows from Eqs. (3.1.6) and (4.3.15) as

$$\psi(r) = -\frac{ez_p}{2\pi\epsilon_0 \epsilon \ell} \ln\left(\frac{r}{a} \right). \tag{4.3.17}$$

The electric potential is an arbitrary constant at the cylinder surface ($r = a$), and it must be emphasized that there is no mathematical divergence in the potential, for physically relevant values of r, at the surface of the cylinder and its neighborhood in the solution, for cylinders of finite radius. For distances near the cylinder, the potential varies logarithmically, reflecting the two-dimensional spatial symmetry around an infinite cylinder.

The ion cloud in the solution modifies the result of Eq. (4.3.17). Solving the Debye–Hückel equation, Eq. (4.3.13) with the boundary conditions that the electric field vanishes far away from the cylinder and that it is given by Eq. (4.3.16) at the surface of the cylinder, the result is [Daune (1999)]

$$\frac{ez_c\psi(r)}{k_BT} = 2z_pz_c\frac{\ell_B}{\ell}\frac{1}{\kappa a}\frac{K_0(\kappa r)}{K_1(\kappa a)},\qquad(4.3.18)$$

where $K_n(r)$ is the modified Bessel function of nth order [Abramowitz & Stegun (1965)]. Noting that the strength of the electrostatic interaction compared to the thermal energy is given by the Bjerrum length, we define a dimensionless Coulomb strength parameter Γ as the ratio of the Bjerrum length to the charge separation distance ℓ along the cylinder axis,

$$\Gamma \equiv \frac{\ell_B}{\ell} = \frac{e^2}{4\pi\epsilon_0\epsilon k_BT\ell}.\qquad(4.3.19)$$

We shall refer to Γ as the charge density parameter or Coulomb strength parameter interchangeably. The electric potential follows from Eqs. (4.3.18) and (4.3.19) as

$$\frac{ez_c\psi(r)}{k_BT} = \frac{2z_pz_c\Gamma}{\kappa a}\frac{K_0(\kappa r)}{K_1(\kappa a)},\qquad(4.3.20)$$

depending on the two dimensionless parameters κa and Γ.

The spatial variation of the electric potential given by the above equation is presented in Fig. 4.11b, where $e\psi(r)/2z_p\Gamma k_BT$ is plotted against r/a for different values of the ratio of the Debye length to the radius of the cylinder. As an illustration, let us assume that a ds-DNA molecule can be modeled as a cylinder of radius $a = 0.9$ nm and charge separation length $\ell = 0.17$ nm, so that $\Gamma = 4.2$ at 20°C. If the concentration of monovalent salt is 0.1 M, 0.5 M, and 1 M, the corresponding Debye lengths are 0.98 nm, 0.42 nm, and 0.3 nm, respectively. The variations in the electric potential for these three concentrations are given in Fig. 4.11b. For low salt concentrations, the range of the potential can be substantially longer in comparison with the radius of the cylinder. However, as the salt concentration increases, the electric potential away from the cylinder axis decays more sharply.

If the Debye length is very large as in the case of extremely dilute electrolyte solutions, and if the radius of the cylinder is very small, the arguments of the modified Bessel functions in the above equations can become small. For very small values of the argument, the limiting behaviors of the modified Bessel functions [Abramowitz & Stegun (1965)] are

$$K_0(\kappa r) \simeq -\ln(\kappa r)\qquad(4.3.21)$$

and

$$K_1(\kappa a) \simeq \frac{1}{\kappa a}. \tag{4.3.22}$$

Therefore in the limit of $\kappa a \to 0$, the electric potential is given by

$$\psi(r) \simeq -\frac{2k_B T z_p \Gamma}{e}[\ln(\kappa r)]. \tag{4.3.23}$$

Substitution of Eq. (4.3.19) in the above equation yields

$$\psi(r) = -\frac{e z_p}{2\pi\epsilon_0\epsilon\ell}(\ln r - \ln \xi_D). \tag{4.3.24}$$

The first term ($\ln r$) in the above equation corresponds to the potential due to the charges inside the cylinder, and the second term ($-\ln \xi_D$) is the potential due to the ion cloud essentially located at the Debye length ξ_D.

The above formula, Eq. (4.3.24), is valid only for large Debye lengths. For the usual electrolyte concentrations used in experiments, Eq. (4.3.20) should be used. In general, the electric potential around a cylinder is directly proportional to the charge density parameter Γ, defined in Eq. (4.3.19). If Γ is large, the potential can become so large that the assumption of linearization of the Poisson–Boltzmann equation ($e z_c \psi / k_B T \ll 1$) breaks down. Therefore, for large Γ values, the Debye–Hückel description is valid only for large distances away from the cylinder where the potential becomes weak enough to satisfy the assumption made in the derivation of Eq. (4.3.20). Also, similar to the case of planar interfaces, for very short distances from the surface, the counterions adsorb essentially into a monolayer. Thus, we can imagine three regimes [Hiemenz & Rajagopalan (1997)] for the variation of the electric potential as a function of the radial distance from the cylinder axis: (a) the Debye–Hückel regime at large distances, (b) Gouy regime at closer distances to the cylinder, where the nonlinear Poisson–Boltzmann description is needed, and (c) a monolayer of adsorbed counterions, called the Stern regime. Although the quantitative details of the crossover behaviors between these regimes are rather intricate, the basic feature is that the counterions accumulate near the charged surface and the electric potential decays into the solution with a characteristic distance roughly given by the Debye length. The dependencies of the electric potential and ion distribution on the radial distance are smooth as also evident from the exact solutions for salt-free solutions around thin charged cylinders discussed in Section 4.3.1.

4.4   Line Charge and Manning Condensation

Let us consider an infinitely thin and infinitely long line of charges in a medium of dielectric constant ϵ, since this model is the simplest that might mimic a long rigid polyelectrolyte. For this model, an argument, called Manning condensation, is made to derive a simple relation for the accumulation of counterions near the line charge. The

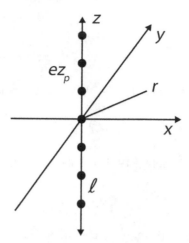

Figure 4.12 Line charge used in deriving Manning condensation [Muthukumar (2011)].

geometry and the parameters of the model are given in Fig. 4.12. Let the separation distance between the consecutive uniform charges be ℓ and each charge be ez_p, as in the preceding section. The difference now is that the line charge does not have any radius. The line charge is along the z-direction and r is the radial distance from the line charge in a plane perpendicular to the line charge.

As derived in Appendix 1, the number of counterions of valency z_c inside a cylinder of radius r_0 (of length ℓ) is

$$n_{r_0} \sim \left. \frac{1}{r^{2(\Gamma z_p z_c - 1)}} \right|_0^{r_0}, \tag{4.4.1}$$

showing that n_{r_0} diverges at $r \to 0$ for $z_p z_c \Gamma > 1$. This apparent divergence can be mathematically avoided by assuming that $z_p z_c \Gamma$ is never allowed to be greater than unity. In other words, we imagine that enough counterions condense on the line charge and reduce the charge density parameter to be Γ_{eff} so as to make $z_p z_c \Gamma_{\text{eff}}$ to become unity. Hence, the effective charge separation ℓ_{eff} is imagined to be greater than ℓ.

The above argument, constructed to avoid the divergence in the number of counterions on an infinitely thin and infinitely long line charge, is referred to as the Manning condensation. According to this hypothesis, a plot of Γ_{eff} against Γ is given in Fig. 4.13a. By accounting for the counterion condensation, the charge fraction α of the line charge is defined as the ratio of ℓ to ℓ_{eff},

$$\alpha \equiv \frac{\ell}{\ell_{\text{eff}}} = \frac{\Gamma_{\text{eff}}}{\Gamma}. \tag{4.4.2}$$

For $\Gamma z_p z_c < 1$, $\ell_{\text{eff}} = \ell$ so that $\alpha = 1$. For $\Gamma z_p z_c > 1$, $\Gamma_{\text{eff}} z_p z_c = 1$ so that $\alpha = 1/\Gamma z_p z_c$,

$$\alpha = \begin{cases} 1, & \Gamma z_p z_c < 1 \\ \frac{1}{\Gamma z_p z_c}, & \Gamma z_p z_c > 1 \end{cases} \tag{4.4.3}$$

as sketched in Fig. 4.13b.

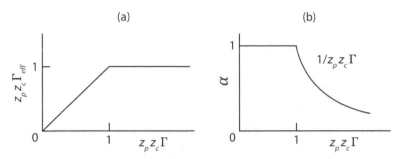

Figure 4.13 Counterions condense on the line charge for $z_p z_c \Gamma > 1$. (a) Effective charge density parameter. (b) Fraction of uncondensed charge density, $\alpha = \ell / \ell_{\text{eff}}$. This picture with discontinuity is to be contrasted with the exact results in Fig. 4.10, where various quantities are continuous [Muthukumar (2011)].

The simplicity and consequences of the above results can be illustrated by considering a line charge model of dsDNA. Let the charge separation length ℓ be 0.17 nm, so that $\Gamma = 4.2$ at 20°C. Let $z_p = 1$ and $z_c = 1$. Therefore, for conditions where counterions would condense ($\Gamma z_p z_c > 1$), the effective charge fraction α follows from Eq. (4.4.3) as

$$\alpha = \frac{1}{\Gamma} = 0.24. \qquad (4.4.4)$$

Therefore the fraction of condensed counterions, $1 - \alpha$, is about three quarters. The effective charge of each phosphate group depends on the valency of the counterion,

$$(ez_p)_{\text{eff}} = \frac{e}{z_c \Gamma}. \qquad (4.4.5)$$

There has been extensive discussion about the applicability of the Manning condensation to polyelectrolytes [Oosawa (1957), Rice & Nagasawa (1961), Manning (1969), Manning (1978), Hoagland (2003), Muthukumar (2004)]. While the above argument is exact for the particular model of an infinitely thin and infinitely long one-dimensional line charge, the results of Eq. (4.4.5) cannot be applied to experimental systems involving flexible and semiflexible polyelectrolyte molecules [Beer et al. (1997), Holm et al. (2004), Muthukumar (2004)]. The discontinuity of α shown in Fig. 4.13 is also not to be expected for these experimental systems [Alfrey et al. (1951), Fuoss et al. (1951)]. As described in Section 4.3.1, it must be emphasized that the Manning condensation criterion is only a mathematical convenience, since the exact solutions for the potential and counterion distribution, for the line charge model with a finite radius, do not show any such phase-transition-like discontinuities [Alfrey et al. (1951), Fuoss et al. (1951)]. The three major objections to the line-charge model of flexible polyelectrolyte are: zero thickness, infinite length, and no chain flexibility.

4.5 Dielectric Mismatch

So far in this chapter, we have taken the dielectric constant across the various interfaces to be the same (except in Section 4.2.2). In general, this condition is not true in

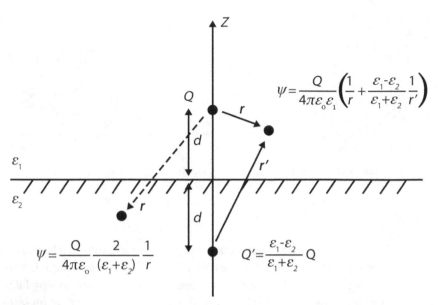

Figure 4.14 Effects of dielectric mismatch across the interface in the (x, y) plane at $z = 0$ in the Cartesian coordinate system. The image charge is Q'. The potential due to the charge Q at a distance r depends on whether r is in medium of ϵ_1 or medium of ϵ_2 and how far r is from the interface [Muthukumar (2011)].

charged macromolecular systems. In the example of an oily pocket immersed inside an aqueous electrolyte solution or an ion channel embedded within a lipid membrane, the dielectric constants are different across the interfaces. The dielectric mismatch across interfaces can have profound effects in terms of the quantitative values of forces exerted on charged species in the neighborhood of interfaces [Griffiths (1999), Jackson (1999)]. As an illustrative example, consider a point charge Q inside a semi-infinite medium of dielectric constant ϵ_1 at a distance d away from a planar interface that separates the first medium from another semi-infinite medium of dielectric constant ϵ_2. Let the interface be the (x, y) plane at $z = 0$, as shown in Fig. 4.14. The electric potential at $\mathbf{r}$ due to the charge Q is no longer given by Eq. (3.2.13), which is valid for a homogeneous infinite medium. The dielectric mismatch alters the potential. The standard arguments [Griffiths (1999), Jackson (1999)] in electrostatics show that the effect of dielectric mismatch is equivalent to placing an image charge Q' at a distance d below the interface and considering the effects of Q and Q' in a medium of uniform dielectric constant ϵ_1. The result is

$$\psi = \begin{cases} \frac{1}{4\pi\epsilon_0\epsilon_1}\left(\frac{Q}{r} + \frac{Q'}{r'}\right), & z > 0 \\ \frac{1}{4\pi\epsilon_0}\frac{2}{(\epsilon_1+\epsilon_2)}\frac{Q}{r}, & z < 0 \end{cases} \qquad (4.5.1)$$

as given in Fig. 4.14, where the image charge Q' is

$$Q' = \frac{\epsilon_1 - \epsilon_2}{\epsilon_1 + \epsilon_2}Q. \qquad (4.5.2)$$

The distances r and r' are, respectively, $(x^2+y^2+(z-d)^2)^{1/2}$ and $(x^2+y^2+(z+d)^2)^{1/2}$, as marked in Fig. 4.14.

Depending on the ratio ϵ_1/ϵ_2, the image charge may attract or repel the charge Q. For $\epsilon_1 \gg \epsilon_2$, the image charge has the same sign and magnitude as Q, and the charge Q is repelled by the interface. On the other hand, if $\epsilon_2 \gg \epsilon_1$, the charge in a medium of dielectric constant ϵ_1 is attracted by the medium of dielectric constant ϵ_2. As an example of the net effect from the charge Q and its image charge Q', let us consider the electric potential at a distance r much larger than the distance d of the charge from the interface. This follows from the above equations as

$$\psi = \frac{Q}{4\pi\epsilon_0} \frac{2}{(\epsilon_1 + \epsilon_2)} \frac{1}{r}. \tag{4.5.3}$$

Therefore, an effective dielectric constant ϵ_{eff} may be identified in describing the electric potential at distances far away from Q, by writing the above equation as

$$\psi = \frac{Q}{4\pi\epsilon_0 \epsilon_{\text{eff}} r}, \tag{4.5.4}$$

where

$$\epsilon_{\text{eff}} = \frac{\epsilon_1 + \epsilon_2}{2}. \tag{4.5.5}$$

If the dielectric constants of an aqueous medium and an oily enclosure, such as the interior of a protein or a hydrophobic polymer, are taken to be 80 and 3, respectively, then the effective dielectric constant at large distances is roughly 40. Thus the apparent dielectric constant can be different by a factor of two. Only for distances much less than the distance between the charge and the interface, the dielectric constant is effectively the local value. The complementary problem of how a charge buried in an oily enclosure is subjected to an attractive force by its image charge, which is now present in the aqueous medium with higher dielectric constant, can be readily addressed from the above equations.

Analogous calculations of the effects of dielectric mismatch due to curved interfaces become highly technical. Nevertheless, if one is interested in very accurate estimates of various forces near interfaces with dielectric mismatch, it is necessary to account for the presence of dielectric mismatch across interfaces.

5 Dilute Solutions of Charged Macromolecules

5.1 Introduction

When charged macromolecules are sparsely and uniformly distributed in an aqueous salt solution, several questions emerge. How do they assume their net electrical charge and to what values? What are their sizes and internal structures? What are the colligative properties of their solutions? The primary goal of this chapter is to address these questions for electrolyte solutions at very low polymer concentrations such that the average distance between any two macromolecules is much larger than their average diameter. We have invested in the previous chapters in laying down the necessary foundation to treat charged macromolecules. The two major features of charged macromolecules are the topological correlation emerging from chain connectivity and the electrostatic interactions among charged species. In order to understand the nature of both of these correlations, the behaviors of macromolecules without any charges and those of solutions of ionic species were treated as separate entities in Chapters 2 and 3, respectively. We are now ready to put the concepts underlying these two aspects together and address the combined behavior of charged macromolecules.

Consider a dilute aqueous solution containing a certain amount of a simple electrolyte, such as sodium chloride. The thermodynamic and dynamic properties of this solution are dictated by the electrostatic interactions among all ions in the solution. Let us now add a small amount of a nonpolar polymer, such as poly(styrene), into this aqueous electrolyte solution. The polymer does not dissolve and it separates into an immiscible oil-like phase. This is due to the short-ranged van der Waals interactions among the hydrophobic monomer units making up the nonpolar polymer. In contrast, consider a small amount of a polyelectrolyte salt, such as deoxyribonucleic acid sodium salt or sodium poly(styrene sulfonate), dropped into this electrolyte solution. In due course, the polymer salt dissolves completely into a homogeneous solution, which consists of the charged polymer molecules, their dissociated counterions, and the cations and anions of the simple electrolyte. The behavior of charged macromolecules, in the presence of their counterions and the electrolyte ions in the background solution, becomes unavoidably rich due to the confluence of electrostatic correlation among all charged species in the solution and the topological correlation arising from chain connectivity of the macromolecule [Rice & Nagasawa (1961), Mandel (1988), Dautzenberg *et al.* (1994), Förster & Schmidt (1995), Hoagland (2003), Muthukumar (2017)].

At the simplest level, as originally envisaged by Katchalsky *et al.* (1950), consider a uniformly charged flexible polyelectrolyte of N segments, each with charge qe (e is the electronic charge), with no other charged species in the sphere of influence of the chain. By recapitulating the essential aspects of chain connectivity (Chapter 2) and electrostatic interaction (Chapter 3), the free energy $F(R)$ of the chain with its end-to-end distance at R is given by Katchalsky *et al.* (1950)

$$\frac{F(R)}{k_B T} = \frac{3R^2}{2N\ell^2} + \frac{\ell_B (qN)^2}{4R}, \tag{5.1.1}$$

where the first term on the right-hand side is due to conformational entropy of the Gaussian chain (Eq. (2.1.11)) and the second term represents the electrostatic repulsion (Eqs. (3.6.1) and (3.6.7)). Here ℓ is the Kuhn segment length, ℓ_B is the Bjerrum length (Eq. (3.6.7)), and $k_B T$ is Boltzmann's constant times the absolute temperature. The second term follows from Eq. (3.6.1) by imagining [Katchalsky *et al.* (1950)] that the total charge qeN of the chain is distributed such that half of the total charge is at one end and the other half of the total charge is at the other end and these ends are separated by R. By minimizing $F(R)$ with respect to R, that is using the condition $\partial F(R)/\partial R = 0$, we get

$$\frac{R}{\ell} \sim \left(\frac{q^2 \ell_B}{\ell} \right)^{1/3} N. \qquad \text{rod-like; salt-free} \tag{5.1.2}$$

The numerical prefactor 0.44 is not expressed explicitly in this proportionality equation since our focus is on the relation between R and the key experimental variables q, ℓ_B, and N. The linear dependence of the end-to-end distance on the number of segments N suggests that an isolated flexible polyelectrolyte chain would adopt a rod-like conformation if the chain were to be alone. On the other hand, if the polyelectrolyte chain is surrounded by counterions and low molar mass salt ions, the electrostatic interaction is screened and the interaction becomes short-ranged (Eq. (3.8.16)). Under this salty condition, the strength of the electrostatic interaction is $4\pi q^2 \ell_B / (\kappa^2 \ell^3)$ (as will be derived later in this chapter, Section 5.3.2.3), where κ is the inverse Debye length (Eqs. (3.8.17) and (3.8.18)). Now $F(R)$ is given by (as for a chain with short-ranged van der Waals excluded volume interactions, Eq. (2.5.3))

$$\frac{F(R)}{k_B T} = \frac{3R^2}{2N\ell^2} + \frac{1}{2} \frac{4\pi q^2 \ell_B}{\kappa^2} \frac{N^2}{R^3}. \tag{5.1.3}$$

Minimization of $F(R)$ with respect to R gives [Flory (1953b)]

$$\frac{R}{\ell} \sim \left(\frac{4\pi q^2 \ell_B}{\kappa^2 \ell^3} \right)^{1/5} N^{3/5}. \qquad \text{swollen coil in good solvents; high salt} \tag{5.1.4}$$

Thus, we anticipate right away that a flexible chain is going to be swollen to an average conformation inbetween the swollen coil (as in good solvents) and rod-like (as in semiflexible chains) conformations, depending on the salt concentration. In this qualitative description, and in general, we need to know the charge of the polymer qeN. What is the net charge of a well-defined charged macromolecule? The answer depends on the experimental conditions presented to the macromolecule.

As in the case of rigid charged particles in electrolyte solutions, described in Chapter 4, an ion cloud would surround a polyelectrolyte chain, due to an optimization between the attractive interaction between the polymer charges and the oppositely charged counterions and the loss of translational freedom of the dissociated free counterions upon binding with the macromolecule. Unlike rigid bodies, a flexible polymer chain has an incipient capacity to adopt an enormous number of conformations due to its conformational flexibility. As a result, the situation with flexible charged macromolecules becomes much richer in comparison with rigid objects.

In general, the nature of polymer conformations is influenced by the inter-segment interactions arising from both the long-ranged electrostatic interactions among charged segments and the short-ranged van der Waals forces among the hydrophobic groups, counterion binding on the polymer, and the conformational entropy associated with chain connectivity. The inter-segment electrostatic interaction is mediated through all charged species in the system, which include charged segments, dissociated ions from the macromolecule and the added salt, and the dipolar ion pairs formed by binding of counterions on the macromolecule. These contributing factors are cartooned in Fig. 5.1. The polymer conformations, in turn, influence the spatial distribution of the small ions. After a short time elapse, the polymer conformations would change even in equilibrium due to thermal noise from solvent molecules. Now the ion cloud surrounding the polymer skeleton will contain different configurations of the counterions, with some of the original counterions replaced by new ones at new locations. Thus, on average, we imagine a counterion cloud looking like a worm around the backbone of the polymer chain. The worm-like counterion cloud is

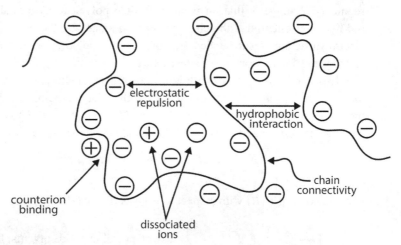

Figure 5.1 Self-consistent coupling between chain conformation and ionization. Electrostatic repulsion between the charges of the repeat units, mediated by all charged species (polymer segments, dissociated ions from the macromolecule and salt, and ion pairs) in the system, counterion adsorption equilibria, hydrophobic interaction among uncharged moieties, and chain connectivity control the conformations of the macromolecule which in turn dictate the distribution of counterions and electrolyte ions in the solution. The net effect is self-regularization of the charge of the macromolecule. The charge is not a fixed quantity for a charged macromolecule, but changes as the experimental conditions change.

dynamic, with the counterions continuously binding and unbinding with the polymer backbone at random locations, but maintaining an average number inside the cloud. As a result, the effective charge of the polyelectrolyte is not the same as its fully ionized chemical charge. The net charge of a macromolecule is less than its chemical charge due to counterion binding. The effective charge is unique to the average polymer conformation, and it self-regulates with changes in polymer conformations accompanying changes in experimental conditions.

In addition to the net charge, the other attributes of the macromolecule are its size (radius of gyration, mean square end-to-end distance) and structure (monomer density profile). Due to the presence of charges of the same sign in a polyelectrolyte chain, we expect intra-chain segment–segment repulsion. The net repulsion among the effective charges on the polymer backbone is manifest as electrostatic swelling for flexible polymers, or equivalently as chain stiffening for semiflexible polymers. Furthermore, we can readily expect nonideal behavior in the colligative properties of solutions of charged macromolecules arising from long-ranged topological and electrostatic correlations. In addition, due to the fact that not all charges of the repeat units of the macromolecule are visible under the experimental conditions of finite concentrations (which can prohibit full dissociation of counterions), the colligative properties of dilute solutions of charged macromolecules deviate substantially from ideal behaviors. The main focus of this chapter is to combine the key concepts of electrostatic interactions (Chapter 3) and the various polymer models (Chapter 2), toward a description of the equilibrium properties of charged macromolecules in dilute solutions.

The outline of this chapter is the following. (1) We shall exhibit the importance of the omnipresent self-regularization of macromolecular charge by describing experimental titration curves. (2) After introducing a minimal model of a charged polymer where its effective charge is uniformly smeared over its backbone, we shall present major concepts as guidelines toward an understanding of the physics of charged macromolecules. (3) Equipped with a general conceptual framework, we will next discuss key experimental and simulation results. (4) We will then describe several theoretical treatments of electrostatic effects on charged macromolecules including scaling arguments, analytical theories and self-consistent charge regularization of macromolecules. (5) Next, we will present a discussion of the osmotic pressure of dilute solutions and the very important aspects of the Donnan equilibrium, which is of common occurrence in a variety of phenomena involving charged systems. (6) Finally, building on the above concepts, a brief discussion of polyampholytes, polyzwitterions, and intrinsically disordered proteins will be presented.

5.2 Titration Curves and Charge Regularization

Consider first a low molar mass acid HA undergoing dissociation into the anion A^- and the proton H^+,

$$HA \rightleftharpoons H^+ + A^-. \tag{5.2.1}$$

The acid dissociation constant K_a is defined by

$$K_a = \frac{[\text{H}^+][\text{A}^-]}{[\text{HA}]},$$ (5.2.2)

where the square brackets denote the concentrations of the various species. The dissociation constant can also be expressed in terms of the free energy change ΔG_0 associated with the dissociation equilibrium,

$$K_a = \exp\left[-\frac{\Delta G_0}{k_B T}\right].$$ (5.2.3)

Let α be the degree of ionization (dissociation),

$$\alpha = \frac{[\text{A}^-]}{[\text{A}^-] + [\text{HA}]}.$$ (5.2.4)

Combining Eqs. (5.2.2) and (5.2.4), the concentration of protons in the solution is given by

$$[\text{H}^+] = K_a \frac{(1-\alpha)}{\alpha}.$$ (5.2.5)

Taking the negative common logarithm of this equation, we get

$$\text{pH} = \text{pK}_a + \log_{10}\left(\frac{\alpha}{1-\alpha}\right),$$ (5.2.6)

where $\text{pH} = -\log_{10}[\text{H}^+]$ and $\text{pK}_a = -\log_{10} K_a$. This equation is the classic Henderson–Hasselbalch equation for ionization equilibria, giving the dependence of the degree of ionization on the pH of the medium and vice versa [Daune (1999)]. For $\alpha = 1/2$, $\text{pH} = \text{pK}_a$. At this point of α, pK_a is the free energy change $\Delta G_0/(2.303 k_B T)$ associated with the dissociation of the acid, as seen from Eq. (5.2.3).

The Henderson–Hasselbalch equation can be written equivalently by rearranging the above equation as

$$\alpha = \frac{1}{\left[1 + e^{-2.303(\text{pH}-\text{pK}_a)}\right]}.$$ (5.2.7)

The degree of ionization of low molar mass acids can be tuned by changing pH in accordance with the above equation. According to this equation, α rises sharply from zero to unity over a narrow pH range around the pK_a of the particular acid. This is illustrated by the titration curve for acrylic acid given in Fig. 5.2, where $\text{pK}_a = 4.76$ [Lagueci et al. (2006)].

For polyelectrolytes, the titration curves are quite different from those of the corresponding monomers. This is illustrated in Fig. 5.2, where the titration curves of poly(acrylic acid) at different degrees of polymerization ($N = 25$, 70, and 700) are contrasted with the monomeric acrylic acid. It is clear that the Henderson–Hasselbalch equation for the monomer is not applicable to its polymer. The value of pK_a increases with molecular weight of the polyelectrolyte as seen in Fig. 5.2. In view of Eqs. (5.2.3) and (5.2.6), the free energy associated with ionization changes as the molecular weight of the polyelectrolyte changes. When the polymer size changes, the local electric potential is modified [Luo et al. (2013)], which in turn influences the dissociation

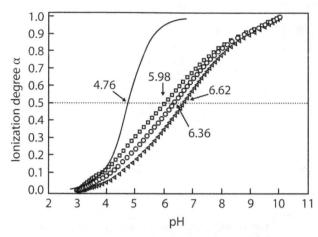

Figure 5.2 Titration curves of acrylic acid and poly(acrylic acid). The solid curve is for acrylic acid given by Eq. (5.2.7) with $pK_a = 4.76$. The data are for poly(acrylic acid) with degree of polymerization $N = 25$ (square), 70 (circle) and 700 (triangle). The values of pK are indicated on the titration curves [Adapted from Lagueci *et al.* (2006)].

of a proton from its monomeric acid. Now, we need to do extra electric work, resulting in ΔG_e, to transfer H^+ from the polyion (instead of an isolated monomer) into the bulk solution against the attraction of the negatively charged polyanion. Therefore, we write the net free energy change associated with ionization as [Katchalsky *et al.* (1954)]

$$\Delta G = \Delta G_0 + \Delta G_e(\alpha, c_s, \epsilon, T), \qquad (5.2.8)$$

where ΔG_e is the additional electrostatic part, which depends on the average degree of ionization α for the polyelectrolyte chain, salt concentration c_s, local dielectric constant ϵ, and temperature T. This extra ΔG_e can be equivalently written in terms of an average electric potential $\overline{\psi}_{\text{site}}$ at the site of ionization as

$$\Delta G_e = e\overline{\psi}_{\text{site}}, \qquad (5.2.9)$$

where e is the electronic charge.

Since ΔG_0 is no longer sufficient to describe the titration curve of polyacids, we define an apparent ionization constant K as

$$K = \exp\left(-\frac{\Delta G}{k_B T}\right) = K_a \exp\left(-\frac{\Delta G_e}{k_B T}\right), \qquad (5.2.10)$$

where Eqs. (5.2.3) and (5.2.8) are used. Using $K = [H^+][A^-]/[HA] = \exp\left(-\frac{\Delta G}{k_B T}\right)$, the analog of the Henderson–Hasselbalch equation (Eq. (5.2.6)) for polyacids is

$$pH = pK + \log_{10}\left(\frac{\alpha}{1 - \alpha}\right), \qquad (5.2.11)$$

with

$$pK = pK_a + \frac{1}{2.303}\frac{\Delta G_e(\alpha, c_s, \epsilon, T)}{k_B T}. \qquad (5.2.12)$$

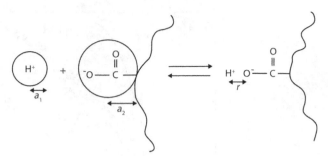

Figure 5.3 Ionization equilibrium of a monomer of poly(acrylic acid).

Several theoretical models to get $\overline{\psi}_{site}$ have been attempted in the past. The computation of $\overline{\psi}_{site}$ is a nontrivial task due to the necessity to perform self-consistent charge regularization accompanying changes in polymer conformations. We shall address this issue in Section 5.7. Nevertheless, useful insight can be gleaned by considering the ionization equilibrium based on the Born model described in Section 3.5.

As a specific example, consider the binding equilibrium of a counterion (say H^+) to the acrylate A^- group of a monomer, as in poly(acrylic acid), in water of dielectric constant ϵ (Fig. 5.3).

The free energy associated with charging a proton of radius a_1 to its full charge in a medium of dielectric constant ϵ is given by the Born energy, $e^2/(8\pi\epsilon_0\epsilon a_1)$, from Eq. (3.5.3). Similarly, the Born energy for the acrylate group of radius a_2 is $e^2/(8\pi\epsilon_0\epsilon_\ell a_2)$. Here, the dielectric constant ϵ_ℓ in the neighborhood of the acrylate group must be significantly lower than that of the bulk polar solvent as the group is permanently attached to oil-like nonpolar backbone. As seen in Section 3.5, the dielectric constant changes from a very low value of about 5 near the chain to the bulk value of about 80 at a distance of 2 nm from the chain backbone [Mehler & Eichele (1984)]. Here, let us treat ϵ_ℓ as a phenomenological parameter. The ion-pair energy for a bound H^+ ion to the acrylate group at a distance r is the Coulomb energy $-e^2/(4\pi\epsilon_0\epsilon_\ell r)$ (see Eq. (3.6.1)). Therefore the free energy of formation of the ion pair is

$$\frac{\Delta G}{k_B T} = -\frac{e^2}{4\pi\epsilon_0\epsilon_\ell r} - \frac{e^2}{8\pi\epsilon_0}\left(\frac{1}{\epsilon a_1} + \frac{1}{\epsilon_\ell a_2}\right) \equiv -\frac{e^2}{4\pi\epsilon_0\epsilon_\ell d}, \qquad (5.2.13)$$

where $1/d = (1/r + 1/(2a_2) + \epsilon_\ell/(2\epsilon a_1))$. Here ϵ_ℓ and d are unknown and their values depend on the specifics of the polyelectrolyte backbone and chemical identities of the various ions in the solution. These unknowns can be parametrized into one parameter δ defined as

$$\delta = \frac{\epsilon\ell}{\epsilon_\ell d}, \qquad (5.2.14)$$

where ℓ is now taken as the charge separation distance along the chain contour. Using the definition of the Bjerrum length ℓ_B (Eq. (1.3.2)), we can rewrite Eq. (5.2.13) in terms of the parameter δ as [Muthukumar (2004)]

$$\frac{\Delta G}{k_B T} = -\frac{\ell_B}{\ell}\delta, \tag{5.2.15}$$

The parameter δ is related to the ionization equilibrium constant K through

$$pK = -\log_{10} K = \frac{1}{2.303}\frac{\ell_B}{\ell}\delta. \tag{5.2.16}$$

The fact that the ionization equilibrium of a repeat unit as a part of the polyelectrolyte chain is different from that of the unit if it were to be dispersed into the polar medium as simply a monomer, as seen in Fig. 5.2, can be represented using the Born model given by Eq. (5.2.16).

Although the above description is couched with the specific example of poly(acrylic acid), which is a weak acid, the conclusions are general to all polyelectrolytes, including strongly ionizing systems such as poly(styrene sulfonate) salts. The local polarizability of the medium controlling the ionization of a specific monomer is influenced by all other monomers of the chain. Therefore, ΔG and α discussed above can only be averaged quantities. For the whole chain, made of an ensemble of ionization equilibria for all ionizable repeat units, an average degree of ionization α may be ascribed to the whole chain for a particular set of experimental conditions. It is clear that the degree of ionization of each charged repeat unit is less than unity [Förster & Schmidt (1995)]. Although, in reality, the chain should be treated as a heteropolymer made of unionized groups and ionized groups, with the sequences changing dynamically, it is a good approximation to take the sequences as "annealed," such that the chain is uniformly charged with each repeat unit of degree of ionization α. The value of α depends on all nonuniversal chemical specifics of the polymer, counterion, salt ions, and solvent, and the physical conditions such as the temperature, polyelectrolyte concentration, and salt concentration. As the chain conformations change, so does the value of α, which is not a fixed number for any polyelectrolyte solution as the experimental conditions change. The resulting charge regularization is uniquely specific to a particular set of experimental variables (temperature, polymer concentration, salt concentration, solvent, etc.).

As α increases, $\Delta G_e (= e\overline{\psi}_{site})$ increases and hence the pK is expected to increase (Eq. (5.2.12)). When the concentration of added salt increases, the electrostatic interaction is screened and we expect a decrease in ΔG_e so that pK decreases. These trends of pK with α and c_s are seen in experiments [Mandel (1988)]. As a further example of self-consistent charge regularization when a polyelectrolyte chain undergoes a conformational change, the titration curve for poly(methacrylic acid) in water at 5°C is given in Fig. 5.4 [Mandel et al. (1967)]. Here, the dependence of pK on α is nonmonotonic due to a conformational transition of poly(methacrylic acid) in the range of $0.15 < \alpha < 0.30$. In general, the degree of ionization of a polyelectrolyte adjusts itself concomitantly with the conformation of the polyelectrolyte. As the experimental conditions (such as temperature, salt concentration, and polyelectrolyte concentration) change, the degree of ionization changes too.

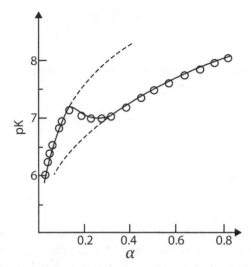

Figure 5.4 Nonmonotonic dependence of pK on degree of ionization α for poly(methacrylic acid) in water at 5°C accompanying a conformational transition in the range $0.15 < \alpha < 0.30$. Dashed portions indicate extrapolations for the two different conformations [Adapted from Mandel *et al.* (1967)].

5.3   Polyelectrolyte Model Chain and Basic Concepts

Firstly, before describing the key concepts, we consider the parameters that reflect the various experimental variables such as the molecular weight, charge, and concentration of polymer molecules, and the concentration of added salt in a polyelectrolyte solution. Consider a dilute solution of volume V containing n polyelectrolyte chains of molecular weight M at added salt concentration c_s. For illustrative purposes, let the polyelectrolyte be a homopolymer with each repeat unit carrying one ionizable group. Upon ionization in a polar solvent, each repeat unit is assumed to carry one monovalent charge e. If the charge separation between any two consecutive monomeric charges along the chain backbone is ℓ_0 (comparable to a couple of bond lengths), the chemical charge density q_0 along the chain contour is e/ℓ_0. Since the basic model for chain connectivity is the Kuhn model with N Kuhn segments (Section 2.1) of segmental length ℓ, we write the chemical charge density as ez_p/ℓ,

$$q_0 = \frac{e}{\ell_0} = \frac{ez_p}{\ell}, \tag{5.3.1}$$

where z_p is the average number of ionizable monomers per one Kuhn length, in accordance with the chemical formula of the polyelectrolyte. The total chemical charge of one polymer molecule is eN_0 (with N_0 being the number of ionizable monomers per chain), or equivalently ez_pN, where N is the number of Kuhn segments per chain. Although all repeat units of the chain are ionizable, it turns out that only a certain fraction of these actually dissociate and consequently bear the monomer charge. As we shall see below, this fraction depends on the experimental conditions. Let α be the average degree of ionization. In view of this, and within the assumption of smearing the charges uniformly throughout the backbone of the chain, we take the effective charge of a segment as $e\alpha z_p$, so that the effective charge density q of the polyelectrolyte chain is

$$q = \frac{e\alpha z_p}{\ell}.$$ (5.3.2)

Let the charge of each counterion be ez_c. The number of counterions n_c in a solution containing n polyelectrolyte chains of uniform contour length with an average degree of ionization α is given by the electroneutrality condition as

$$n_c = \left|\frac{z_p}{z_c}\right|\alpha n N.$$ (5.3.3)

Furthermore, there are cations and anions from the simple salt, with their charges and numbers represented by ez_a and n_{a0} for the a-type ions. The number density of counterions is n_c/V, and that of a-type salt ions is n_{a0}/V, which is proportional to the added salt concentration c_s. The average segment number density of the polymer in the solution is nN/V.

In addition to the charges, the repeat units of the macromolecule interact among themselves and the solvent molecules through the short-ranged van der Waals-type interactions, which are collectively called excluded volume interactions. Furthermore, in general, the charges on the polymer may be quenched on specific segments representing the sequences of charged-uncharged heteropolymers, although we have described above homopolymers for illustrative purposes. By following the same procedure outlined in Section 2.4, the probability distribution function for the end-to-end distance $\mathbf{R}$ of a Kuhn chain with N segments given by Eq. (2.4.6) can be generalized to include the presence of charges as

$$G(\mathbf{R}, \mathbf{0}; N) = \int_0^{\mathbf{R}} \mathcal{D}[\mathbf{R}(s)] \exp\left[-\left(\frac{3}{2\ell^2}\right)\int_0^N ds \left(\frac{\partial \mathbf{R}(s)}{\partial s}\right)^2\right.$$
$$\left. -\frac{1}{2k_B T}\int_0^N ds \int_0^N ds' u\left[\mathbf{R}(s) - \mathbf{R}(s')\right]\right],$$ (5.3.4)

where

$$u\left[\mathbf{R}(s) - \mathbf{R}(s')\right] = u_{\text{exc}}\left[\mathbf{R}(s) - \mathbf{R}(s')\right] + q_s q_{s'} u_{\text{elec}}\left[\mathbf{R}(s) - \mathbf{R}(s')\right].$$ (5.3.5)

The first term in the argument of the exponential in Eq. (5.3.4) represents the conformational entropy arising from chain connectivity. The first term on the right-hand side of Eq. (5.3.5) is the same as in Eq. (2.4.2) denoting the short-ranged excluded volume interaction between the sth and s'th segments. The second term denotes the effective electrostatic interaction between the sth and s'th segments mediated by all charged species in the system. The variable q_s denotes whether the sth monomer is charged ($q_s = 1$) or not ($q_s = 0$). Similarly $q_{s'}$ is 0 or 1 depending on whether the s'th monomer is uncharged or charged. The sign of the charge is included in the factor u_{elec}.

We now address these two kinds of inter-segment interactions and chain connectivity, along with some key concepts pertinent to the equilibrium properties of solutions of charged macromolecules.

5.3.1   Excluded Volume Interaction (Hydrophobic Interaction)

As discussed in Section 2.4, there is an effective potential interaction between any two segments of the polymer mediated by the solvent molecules, even in the absence of charges. We call all of these short-range excluded volume interactions as the van der Waals interaction, or loosely the hydrophobic interaction. As described in Section 2.4, this is parametrized as the excluded volume parameter v, or the Flory–Huggins χ parameter, through Eqs. (2.4.2) and (2.4.3) as

$$\frac{u_{\text{exc}}\left[\mathbf{R}(s) - \mathbf{R}(s')\right]}{k_B T} = v\ell^3 \delta\left[\mathbf{R}(s) - \mathbf{R}(s')\right] = (1 - 2\chi)\,\ell^3 \delta\left[\mathbf{R}(s) - \mathbf{R}(s')\right], \quad (5.3.6)$$

where $\delta\left[\mathbf{R}(s) - \mathbf{R}(s')\right]$ is the Dirac delta function.

The chemical specificities of the uncharged part of the monomer in the polyelectrolyte and the solvent are presumed to be adequately captured by the single parameter $v = (1 - 2\chi)$. For ideal conditions corresponding to the Flory temperature, the excluded volume parameter v is zero ($\chi = 1/2$), due to a cancelation between the net attractive and repulsive interactions among the solvent molecules and the polymer segments. If $v < 0$ ($\chi > 1/2$), polymer segments attract themselves more than the attraction between the polymer and solvent. As a result, the polymer contracts into globular conformations, as seen in Section 2.6. Since most of the polyelectrolyte backbones are immiscible with the polar solvents typically used in polyelectrolyte solutions, the chemical mismatch parameter χ is about 0.5 or higher ($v < 0$). As already noted in Section 2.4, depending on the chemical specificity of the polymer backbone and solvent organization around it, χ can become higher than 0.5 by either lowering the temperature (UCST behavior) or increasing the temperature (LCST behavior), or by changing the solvent quality.

5.3.2   Electrostatic Interaction

As discussed in Chapter 3, there are two basic length scales describing the electrostatic interactions among ions, namely the Bjerrum length and the Debye length. While the Bjerrum length is a measure of the strength of the Coulomb interaction among charged species, the Debye length is a measure of the range of the electrostatic interaction. In the presence of polymers, the Debye length is modified significantly from the standard formula (Eq. (3.8.19)) used in electrolyte solutions, as described below.

5.3.2.1   Coulomb Strength

The Bjerrum length is the distance between two ions at which their electrostatic energy equals the thermal energy. Dividing the Bjerrum length by the Kuhn length, and absorbing the magnitudes of the two charges (z_1 and z_2), we define the Coulomb strength parameter Γ as in Eq. (4.3.19),

$$\Gamma = |z_1 z_2| \frac{\ell_B}{\ell} = |z_1 z_2| \frac{e^2}{4\pi\epsilon_0\epsilon k_B T \ell}, \tag{5.3.7}$$

where the definition of ℓ_B from Eq. (3.6.7) is used. We shall call Γ as the Coulomb strength parameter. This dimensionless parameter is a measure of the strength of the attractive (or repulsive) interaction between two ions separated by distance ℓ.

5.3.2.2   Debye Length for Polyelectrolyte Solutions

The range of electrostatic interaction between two ions in an electrolyte solution is given by the Debye length (Section 3.8.2). For electrolyte solutions, all types of ions contribute to the Debye length ξ_D, as given by Eq. (3.8.19). We make a slight modification for polyelectrolyte solutions. When an electrolyte solution contains long polymers, which are gigantic in size in comparison with the small ions from the simple electrolyte, we expect the time scales for the rearrangements of the small ions and polymer conformations to be widely separated. Therefore, it is convenient to imagine [de Gennes *et al.* (1976), Muthukumar (1996a)] that the polymer molecules are present in an effective medium where the degrees of freedom of all small ions are accounted for. The effective medium can be assumed to obey the Debye–Hückel description, with the Debye length arising from only the counterions and the ions from the simple electrolyte,

$$\xi_D \equiv \kappa^{-1} \equiv \left[4\pi\ell_B \left(z_c^2 \frac{n_c}{V} + \sum_a z_a^2 \frac{n_{a0}}{V} \right) \right]^{-1/2}. \tag{5.3.8}$$

Using the conversion factor discussed in Section 3.8.3, the Debye length can be expressed in terms of the molarities of simple salt ions and counterions as

$$\xi_D^{-2} \equiv \kappa^2 = 4000\pi\ell_B N_A \left(z_c^2 c_{c0} + \sum_a z_a^2 c_{a0} \right), \tag{5.3.9}$$

where N_A is the Avagadro number and c_{a0} is the concentration of small ions of a-type in molarity. c_{c0} is the counterion concentration in molarity and is related to the monomer concentration c_{p0} in molarity according to

$$c_{c0} = \left| \frac{z_p}{z_c} \right| \alpha c_{p0}. \tag{5.3.10}$$

The polyelectrolyte chains exist in the effective medium created by the counterions and ions from the simple salt. The interactions between the charged polymer segments amongst themselves are mediated by the neutralizing plasma that constitutes the background. We further assume that the mediation by the background is adequately described by the Debye–Hückel theory. This assumption allows us to deduce the key features, without resorting to heavy numerical work that will be needed to solve the Poisson–Boltzmann equations for such topologically correlated objects as flexible polymers.

5.3.2.3   Electrostatic Excluded Volume

As described in Section 2.4, the polymer segments are subjected to excluded volume interactions when two nonbonded segments are in close proximity to each other. When the segments carry similar charges, the inter-segment interaction is repulsive and can be long-ranged. As a result, the chain is expected to expand if the intra-chain electrostatic interaction is not fully screened by the salt ions and counterions. Analogous to the excluded volume interaction treated in Section 2.4 for uncharged polymers, we now consider the additional excluded volume interaction due to electrostatic repulsion between the segments. According to the Debye–Hückel theory, the electrostatic interaction energy between two segments i and j, each of charge $e\alpha z_p$, separated by the distance r_{ij}, follows from Eq. (3.8.16) as

$$\frac{u_{\text{elec}}(r_{ij})}{k_B T} = \alpha^2 z_p^2 \ell_B \frac{e^{-\kappa r_{ij}}}{r_{ij}}. \qquad (5.3.11)$$

Here, the inverse Debye length κ arises only from the counterions and salt ions, and not from the charged monomers, as given by Eq. (5.3.8).

For sufficiently large values of κ, such that $\kappa R_g > 1$ (R_g being the radius of gyration of the polymer), the screened Coulomb interaction energy becomes short-ranged (Appendix 2). In this limit, the result of Eq. (5.3.11) can be written exactly as (Eq. (2.8))

$$\frac{u_{\text{elec}}(r_{ij})}{k_B T} = \frac{4\pi \alpha^2 z_p^2 \ell_B}{\kappa^2} \delta(r_{ij}), \qquad (5.3.12)$$

where $\delta(r_{ij})$ is the Dirac delta function. On the other hand, for $\kappa = 0$, we have the Coulomb result,

$$\frac{u_{\text{elec}}(r_{ij})}{k_B T} = \alpha^2 z_p^2 \ell_B \frac{1}{r_{ij}}. \qquad (5.3.13)$$

The situation of $\kappa = 0$ is unphysical, because there are always counterions present in a solution of finite volume to meet the electroneutrality condition. Nevertheless, the behavior of Eq. (5.3.13) is approachable in the asymptotic limit of the Debye length being larger than the radius of gyration of the polymer, $\kappa R_g \ll 1$. Combining the above two limits,

$$\frac{u_{\text{elec}}(r_{ij})}{k_B T} = \begin{cases} \frac{4\pi \alpha^2 z_p^2 \ell_B}{\kappa^2} \delta(r_{ij}), & \kappa R_g \gg 1 \\ \alpha^2 z_p^2 \ell_B \frac{1}{r_{ij}}, & \kappa R_g \ll 1. \end{cases} \qquad (5.3.14)$$

We shall use these two limits as guides in interpreting the experimental data. For intermediate values of κR_g, the full form of Eq. (5.3.11) is needed.

The convenient result of the above argument is that the inter-segment electrostatic interaction energy becomes short-ranged as for the uncharged polymers, provided the solutions are at high enough salt concentrations. For such conditions, the electrostatic contribution can be simply added to the contribution from the excluded volume interaction. Therefore, by combining Eqs. (2.4.2) and (5.3.12),

$$\frac{u(r_{ij})}{k_B T} = \frac{u_{\text{exc}}(r_{ij})}{k_B T} + \frac{u_{\text{elec}}(r_{ij})}{k_B T} = \left(v\ell^3 + \frac{4\pi\alpha^2 z_p^2 \ell_B}{\kappa^2}\right)\delta(r_{ij}). \qquad (5.3.15)$$

The first term is due to the uncharged contribution and the second term is due to the electrostatic contribution. We call the term $4\pi\alpha^2 z_p^2 \ell_B / \kappa^2$ as the electrostatic excluded volume parameter, in the limit of high salt concentrations.

5.3.2.4 Topological Correlation: Segment Density, Counterion Distribution, and Electric Potential

The electric potential in and around a flexible polyelectrolyte chain is directly dependent on how the monomers are spatially correlated. Unlike the situations in Chapter 4, a flexible polymer molecule is not a sphere or cylinder with all its charges on the outer surface. Instead, the polymer is essentially a statistical fractal, with its monomer density falling off algebraically with the radial distance from its center, as illustrated in Fig. 5.5a. The data are from numerical simulations of salt-free flexible polyelectrolyte chains of $N = 60$ segments at the Coulomb strength parameter $\Gamma = 2.8$ (corresponding to aqueous solutions at room temperature of sodium poly(styrene sulfonate) type polyelectrolytes and monovalent salts), and the volume of the system corresponding to the average segment density of $5.63 \times 10^{-6}\ell^{-3}$. As evident from this figure, the monomer density is strongly correlated for distances within the radius of gyration R_g of the chain. The segmental correlation seen in Fig. 5.5a is a direct representation of the concomitant occurrence of the topological correlations of the chain arising from chain connectivity and the long-ranged electrostatic correlations among all charged species. Indeed, the monomer density profile is set up self-consistently by the mutual correlations between the chain connectivity and electrostatic interactions among all

(a) (b)

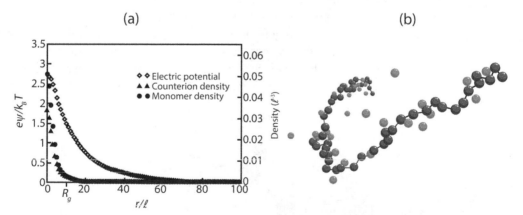

Figure 5.5 (a) Dependence of spherically averaged monomer density, counterion density, and the electric potential on the radial distance from the center-of-mass of a flexible polyelectrolyte chain. The results are from numerical simulations of uniformly charged flexible polyelectrolyte for $N = 60$ at the Coulomb strength parameter $\Gamma = 2.8$ and the average segment concentration $5.63 \times 10^{-6}\ell^{-3}$. (b) A snapshot of the chain with its dress of counterions [Muthukumar (2011)].

charged monomers and small ions. As an example, a snapshot from the above numerical simulation of an equilibrated single flexible polyelectrolyte in salt-free conditions is given in Fig. 5.5b. Many counterions are seen to hover over the chain backbone. The spatial profile of the counterion distribution is included in Fig. 5.5a. It is obvious that a counterion cloud shrouds around the segment density by partially dressing the flexible chain with oppositely charged ions. The simulations reveal that the identities of the counterions around the chain backbone are constantly changing as the polymer conformations fluctuate. The electric potential (ψ in units of $k_B T/e$) at a radial distance r from the center-of-mass of the chain is also given in Fig. 5.5a, along with the average density profiles of the monomers and the counterions. The electric potential is strongly correlated with the monomer distribution and is significant even at distances comparable to four times the radius of gyration of the chain. It must be noted that there are no mathematical divergences (as encountered in the Manning theory of counterion condensation, Chapter 4) as long as we do not consider the unphysical situation of segments with zero radius.

5.3.2.5   Worm-Like Counterion Cloud

The number of counterions adsorbed on the chain backbone is a difficult quantity to measure experimentally. However, as demonstrated in Fig. 5.5, computer simulations have helped to gain an understanding of the counterion cloud around the chain. Since the strongest attractive electric potential for the counterions is generally near the contour of the chain, the ion cloud dresses the chain along its contour. As a result, the chain with its counterions would look like a worm.

In order to obtain a measure of the number of counterions inside this worm, the following construction is usually made. Consider the skeletal chain model made of the repeat units treated as united atoms as in Fig. 5.6. We construct a tube around the chain backbone from the position coordinates of the united atoms. The tube is the nonoverlapping superposition of spheres of fixed radius r_c centered at each united atom, as shown in Fig. 5.6. All ions other than the repeat units inside this tube constitute the ion cloud. Most of these ions are the counterions, and so we shall call the tube the "counterion worm" for a particular polymer conformation. Knowing the number and charges of all ions inside this worm and adding the charges of all monomers of

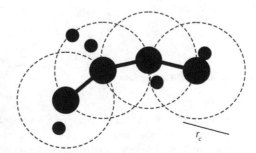

Figure 5.6 Counterion cloud looking like a worm, constructed from the nonoverlapping superposition of spheres of radius r_c centered at each repeat unit (big filled circle) of a skeletal chain. Small filled circles represent counterions [Muthukumar (2011)].

the chain, the net charge is obtained. Averaging over many conformations in equilibrium gives the average polymer charge Q_{eff} and the average degree of ionization of the polymer chain,

$$\alpha = \frac{Q_{eff}}{eN_0} = \frac{Q_{eff}}{ez_p N}, \qquad (5.3.16)$$

where N_0 is the number of ionizable repeat units and N is the number of Kuhn segments per chain. The value of Q_{eff} depends on the choice of the cutoff radius r_c. One of the convenient choices of r_c is the value r_0 at which the electrostatic energy of a pair of monovalent ions ($\ell_B k_B T / r_0$) is comparable to the kinetic energy ($3 k_B T / 2$) of an ion. We shall use this choice for r_c in discussing the simulation results for the effective polymer charge in Section 5.5.

The net result of the ion cloud around the polymer is that the effective polymer charge is different from the nominal chemical charge of the polymer. The extent of counterion adsorption is affected by many factors including the nature of the counterion (size and valency), concentration of added salt, polymer concentration, and temperature. Another factor that significantly affects the extent of counterion adsorption on the chain is the dielectric constant of the medium, as can be readily recognized by the definition of the Bjerrum length. However, the dielectric constant of the solution is not uniform, due to the backbone structure of large polymer chains being oil-like and the solvent being polar. As described in Section 5.2, the dielectric constant in the region of binding of counterions to the pendant charged groups of the polymer can be quite different from that in the solvent. A typical local conformation of a few monomers of poly(styrene sulfonate) to which the sodium counterion binds is sketched in Fig. 5.7a. In this region, the local dielectric constant can be substantially smaller, due to the fact that most of the materials made out of the polymer backbone without charges have dielectric constants in the range of 2–3. In fact, for biological macromolecules, it has been recognized [Mehler & Eichele (1984), Lamm & Pack

(a)   (b)

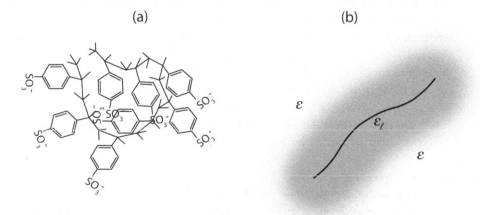

Figure 5.7 (a) Sketch of the local environment of pendant groups at oily backbone against which the counterion binds. (b) The local dielectric constant inside the counterion cloud ϵ_ℓ is lower than the bulk value ϵ. The thicker curve denotes the chain backbone.

(1997), Gong *et al.* (2008)] that the effective dielectric constant varies sharply from a low value near the chain backbone to the high bulk value. As a result, we imagine that the dielectric constant inside the counterion worm is ϵ_ℓ which is substantially lower than the bulk value. This is sketched in Fig. 5.7b, where the dielectric constant is ϵ_ℓ inside the worm and is ϵ outside the worm. Furthermore, the distance d between the bulky charged group of the monomer and the counterion, when an ion pair is formed in the counterion adsorption step, is usually different from the optimum distance between two small ions in the solvent. Therefore, as discussed in Section 5.2, the adsorption energy for one monomer–counterion pair is

$$u_{\text{local}}(r) = -\frac{e^2}{4\pi\epsilon_0\epsilon_\ell d}. \qquad (5.3.17)$$

It must be cautioned that even the notion of the dielectric constant at such nanoscopic length scales is not accurate and it is necessary to compute the polarization forces. However such calculations are yet to be performed for the ill-structured heterogeneous suspensions of charged macromolecules. Therefore, ϵ_ℓ is taken to be different and smaller than the bulk value, in recognizing the existence of dielectric heterogeneity in these solutions. Since we do not know the value of ϵ_ℓ and the ion-pair distance d inside the worm, we combine these two quantities and define the dielectric mismatch parameter $\delta = \epsilon\ell/(\epsilon_\ell d)$ as given in Eq. (5.2.14).

The parameter δ is the enhancement factor for the formation of ion pairs near the polymer backbone, as the ion-pair energy follows from Eqs. (5.2.14) and (5.3.17) as the product of the Bjerrum length and δ,

$$\frac{u_{\text{local}}(r)}{k_B T} = -\frac{\ell_B}{\ell}\delta. \qquad (5.3.18)$$

In view of the discussion in Section 3.5 and Fig. 3.8, the value of ϵ_ℓ in the region of counterion binding to the pendant charged groups of the polymer in aqueous solutions is close to 30 [Lamm & Pack (1997)]. Furthermore, the distance d between the charged monomer and the counterion in an ion pair is comparable to the distance between two consecutive ionizable monomers on the chain backbone. As a result of both of these tendencies, the value of the dielectric mismatch parameter δ is expected to be larger than unity, but of order unity. In view of the lack of an adequate understanding of the polarization forces at short distances for polyelectrolyte solutions, the dielectric mismatch parameter δ can be taken only as an empirical parameter. In addition to parametrizing the dielectric heterogeneity in the solutions, δ reflects the specificity of the counterions, because the ion-pair distance d depends on the hydrated ionic radii of the counterions involved in the ion-pair formation.

5.3.2.6   Dipole–Dipole Interactions

In addition to the electrostatic repulsion among the charged monomers (that are free from counterion adsorption), there are electrostatic interactions between the ion pairs formed by monomer–counterion binding. Since these ion pairs are constantly changing their locations along the chain contour as the chain itself adopts many random conformations, the interaction energy between a pair of ion pairs can be assumed to be that of a freely rotating pair of dipoles. The interaction energy between two freely rotating dipoles is short-ranged [Israelachvili (2011)] and attractive. The dipole–dipole interaction can be stronger than the thermal energy $k_B T$. As a specific example, for two freely rotating dipoles $\mathbf{p}_1$ and $\mathbf{p}_2$ of unit charge separated by a distance r in an electrolyte solution, the angularly averaged interaction energy is attractive as given by (Appendix 3)

$$\frac{u_{\text{dipole–dipole}}(r)}{k_B T} = -\frac{p_1^2 p_2^2 \ell_B^2}{3r^6} \left[1 + 2\kappa r + \frac{5}{3}(\kappa r)^2 + \frac{2}{3}(\kappa r)^3 + \frac{1}{6}(\kappa r)^4 \right] \exp\left(-2\kappa r\right).$$

$$(5.3.19)$$

For typical physiological values of monovalent salt concentration of about 150 mM, and typical lengths of ion pairs, the effective pairwise interaction energy is about $10k_B T$.

The contribution from the dipolar attraction among the ion pairs becomes significant when the polymer chain collects a sufficient number of counterions and consequently contracts in size. When the polymer coil shrinks, more counterions bind due to the lower local dielectric constant. Thus, there is a cascade mechanism by which counterion adsorption escalates as the polymer collapses due to the attraction arising from ion-pair interactions [Khokhlov & Kramarenko (1994), Brilliantov et al. (1998), Kundagrami & Muthukumar (2010)]. The situation becomes quite complicated when multivalent counterions are involved in the formation of ion pairs [Huber (1993), de la Cruz et al. (1995), Wittmer et al. (1995), Nguyen et al. (2000), Kundagrami & Muthukumar (2008)]. For the monovalent counterions, the effect from the ion-pair interactions can be treated [Muthukumar (2004)] as the short-ranged excluded volume effect by adding a negative term to the excluded volume parameter v in Eq. (5.3.15) that is dependent on $\ell_B, \delta,$ and d.

5.3.3   Electrostatic Persistence Length

Many polyelectrolyte molecules possess intrinsic stiffness along their backbones. An example is dsDNA with its intrinsic stiffness mainly arising from hydrogen bonding between its base-pairing strands. When such molecules carry charges, the chain stiffness can be enhanced due to the repulsion between charged repeat units. Consider a rod-like chain of length L which is shorter than its intrinsic persistence length ℓ_p (Section 2.2). Also, let the charge separation distance ℓ along the chain be smaller than the Debye length κ^{-1}, which in turn is taken to be smaller than the persistence length ℓ_p (Fig. 5.8a). Let this chain bend slightly with a constant curvature radius so that the directions of the ends subtend an angle θ ($\ll 1$).

If the chain is uncharged, the bending energy can be expressed using the theory of elasticity of rods as discussed in Section 2.2 [Landau & Lifshitz (1980)]. The energy U_b to bend a rod, per unit length, is half the bending force constant ϵ_b times the inverse square of the local radius of curvature R_c,

$$\frac{U_b}{L} = \frac{\epsilon_b}{2} \left(\frac{1}{R_c}\right)^2.$$

$$(5.3.20)$$

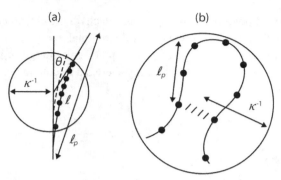

Figure 5.8 (a) Electrostatic repulsion leads to additional stiffening of rod-like molecules, resulting in electrostatic persistence length, which is in addition to the intrinsic persistence length of the molecule. (b) For very long chains, electrostatic swelling leads to more swollen semiflexible coil [Adapted from Muthukumar (2011)].

Since $R_c = L/\theta$ (from geometry), the bending energy for the whole length L is

$$\frac{U_b}{k_B T} = \frac{1}{2} \frac{\epsilon_b}{k_B T} \frac{\theta^2}{L}. \tag{5.3.21}$$

Since $\epsilon_b/(k_B T)$ is the intrinsic persistence length ℓ_p of the uncharged polymer (Eq. (2.2.2)), the bending energy is given as

$$\frac{U_b}{k_B T} = \frac{\ell_p}{2} \frac{\theta^2}{L}. \tag{5.3.22}$$

If the chain is uniformly charged, consideration [Odijk (1977), Skolnick & Fixman (1977)] of electrostatic repulsion within the Debye–Hückel approximation leads to an additional term for $U_b/k_B T$, which is proportional to $\theta^2/2L$. Therefore, the coefficient of the additional contribution can be identified as an extra persistence length,

$$\frac{U_b}{k_B T} = \frac{\ell_p}{2} \frac{\theta^2}{L} + \frac{\ell_{pe}}{2} \frac{\theta^2}{L}, \tag{5.3.23}$$

where ℓ_{pe} is called the electrostatic persistence length. It is given by the Odijk–Skolnick–Fixman theory [Odijk (1977), Skolnick & Fixman (1977)] as

$$\ell_{pe} = \frac{\ell_B}{4} \frac{\xi_D^2}{\ell^2}, \tag{5.3.24}$$

where ℓ_B, ξ_D, and ℓ are the Bjerrum length, Debye length, and the charge separation length along the rod. The total persistence length $\ell_{p,\text{eff}}$ is given by

$$\ell_{p,\text{eff}} = \ell_p + \ell_{pe}. \tag{5.3.25}$$

It is difficult to experimentally observe the proportionality between the electrostatic persistence length and the inverse salt concentration, because the result of Eq. (5.3.23) is only a perturbation theory in the limit $\ell_p > L$. For semiflexible chains with L larger than ℓ_p (Fig. 5.8b), the role of intrachain electrostatic interactions can be conveniently treated as electrostatic swelling in the vein of Section 5.3.2.3, as shown in Section 5.6.5.

All of the above concepts must be addressed in interpreting experimental results on various measures of the equilibrium structures of polyelectrolyte chains in dilute solutions. Before implementing the basic concepts introduced in this section to derive closed-form analytical formulas for size, structure, and charge of polyelectrolyte chains in dilute solutions, it is prudent to be cognizant of experimental facts. In addition, computer simulations have revealed the richness of consequences of the coexisting topological and electrostatic forces on the charged macromolecule. Therefore, we first address the experimental and simulation results in the following two sections, and then we return to theoretical formulations of polyelectrolyte behavior in dilute solutions.

5.4   Experimental Results

We first review the experimental data on the size of a polyelectrolyte molecule in dilute solutions in terms of its dependence on chain length and salt concentration. The main methodology for the determination of polymer size in solutions is the use of light scattering. Most of the experiments dealing with polyelectrolyte solutions are at polymer concentrations not low enough to avoid the long-ranged interference from other chains [Rice & Nagasawa (1961), Dautzenberg et al. (1994)]. Furthermore, interpretation of data from light scattering measurements on polyelectrolyte solutions at very low salt concentrations is quite difficult [Förster & Schmidt (1995), Drifford & Dalbiez (1984,1985), Volk et al. (2004)]. Therefore, there has been a scarcity of reliable experimental data on polyelectrolyte size in infinitely dilute solutions. It is necessary to choose polymer concentrations low enough and the salt concentration high enough to minimize long-ranged inter-chain correlations. As an example, precise light scattering results [Beer et al. (1997)] on the radius of gyration R_g of ethyl-poly(vinylpyridinium)-bromide (Et-PVP-Br) in very dilute solutions are given in Fig. 5.9a in terms of the polymer contour length L and concentration c_s of the added salt NaBr.

As expected, R_g increases as the chain length is increased at a fixed salt concentration. Also, for a fixed chain length, R_g decreases as the salt concentration increases. This is consistent with our expectation based on electrostatic expansion. At lower salt concentrations, there is less screening of the intra-chain electrostatic repulsion, and as a result the chain swells more. One could intuitively expect, in accordance with Eq. (5.1.2), that a flexible polyelectrolyte molecule adopts a rod-like conformation due to repulsive intra-chain electrostatic interactions in salt-free solutions because the Debye length for such conditions can be larger than the radius of gyration. However, this limit seems never to be approached in experiments measuring R_g in dilute solutions due to the omnipresence of counterions and counterion adsorption to the polymer. It is seen from the experimental data in Fig. 5.9a that the radius of gyration increases only by a factor of about three when the salt concentration is reduced by three orders of magnitude, for each of the chain lengths. The solid lines in this figure is an analytical prediction,

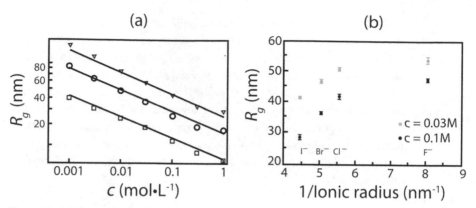

Figure 5.9 (a) Dependence of radius of gyration R_g of Et-PVP-Br on concentration of NaBr. $L = 400$ nm ($\square$), 1000 nm ($\bigcirc$), and 1800 nm ($\triangledown$). The error bars are smaller than the symbols. The solid lines are from Eq. (5.6.11) with $\alpha = 0.17$ and $v_1 = -0.03$. (b) Dependence of R_g on the specificity of the monovalent counterion (I^-, Br^-, Cl^-, and F^-, represented in terms of their ionic radii) [Adapted from Beer *et al.* (1997)].

Eq. (5.6.11), derived in Section 5.6.1 for the so called "high salt limit." The agreement between the solid lines and the experimental values is reasonably good, given the complexity of the problem. Therefore, the applicability of Eq. (5.6.11) derived for high salt concentration limit, for even such low salt concentration as 0.001M, suggests that the "high salt limit" can be assumed even for 0.001 M monovalent salt concentration.

Furthermore, the specificity of the counterions plays a significant role in determining the polymer size. For example [Beer *et al.* (1997)], the dependence of R_g of ethyl-poly(vinylpyridinium)-bromide (Et-PVP-Br) in very dilute solutions on the specificity of the counterions (I^-, Br^-, Cl^-, and F^-) is given in Fig. 5.9b, where the identity of the counterion is represented by its ionic radius [Marcus (2015)]. The ability of the counterions to compactify the chain increases in the order $F^- <$ $Cl^- < Br^- < I^-$, although all of these ions are monovalent. Thus it is clear that the effective linear charge density of the polymer decreases with increasing polarizability of the counterions. A proper treatment of the polarizability of counterions is beyond the scope of the current theories of charged macromolecules. Furthermore, the effect of multivalent counterions on the polymer size is much stronger than that from monovalent counterions. This is due to a combination of the fact that a multivalent counterion is a stronger adsorbing ion to the polymer than a monovalent ion and the ability of multivalent ions to bridge multiple monomers to make transient crosslinks and, as a result, collapsing the polymer dramatically [Huber (1993), Ikeda *et al.* (1998), Prabhu *et al.* (2004)]. If the binding of multivalent counterions to the polymer is strong, even the net charge of the polymer can be reversed [Besteman *et al.* (2007)]. It is also experimentally known that the effective charge density of the polymer decreases with increasing hydrophobicity of the polymer backbone [Essafi *et al.* (1995)].

5.5   Simulation Results

As mentioned above, experimental protocols that directly probe isolated polyelectrolyte chains, their sizes, counterion distributions, and electric potential variations inside and outside the coils, are challenging. These features are sometimes deduced from measurements of other quantities, such as the electrophoretic mobility. The interpretation of data in these indirect measurements also depends heavily on reliable theories. The theoretical formulation of the internal correlations within a polyelectrolyte molecule is also difficult, as discussed in Section 5.3. Faced with such challenges, computer simulations have played a significant role in helping to understand the basic features of charge correlations in polyelectrolytes.

There have been several simulation techniques implemented in modeling polyelectrolyte chains, including Molecular Dynamics, Brownian Dynamics, Langevin Dynamics, and Monte Carlo [Severin (1993), Stevens & Kremer (1995), Winkler *et al.* (1998), Liu & Muthukumar (2002), Liu *et al.* (2003), Holm *et al.* (2004)]. The fundamental concepts derived from these simulations are naturally uniform, independent of the particular technique used. Since we are presently interested in key concepts about the polyelectrolyte molecule and not in simulation techniques, we shall only illustrate the results from a generic model [Liu & Muthukumar (2002), Liu *et al.* (2003), Ou & Muthukumar (2005)] of a polyelectrolyte chain and its properties as obtained from Langevin Dynamics simulations. The simulation system is made of n freely jointed chains each with N spherical beads of point unit electric charge $-e, nN/z_c$ counterions (z_c being the valency of the counterion), n_+ cations of added salt with valency z_+ and n_+z_+/z_- anions of added salt with valency z_-, all placed in a cubic medium of permittivity $\epsilon_0\epsilon$ and volume V. The bond length between any two successive beads along a chain is allowed to fluctuate about the equilibrium value ℓ. The beads are allowed to interact among themselves with excluded volume and electrostatic interactions. The excluded volume contribution is the non-electrostatic part of the potential interaction between nonbonded beads of the chain and is taken as a purely repulsive Lennard-Jones (LJ) potential,

$$u_{\text{LJ}} = \begin{cases} \epsilon_{\text{LJ}}\left[\left(\frac{\sigma}{r}\right)^{12} - 2\left(\frac{\sigma}{r}\right)^6 + 1\right] & ,r \leq \sigma \\ 0 & ,r > \sigma \end{cases} \tag{5.5.1}$$

where ϵ_{LJ} is the strength, σ is the hard-core distance at contact, and r is the distance between two nonbonded beads. With the repulsive LJ potential, polymer collapse due to the hydrophobic effect is not addressed. The same form of Eq. (5.5.1) can also be used to capture the non-electrostatic excluded volume interactions among the polymer beads and counterions, the difference appearing in the choice of the hard-core distance σ, and the strength parameter ϵ_{LJ}. The electrostatic interaction among the charged beads and ions is taken as the Coulomb energy,

$$u_c\left(r_{ij}\right) = \frac{z_i z_j e^2}{4\pi\epsilon_0\epsilon r_{ij}} = \frac{z_i z_j \ell_B k_B T}{r_{ij}}, \tag{5.5.2}$$

where r_{ij} is the distance between the ions i and j, z_k is the valency of the kth ion, and ℓ_B is the Bjerrum length.

The key control parameter in these simulations is the Coulomb strength parameter Γ defined in Eq. (5.3.7),

$$\Gamma = |z_p z_c|\frac{\ell_B}{\ell} = |z_p z_c|\frac{e^2}{4\pi\epsilon_0\epsilon k_B T\ell} = |z_c|\frac{\ell_B}{\ell}, \qquad (5.5.3)$$

where z_p is taken as -1. As noted already, the experimentally relevant range of Γ is $3.2 > \Gamma > 2.4$ for aqueous solutions ($0°C < T < 100°C$, respectively) of flexible polyelectrolytes with chemical charge separation along the chain backbone of about 0.25 nm and $z_c = 1 = z_p$. Since the vast majority of experiments on polyelectrolytes use water as the solvent, $\Gamma \approx 3$. For multivalent counterions, this range of Γ is expanded by the multiple of z_c. Naturally, the values of Γ outside the above range represent solvents other than water and charge separations along the chain much different from ≈ 0.25 nm. Due to the temperature dependence of the permittivity of the polyelectrolyte solution, and since the product $T\epsilon$ appears in the definition of ℓ_B (and incidentally, Γ), we must consider $T\epsilon$ as the temperature variable instead of T alone.

5.5.1   Radius of Gyration

The radius of gyration of a freely jointed chain of $N = 100$ beads in the absence of any added salt is given in Fig. 5.10 as a function of the Coulomb strength parameter Γ. The cases of monovalent counterions and multivalent counterions are presented in Figs. 5.10a and 5.10b, respectively. When the chain is uncharged and only has repulsive LJ interactions, data for R_g are included in Fig. 5.10a, to illustrate the role of electrostatics on R_g. For very high values of $T\epsilon$ ($\Gamma \to 0$), only weak electrostatic repulsion is present. Consequently, R_g is slightly higher than that for the LJ chain.

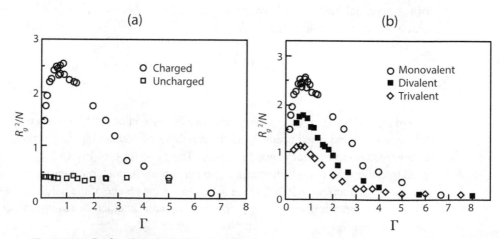

Figure 5.10 R_g for $N = 100$. (a) Freely jointed polyelectrolyte chain with monovalent counterions in comparison with an athermal uncharged chain. (b) Comparison of electrostatic swelling for monovalent, divalent, and trivalent counterions [From Liu & Muthukumar (2002)].

(a) (b)

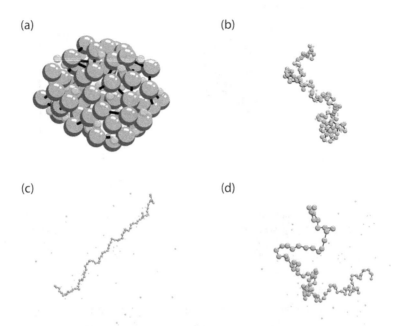

(c) (d)

Figure 5.11 Snapshots from the simulations for $N = 100$. (a) $\Gamma = 20.0$; (b) $\Gamma = 7.0$; (c) $\Gamma = 1.0$; (d) $\Gamma = 0.13$. (Pictures are not in the same scale.) [From Liu & Muthukumar (2002)].

As the value of $T\epsilon$ is lowered, the electrostatic repulsion between the beads becomes stronger, and consequently R_g begins to increase with Γ. As $T\epsilon$ is decreased even further (i.e., Γ approaches about 0.5), the intra-chain electrostatic repulsion is attenuated by electrostatic attractions between beads and counterions. The rate of chain swelling with a decrease in $T\epsilon$ begins to decrease.

As the value of $T\epsilon$ is lowered even more (i.e., Γ goes above roughly 1) there are significant numbers of counterions close to the chain backbone, creating many dipoles. The interaction between these dipoles leads to intrachain attraction, working against the intrachain swelling arising from the uncompensated charges on the chain backbone. The net result is that R_g decreases as Γ increases (i.e., ϵT decreases). Yet, until $\Gamma \approx 5.0$, R_g of the polyelectrolyte chain is bigger than the value expected for an uncharged chain in good solvents. The chain begins to collapse as Γ is increased beyond 5. If the counterions are multivalent, the chain expansion is substantially weaker, as illustrated in Fig. 5.10b. Typical configurations of the chain with monovalent counterions are given in Fig. 5.11 at different values of Γ.

Another measure of chain size, in addition to the radius of gyration, is the hydrodynamic radius R_h defined as

$$R_h = \left(\frac{1}{N^2} \sum_{i=1}^{N} \sum_{j>i} < \frac{1}{|\mathbf{R}_i - \mathbf{R}_j|} > \right)^{-1}, \tag{5.5.4}$$

where $\mathbf{R}_i$ is the position vector of the ith bead and the angular brackets denote averaging over chain conformations. The hydrodynamic radius is measured using dynamic

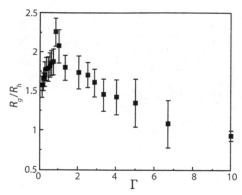

Figure 5.12   The shape factor R_g/R_h for $N = 100, z_c = 1$ and $c_p = 8 \times 10^{-4} \ell^{-3}$. The chain is anisotropic at $\Gamma \approx 1$ with an effective size exponent $\nu = 0.98$. The shape is self-avoiding-walk-like for $\Gamma \approx 3$ and globule-like for $\Gamma \approx 10$ [From Liu & Muthukumar (2002)].

light scattering, and its origin lies in the hydrodynamic interactions in the solution, as will be discussed in Chapter 7. The ratio R_g/R_h is a measure of the anisotropy of the average shape of the chain, and is called the shape factor. The values of R_g/R_h are 0.77, 1.3, and about 4, for compact globules, Gaussian chains, and rod-like chains, respectively. For the present situation of monovalent counterions, the shape factor is plotted versus Γ in Fig. 5.12. It is obvious from Figs. 5.11 and 5.12 that the chain is highly anisotropic for Γ values around unity. In Figs. 5.10a, 5.11, and 5.12, $N = 100, z_c = 1$ and the monomer density is $8 \times 10^{-4} \ell^{-3}$. By monitoring the N-dependence of R_g, the effective size exponent ν is reported to be 0.98, 0.85, 0.59, and 0.33 for $\Gamma = 1.0, 3.0, 5.0$, and 20.0, respectively [Liu & Muthukumar (2002)]. In spite of the exponents approaching the rod-like and globule-like limits, the actual size of the chain is less stretched than a rod (at $\Gamma = 1.0$) and less compact than a compact sphere (at $\Gamma = 20.0$).

5.5.2   Counterion Adsorption and Effective Charge

When the Coulomb interaction parameter is large enough such that $\Gamma \geq 0.5$, attraction between the charged beads and some counterions begins to contribute significantly. Close examination of the positions of counterions near the chain backbone reveals that these counterions are not frozen at fixed positions. Instead they are found to undergo dynamics by which they bind and unbind, move along the chain backbone, and eventually exchange with other counterions which were not originally in the proximity of the chain backbone. A dynamic equilibrium of counterion distribution around the polyelectrolyte is maintained with a higher average density of counterions around the polyelectrolyte backbone than in the bulk.

By following the procedure described in Section 5.3.2.5 to construct the counterion worm, the average degree of ionization α as defined by Eq. (5.3.16) is computed. The dependence of α on the electrostatic interaction parameter Γ is given in Fig. 5.13 for

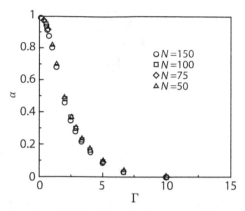

Figure 5.13 Charge fraction of a single flexible polyelectrolyte chain due to adsorption of monovalent counterions in salt-free solutions [From Liu & Muthukumar (2002)].

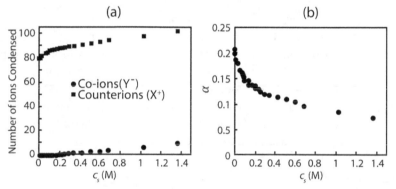

Figure 5.14 Decrease in the effective charge of the polymer with increasing salt concentration for $N = 100$ at $\Gamma = 3$. (a) Number of counterions and co-ions inside the counterion worm and (b) net degree of ionization [From Liu *et al.* (2003)].

several values of N, by fixing the monomer concentration at $c_p = 8 \times 10^{-4}\ell^{-3}$, and $z_c = 1$. As seen in this figure, α decreases smoothly, without any discontinuity, as Γ is increased. For aqueous solutions of flexible polyelectrolytes at room temperature, Γ is about 3, and α is around 0.2, due to the mechanism of counterion adsorption on the polymer backbone. In spite of copious numbers of references in the literature to the Manning model (Section 4.4) of counterion condensation on an infinitely long and infinitely thin rod-like line charge, to be presumably valid even for flexible coil-like chains, the discontinuity of Fig. 4.13a is not seen.

When a fixed quantity of salt is added to an equilibrated polyelectrolyte chain with a monovalent counterion, there is an additional counterion adsorption and α decreases further. This result is illustrated in Fig. 5.14, where the monovalent counterion is labeled as X^+ and the monovalent added salt is XY (Y^- being the co-ion to the polymer). In these figures, $\Gamma = 3, N = 100$ and the monomer concentration $c_p = 8 \times 10^{-4}\ell^{-3}$. The numbers of X and Y ions inside the counterion worm are plotted in Fig. 5.14a, as the salt concentration c_s of the XY salt is increased from 0 to 1.36 M. As already noted in Fig. 5.13 for salt-free solutions, about 80% of the counterions

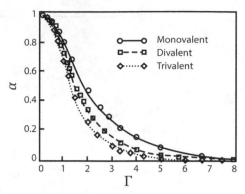

Figure 5.15 Dependence of α on the valency of the counterion in salt-free solutions for $N = 100$ [From Liu & Muthukumar (2002)].

(X^{+1}) are adsorbed inside the counterion worm for $\Gamma = 3$. As the salt concentration increases, the number of adsorbed counterions around the polymer backbone increases. At higher salt concentrations, as the counterion concentration inside the counterion worm increases, increasing numbers of co-ions (Y^-) are also brought inside the worm in an effort to maintain electroneutrality even at local scales. By adding the numbers of X^+ and Y^- ions and their charges and combining this quantity with the bare polymer charge, the effective degree of ionization of the polymer is obtained, as given in Fig. 5.14b. The degree of ionization decreases monotonically with the salt concentration.

When the counterion is multivalent, the nature of counterion adsorption and the consequent chain contraction are more drastic. The results for salt-free solutions are given in Fig. 5.15 for the average charge fraction. As expected, the degree of ionization of the chain decreases with the valency of the counterions at all values of Γ. In addition to the contribution from z_c, this decrease is contained in the definition of $\Gamma = |z_p z_c| \ell_B / \ell_0$.

When a salt of type AY_2 (A^{2+} is the counterion and Y^- is the co-ion) is added to an equilibrated flexible polyelectrolyte chain with a monovalent counterion (X^+), the divalent counterion competitively adsorbs on the polymer backbone by displacing the already adsorbed monovalent counterions. Typical results of this competition and the net effective charge of the polymer are given in Fig. 5.16 as functions of the salt concentration c_s, for $N = 100$ and $c_p \ell^3 = 8 \times 10^{-4}$. Whenever small amounts of divalent counterions are introduced into the system, these effectively replace the adsorbed monovalent counterions. This is shown in Fig. 5.16a, where the numbers of adsorbed divalent counterions, adsorbed monovalent counterions, and co-ions inside the counterion worm are plotted against the salt (AY_2) concentration. As a net result, α is reduced sharply by the divalent ions of the salt in comparison with the case of monovalent ions. In fact, depending on the value of Γ, ion size, and multivalent salt concentration, the counterion adsorption can lead to overcharging with a charge reversal of the net charge of the polymer.

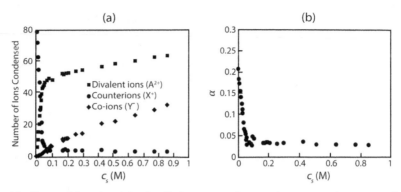

Figure 5.16 Competitive adsorption by divalent counterions against monovalent counterions. (a) Partitioning of monovalent counterions, divalent counterions, and monovalent co-ions inside the counterion worm as a function of the concentration of the AY_2-type salt. (b) The net charge fraction of the polymer [From Liu *et al.* (2003)].

5.6 Electrostatic Swelling with Fixed Polymer Charge

We shall now consider mean field arguments, resulting in analytical formulas, to capture the various experimental and simulation results seen above. One set of arguments assumes that the polymer charge is a fixed quantity. This approach facilitates simple derivations of analytical formulas for the sizes of charged macromolecules. The other set of arguments allows the polymer charge to self-regulate as the experimental conditions change, which will be presented in Section 5.7.

In view of the counterion adsorption that is always present in a polyelectrolyte solution with finite total volume, and because of the continuous exchange of counterions within the counterion worm, we assume that the adsorbed counterions are uniformly smeared along the polymer contour, as already mentioned several times. Thus, the average degree of ionization of the polymer α is presumed to be the same throughout the chain contour. Equivalently, we assume that the charge of a segment is $e\alpha z_p$, instead of the bare chemical charge ez_p (Section 5.3). In other words, the effective charge density along the polymer is q given by Eq. (5.3.2), and the electrostatic interaction energy between two segments is given by Eq. (5.3.11) as

$$\frac{u_{ij}(r_{ij})}{k_B T} = \alpha^2 z_p^2 \ell_B \frac{e^{-\kappa r_{ij}}}{r_{ij}}. \tag{5.6.1}$$

In view of the arguments leading to the two limits of Eq. (5.3.14), we shall consider the two limits of (a) high salt ($\kappa R_g \gg 1$) and (b) low salt ($\kappa R_g \ll 1$) as follows.

5.6.1 High Salt Limit

As discussed in Section 5.3.2.3, the intra-chain electrostatic interaction becomes short ranged for high salt concentrations and its strength can be simply added to the excluded volume parameter without charges. Recalling Eq. (5.3.15), an effective excluded volume parameter v_{eff} can be defined as

$$v_{\text{eff}} = v + \frac{4\pi\alpha^2 z_p^2 \ell_B}{\kappa^2 \ell^3}. \tag{5.6.2}$$

In view of Eq. (2.4.3), v_{eff} can be equivalently expressed in terms of an effective Flory–Huggins chi parameter, χ_{eff}, defined as

$$v_{\text{eff}} = (1 - 2\chi_{\text{eff}}), \tag{5.6.3}$$

where

$$\chi_{\text{eff}} = \chi - \frac{2\pi\alpha^2 z_p^2 \ell_B}{\kappa^2 \ell^3}. \tag{5.6.4}$$

This result shows that the effective Flory–Huggins χ parameter becomes lower (meaning more thermodynamic miscibility between the polymer and solvent) as the concentration of added salt is decreased. Hence, liquid–liquid phase separation can be induced by adding salt (see Chapter 8).

It is evident from Eqs. (5.3.15) and (5.6.2) that the only consequence of electrostatic interactions in the high salt limit is to simply renormalize the excluded volume parameter. Therefore, one needs simply to copy the results of Section 2.5, and substitute v_{eff} for v to derive the dependence of polymer size on Bjerrum length, Debye length, and chain length in the high salt limit. By adding the free energy contributions from chain connectivity and the sum of excluded volume effects from hydrophobicity and electrostatic repulsion, we rewrite Eq. (2.5.5) as

$$R_g \sim \ell \left[v + \frac{4\pi\alpha^2 z_p^2 \ell_B}{\kappa^2 \ell^3} \right]^{1/5} N^{3/5}, \tag{5.6.5}$$

where the radius of gyration is proportional to the Flory radius (described in Section 2.5) for the present situation. If we consider only the electrostatic contribution, R_g depends on the salt concentration c_s as

$$R_g \sim \frac{N^{3/5}}{c_s^{1/5}}, \tag{5.6.6}$$

where Eqs. (3.8.17) and (3.8.28), valid for monovalent salt, have been used. Thus, the radius of gyration of a polyelectrolyte chain in dilute salty solutions depends on the chain length with the Flory exponent 3/5, as for an uncharged polymer in good solvents, and is inversely proportional to $c_s^{1/5}$.

Analogous to the derivation of Eq. (2.5.7) for an uncharged polymer, a crossover formula for the radius of gyration of a flexible polyelectrolyte in a solution with high enough salt is obtained as follows. Substituting Eq. (2.1.14) for the free energy of chain connectivity and Eq. (2.5.2) (with v replaced by v_{eff}) for the excluded volume effect in Eq. (2.5.1), we get (Appendix 2)

$$\frac{F(R)}{k_B T} = \frac{3}{2}\left(\frac{R^2}{N\ell^2} - 1 - \ln\frac{R^2}{N\ell^2}\right) + \frac{4}{3}\left(\frac{3}{2\pi}\right)^{3/2}\left[v + \frac{4\pi\alpha^2 z_p^2 \ell_B}{\kappa^2 \ell^3}\right]\frac{N^2 \ell^3}{R^3}, \tag{5.6.7}$$

where $F(R)$ is the free energy of a chain with its root mean square end-to-end distance at R. Minimization of $F(R)$ with respect to R yields

$$\frac{\partial F(R)}{\partial R} = 3\left(\frac{R}{N\ell^2} - \frac{1}{R}\right) - 4\left(\frac{3}{2\pi}\right)^{3/2}\left[v + \frac{4\pi\alpha^2 z_p^2 \ell_B}{\kappa^2 \ell^3}\right]\frac{N^2\ell^3}{R^4} = 0, \qquad (5.6.8)$$

which upon slight rearrangement yields

$$(\frac{R^2}{N\ell^2})^{5/2} - (\frac{R^2}{N\ell^2})^{3/2} = \frac{4}{3}(\frac{3}{2\pi})^{3/2}\left[v + \frac{4\pi\alpha^2 z_p^2 \ell_B}{\kappa^2 \ell^3}\right]\sqrt{N}, \qquad (5.6.9)$$

for the high salt limit (as derived in Appendix 2).

For the radius of gyration R_g, the above equation is modified by replacing the factor 4/3 by 134/105. The factor 4/3 in Eq. (5.6.9) for root mean square end-to-end distance and the factor 135/105 for R_g are based on the exact first order perturbation theory of chain swelling [Yamakawa (1971)]. Defining the expansion factor α_S by which the R_g of the chain swells over its value at the Flory temperature ($R_{g0} = (N\ell^2/6)^{1/2}$),

$$\alpha_S \equiv \frac{R_g}{R_{g0}}, \qquad (5.6.10)$$

Eq. (5.6.9) can be rewritten for R_g as [Beer *et al.* (1997)]

$$\alpha_S^5 - \alpha_S^3 = \left[v_1 + \frac{134}{35}\sqrt{\frac{6}{\pi}\frac{\alpha^2 z_p^2}{\kappa^2 \ell^2}\frac{\ell_B}{\ell}}\right]\sqrt{N}, \qquad (5.6.11)$$

where v_1 is proportional to v ($v_1 = (67/35)(3/2\pi^3)^{3/2}v$).

The theory given above, based on the simplifying assumption of uniform radial expansion of the coil due to fixed polymer charge, appears to capture the experimental data of Fig. 5.9a. By taking v_1 and α as fitting parameters, the curves in Fig. 5.9a are given by Eq. (5.6.11). One set of values, $\alpha = 0.17$ and $v_1 = -0.03$, seems to fit the data for various chain lengths and salt concentrations over three decades. The slightly negative value for v_1 is reasonable, since the polymer backbone is immiscible with water and therefore the excluded volume parameter must be negative. The degree of ionization around 0.2 is also consistent with simulation results. Nevertheless, the degree of ionization α is taken here only as a fitting parameter, and we shall return to this issue in Section 5.7.

5.6.2   Low Salt Limit

For solutions of polyelectrolyte chains without added salt, the electrostatic interaction is relatively long-ranged, although the Debye length is finite due to the omnipresence of counterions in finite volumes. In order to assess the maximum limit for chain swelling due to intrachain electrostatic repulsion, let us assume that $\kappa = 0$, corresponding to the scenario where all counterions have left the polymer to explore their translational degrees of freedom in an infinite volume. In this limit, the electrostatic contribution to the free energy of a chain with its end-to-end distance at R follows from Eq. (5.3.13) as

$$\frac{F_{\text{elec}}}{k_B T} = \frac{\alpha^2 z_p^2 \ell_B}{2} \sum_i \sum_{j \neq i} \frac{1}{r_{ij}}, \tag{5.6.12}$$

where r_{ij} is the distance between the segments i and j. α is the uniform degree of ionization. On dimensional grounds, the double sum is proportional to N^2/R, where R is a typical length characterizing the polymer radius. Combining this result with Eq. (2.5.3) yields the scaling form,

$$\frac{F(R)}{k_B T} = \frac{3}{2\ell^2} \frac{R^2}{N} + \frac{v\ell^3}{2} \frac{N^2}{R^3} + \frac{\alpha^2 z_p^2 \ell_B}{2} \frac{N^2}{R}. \tag{5.6.13}$$

The first term, due to chain entropy, favors a lower value of R. The electrostatic repulsion favors larger values of R, even more strongly than the short-ranged excluded volume effect. The optimum value of R, $R^\star$, is obtained by minimizing $F(R)$ with respect to R,

$$\left. \frac{dF(R)/k_B T}{dR} \right|_{R^\star} = \frac{3R^\star}{N\ell^2} - \frac{\alpha^2 z_p^2 \ell_B N^2}{2\,(R^\star)^2} = 0 \tag{5.6.14}$$

so that

$$\frac{R^\star}{\ell} = \left(\frac{\alpha^2 z_p^2 \ell_B}{6\ell} \right)^{1/3} N. \tag{5.6.15}$$

In the above results, we have taken $v = 0$, to highlight the dominant electrostatic effect.

By including the correct prefactors and the correct expression for chain entropy (Eq. (2.1.14)), $F(R)$ (where R is now specifically the root mean square end-to-end distance) becomes (Appendix 2)

$$\frac{F(R)}{k_B T} = \frac{3}{2} \left[\frac{R^2}{N\ell^2} - 1 - \ln\left(\frac{R^2}{N\ell^2} \right) \right] + \frac{4}{3} \left(\frac{3}{2\pi} \right)^{3/2} \frac{v\ell^3 N^2}{R^3} + \frac{8}{5\sqrt{3}} \left(\frac{1}{2\pi} \right)^{1/2} \frac{\alpha^2 z_p^2 \ell_B N^2}{R}. \tag{5.6.16}$$

The optimum value for the end-to-end distance is

$$\frac{R^\star}{\ell} = \left(\frac{8}{15\sqrt{6\pi}} \frac{\alpha^2 z_p^2 \ell_B}{\ell} \right)^{1/3} N. \tag{5.6.17}$$

The prefactor $\left(8/15\sqrt{6\pi} \right)^{1/3} \left(\alpha^2 z_p^2 \ell_B/\ell \right)^{1/3}$ is smaller than the value of unity valid for a perfect rod. Therefore, the chain is not a rod even in the asymptotic limit of $\kappa = 0$. However, in view of the proportionality of $R^\star$ to N, the polyelectrolyte conformation in the zero-salt limit is called rod-like. In view of Eq. (5.6.17), the contribution from the hydrophobic interaction becomes negligible as the second term of Eq. (5.6.16) is of order $1/N$, whereas the other terms are proportional to N. Similar to the dimensionless excluded volume parameter (Fixman parameter) identified in Section 2.5, the strength of the intrachain electrostatic interaction in salt-free solutions is given by the dimensionless electrostatic excluded volume parameter $z_{\text{el,low}}$,

$$z_{\text{el,low}} \equiv \frac{\alpha^2 z_p^2 \ell_B N^{3/2}}{\ell}, \tag{5.6.18}$$

which may be identified from Eq. (5.6.13) as the strength of electrostatic interaction among segments when the coil radius is the Gaussian chain radius proportional to $\sqrt{N}$. The subscript "low" denotes the low salt condition. It is the above combination of α, z_p, ℓ_B, and N that dictates the strength of the interaction, rather than any one of these parameters individually. When the value of this parameter $z_{el,low}$ is large, the third term in Eq. (5.6.16) dominates so that $R \sim N$; otherwise, $R \sim \sqrt{N}$. Therefore, the electrostatic interaction is considered weak if $z_{el,low}$ is less than unity,

$$\frac{\alpha^2 z_p^2 \ell_B N^{3/2}}{\ell} < 1. \tag{5.6.19}$$

In other words, there exists a threshold value of the charge fraction $\alpha^\star$,

$$\alpha^\star \equiv \frac{1}{z_p N^{3/4} \left(\frac{\ell_B}{\ell}\right)^{1/2}}, \tag{5.6.20}$$

to delineate regimes of weak and strong electrostatic interactions. If $\alpha < \alpha^\star$, the intrachain electrostatic interactions are weak. Instead of defining the threshold in terms of α, it can be equivalently defined in terms of N. The electrostatic interaction is weak if the number of Kuhn segments in the chain is smaller than the threshold value $N^\star$,

$$N^\star = \left(\frac{\ell}{\alpha^2 z_p^2 \ell_B}\right)^{2/3}. \tag{5.6.21}$$

In other words, if $N < N^\star$, the chain does not have a significant contribution from electrostatic swelling.

5.6.3   Electrostatic Blob

It is sometimes convenient to couch the results of Section 5.6.2 in terms of a graphical representation. Let us consider a chain of N segments. As seen from Eq. (5.6.21), the electrostatic interaction among segments within a small section along the chain is expected to be weak. Therefore, we can imagine the chain to be comprised of several contiguous sections within which the electrostatic interaction is weak and beyond which the electrostatic interaction is significant. Let us call these sections as "electrostatic blobs" (Fig. 5.17). Let the number of segments in each of the blobs be g and the typical linear size of the blob be ξ_e. We choose g and ξ_e by stipulating that the electrostatic energy of one blob is comparable to $k_B T$. It follows from the third term on the right-hand side of Eq. (5.6.16) that the electrostatic energy of g monomers in a blob of size ξ_e in units of $k_B T$ is $\alpha^2 z_p^2 \ell_B g^2 / \xi_e$. Therefore, the condition in which the blob energy is comparable to $k_B T$ is given by

$$\left(\alpha z_p g\right)^2 \frac{\ell_B}{\xi_e} \simeq 1. \tag{5.6.22}$$

Within the blob, Gaussian statistics is applicable for the case of $v = 0$, so that

$$\xi_e^2 \sim g \ell^2. \tag{5.6.23}$$

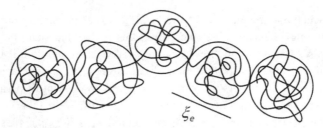

Figure 5.17 A chain is imagined to be made of electrostatic blobs of linear size ξ_e. Within the blob, electrostatic interactions are weak and the chain statistics appropriate for the uncharged backbone is applicable. Outside the blobs, electrostatic repulsion dominates and the chain is extended into a rod-like conformation.

Combining Eqs. (5.6.22) and (5.6.23) yields

$$\frac{\xi_e}{\ell} \sim \left(\frac{\alpha^2 z_p^2 \ell_B}{\ell}\right)^{-1/3}, \tag{5.6.24}$$

and

$$g \sim \left(\frac{\alpha^2 z_p^2 \ell_B}{\ell}\right)^{-2/3}. \tag{5.6.25}$$

Since the electrostatic interaction between the blobs is assumed to be strong, we expect the chain to be stretched into a rod-like conformation with N/g blobs so that the end-to-end distance of the chain is

$$R \sim \left(\frac{N}{g}\right)\xi_e. \tag{5.6.26}$$

Substitution of Eqs. (5.6.24) and (5.6.25) in Eq. (5.6.26) gives

$$\frac{R}{\ell} \sim \left(\frac{\alpha^2 z_p^2 \ell_B}{\ell}\right)^{1/3} N, \tag{5.6.27}$$

which is the same result as Eqs. (5.1.2) and (5.6.17), within the numerical prefactor.

The simplicity of the geometrical derivation of the scaling form, Eq. (5.6.27), is appealing. Although the particular scaling law derived here is not realizable due to experimental difficulty in reaching the limit of $\kappa \to 0$, we have taken this opportunity to introduce the notion of thermal blobs. This illustrates an alternate method to derive scaling laws [de Gennes *et al.* (1976), de Gennes (1979)], instead of minimizing derived free energy expressions. We shall take advantage of this method in the following chapter in deriving scaling laws for semidilute polyelectrolyte solutions.

5.6.4 Crossover Formula

The limits of high salt and low salt have been addressed in Sections 5.6.1 and 5.6.2, respectively. In experiments, the salt concentrations are not necessarily in such extreme limits. Similarly the dimensionless excluded volume parameter $v\sqrt{N}$ and the electrostatic excluded volume parameter defined in Eq. (5.6.18) can assume intermediate values, instead of being either zero or very large. The crossover formula for the free energy that recovers the limits of Eqs. (5.6.7) and (5.6.16) is [Muthukumar (1987), Appendix 2]

$$\frac{F(R)}{k_B T} = \frac{3}{2} \left(\tilde{\ell}_{\text{eff}} - 1 - \ln \tilde{\ell}_{\text{eff}} \right) + \frac{4}{3} \left(\frac{3}{2\pi} \right)^{3/2} \frac{v \sqrt{N}}{\tilde{\ell}_{\text{eff}}^{3/2}} + 2 \sqrt{\frac{6}{\pi}} \frac{\alpha^2 z_p^2 \tilde{\ell}_B N^{3/2}}{\sqrt{\tilde{\ell}_{\text{eff}}}} \Theta_0 (\tilde{a}),$$

(5.6.28)

where

$$\Theta_0 (\tilde{a}) = \frac{\sqrt{\pi}}{2} \left(\frac{2}{\tilde{a}^{5/2}} - \frac{1}{\tilde{a}^{3/2}} \right) e^{\tilde{a}} \text{erfc} \left(\sqrt{\tilde{a}} \right) + \frac{1}{3\tilde{a}} + \frac{2}{\tilde{a}^2} - \frac{\sqrt{\pi}}{\tilde{a}^{5/2}} - \frac{\sqrt{\pi}}{2\tilde{a}^{3/2}}.$$

(5.6.29)

Here $\tilde{\ell}_{\text{eff}}$ is defined as the square of the expansion factor $R^2/N\ell^2$,

$$\tilde{\ell}_{\text{eff}} \equiv \frac{\ell_{\text{eff}}}{\ell} \equiv \frac{R^2}{N\ell^2},$$

(5.6.30)

and

$$\tilde{\ell}_B = \frac{\ell_B}{\ell}, \qquad \tilde{a} = \frac{\kappa^2 \ell^2 N \tilde{\ell}_{\text{eff}}}{6},$$

(5.6.31)

and erfc is the complimentary error function [Abramowitz & Stegun (1965)]. The root mean square end-to-end distance R is

$$R = \ell \sqrt{N \tilde{\ell}_{\text{eff}}},$$

(5.6.32)

where $\tilde{\ell}_{\text{eff}}$ is given by the crossover formula [Muthukumar (1987)]

$$\tilde{\ell}_{\text{eff}}^{5/2} - \tilde{\ell}_{\text{eff}}^{3/2} = \frac{4}{3} \left(\frac{3}{2\pi} \right)^{3/2} v \sqrt{N} + \frac{4}{3} \sqrt{\frac{6}{\pi}} \alpha^2 z_p^2 \tilde{\ell}_B \tilde{\ell}_{\text{eff}} N^{3/2} \Theta (\tilde{a}),$$

(5.6.33)

with Θ being

$$\Theta (\tilde{a}) = \frac{\sqrt{\pi}}{2\tilde{a}^{5/2}} \left[\left(\tilde{a}^2 - 4\tilde{a} + 6 \right) e^{\tilde{a}} \text{erfc} \left(\sqrt{\tilde{a}} \right) - 6 - 2\tilde{a} + \frac{12}{\sqrt{\pi}} \sqrt{\tilde{a}} \right].$$

(5.6.34)

With the assumption of uniform expansion used in deriving the above crossover formula, the radius of gyration R_g is $\ell \sqrt{N \tilde{\ell}_{\text{eff}}/6}$. Therefore Eq. (5.6.33) can be used to calculate the radius of gyration of a flexible polyelectrolyte at all values of the Bjerrum length, Debye length, chain length, and the excluded volume parameter for a fixed polymer charge.

5.6.5 Electrostatic Stretching of Semiflexible Chains

As described in Section 2.2, a semiflexible chain such as dsDNA is modeled as a worm-like (Kratky–Porod) chain with an intrinsic persistence length ℓ_p arising from the chemical nature of the chain backbone. When ℓ_p is larger than the chain's contour length L, the chain conformation is rod-like, so that the mean square end-to-end distance $\langle R^2 \rangle_0$ is L^2. If L is much longer than ℓ_p, then $\langle R^2 \rangle_0$ is $L\ell$ (where $\ell = 2\ell_p$),

$$\langle R^2 \rangle_0 = \begin{cases} L^2, & \ell_p \gg L \text{(rod-like)} \\ L\ell, & \ell_p \ll L \text{(coil-like)} \end{cases} \qquad (5.6.35)$$

where the subscript "0" indicates the absence of inter-segment excluded volume and electrostatic interactions. As shown in Section 5.3.3 for a semiflexible chain in the limit of $\ell_p > L$, presence of charges along the chain backbone amplifies the intrinsic persistence length with an additional contribution from the electrostatic persistence length ℓ_{pe}. In this rod-like limit, ℓ_{pe} is inversely proportional to the concentration of the added salt in the medium (Eq. (5.3.24)). On the other hand, in the coil-like limit valid for $\ell_p \ll L$, the semiflexible chain undergoes swelling due to excluded volume and electrostatic interactions. The mean square end-to-end distance $\langle R^2 \rangle$ of a semiflexible chain in these two limits is given by

$$\langle R^2 \rangle = \begin{cases} L^2, & \ell_p \gg L \text{(rod-like)} \\ L\ell_{\text{eff}}, & \ell_p \ll L \text{(swollen coil)} \end{cases} \qquad (5.6.36)$$

where $\ell_{\text{eff}} = \langle R^2 \rangle / N\ell$ is described in Sections 5.6.1, 5.6.2, and 5.6.4. In particular, the radius of gyration R_g of a semiflexible chain with $L \gg \ell_p$ depends on the salt concentration as

$$R_g \sim \frac{1}{c_s^{1/5}}. \qquad (5.6.37)$$

Here, the role of excluded volume is ignored. In general, the dependence of R_g on c_s is a crossover function as given in Eq. (5.6.33). Although dsDNA can be modeled as a semiflexible chain, the above coil-like electrostatic swelling is applicable if its contour length is much longer than its intrinsic persistence length of 35 nm [Brinkers *et al.* (2009)], as well as ssDNA. The full crossover behavior at intermediate values of ℓ_p and c_s for semiflexible polyelectrolytes remains as an active subject [Ghosh *et al.* (2001), Innes-Gold *et al.* (2021)].

5.7   Self-regularization of Polymer Charge

As seen in Section 5.5, computer simulations show, for parameters representing dilute aqueous solutions of flexible polyelectrolytes, that there is a continuous exchange of counterions between the neighborhood of the polymer and the background, and that any counterion adsorbed on the polymer can move along the polymer backbone. The counterion condensation around a polymer is reminiscent of adsorption of a gas on a lattice, except that now the lattice is a topologically correlated polymer chain. The optimization between the translational entropy of the counterions (resulting in less adsorption) and the electrostatic attraction of the counterions by the polymer (resulting in more adsorption) is influenced by the conformations of the polymer. As a result, the polymer charge self-regulates as experimental conditions change, and its value is dictated self-consistently by the compatibility between polymer conformations and counterion adsorption.

The net polymer charge is unique to a polymer conformation [Muthukumar (2004)]. For example, a polymer chain cannot be assumed to bear the same net charge as it undergoes a coil–globule transition, a common assumption prevalent in the polyelectrolyte literature. As discussed in Sections 5.4 and 5.5, the extent of conformation-dependent counterion condensation is affected by many factors including the nature of the counterion (size and valency), concentration of added salt, polymer concentration, solvent dielectric constant, and temperature. This effect has been addressed theoretically in the contexts of isolated chains [Khokhlov (1980), Muthukumar (2004)], polyelectrolyte brushes [Pincus (1991)], and gels [Khokhlov & Kramarenko (1994)]. We give a brief discussion of self-regularization of polymer charge below by addressing all of the concepts introduced in Section 5.3.

Consider a single chain of N monomers in volume V. Each monomer is monovalently charged and ℓ is now the distance between two successive monomers along the polymer. Due to the electroneutrality condition, there are N monovalent counterions. Let M be the number of counterions adsorbed on the polyelectrolyte so that M/N is the degree of counterion adsorption and $\alpha = 1 - (M/N)$ is the degree of ionization of the polyelectrolyte. In addition, let c_s be the number concentration of an added salt which is fully dissociated into n_+ counterions and n_- co-ions ($c_s = n_+/V = n_-/V$). For the sake of simplicity, the dissolved ions are assumed to be monovalent and the counterion from the salt is chemically identical to that of the polymer. The coupling between α and the radius of gyration R_g is calculated self-consistently as follows [Muthukumar (2004)].

The free energy F has six contributions, related, respectively, to (i) entropy of adsorbed counterions on the polymer backbone, (ii) translational entropy of unadsorbed counterions and all co-ions (except the polymer) distributed in volume V, (iii) fluctuations in distributions of all dissociated ions arising from their electrostatic interactions, (iv) gain in energy due to the formation of ion pairs accompanying counterion adsorption, (v) free energy of the polymer with $N - M$ charges and M dipoles on its backbone interacting with the neutralizing background composed of $N - M$ counterions and salt ions in a solution of average monomer density $\rho = N/V$, and (vi) correlations among the ion pairs on the polymer.

(i) Entropy of adsorbed ions: Since $(1 - \alpha)N$ counterions are adsorbed to each chain on average and there are $N!/[((1 - \alpha)N)!][(\alpha N)!]$ ways of formation of ion pairs, the free energy associated with the entropy of adsorbed counterions, with the Stirling approximation, is

$$\frac{F_1}{k_B T} = N \left[\alpha \ln \alpha + (1 - \alpha) \ln(1 - \alpha)\right]. \tag{5.7.1}$$

All nonlinear effects resulting from possible cooperative features associated with the placement of counterions along the polymer contour are ignored.

(ii) Entropy of unadsorbed ions: The free energy due to the translational entropy associated with $(N - M + n_+)$ unadsorbed counterions and n_- co-ions in volume V is the familiar entropy of ideal mixing

$$\frac{F_2}{k_B T} = V \left[(\alpha \tilde{\rho} + \tilde{c}_s) \ln (\alpha \tilde{\rho} + \tilde{c}_s) + \tilde{c}_s \ln \tilde{c}_s - (\alpha \tilde{\rho} + 2\tilde{c}_s) \right], \qquad (5.7.2)$$

where $\tilde{c}_s$ and $\tilde{\rho}$ are the dimensionless salt concentration and monomer density, respectively,

$$\tilde{c}_s \equiv c_s \ell^3, \qquad \tilde{\rho} \equiv \rho \ell^3. \qquad (5.7.3)$$

(iii) Correlations among dissociated ions: Using the Debye–Hückel result given by Eq. (3.8.42), we get

$$\frac{F_3}{k_B T} = -V \frac{\kappa^3}{12\pi}, \qquad (5.7.4)$$

where

$$\kappa^2 = 4\pi \ell_B (\alpha \rho + 2c_s). \qquad (5.7.5)$$

This contribution is due to the Coulomb interactions among the ions in the counterion cloud around the whole polymer.

(iv) Energy of adsorbed ions: The gain in energy due to an adsorbed ion on the chain backbone depends on microscopic details such as the ionic radii and the local dielectric constant, as discussed in Section 5.3.2.5. Using the phenomenological dielectric mismatch parameter defined in Eq. (5.2.14), the free energy associated with the formation of ion pairs follows from Eq. (5.3.18) as

$$\frac{F_4}{k_B T} = -N \left[(1 - \alpha) \frac{\ell_B}{\ell} \delta \right]. \qquad (5.7.6)$$

(v) Chain free energy: The free energy of a flexible chain with degree of ionization α and radius of gyration $R_g = \ell \sqrt{N \tilde{\ell}_{\text{eff}}/6}$ is given by Eq. (5.6.28).

(vi) Interaction between ion pairs: As discussed in Section 5.3.2.6, the ion pairs resulting from the adsorbed counterions may be assumed to be randomly distributed along the chain backbone with random orientations. The interaction among these ion-pairs leads to a short-ranged attractive contribution, and can be absorbed into the two-body excluded volume interaction parameter v as long as the polymer has not substantially collapsed into a globule (Section 5.3.2.6).

Since the contribution from ion-pair interactions can be absorbed in the expression for the chain free energy by redefining the excluded volume parameter v, the total free energy F of the chain is

$$F = F_1 + F_2 + F_3 + F_4 + F_5, \qquad (5.7.7)$$

where F_1, F_2, F_3, and F_4 are given by Eqs. (5.7.1)–(5.7.6), and F_5 is given by Eq. (5.6.28). By minimizing F with respect to the degree of ionization α and the expansion factor $\tilde{\ell}_{\text{eff}}$,

$$\frac{\partial F}{\partial \alpha} = 0 = \frac{\partial F}{\partial \tilde{\ell}_{\text{eff}}}, \qquad (5.7.8)$$

the optimum effective charge of the polymer $N\alpha e$ and the optimum radius of gyration of the polymer, $R_g = \ell \sqrt{N \tilde{\ell}_{\text{eff}}/6}$ are obtained [Muthukumar (2004)].

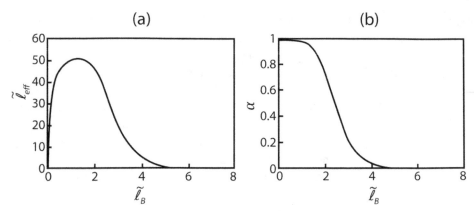

Figure 5.18 Dependence of (a) expansion factor $\tilde{\ell}_{\text{eff}} = R_g^2 / R_{g0}^2$, and (b) degree of ionization α on the Coulomb strength parameter $\tilde{\ell}_B = \ell_B / \ell$ [Muthukumar (2004)].

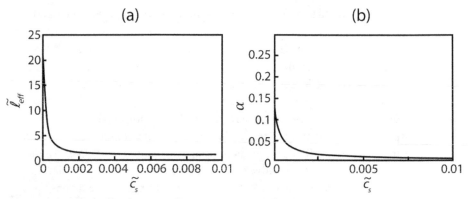

Figure 5.19 Dependence of (a) expansion factor $\tilde{\ell}_{\text{eff}} = R_g^2 / R_{g0}^2$, and (b) degree of ionization α on salt concentration $\tilde{c}_s = c_s \ell^3$ [Muthukumar (2004)].

The results from this self-consistent procedure are given in Fig. 5.18, showing the dependencies of the chain expansion factor $\tilde{\ell}_{\text{eff}}$ and the degree of ionization α on the Bjerrum length. The effect of monovalent salt concentration $\tilde{c}_s \equiv c_s \ell^3$ on $\tilde{\ell}_{\text{eff}}$ and α is given in Fig. 5.19. These results capture all trends observed in computer simulations discussed in Section 5.5 for the polymer size and the effective polymer charge.

A simplification arises in the self-consistent calculation if we were to consider only highly swollen flexible polyelectrolyte molecules. Inspection of numerical values of the various terms in Eq. (5.7.7) shows that the contributions from chain entropy (F_5) and ion correlations (F_3) are weak in comparison with the other three terms [Kundagrami & Muthukumar (2010)]. By taking only F_1, F_2, and F_4 in Eq. (5.7.7), the degree of ionization is given by the formula

$$\alpha = \frac{1}{2\tilde{\rho}} \left[-\left(\tilde{c}_s + e^{-\delta \tilde{\ell}_B}\right) + \sqrt{\left(\tilde{c}_s + e^{-\delta \tilde{\ell}_B}\right)^2 + 4\tilde{\rho} e^{-\delta \tilde{\ell}_B}} \right]. \tag{5.7.9}$$

This closed-form result for α gives its dependence on $\tilde{\ell}_B$ (i.e., $T \epsilon (T)$), the salt concentration c_s, the monomer concentration ρ, and the dielectric mismatch parameter

δ. The value of α given by Eq. (5.7.9) is substituted into Eq. (5.6.33) to calculate the expansion factor $\tilde{\ell}_{\text{eff}}$. A comparison of α and $\tilde{\ell}_{\text{eff}}$ thus calculated using the separation approximation, with the numerical computation with full coupling, reveals that the separation approximation is adequate if the chain is not collapsed below the Gaussian chain size.

The full self-consistent calculation is necessary if the chain undergoes significant conformational changes as the experimental conditions are varied. As an example, the polymer chain collects most of the counterions when it undergoes a coil–globule transition [Kundagrami & Muthukumar (2010)] as seen in experiments [Loh *et al.* (2008)]. This example is discussed in detail in the following section.

5.8   Coil–Globule Transition in Polyelectrolytes

The coil-globule conformational transition can be induced in charged macromolecules by changing the solvent into a poor solvent. This can be experimentally realized either by changing the temperature, by increasing the salt concentration (Section 5.6.1), or by increasing the chemical mismatch (Eq. (2.4.4)) with an increase in the proportion of the non-solvent in a solvent mixture at a fixed temperature. An example of the third strategy is the experiments by Loh *et al.* (2008), who studied the collapse transition of poly-2-vinylpyridine quaternized to 4.3% with ethyl bromide in a 1-propanol/2-pentanone solvent mixture. As the content of the non-solvent 2-pentanone increased, the polymer exhibited a transition from an expanded coil state to a globular state.

As we expected intuitively in Section 5.3.2.6, the local dielectric constant inside the coil decreases when the chain is allowed to shrink. As a consequence, counterion binding would be enhanced. Therefore, the net electrostatic repulsion arising from the lowered number of charged monomers is weakened, generating further shrinkage of the coil. This in turn would lead to further collection of counterions, and subsequent further shrinkage of the coil. This cascade process of counterion collection continues until the globule becomes essentially neutral when all counterions are buried within it. The cartoon in Fig. 5.20 depicts the cascade process of chain collapse by simultaneous collection of counterions and conformational adjustments. The transition from the expanded coil state to the collapsed globular state occurs through a series of intermediate states. An example of one intermediate state is shown in the figure as the Gaussian coil state.

The self-consistent theoretical treatment of the coil–globule transition in charged macromolecules can be performed by following the same procedure as in Section 5.7, except that a three-body excluded volume interaction term, $w\ell^6 N^3/R^6$ (as in Eq. (2.6.1)), should be added on the right-hand side of Eq. (5.6.28). Minimizing the resulting free energy with respect to the size expansion factor $\tilde{\ell}_{\text{eff}}$ and degree of ionization α, we get the optimum values of $\tilde{\ell}_{\text{eff}}$ and α from Eq. (5.7.8).

The coil–globule transition is given by computing the temperature dependence of $\tilde{\ell}_{\text{eff}}$ and α so obtained [Kundagrami & Muthukumar (2010)]. The two key variables in Eq. (5.6.28) that depend on temperature are the two-body excluded volume parameter v ($= 1 - 2\chi$) and the Bjerrum length ℓ_B, assuming that the three-body repulsive

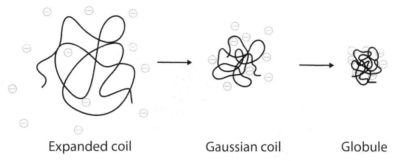

Expanded coil   Gaussian coil   Globule

Figure 5.20   Schematic of the cascade of collection of counterions during coil-to-globule transition.

excluded volume parameter w is positive and independent of temperature (Section 2.6). Since ℓ_B is inversely proportional to $T\epsilon(T)$, define a dimensionless reduced temperature t as

$$t \equiv \frac{\ell}{4\pi\ell_B}. \qquad (5.8.1)$$

In view of the discussion in Section 2.4, the Flory–Huggins χ (proportional to $1/T$ for UCST systems) can be written in terms of the reduced temperature t as

$$\chi = \frac{\theta_F}{2T} \equiv \frac{a_\chi}{20\pi t}, \qquad (5.8.2)$$

where θ_F is the Flory temperature. The parameter a_χ ($= 10\pi\ell\epsilon_0\epsilon k_B\theta_F/e^2$) reflects the specific value of the Flory temperature and is taken as a measure of the van der Waals interactions between the polyelectrolyte backbone and the solvent for a particular system.

The self-consistently calculated dependencies of the degree of ionization α and the size expansion factor $\tilde{\ell}_{\text{eff}} = R_g^2/R_{g0}^2$ on the reduced temperature t are given in Fig. 5.21, for different values of the three-body excluded volume parameter w (for $N = 1000, \rho\ell^3 = 0.0005$, and $\delta = 3.0$). We see from Fig. 5.21 that for w smaller than a critical value $w^\star$, the charged macromolecule undergoes a coil–globule transition. As an example of the transition, consider the case of $w = 0.2$. As the reduced temperature is lowered from a high value ($t = 0.023$), the chain in the coil state shrinks in size. When the reduced temperature is about 0.020, the coil undergoes a discontinuous transition into a globule. For temperatures below the transition temperature, the chain remains in the globular state with its conformational entropy associated with packing progressively decreasing. Simultaneous to the change in the chain conformation, the degree of ionization changes. As seen in Fig. 5.21, α decreases gradually as the temperature is lowered above the transition temperature, and precipitously collapses at the transition temperature. For temperatures below the transition temperature, α continues to decrease until it approaches zero. The discontinuous transition for both α and $\tilde{\ell}_{\text{eff}}$ occurs at the same transition temperature, demonstrating the self-regulation of the charge of the macromolecule accompanying conformational changes. The coil–globule transition temperature, based on this theoretical model,

(a)

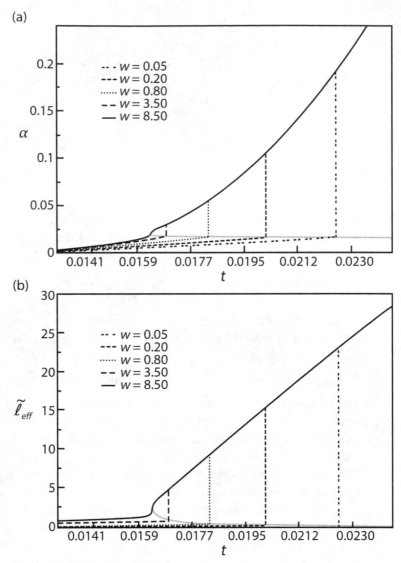

(b)

Figure 5.21 Coil collapse and collection of counterions occur simultaneously during coil-to-globule transition. (a) Degree of ionization versus the reduced temperature t for different values of the three-body excluded volume parameter w. (b) Size expansion parameter $\tilde{\ell}_{\mathrm{eff}}$ versus t for different values of w [Adapted from Kundagrami & Muthukumar (2010)].

depends on the three-body excluded volume parameter w. The transition temperature decreases as w increases until a critical point is reached. For the particular choice of $N = 1000, \rho\ell^3 = 0.0005$, and $\delta = 3.0$, the critical point is given by $t^\star = 0.01634$ (or equivalently $\ell_B^\star = 4.867$), $v^\star = -1.164, w^\star = 8.5, \tilde{\ell}_{\mathrm{eff}}^\star = 2.3$, and $\alpha^\star = 0.02$ [Kundagrami & Muthukumar (2010)].

The main conclusion from Fig. 5.21 is that the chain collapse and the precipitous collection of counterions by the macromolecule occur simultaneously. Consequently,

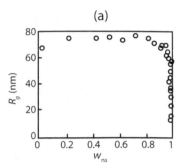
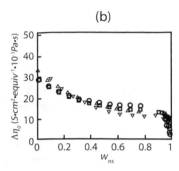

Figure 5.22 (a) Radius of gyration of quaternized poly(2-vinylpyridine) in a mixed solvent of 1-propanol and 2-pentanone (non-solvent), measured at room temperature using light scattering under salt-free conditions, as a function of the weight fraction w_{ns} of the non-solvent. (b) The dependence of the Walden product of equivalent conductivity Λ and the viscosity of the solvent mixture η_0 on w_{ns} [Adapted from Loh *et al.* (2008)].

there cannot be a coexistence of swollen and collapsed states for the same effective charge of the macromolecule.

The mutually cooperative coupling between α and R_g has been demonstrated in the aforementioned experiments by Loh *et al.* (2008) on the coil–globule transition of quaternized poly(2-vinylpyridine) in the mixed solvent of 1-propanaol and 2-pentanone (non-solvent). The measured radius of gyration of the polymer chain is given in Fig. 5.22a as a function of the weight fraction w_{ns} of the non-solvent. As seen in the figure, the chain undergoes a coil–globule transition, where R_g shrinks sharply from about 60 nm to about 10 nm as w_{ns} is changed from about 0.95 to 0.995.

Concurrent with this structural transition, the equivalent conductivity Λ corrected by the solvent viscosity η_0, expressed as the Walden product $\Lambda\eta_0$, decreases sharply to zero in the range $0.95 < w_{ns} < 0.995$, as shown in Fig. 5.22b. The polyelectrolyte chain is essentially uncharged in its collapsed state. The fact that the measured ionic conductivity of the solution approaches zero when the chain becomes globular indicates that the counterions are progressively collected by the collapsing coil during the coil–globule transition [Loh *et al.* (2008)].

The above description is for polyelectrolyte solutions without added salt. As pointed above, the coil–globule transition can be facilitated by increasing the concentration of added salt. The role of multivalent counterions and salt ions can be addressed in an analogous manner, and the behavior of polyelectrolyte chains becomes very rich [de la Cruz *et al.* (1995), Wittmer *et al.* (1995), Schiessel (1999), Nguyen *et al.* (2000), Kundagrami & Muthukumar (2008)].

It is evident from Figs. 5.21 and 5.22 that charge regularization must be considered in interpreting experimental results obtained with wide variations in temperature, solvent quality, and salt concentration. Obviously, caution must be exercised in using theories based on scaling arguments, electrostatic blobs, pearl-like structures, etc., which assume a *fixed* value of α over the entire ranges of the various experimental variables. Exceptions might arise under conditions where α does not vary too much, as perhaps in the situation of homogeneous single phases of polyelectrolyte solutions in a narrow temperature range.

5.9   Apparent Molar Mass, Apparent Radius of Gyration, and Structure Factor

For a two-component dilute solution of uncharged macromolecules in a solvent, the degree of polymerization N of the macromolecule, its radius of gyration R_g, and the size exponent ν are usually determined using light scattering techniques. In infinitely dilute solutions, the scattered intensity $I(\mathbf{k})$ at the scattering wave number $k = (4\pi/\lambda)\sin(\theta/2)$ (where λ is the wavelength of the incident radiation and θ is the scattering angle) is proportional to the form factor $P(\mathbf{k})$ [Yamakawa (1971), Higgins & Benoit (1994)]

$$I(\mathbf{k}) \sim \frac{R_\theta}{Kc_p} = NP(\mathbf{k}) = \frac{1}{N}\sum_{i=1}^{N}\sum_{j=1}^{N}\langle \exp\left(-i\mathbf{k}\cdot\mathbf{R}_{ij}\right)\rangle, \tag{5.9.1}$$

where R_θ is known as the Rayleigh ratio, c_p is the monomer number concentration, and K is a factor depending on the refractive index of the solution. On the right-hand side of Eq. (5.9.1) N is the number of monomers per chain, $\mathbf{R}_{ij}$ is the distance vector between the ith and jth monomers of the macromolecule, and the angular brackets denote the averaging over all conformations of the macromolecule. As seen in Section 2.9, $P(\mathbf{k})$ for two-component solutions of uncharged polymers is given by

$$P(\mathbf{k}) = \begin{cases} 1 - \frac{k^2 R_g^2}{3} + \cdots, & kR_g \ll 1 \\ \sim \frac{1}{Nk^{1/\nu}}. & \frac{2\pi}{\ell} > k > \frac{2\pi}{R_g} \end{cases} \tag{5.9.2}$$

The above relation between R_θ/Kc_p and the form factor cannot be used for solutions of charged macromolecules, which contain more than two species (macromolecules, counterions, salt ions, and solvent). It is necessary to use the correct Rayleigh ratio valid for multi-component systems [Brinkman & Hermans (1949), Kirkwood & Goldberg (1950), Stockmayer (1950)], by obtaining expressions for the chemical potentials of all components in the solution [Muthukumar (2012a)]. Furthermore, the strong electrostatic correlations and charge regularization of macromolecules due to counterion binding must be included in the expression for the Rayleigh ratio. When there are correlations between several macromolecules due to inter-chain interactions, the form factor of Eq. (5.9.1) is replaced by the structure factor $S(\mathbf{k})$ [Yamakawa (1971), Higgins & Benoit (1994), McQuarrie (1976)],

$$S(\mathbf{k}) = P(\mathbf{k})\left[1 - \frac{c}{N}Q(\mathbf{k})\right], \tag{5.9.3}$$

where $Q(\mathbf{k})$ is the Fourier transform of the pair correlation function for a pair of macromolecules,

$$Q(\mathbf{k}) = \int d\mathbf{r}\left[1 - \exp\left(-\frac{U(\mathbf{r})}{k_B T}\right)\right]\exp\left(i\mathbf{k}\cdot\mathbf{r}\right), \tag{5.9.4}$$

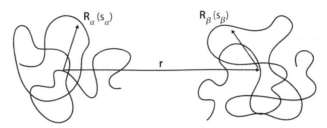

Figure 5.23 The inter-chain potential between a pair of chains is calculated by considering all pairwise interactions among monomers of both chains. The separation distance between the centers-of-mass of αth and βth chains is $\mathbf{r}$, and $\mathbf{R}_\alpha(s_\alpha)$ and $\mathbf{R}_\beta(s_\beta)$ are the position vectors of monomers from the center of mass of the respective chains.

with $U(\mathbf{r})$ being the inter-chain potential as a function of the distance $\mathbf{r}$ between the centers of the chains, as depicted in Fig. 5.23.

The inter-chain potential is the sum of the excluded volume and electrostatic contributions,

$$U(\mathbf{r}) = U_{\text{exc}}(\mathbf{r}) + U_{\text{el}}(\mathbf{r}), \tag{5.9.5}$$

with

$$\frac{U_{\text{exc}}(\mathbf{r})}{k_B T} = v\ell^3 \int_0^N ds_\alpha \int_0^N ds_\beta \delta\left[\mathbf{R}_\alpha(s_\alpha) - \mathbf{R}_\beta(s_\beta)\right], \tag{5.9.6}$$

and

$$\frac{U_{\text{el}}(\mathbf{r})}{k_B T} = \alpha^2 z_p^2 \ell_B \int_0^N ds_\alpha \int_0^N ds_\beta \frac{e^{-\kappa|\mathbf{R}_\alpha(s_\alpha) - \mathbf{R}_\beta(s_\beta)|}}{|\mathbf{R}_\alpha(s_\alpha) - \mathbf{R}_\beta(s_\beta)|}. \tag{5.9.7}$$

Here the positions of the monomers s_α and s_β belonging to the αth and βth chains are $\mathbf{R}_\alpha(s_\alpha)$ and $\mathbf{R}_\beta(s_\beta)$ with respect to their centers of mass. Choosing the origin of the coordinate system as the center-of-mass of the αth chain, the positions of s_α and s_β are $\mathbf{R}_\alpha(s_\alpha)$ and $\mathbf{R}_\beta(s_\beta) + \mathbf{r}$ (Fig. 5.23).

Let us assume, for the sake of analytical tractability, that the monomer density profile $\rho_p(\mathbf{r}')$ at the spacial location $\mathbf{r}'$ inside the volume occupied by the chain can be approximated by the Gaussian distribution function,

$$\rho_p(\mathbf{r}') = \int_0^N ds_\alpha \delta\left(\mathbf{R}_\alpha(s_\alpha) - \mathbf{r}'\right) = N\left(\frac{3}{2\pi R_g^2}\right)^{3/2} e^{-\frac{3}{2}\frac{r'^2}{R_g^2}}. \tag{5.9.8}$$

Substituting this approximation in Eq. (5.9.6), we get

$$\frac{U_{\text{exc}}(\mathbf{r})}{k_B T} = v\ell^3 N^2 \left(\frac{3}{4\pi R_g^2}\right)^{3/2} e^{-\frac{3}{4}\frac{r^2}{R_g^2}}. \tag{5.9.9}$$

This is known as the Flory–Krigbaum potential between two uncharged chains [Flory & Krigbaum (1950), Yamakawa (1971)]. Using the same Gaussian approximation for monomer density profile as in Eq. (5.9.8), the electrostatic contribution to the inter-chain potential becomes [Muthukumar (2012a)]

$$\frac{U_{el}(\mathbf{r})}{k_B T} = \frac{1}{2} \frac{\alpha^2 z_p^2 \ell_B N^2}{r} e^{k^2 R_g^2/3} \left[e^{-\kappa r} \text{erfc} \left(\frac{\kappa R_g}{\sqrt{3}} - \frac{\sqrt{3}}{2} \frac{r}{R_g} \right) - e^{\kappa r} \text{erfc} \left(\frac{\kappa R_g}{\sqrt{3}} + \frac{\sqrt{3}}{2} \frac{r}{R_g} \right) \right],$$

(5.9.10)

where erfc is the complementary error function [Abramovitz & Stegun (1965)]. Note that, for $\kappa R_g < 3r/R_g$, this inter-chain potential becomes the screened Coulomb potential between two chains separated by distance r,

$$\frac{U_{el}(r)}{k_B T} = \alpha^2 z_p^2 \ell_B N^2 \frac{e^{-\kappa r}}{r}, \qquad r > 2R_g.$$

(5.9.11)

In order to interpret small angle scattering experiments ($kR_g \ll 1$), $Q(\mathbf{k})$ of Eq. (5.9.4) is expanded as a Taylor series in k,

$$Q(\mathbf{k}) = v_0 - v_2 k^2 + \cdots,$$

(5.9.12)

where v_0 and v_2 depend on $U(\mathbf{r})$ given above. v_0 is twice the second virial coefficient for polymer chains (discussed in Section 5.10) and v_2 is proportional to R_g^5. By addressing inter-chain interaction as approximated above and charge regularization arising from self-consistent coupling between counterion adsorption and chain conformations, the Rayleigh ratio can be obtained from Eqs. (5.9.1)–(5.9.3) and (5.9.12) as [Muthukumar (2012a)]

$$\frac{R_\theta}{K c_p} = N_{app}^{(0)} \left(1 - \frac{c_p v_0}{N} \right) \left\{ 1 - k^2 \left[\frac{R_g^2}{3} - \frac{c_p v_2}{(N - c_p v_0)} \right] + 0(k^4) \right\},$$

(5.9.13)

where $N_{app}^{(0)}$ is due to counterion binding and thermodynamic coupling among various components by the Gibbs–Duhem equation, but without the interchain correlations. By defining apparent values for N and R_g, Eq. (5.9.13) can be equivalently written in the form of Eq. (5.9.2) in the small k limit as

$$\frac{R_\theta}{K c_p} = N_{app} \left(1 - \frac{k^2 R_{g,app}^2}{3} + \cdots \right), \qquad kR_g \ll 1$$

(5.9.14)

where the apparent values of degree of polymerization (N_{app}) and radius of gyration ($R_{g,app}$) obtained from the above equation are significantly different from their true values. For example, N_{app} obtained from light scattering in salt-free conditions may even be an order of magnitude smaller than the true N, and $R_{g,app}^2$ can even be negative [Saha et al. (2013)]. The resolution of the discrepancies between the apparent values and true values is described in the following subsections.

5.9.1   Apparent Molar Mass

According to the counterion-adsorption theory [Muthukumar (2012a)], the difference between the experimentally measured N_{app} and the true N is given by

$$\frac{1}{N_{app}} = \frac{1}{N} + f(c_p, c_s, q_0, K_i),$$

(5.9.15)

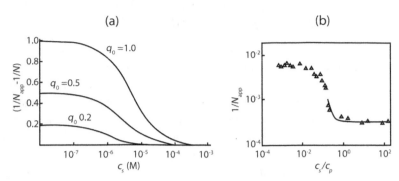

Figure 5.24 The apparent molar mass determined using light scattering is anomalously low at low ionic strengths. Only at sufficiently high salt concentrations can true molar mass be determined from the Zimm plot. (a) Theoretical calculation of the dependence of the apparent degree of polymerization as a function of salt concentration for different values of the maximum fraction q_0 of ionizable groups per chain [Muthukumar (2012a)]. (b) Experimental results on the dependence of the apparent degree of polymerization of poly-2-vinylpyridine (with $q_0 = 0.35$) in propanol on c_s/c_p. The solid curve is from the counterion-binding theory [Saha *et al.* (2013)].

where f is a derived function of the indicated variables (q_0 is the maximum allowed degree of ionization per molecule and $K_i = \exp(-\delta \ell_B/\ell)$ is the ionization equilibrium constant for a charged monomer-counterion pair). For salt-free, infinitely dilute solutions, Eq. (5.9.15) becomes

$$\frac{1}{N_{\text{app}}} = \frac{1}{N} + q_0, \qquad (c_s = 0, c_p \to 0). \qquad (5.9.16)$$

Since q_0 is of order unity and $N \gg 1$, the measured value of N_{app} is only of order unity in this limit, without reflecting the true value of N. In the presence of salt, the function f in Eq. (5.9.15) is a decreasing function, as shown in Fig. 5.24a for $c_p = 10^{-5}$ M and $K_i = e^{-9}$. The apparent degree of polymerization is $1/q_0$ in the limit of $c_s = 0$, and it approaches the true N at higher salt concentrations. As seen in Fig. 5.24a in order to measure the true molar mass of the charged macromolecule, the required salt concentration depends on the maximum allowable degree of ionization. The apparent molar mass of poly-2-vinylpyridine (with degree of quaternization $q_0 = 0.35$) in propanol measured by light scattering at low ionic strengths (10^{-6} M $< c_s < 10^{-3}$ M) is given in Fig. 5.24b as a function of c_s/c_p [Saha *et al.* (2013)]. The N_{app} changes from a small value of about 167 to its true value 3200, as c_s/c_p is increased. The solid curve in Fig. 5.24b is the prediction from the self-consistent counterion adsorption theory with $K_i = 5 \times 10^{-8}$ M (see Section 5.2). It is evident from Figs. 5.24a and 5.24b that the true molar mass of charged macromolecules can be measured from light scattering (using the Zimm plot), only if a sufficiently large concentration of added salt is guaranteed. A safe assumption for this criterion is that the salt concentration is higher than the monomer concentration (as seen in Fig. 5.24(b)).

5.9.2   Correlation Hole and Apparent Radius of Gyration

In view of the negative term inside the square brackets in Eq. (5.9.13), arising from electrostatic correlations among chains, the scattered intensity at small k values can show a down-turn, as shown in Fig. 5.25a, instead of the monotonic behavior of Eq. (5.9.2) seen in two-component solutions of uncharged macromolecules. The data in Fig. 5.25a are from salt-free propanol solution of quaternized poly-2-vinylpyridine (with 20 % quaternization) at the monomer concentration c_p = 10 mg/L (filled circles), 20 mg/L (open circles), and 27 mg/L (stars). The region of diminished scattered intensity, instead of the expected higher value at small scattering wave vectors, is referred to as the "correlation hole." The emergence of the correlation hole in the scattering intensity is due to long-ranged electrostatic interactions. The location of the maximum scattering intensity k_m increases with polymer concentration. In fact, $k_m \sim \sqrt{c_p}$ as will be derived in Section 6.4. The correlation hole occurs only at low salt concentrations. At high enough salt concentrations, where the long-ranged electrostatic interaction is screened, the correlation hole in the scattered intensity is absent and the behavior is identical to that of uncharged dilute solutions (Eq. (5.9.2)).

The occurrence of the correlation hole presents difficulties in determining the radius of gyration from light scattering with the use of the Zimm plot. If the scattering data at very small wave vectors were analyzed using the Zimm plot, the resulting R_g^2 can be negative. For solutions of charged macromolecules at low ionic strengths, such an analysis is obviously invalid. The full theory which addresses the multicomponent nature of the system, electrostatic correlations, and charge-regulation due to counterion binding, must be implemented in determining the true R_g. An example of the discrepancy between the true radius of gyration and the apparent radius of gyration, which is measured from the initial slope of Kc_p/R_θ versus k^2 is provided in Fig. 5.25b,

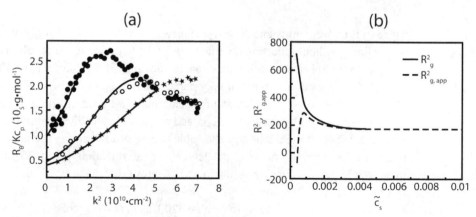

Figure 5.25   (a) Emergence of correlation hole due to electrostatic correlations among chains even in infinitely dilute solutions at small scattering wavevectors [Saha *et al.* (2013)]. The monomer concentration c_p = 10 mg/L (filled circles), 20 mg/L (open circles), and 27 mg/L (stars). (b) The straight forward deduction of the radius of gyration from the Zimm analysis can be misleading at low ionic strengths (dashed curve). Using the counterion-adsorption theory, the correct R_g can be obtained from experimental data (solid curve) [Muthukumar (2012a)].

where the square of the radius of gyration is plotted against the dimensionless salt concentration $\tilde{c}_s = c_s \ell^3$. The values of the parameters used in this illustration are $N = 1000, \ell_B/\ell = 3, \delta = 3.5, v = 0$, and $c_p \ell^3 = 0.0005$ [Muthukumar (2012a)]. If the salt concentration is sufficiently high, this discrepancy is absent and R_g can be determined accurately using the conventional Zimm plot.

5.9.3 Internal Structure

The details of how local monomer concentration is correlated in space is obtained by measuring the scattered intensity at scattering wave numbers in the range $2\pi/\ell > k > 2\pi/R_g$. In general, the form factor as defined in Eq. (5.9.1) is given in this range of scattering wave numbers by

$$P(k) \sim \frac{1}{k^{1/v}}, \qquad \left(\frac{2\pi}{\ell} > k > \frac{2\pi}{R_g}\right) \qquad (5.9.17)$$

where v is the effective size exponent. As described in Section 5.6, the range of v is from 1/3 to 1. A closed-form formula for $P(k)$ valid for all k values is not available. However, a convenient interpolation formula which captures the limits of $kR_g < 1$ and $kR_g > 1$, (Eq. (5.9.2)), is

$$P(k) \simeq \frac{1}{\left[1 + \frac{2v}{3}\left(kR_g\right)^2\right]^{1/2v}}, \qquad (5.9.18)$$

where $R_g^{1/v} \sim N$ is recognized. For example, consider a dilute aqueous solution containing a flexible polyelectrolyte chain of $R_g = 50$ nm in 0.1 M NaBr. As can be inferred from Fig. 5.9, the size exponent for such a condition is $v = 0.6$. With this value of v, the dependence of $P(k)$ on k for $N = 1000$, as predicted by theory, is presented in Fig. 5.26 as a double logarithmic plot. The expected slope $-5/3$ is indicated on the plot. In general, the slope is $-1/v$ for any self-similar statistical fractal-like macromolecule, and $1/v$ is known as the fractal dimension (Chapter 2). The size exponent v can be experimentally determined from such plots at $kR_g > 1$, which would

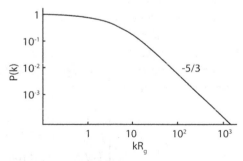

Figure 5.26 Plot of the form factor, proportional to the scattering intensity, versus kR_g, for $N = 1000$. The slope in the $kR_g \gg 1$ regime is the negative of the inverse of the size exponent v of the macromolecule. The size exponent $v = 0.6$ is illustrated in this figure.

complement the deduction of ν from a table of R_g versus N for a series of molecular weights measured at $kR_g < 1$.

5.10   Osmotic Pressure and Donnan Equilibrium

When a solution of macromolecules and its solvent are separated by a membrane that is permeable to the solvent, there is a net flow of the solvent into the solution because the chemical potential of the solvent in the solution is always lower than in the pure solvent at a given pressure. The osmotic pressure Π of the solution is the pressure that must be applied to the solution to prevent the net diffusion of solvent through the semipermeable membrane into the solution. For infinitely dilute solutions under ideal conditions where intermolecular interactions can be ignored, the osmotic pressure is given by the van't Hoff law [McQuarrie (1976), Hill (1986)],

$$\frac{\Pi}{k_B T} = \sum_j \rho_j, \tag{5.10.1}$$

where ρ_j is the number concentration of the jth solute in the solution [Hill (1986)]. When inter molecular interactions are taken into account, the osmotic pressure of a dilute two-component solution of an uncharged macromolecule is given by the virial series,

$$\frac{\Pi}{k_B T} = \rho + B_2 \rho^2 + \cdots, \tag{5.10.2}$$

where ρ is the number concentration of the macromolecule (n/V). B_2 is the second virial coefficient given by [Hill (1986)]

$$B_2 = \frac{1}{2} \int_0^\infty \left[1 - \exp\left(-\frac{U(r_{ij})}{k_B T} \right) \right] 4\pi r_{ij}^2 \, dr_{ij}, \tag{5.10.3}$$

where $U(r_{ij})$ is the pairwise inter molecular interaction energy as a function of the separation distance r_{ij} between the centers of mass of spherically averaged molecules i and j. If the molecules are spherical proteins with essentially no electrostatic interactions, they can be treated as hard spheres of radius R. For this case, $B_2 = 16\pi R^3/3$, which is proportional to the volume of the molecule. In general, for flexible chains, B_2 is proportional to the volume of the macromolecule, $B_2 \sim R_g^3$ [Hill (1986), Yamakawa (1971)]. If the macromolecule can be treated as a hard rigid rod of diameter d and length L, B_2 is $\pi dL^2/4$ for $L \gg d$ [Onsager (1949), Yamakawa (1971)]. Summarizing the results for the second virial coefficient B_2,

$$B_2 = \begin{cases} \frac{16}{3}\pi R^3, & \text{spheres} \\ \beta R_g^3, & \text{flexible chains} \\ \frac{\pi}{4} dL^2, & \text{rigid rods.} \end{cases} \tag{5.10.4}$$

where β depends on excluded volume interactions and persistence length of the macromolecule.

It is customary to rewrite Eq. (5.10.2) in terms of monomer number concentration $c = nN/V = \rho N$, so that

$$\frac{\Pi}{k_B T} = \frac{c}{N} + A_2 c^2 + \cdots ,\tag{5.10.5}$$

where the redefined second virial coefficient $A_2 = B_2/N^2$ is proportional to R_g^3/N^2 for flexible chains. If the monomer concentration c is in units of molarity ($c = nM/V$, with M being the molar mass), instead of the number concentration in Eq. (5.10.2), then this equation is rewritten as

$$\frac{\Pi}{RT} = \frac{c}{M} + A_2 c^2 + \cdots ,\tag{5.10.6}$$

where R is the gas constant $k_B N_A$, where N_A is the Avogadro number, and A_2 is $N_A B_2/M^2$, which is proportional to R_g^3/M^2. Since $R_g \sim N^\nu$, we get

$$A_2 \sim N^{3\nu-2} \sim M^{3\nu-2}.\tag{5.10.7}$$

The above result is valid for uncharged flexible polymers.

Let us now consider the implications of van't Hoff's law for the osmotic pressure of a solution of charged macromolecules. We shall address three situations as shown in Fig. 5.27. In Fig. 5.27a, a solution of charged macromolecules and low molar mass electrolyte ions is in compartment "in," which is separated from pure solvent in compartment "out" by a semipermeable membrane. The membrane is permeable only to the solvent, but not to the macromolecules and electrolyte ions. In this first situation (Fig. 5.27a), the osmotic pressure is given by a generalization of Eq. (5.10.6). The second virial coefficient for charged macromolecules can be intricate (Section 3.8.7) and there are also $c^{3/2}$ terms, arising from electrostatic correlations, in Eq. (5.10.6) [Hill (1986), Muthukumar (2012a)].

In the second situation (Fig. 5.27b), the "in" compartment is a solution of charged macromolecules and a low molar mass electrolyte; the "out" compartment is a solution of the same electrolyte as in the compartment "in." The semipermeable membrane in Fig. 5.27b is permeable to both the solvent and the electrolyte ions, but not to

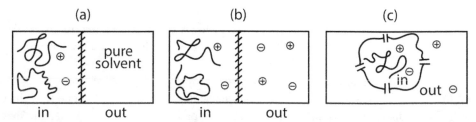

(a)   (b)   (c)

in   out   in   out

Figure 5.27 A solution of charged macromolecules and low molar mass electrolytes contained inside the "in" compartment, which is separated by a semipermeable membrane from the outside ("out") compartment. (a) The "out" compartment contains pure solvent, and the semipermeable membrane is permeable only to the solvent and not the electrolyte ions or macromolecules. (b) and (c) The "out" compartment is an electrolyte solution and the membrane allows passage of solvent, counterions, and electrolyte ions, but not the macromolecule.

the macromolecules. Further, to simplify the argument, let the counterion from the macromolecule be the same as that from the electrolyte. In this situation (Fig. 5.27b), the electrolyte concentrations in the equilibrated "in" and "out" compartments can be different, and as a result, a difference in osmotic pressure is established between the two compartments. This is known as the Donnan effect.

In the third situation (Fig. 5.27c), let the "in" compartment be the inside of a membrane-bound sac of charged macromolecules and electrolyte ions, and the "out" compartment be the outer solution containing electrolyte ions not necessarily the same ions as in the "in" compartment. If the membrane enveloping the "in" compartment has ion channels allowing passage of electrolyte ions and water, but not the macromolecule, a difference in electric potential is generated across the membrane accompanying the Donnan effect. This potential difference is known as the Donnan membrane potential. We describe these three situations sketched in Fig. 5.27 as follows.

5.10.1   Osmotic Pressure of a Dilute Polyelectrolyte Solution

Consider n charged macromolecules each of N segments and a net charge $Q = \alpha z_p N$ (α is the degree of ionization and z_p is the number of charges per segment before counterion adsorption) in volume V. In the "in" compartment, there are n chains and $\alpha z_p N n$ monovalent counterions. Using the ideal law of van't Hoff, Eq. (5.10.1) gives

$$\left(\frac{\Pi}{k_B T}\right)_{\text{ideal}} = \frac{n}{V} + \alpha z_p N \frac{n}{V}. \tag{5.10.8}$$

Using the definition of monomer number concentration, $c = nN/V$, we get

$$\left(\frac{\Pi}{k_B T}\right)_{\text{ideal}} = \frac{c}{N}\left(1 + \alpha z_p N\right) = \frac{c}{N}\left(1 + Q\right). \tag{5.10.9}$$

In the presence of monovalent salt at the number concentration c_s, Eq. (5.10.9) becomes

$$\left(\frac{\Pi}{k_B T}\right)_{\text{ideal}} = \frac{c}{N}\left(1 + \alpha z_p N\right) + 2c_s. \tag{5.10.10}$$

In addition, there are corrections to the above ideal law due to intermolecular interactions as manifest in the virial coefficients for the osmotic pressure, and electrostatic correlations as in the Debye–Hückel theory (Section 3.8.7).

Even in the ideal limit, we see from Eq. (5.10.9) that counterions dominate the osmotic pressure over the macromolecules since $\alpha z_p N$ is much larger than unity. For uncharged systems, the intercept of a plot of Π/c versus c is used to determine the degree of polymerization, or equivalently the molecular weight, of macromolecules. Also, this intercept, $\partial \Pi/\partial c$, is related to the inverse of the scattered intensity in the limit of zero scattering wave vector; hence, the molecular weight is easily measured using scattering techniques. However, this procedure is not feasible for determining the molecular weight of charged macromolecules due to the factor $(1 + \alpha z_p N)$ instead of unity in the case of uncharged systems. Since $\alpha z_p N$ can be much greater than unity,

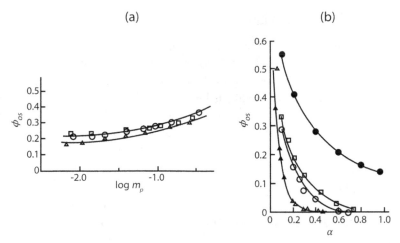

Figure 5.28 (a) Osmotic coefficient ϕ_{os} versus monomolal concentration m_p of aqueous poly(styrene sulfonate) solutions for three different counterions: H^+ (circle); Na^+ (square); Cs^+ (triangle). Molar mass of NaPSS is 4×10^4 g/mol. [Kozak et al. (1971)]. (b) Osmotic coefficient versus degree of ionization α for NaPAA in water at the polymer concentration $c_p = 0.01$ M (filled circle) and in methanol at $c_p = 8.7 \times 10^{-2}$ (triangle); ϕ_{os} of LiPAA in methanol at $c_p = 8.7 \times 10^{-2}$ (square) and $c_p = 1.7 \times 10^{-2}$ (open circle) [Klooster et al. (1984)].

the molecular weight measured from $\partial\Pi/\partial c$ is only an apparent molecular weight, which is much lower than the true molecular weight as seen in Section 5.9.1.

Furthermore, due to nonideal interactions and correlations among the various species in the solution, there are strong deviations from the ideal osmotic pressure given by Eq. (5.10.10). These deviations are collectively expressed as the osmotic coefficient ϕ_{os} defined as

$$\Pi = \phi_{os}\Pi_{ideal}. \tag{5.10.11}$$

In general, ϕ_{os} is smaller than unity, since the electrostatic correlations and inter molecular interactions contribute to reduce the osmotic pressure, in comparison with the expectation from the ideal van't Hoff's law. The value of ϕ_{os} depends on the speci ficity of the counterion (Fig. 5.28a), the degree of ionization (Fig. 5.28b) and the dielectric constant of the solvent (Fig. 5.28b). The experimentally measured ϕ_{os} for aqueous poly(styrene sulfonate) (PSS) solutions is given in Fig. 5.28a, as a function of the logarithm of monomermolal concentration (mol/kg) for three different counterions (H^+, Na^+, and Cs^+). In this figure, the molar mass of NaPSS is 4×10^4 g/mol. The role of specificity of the counterion on the deviation of osmotic pressure from the ideal value is evident in Fig. 5.28a.

The dependence of the osmotic coefficient on the degree of ionization for sodium poly(acrylic acid) (NaPAA) in water ($\epsilon \simeq 80$) and in methanol ($\epsilon \simeq 33$), and for lithium poly(acrylic acid) (LiPAA) in methanol is given in Fig. 5.28b. In general, the osmotic coefficient significantly decreases as the degree of ionization is increased reflecting the enhanced electrostatic effects at higher degrees of ionization. As the dielectric constant is decreased, the stronger counterion adsorption leads to lower ϕ_{os}.

5.10.2 Donnan Equilibrium

Corresponding to Fig. 5.27b, let us consider a specific example of each of the macromolecules in the "in" compartment to be positively charged with a net charge $Q = \alpha z_p N$, and let $c = nN/V$ be the monomer number concentration. A monovalent electrolyte is equilibrated between "in" and "out" compartments. The negatively charged counterion from the macromolecule is taken to be chemically identical to the anion of the electrolyte. The semipermeable membrane allows passage of counterions, electrolyte ions, and water, but not the macromolecules. Let c_s be the number concentration of the monovalent electrolyte ions in the "out" compartment at equilibrium.

As derived in Appendix 4, the difference in the osmotic pressure between the "in" and "out" compartments, called the Donnan pressure Π_D, is given by

$$\frac{\Pi_D}{k_B T} = \frac{c}{N} + \sqrt{\left(\frac{Qc}{N}\right)^2 + 4c_s^2} - 2c_s, \qquad (5.10.12)$$

where the activity of the electrolyte is approximated by the concentration.

In the zero-salt limit, this equation recovers the osmotic pressure of the solution given by Eq. (5.10.9) as

$$\frac{\Pi_D}{k_B T} = \frac{c}{N}(1 + Q). \qquad (5.10.13)$$

In the high salt limit, upon an expansion of the right-hand side of Eq. (5.10.12) as a series in $Qc/(2Nc_s)$, we get

$$\frac{\Pi_D}{k_B T} = \frac{c}{N}\left[1 + \frac{Q^2 c}{4Nc_s} + \cdots\right] \qquad \text{(high salt)}. \qquad (5.10.14)$$

The second term inside the brackets is the contribution to the second virial coefficient arising from the Donnan equilibrium. In addition, the second virial coefficient depends on the inter-segment excluded volume and electrostatic interactions.

For positively charged macromolecules in the "in" compartment, the amounts of cations and anions in the "in" compartment are depleted and enhanced, respectively, with respect to the amounts in the "out" compartment (see Eqs. (A4.6) and (A4.7)). For example, the relative distribution coefficient of the salt, $\Gamma_{\text{rel}} \equiv (c_s - c_+^{\text{in}})/(Qc/N)$ (where c_+^{in} is the number concentration of cations in the "in" compartment at equilibrium), follows from Eq. (A4.6) as

$$\Gamma_{\text{rel}} \equiv \frac{(c_s - c_+^{\text{in}})}{(Qc/N)} = \frac{1}{2y}\left[1 + y - \sqrt{1 + y^2}\right], \qquad (5.10.15)$$

where $y = Qc/(2Nc_s)$. In the limit of high salt concentration, $y \ll 1$, we get

$$\Gamma_{\text{rel}} \simeq \frac{1}{2}\left[1 - \frac{Qc}{4Nc_s} + \cdots\right] \qquad \text{(high salt)}. \qquad (5.10.16)$$

The experimental values of the relative Donnan equilibrium distribution coefficient Γ_{rel} at infinitely dilute aqueous DNA solutions containing the electrolyte MBr (M = Li, Na, K, or tetramethyl ammonium) are given in Fig. 5.29. The asymmetry in the

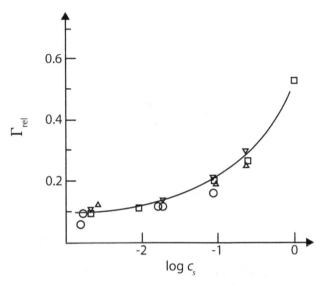

Figure 5.29 The relative Donnan equilibrium distribution coefficient Γ_{rel} at infinitely dilute aqueous DNA/MBr solutions as a function of salt concentration c_s in mol/L. M represents Li (circle), Na (square), K (up triangle), and tetramethyl ammonium (down triangle) [Adapted from Strauss *et al.* (1967)].

concentration of cations between the "out" and "in" compartments increases with the salt concentration until it reaches the asymptotic value of 0.5, in accordance with Eq. (5.10.16).

5.10.3   Donnan Membrane Potential

In this section, we present the electric potential difference generated across the membrane separating the "in" and "out" compartments (Fig. 5.27c), due to the Donnan equilibrium. As discussed in Chapter 3, the local concentration of the ith species of charge $z_i e$ at location $\mathbf{r}$ is given by the Boltzmann distribution,

$$c_i(\mathbf{r}) = c_{i0} \exp\left[-\frac{z_i e \psi(\mathbf{r})}{k_B T}\right], \qquad (5.10.17)$$

where $\psi(\mathbf{r})$ is the electric potential at $\mathbf{r}$ and c_{i0} is the concentration when $\psi(\mathbf{r}) = 0$. Assuming that the electric potential in the "in" compartment is uniformly ψ^{in} and that in the "out" compartment is uniformly ψ^{out}, the ratio of concentrations of the ith species in the "in" compartment to that in the "out" compartment follows from Eq. (5.10.17) as

$$\frac{c_i^{in}}{c_i^{out}} = \exp\left[-\frac{z_i e}{k_B T}\left(\psi^{in} - \psi^{out}\right),\right]. \qquad (5.10.18)$$

Therefore, the difference in the potential between the two compartments, $\Delta\psi = \psi^{in} - \psi^{out}$, is given by

$$\Delta\psi = \psi^{in} - \psi^{out} = \frac{k_B T}{z_i e} \ln \frac{c_i^{out}}{c_i^{in}}, \qquad (5.10.19)$$

known as the Nernst potential. For the Donnan equilibrium with monovalent electrolyte described above, we get from Eq. (5.10.18)

$$\frac{c_+^{in}}{c_S} = e^{-\tilde{\psi}}, \qquad \frac{c_-^{in}}{c_S} = e^{\tilde{\psi}}, \qquad (5.10.20)$$

where $\tilde{\psi} = \Delta\psi e/k_B T$. Substituting Eq. (5.10.20) in the electroneutrality condition, $\frac{c}{N}Q + c_+^{in} = c^{out}_-$, we get

$$\frac{c}{N}Q + c_S e^{-\tilde{\psi}} - c_S e^{\tilde{\psi}} = 0. \qquad (5.10.21)$$

For monovalent electrolytes, the solution of this quadratic equation in $e^{-\tilde{\psi}}$ yields

$$\Delta\psi = \frac{k_B T}{e} \ln \left[\frac{Qc}{2Nc_S} + \sqrt{1 + \left(\frac{Qc}{2Nc_S}\right)^2} \right]. \qquad (5.10.22)$$

The membrane potential set up by the Donnan equilibrium, given by Eq. (5.10.22), is called the Donnan membrane potential. Although the above derivation applies to ideal dilute conditions, it shows that the membrane potential can be tuned by controlling the charge and the concentration of charged macromolecules and electrolyte concentration, as given by Eq. (5.10.22).

The above results in Sections 5.10.1–5.10.3 are general for dilute solutions of charged macromolecules and depend only on the effective charge and the number concentration of the macromolecule, and salt concentration. The same results are valid if the system is made of charged brushes or charged gels, instead of simple charged macromolecules, as we shall see in Chapters 8 and 10. The architecture of the macromolecular systems affects only the higher order terms in c in Eq. (5.10.22).

5.11   Polyampholytes, Polyzwitterions, and Proteins

So far, we have described the nature and consequences of electrostatic interactions and chain connectivity in homopolymers that bear charges of only one kind. Heteropolymers (copolymers) composed of both positively charged (basic) monomers and negatively charged (acidic) monomers are referred to as polyampholytes. On the other hand, if every repeat unit of a homopolymer carries an ion-pair (zwitterion), namely a permanent dipole, then the polymer is called a polyzwitterion. The net charge of a polyzwitterion is zero. In addition to these specialized polymers, proteins constituting a very large class of naturally occurring molecular engines in various biological processes bear charges. The chemical sequence of a protein consists of charged monomers, polar monomers, and hydrophobic monomers. The confluence of the various properties of polyelectrolytes, polyampholytes, and polyzwitterions results in a broad spectrum of structural and functional properties manifest in proteins. The physics of charged macromolecules described thus far forms the basis for a fundamental understanding of the organization of these classes of macromolecules. Furthermore, the concepts and theoretical techniques developed so far are pertinent to address the class of "intrinsically disordered proteins" where the molecules remain essentially unfolded with intrinsic structural disorder, but associated with unique biologically significant functional properties.

5.11.1   Polyampholytes

Various copolymerization methods allow synthesis of polyampholytes of varying compositions of opposite charges and molecular weights. As examples, consider copolymers of 2-vinylpyridine and methacrylic acid (VP-MA) and N,N-diethylaminoethyl methacrylate and methacrylic acid (DEAEMA-MA) [Katchalsky & Miller (1954), Alfrey *et al.* (1952)]. These copolymers exist as polycations at low pH, and as polyanions at high pH. The titration curves for three copolymers of VP-MA are given in Fig. 5.30, where α is the degree of ionization of the basic pyridine group [Katchalsky & Miller (1954)]. The proportions of the methacrylic acid monomer in the monomer mixtures are 94%, 76.6%, and 55.1%, respectively, for curves 1, 2, and 3. For comparison, the titration curve for the homopolymer polyvinylpyridine (PVP) is included in the figure. At higher pH, the copolymers exhibit acid-like behavior, and at lower pH, base-like behavior is seen. At some intermediate pH value, a transition from acid-like behavior to base-like behavior occurs. The degree of ionization of the basic group at which the transition occurs depends on the composition of the copolymer. The value

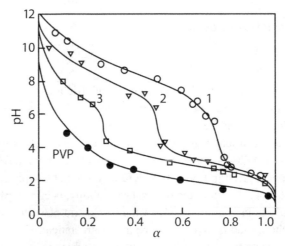

Figure 5.30 Titration curves of copolymers of 2-vinylpyridine and methacrylic acid. The methacrylic acid concentrations in the copolymers are 94% (curve 1), 76.6% (curve 2), and 55.1% (curve 3). The curve PVP denotes the polyvinylpyridine homopolymer [Adapted from Katchalsky & Miller (1954)].

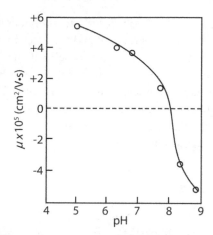

Figure 5.31 Electrophoretic mobility of a typical polyampholyte (copolymer of DEAEMA and MA) as a function of pH [Adapted from Alfrey *et al.* (1952)].

of α at the transition is higher if the content of the acidic group is higher. In fact, the sharp transition from acidic behavior at higher pH to basic behavior at lower pH is associated with the net charge of the copolymer becoming zero. The pH condition at which a charged macromolecule acquires zero net charge is called the isoelectric point. As seen in Fig. 5.30, the pH at the isoelectric point increases with the content of basic groups in the copolymer.

The occurrence of the isoelectric point can be easily monitored by measuring the electrophoretic mobility μ (discussed in Chapter 7). The velocity of a charged macromolecule $\mathbf{v}$ in a solution is proportional to a weak applied electric field $\mathbf{E}$, and the electrophoretic mobility is defined through the linear relation,

$$\mathbf{v} = \mu\mathbf{E}, \tag{5.11.1}$$

where μ is given by the Einstein law

$$\mu = \frac{Q}{\zeta}. \tag{5.11.2}$$

Here Q and ζ are the net charge and friction coefficient of the macromolecule, respectively. The electrophoretic mobility of the macromolecule must be zero at the isoelectric point, because $Q = 0$ at this condition.

The electrophoretic mobility of the copolymer N,N-diethylaminoethyl methacrylate (DEAEMA) and methacrylic acid (MA), with DEAEMA content of 45 mole percent is given in Fig. 5.31 as a function of pH [Alfrey *et al.* (1952)]. The isoelectric point, as reflected by the reversal of mobility, occurs at pH $\simeq 8.0$. Away from the isoelectric point, on either side of the isoelectric pH value, the macromolecule carries a net charge and is expected to be an electrostatically swollen coil. This is seen in Fig. 5.32, where the reduced viscosity, η_{sp}/c, which is proportional to R_g^3/M (where M is the molecular weight, Chapter 7) in the dilute limit, is plotted against pH for the same polyampholyte used in Fig. 5.31. The molecule is swollen on both sides of the isoelectric point. Polyampholytes with a strong net charge of one sign behave as conventional

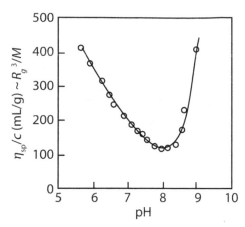

Figure 5.32 Nonmonotonic dependence of the reduced viscosity $\eta_{sp}/c(\sim R_g^3/M)$ of a polyampholyte near its isoelectric point. Away from the isoelectric point, the polyampholyte is swollen behaving like a polyelectrolyte. At the isoelectric point, the molecule shrinks into a globule. The polyampholyte is the same as in Fig. 5.31 [Adapted from Alfrey *et al.* (1952)].

polyelectrolytes with charges of one sign. The minimum of the curve occurs at the isoelectric point when the positive and negative ions would pair to form intramolecular ion pairs (dipoles). Thus, intramolecularly neutralized chains would collapse into globules due to attractive electrostatic interactions manifest as dipoles, as described in Section 5.8. The behavior of the charges inside the globule is similar to that of charges in a small volume of simple electrolytes [Higgs & Joanny (1991)]. Analogous to the Debye–Hückel theory of electrolyte solutions, an effective Debye length ξ_p can be identified as a blob inside the globule. The globule can then be treated as a compact collection of blobs of linear size ξ_p [Higgs & Joanny (1991), Gutin & Shakhnovich (1994), Dobrynin *et al.* (2004)]. Alternatively, polyampholyte globules can be treated as dense dipolar droplets (Section 5.8).

As seen in Section 5.4 (Fig. 5.9), the radius of gyration of a polyelectrolyte decreases upon addition of low molar mass salt. On the other hand, at the isoelectric point, the globule is expected to open up and expand upon addition of salt, since the ion-pair formation would be screened by the added salt. This is shown in Fig. 5.33, where η_{sp}/c (proportional to R_g^3/M) is plotted against the polyampholyte concentration c at the isoelectric point, for salt-free solutions and solutions with 1M NaCl [Ehrlich & Doty (1954)]. The polyampholyte is the same as in Figs. 5.31 and 5.32, except that the DEAEMA content is 5.45% and the isoelectric point is at pH 5.62. The chain expansion due to electrostatic screening of dipolar interactions is clearly visible in the figure, which is the opposite behavior at pH values further from the isoelectric point.

5.11.2 Polyzwitterions

Polyzwitterions are composed of repeat units having permanent charge pair of positive and negative charges (permanent dipoles). A few examples are given in Fig. 5.34.

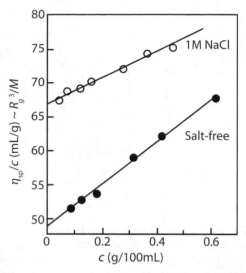

Figure 5.33 Reduced viscosity η_{sp}/c of a polyampholyte solution at its isoelectric point in the absence and presence of added salt [Ehrlich & Doty (1954)].

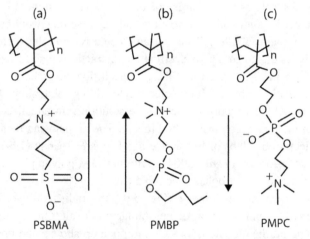

Figure 5.34 Chemical structures of polyzwitterions. (a) Poly(sulfobetaine methacrylate) (PSBMA), (b) n-butyl substituted choline phosphate (PMBP), and (c) poly(2-methacryloyloxyethyl phosphorylcholine (PMPC). The arrows denote the dipole moment vectors.

The dipoles, prescribed by the chemical structure, can be oriented toward the backbone of the polymer chain (as in poly(sulfobetaine methacrylate)) (Fig. 5.34a) and n-butyl substituted choline phosphate (Fig. 5.34b), or away from the backbone (as in poly(2-methacryloyloxyethyl phosphorylcholine) (Fig. 5.34c). In general, polyzwitterions exhibit diversely different solution behaviors compared to polyelectrolytes of one type of charge. In view of this, polyzwitterions have a wide variety of applications in the areas of cosmetics, soft contact lenses, membranes, biolubricants, etc. [Kudaibergenov *et al*. (2006), Lowe & McCormick (2002), Laschewsky (2014), Gitlin *et al*. (2006)].

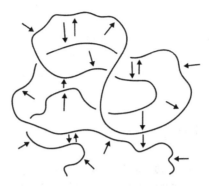

Figure 5.35 Cartoon of cooperative organization of dipoles inside a single chain. The arrows denote the zwitterion monomers of PSBMA (Fig. 5.34a) or PMBP (Fig. 5.34b). Attraction between dipoles results in globule formation. At higher polymer concentrations, many chains intermingle to form dipolar droplets.

The diverse functional properties of polyzwitterions, unprecedented in polyelectrolyte systems, originate from cooperative organization of dipoles of the repeat units of the molecules, as cartooned in Fig. 5.35. Since the dipole–dipole interactions are electrostatic in nature, assembly processes with polyzwitterions can be tuned with experimental variables that affect electrostatics such as ionic strength, pH, and dielectric constant of the solvent medium. In general, polyzwitterion behavior depends on dipole density along the chain, dipole length, chemical identities of the dipoles, orientation of dipoles with respect to the chain backbone, ion specificity of added salt, and local dielectric properties.

As we have seen in Sections 1.3 and 3.7.1, the fundamental property of randomly oriented dipoles is that the pairwise interaction energy between two dipoles is attractive (Eq. (1.3.8)) [Israelachvilli (2011), Muthukumar (1996b), Kumar & Frederickson (2009)]. Consider an isolated polyzwitterion chain of N Kuhn segments, each of segment length ℓ, in an electrolyte solution. Let each segment carry a zwitterion of dipole **p** of unit charge and let the inverse Debye length in the background solution be κ. For the sake of simplicity, let each dipole be freely rotating. The two-body excluded volume interaction among all segments arising from dipolar interactions is obtained by following the same derivation as that given in Section 2.4.

Using Eq. (1.3.8) with $\mathbf{p}_1 = \mathbf{p}_2 = \mathbf{p}$ in Eq. (2.4.1), the contribution to the excluded volume parameter from dipole–dipole intersegment interactions within a chain is written in the form of Eq. (2.4.2) as [Muthukumar (2016b), Muthukumar (2017), Adhikari *et al*. (2018), Appendix 3]

$$\frac{U_{\text{dipole–dipole}}(\mathbf{r}_{ij})}{k_B T} = v_{dd}\ell^3\delta(\mathbf{r}_{ij}), \qquad (5.11.3)$$

where

$$v_{dd} = -\frac{\pi}{9}\frac{\ell_B^2 p^4}{\ell^6}\left[4 + 8\kappa\ell + 4(\kappa\ell)^2 + (\kappa\ell)^3\right]\exp\left(-2\kappa\ell\right). \qquad (5.11.4)$$

Since the two-body excluded volume parameter from dipoles is attractive and stronger than the short-range van der Waals interactions, the chain collapses into a globule in the absence of added salt. The mean field theory for the formation of globules by polyzwitterions can be readily obtained by generalizing Eq. (2.6.1), with v_{dd} as an additional term in the excluded volume parameter v in Eq. (2.4.3). The free energy of a polyzwitterion chain with end-to-end distance R is obtained from Eqs. (2.6.1) and (5.11.3) as

$$\frac{F(R)}{k_B T} = \frac{3}{2}\left(\frac{R^2}{2N\ell^2} - 1 - \ln\frac{R^2}{N\ell^2}\right) + \frac{4}{3}\left(\frac{3}{2\pi}\right)^{3/2}\frac{(v + v_{dd})\ell^3 N^2}{R^3} + \frac{w\ell^6 N^3}{R^6}, \quad (5.11.5)$$

where v is defined in Eq. (2.4.3) and w is again the three-body repulsive excluded volume parameter necessary to stabilize the globular state (Section 2.6). For strong attractive dipolar interactions among the repeat units of the polyzwitterion molecule, the optimum coil radius follows from Eq. (2.6.3) as

$$\frac{R}{\ell} \sim \left(\frac{w}{|v + v_{dd}|}\right)^{1/3} N^{1/3}, \quad (5.11.6)$$

where v_{dd} is defined in Eq. (5.11.4). Thus, isolated polyzwitterion chains in salt-free conditions are expected to be globule-like with an effective size exponent $v = 1/3$.

In the presence of added salt, the dipole–dipole interaction becomes weak due to electrostatic screening (Eq. (5.11.4)). As a result, the chain is swollen in comparison with its globular state. For a given salt concentration, Eq. (5.11.5) should be analyzed to determine the optimum coil radius as a function of added salt by following the same procedure in obtaining Fig. 2.7b. The general result is that a polyzwitterion chain would swell from a globular state upon addition of a low molar mass salt. This feature is exactly the opposite of the behavior exhibited by a polyelectrolyte chain where the chain shrinks in size upon addition of salt. Hence, the dependence of polyzwitterion size on salt concentration is sometimes referred to as the **"anti-polyelectrolyte effect."**

The above theoretical results are observed experimentally for most polyzwitterions. As an example, the dependence of the hydrodynamic radius R_h of the polyzwitterion poly(3-dimethyl acryloyloxyethyl ammonium propiolactone (PDMAEAPL) on NaCl concentration, obtained from dynamic light scattering, is given in Fig. 5.36 [Liaw & Huang (1997)]. The chain dimension increases with an increase in salt concentration.

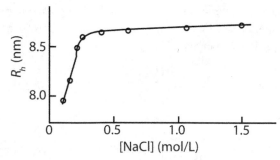

Figure 5.36 Anti-polyelectrolyte effect observed in the increase of the hydrodynamic radius R_h of poly(PDMAEAPL) with NaCl concentration. The polymer concentration is 0.1 g/dl and temperature is 30°C [Liaw & Huang (1997)].

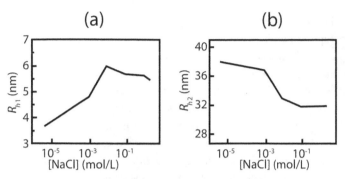

Figure 5.37 (a) Hydrodynamic radius R_{h1} of unaggregated PMBP chain increases with NaCl concentration. (b) Hydrodynamic radius R_{h2} of aggregated PMBP cluster decreases with NaCl concentration [Morozova *et al.* (2016)].

Furthermore, dynamic light scattering measurements on polyzwitterions in dilute solutions show that there are two populations of diffusive modes. One of these is identified with individual molecules and the other can be attributed to aggregates of many chains [Niu *et al.* (2000), Morozova *et al.* (2016)]. The same mechanism of stable quadrupole formation mediated by dipole–dipole attraction, sketched in Fig. 5.35, is present when several chains interpenetrate to enable chain aggregation [Muthukumar (2016b)]. The hydrodynamic radius R_{h1} discerned from the diffusive mode corresponding to unaggregated chains is presented in Fig. 3.37a as a function of NaCl concentration, where the polyzwitterion is PMBP (Fig. 5.34b). Similar to Fig. 5.36, the chain dimension increases with increasing salt concentration due to screening of dipole–dipole interactions. In addition to the screening, the individual cation and anion constituting the zwitterion can each be surrounded by its own counterion cloud arising from the dissociated salt ions. The hydrodynamic radius R_{h2}, corresponding to the aggregates, actually decreases with salt concentration as shown in Fig. 5.37b. This effect is due to the weakening of the quadrupole stability upon an increase in the concentration of added salt (see Section 6.5).

Although the experimental results of Figs. 5.36 and 5.37 are in good agreement with the above theoretical arguments, there are many unpredictable behaviors exhibited by specific polyzwitterions. As an example, the "anti-polyelectrolyte effect" is nonuniversal. To be specific, PMPC (Fig. 5.34c) is insensitive to added salt. Furthermore, it does not aggregate in contrast with the result in Fig. 5.37. A fundamental understanding of the structural organization and functions of polyzwitterions is still at its nascent stages, despite their prevalent usage in medicine, biotechnology, and the oil industry.

5.11.3 Intrinsically Disordered Proteins

A protein molecule is typically a linear heteropolymer where the repeat units are amino acid monomers. There are 20 common amino acids constituting proteins with variable composition [Creighton (1993), Nelson & Cox (2005)]. When taken alone, each amino acid in solution at physiological pH is a zwitterion, $H_3N^+–CHR\text{-}COO^-$, where R is the side group. The side group specifies the uniqueness of a particular amino acid.

The various amino acids in a protein are contiguously connected through formation of peptide bonds between adjacent monomer units. The sequence of the amino acids in a protein molecule is tabulated as the sequence of the side groups of the constituting amino acids. The number of possible sequences that can be generated from 20 different amino acids is astronomically large even for a protein molecule of just 50 repeat units. The great number of different side chains that may be present in a peptide sequence leads to immense versatility of protein molecules.

The side chains of the 20 amino acid repeat units can be broadly classified into acidic, basic, neutral, and hydrophobic. Examples of these groups are aspartic acid, lysine, glutamine, and phenylalanine, respectively. The ionization equilibria of the acidic and basic groups can be varied through pH and local dielectric environment. Therefore, the net charge and charge decoration of a protein molecule depend significantly on experimental and physiological conditions. The combination of net charge and the sequence of charged, neutral, and hydrophobic groups defines the structure and functions of every protein molecule.

In the native state, most of known proteins fold autonomously into well-defined three-dimensional compact structures. When exposed to heat or denaturing agents such as urea, the folded protein opens up into a denatured unfolded state. The conformational transition from the folded state to the unfolded state has been mapped into the globule-to-coil transition discussed in Sections 2.6 and 5.8. Based on polymer physics ideas described so far, we expect the radius of gyration R_g of a protein to depend on the number of repeat units N as $R_g \sim N^{0.6}$ in the unfolded state and as $R_g \sim N^{1/3}$ in the folded state,

$$R_g \sim \begin{cases} N^{3/5}, & \text{(unfolded state)} \\ N^{1/3} & \text{(folded state)}. \end{cases} \tag{5.11.7}$$

All foldable proteins obey these expectations as seen in Figs. 2.6b and 2.8b.

The common paradigm in protein science has been that proteins fold into well-defined three-dimensional compact conformational states with unique interfacial structural features, which are subsequently required for their biological functions [Creighton (1993), Finkelstein & Ptitsyn (2002), Nelson & Cox (2005)]. Not all proteins fold into such well-defined folded structures. Proteins, that possess an inherent resistance to folding, are known as "intrinsically disordered proteins (IDPs)." In fact, a large number of proteins can be classified as IDPs [Dyson & Wright (2005), Tompa (2010)]. Although IDPs do not fold on their own, they may fold into well-defined three-dimensional structures in the presence of binding partners [Schuler et al. (2020)]. In spite of the lack of robust folded structures of their own, IDPs are involved in a wide range of cellular functions, including cell division, signal transduction, intracellular transport, and regulation of protein assembly [Müller-Späth et al. (2010), England & Haran (2010), Das et al. (2015)]. Furthermore, many IDPs as well as misfolded proteins are associated with diseases such as neurodegenerative disorders [Uversky (2002), Uversky et al. (2008)].

(a)

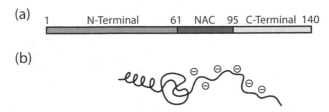

(b)

Figure 5.38 (a) Sequential blocks of amphipathic N-terminal , non-amyloid-β-component (NAC), and polyelectrolytic C-terminal. (b) Sketch of competition among globule formation, chain extension of the C-terminal due to electrostatic repulsion and immiscibility between the globular and charged regions.

The failure of IDPs to fold autonomously is attributed to the composition and sequence of charged and hydrophobic repeat units. IDPs in general are characterized by fewer hydrophobic groups and enrichment of polar and charged amino acid residues. This characteristic of IDPs results in a more dominant role of conformational entropy in comparison with intramolecular van der Waals attractions among the hydrophobic groups. As a consequence, an IDP molecule can adopt many expanded conformational states instead of settling into a single well-defined folded conformational state.

A typical example of IDPs is α-synuclein [Ray *et al.* (2020), Das & Muthukumar (2022)]. As sketched in Fig. 5.38a, the amino acid sequence of α-synuclein (containing 140 residues) can be grouped into three blocks: (i) amphipathic N-terminal region of 60 residues made up of 12 positive charges and 7 negative charges, with the rest being polar or hydrophobic, (ii) hydrophobic non-amyloid-β-component (NAC) region of 35 residues, which mostly bear hydrophobic side groups, and (iii) polyelectrolyte-like C-terminal region of 45 residues with 14 negative charges and 3 positive charges. We expect from the sequences and charges in these three blocks that (i) N-terminal region can form secondary structures dominated by dipolar interactions and hydrogen bonding, (ii) the NAC region would form a globule, and (iii) the C-terminal region is essentially a polyelectrolyte chain with electrostatic repulsion among its residues trying to keep the chain unfolded as sketched in Fig. 5.38b.

Broadly speaking, the amino acid sequences of IDPs are such that the macromolecule contains domains which can easily fold or form globules and domains which remain in the expanded conformational state. This chimeric feature enables IDPs not to fold completely. To model these effects, let f_+ be the fraction of positively charged residues in one molecule, namely the ratio of total number of positively charged residues to the total number of residues in the chain. Similarly, let f_- be the fraction of negatively charged residues. In addition to ionic charges, there can be additional factors arising from hydrogen-bonding and $\pi - \pi$ stacking. Focusing on only the ionic charges on the residues, the net charge fraction $|f_+ - f_-|$, the charge fractions f_+ and f_-, and the degree of hydrophobicity $(1 - f_+ - f_-)$ determine the extent of unfolding in the native state of an IDP. If $|f_+ - f_-|$ is large, the protein behaves like a polyelectrolyte (PE), as indicated in the state-diagram of Fig. 5.39a. When both f_+ and f_- are small, hydrophobicity dominates and the protein would adopt globule-like structure (region

(a)   (b)

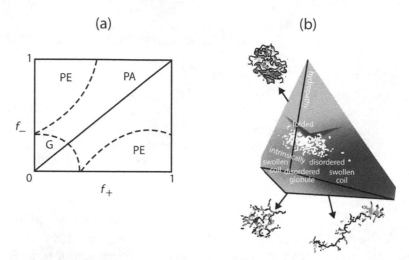

Figure 5.39 (a) Conformational state diagram based on the composition of charged and hydrophobic residues in an IDP. The states of globule, electrostatically swollen polyelectrolyte chain, and polyampholyte chain are denoted as G, PE, and PA, respectively.
(b) Three-dimensional plot of composition–conformation relation in terms of f_+, f_-, and hydropathy. Depending on the relative contributions of polyampholyte effect and hydrophobic effect, disordered globules, intrinsically disordered conformations, and folded states can form. The dots correspond to various IDPs [Adapted from Mao *et al.* (2010)].

G in Fig. 5.39a). If f_+ and f_- are large, but the net charge fraction $|f_+ - f_-|$ is zero, then the protein sequence is similar to a polyampholyte. This situation is represented by the diagonal line in Fig. 5.39a. Even when $|f_+ - f_-|$ is not zero, but small with larger values of f_+ and f_-, the chain would behave like a polyampholyte (marked as region PA in Fig. 5.39a).

The details of the globule-like state in regions G and PA depend on the relative contributions from the hydrophobicity and polyampholyte effects [Mao *et al.* (2010), Hofmann *et al.* (2012), Das & Pappu (2013), van der Lee *et al.* (2014), Schuler *et al.* (2016)]. If the polyampholyte effect is weak, then the globule state is the folded state. For stronger polyampholyte effects (high f_+ and f_-, but $|f_+ - f_-| \simeq 0$), the preferred state is a disordered globule. For an intermediate strength of the polyampholyte effect, the preferred states are intrinsically disordered conformations. These expectations along the PA line in Fig. 5.39a are illustrated in Fig. 5.39b, where the hydropathy index is a measure of the net hydrophobicity of the protein sequence. As seen in Fig. 5.39b, the protein adopts folded conformations for strong hydrophobicity. As the value of the hydropathy index decreases, the chain becomes intrinsically disordered and eventually the resultant conformation is a disordered globular state. The dots in Fig. 5.39b denote various sequences of IDPs [Mao *et al.* (2010), van der Lee *et al.* (2014), Schuler *et al.* (2016)].

The boundaries of the domains sketched in Fig. 5.39 are only sketches. The location of these conformational state boundaries, based only on the composition of IDPs, can be calculated using the results derived in the previous sections. Let f_+, f_-, and f_d denote, respectively, the fractions of the positively charged residues, negatively charged residues, and dipolar residues (such as the charged residues binding with their counterions) in an IDP. As a result of this composition, there are charge–charge, charge–dipole, and dipole–dipole interactions among the various amino acid residues. In addition, there are van der Waals interactions among all residues collectively classified as the hydrophobic excluded volume effect. The discussion of the coil–globule transition in Section 2.6 can be generalized to derive the composition–conformation relation for IDPs by accounting for these interactions. As an illustration, consider a dilute solution of an IDP with finite Debye length κ^{-1} due to presence of counterions and added salt. The excluded volume parameter v appearing in Eq. (2.6.1) becomes

$$v \to v_{\text{IDP}} = v + (f_+ - f_-)^2 \, v_{cc} + 2 \, (f_+ + f_-) \, f_d v_{cd} + f_d^2 v_{dd}, \qquad (5.11.8)$$

where v is the average van der Waals type excluded volume parameter between two residues mediated by solvent including the hydrophobicity of the side groups. v_{cc}, v_{cd}, and v_{dd} are parameters representing the charge–charge, charge–dipole, and dipole–dipole inter-segment interactions, respectively. Within the premise of the Debye–Hückel theory, the charge–charge interaction parameter for segments with unit charge follows from Eqs. (5.3.11)–(5.3.15) as

$$v_{cc} = \frac{4\pi \ell_B}{\kappa^2}. \qquad (5.11.9)$$

As shown in Section 5.4, this form is adequate for treating dilute solutions of macromolecules even for salt concentrations as low as 0.001 M monovalent salt. If it is necessary to consider lower salt concentrations, the full expression in Eq. (5.6.28) must be used. The interaction parameter v_{cd} for a monovalent charge and a randomly rotating dipole $\mathbf{p}$ is given within the Debye–Hückel theory as (Appendix 3)

$$v_{cd} = -\frac{\pi}{3} \frac{\ell_B^2 p^2}{\ell^4} (2 + \kappa\ell) \exp(-2\kappa\ell). \qquad (5.11.10)$$

The interaction parameter for a pair of randomly oriented dipoles $\mathbf{p}_1$ and $\mathbf{p}_2$ is given by (Eq. (5.11.4)) (Appendix 3)

$$v_{dd} = -\frac{\pi}{9} \frac{\ell_B^2 p_1^2 p_2^2}{\ell^6} \left[4 + 8\kappa\ell + 4(\kappa\ell)^2 + (\kappa\ell)^3 \right] \exp(-2\kappa\ell). \qquad (5.11.11)$$

Using Eq. (2.6.2), the radial size of the protein is given by

$$\left(\frac{R^2}{N\ell^2} \right)^{5/2} - \left(\frac{R^2}{N\ell^2} \right)^{3/2} = \frac{4}{3} \left(\frac{3}{2\pi} \right)^{3/2} v_{\text{IDP}} \sqrt{N} + 2w \left(\frac{N\ell^2}{R^2} \right)^{3/2}, \qquad (5.11.12)$$

where v_{IDP} is given by Eq. (5.11.8). Using the same procedure employed in getting Fig. 2.7b, the preferred conformation of an IDP can be calculated in terms of f_+, f_-, and f_d, based only on its composition without any regard to its sequence.

Although analytical calculations along the above line of mean field theory show that conformational states can be directly correlated with the composition of some IDPs, the sequence of charges and hydrophobic groups plays significant additional role [Muthukumar (1996b), Srivastava & Muthukumar (1996)]. For example, even modest alterations of charge placement in the primary sequence, while maintaining the same charge composition, can lead to significant changes in conformation

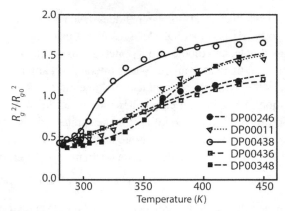

Figure 5.40 Temperature dependence of the square of the chain expansion factor R_g^2/R_{g0}^2 for five sequences from DisProt database. The symbols are from simulations [Das & Pappu (2013)] and the lines are from analytical calculations [Firman & Ghosh (2018)].

[Sawle & Ghosh (2015), Firman & Ghosh (2018), Huihui *et al.* (2018), Huihui & Ghosh (2020)]. Sometimes, the change can be as drastic as a coil–globule transition and vice versa. As an example, the temperature dependence of the expansion factor R_g^2/R_{g0}^2 (where R_{g0} is the radius of gyration under Flory theta conditions) is presented in Fig. 5.40 for five different IDP sequences, with sequence labels from the DisProt database. The symbols are from Monte Carlo simulations [Das & Pappu (2013)] and the lines are from an analytical theory [Firman & Ghosh (2018)] based on the discussion in Chapter 2 and Eqs. (5.3.4) and (5.3.5). It is obvious from Fig. 5.40 that the sequence of charges and hydrophobic groups in IDPs plays a significant role and that the sequence-conformation relation can be quite different from the composition–conformation relation depicted in Fig. 5.39. The role of sequence of the protein must be taken into account for an adequate understanding of IDPs [Firman & Ghosh (2018), Samanta *et al.* (2018)]. A quantitative understanding of sequence-conformation relations for IDPs is only at its early stage of development.

6 Structure and Thermodynamics in Homogeneous Polyelectrolyte Solutions

6.1 Introduction

Every charged macromolecule senses every other charged macromolecule in their solution due to long-ranged electrostatic correlations, even when they are far separated in space. Owing to the combined effects from topological and electrostatic cooperativity, solutions of charged macromolecules exhibit spontaneous formation of certain structures even in thermodynamically stable "homogeneous" single phases. For example, consider a dilute solution of charged macromolecules, such as a solution of folded proteins (Fig. 6.1a). The molecules position themselves at some preferred intermolecular distance Λ in order to optimize the electrostatic interactions between them. When the concentration of solutions of relatively rigid molecules or charged colloidal particles is increased, Λ decreases due to more crowding. This crowding is resisted by electrostatic repulsion between the molecules. Upon a further increase in concentration, Λ reaches a lower limit for a stable homogeneous liquid state, at which point the solution undergoes a phase transition into a stable ordered solid state.

On the other hand, if the macromolecules are flexible polyelectrolyte chains or intrinsically disordered proteins, the chains can interpenetrate upon an increase in concentration c, due to a more favorable entropy gain over mutual electrostatic repulsion. A measure of the condition for interpenetration between chains is the overlap concentration $c^\star$, defined as the concentration at which chains of radius of gyration R_g are imagined to be just touching each other. This situation can be depicted as a hypothetical regular array of spheres of radius R_g, as sketched in Fig. 6.1b [de Gennes (1979)]. In terms of the monomer concentration of the polymer in a solution containing n chains of N monomers per chain, the overlap concentration is generally given by

$$c^\star = \frac{nN}{\left(\frac{4}{3}\pi n R_g^3\right)}, \tag{6.1.1}$$

where the volume of the solution at the overlap concentration is taken as the volume of all n chains, $n(4\pi R_g^3/3)$. In view of the definition of the size exponent ν, $R_g \sim N^\nu$, we get from Eq. (6.1.1),

$$c^\star \sim \frac{N}{R_g^3} \sim N^{1-3\nu}. \tag{6.1.2}$$

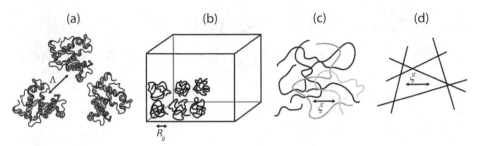

Figure 6.1 (a) Preferred inter-molecule distance Λ in dilute solutions of charged macromolecules, illustrated with folded proteins such as myoglobin. (b) Sketch of imagined arrangement of molecules at the overlap concentration $c^\star$. (c) Flexible polyelectrolyte chains interpenetrate at concentrations above $c^\star$. A characteristic correlation length ξ for monomer concentration fluctuations emerges, which depends on concentration, temperature, and solvent quality. (d) Interpenetration of rod-like molecules arising from gain in orientational entropy.

When $c < c^\star$, we call the solution dilute. For $c > c^\star$, the solution is semidilute or concentrated depending on the value of c. Note that $c^\star$ is only a geometrical estimate and is a rough measure of the onset of interpenetration of chains.

At polymer concentrations above the overlap concentration, chains interpenetrate into each other as depicted in Fig. 6.1c. These interpenetrating structures, albeit disordered and liquid-like, exhibit a certain correlation length ξ for monomer concentration fluctuations. Such mesh-like structures can also emerge for rigid rod-like macromolecules such as dsDNA (Fig. 6.1d). For rod-like molecules, their interpenetration is driven by gain in orientational entropy, in contrast with the conformational entropy in the case of flexible macromolecules.

The intermingling of chains into each other modifies the nature of both the topological correlations (arising from chain connectivity) and electrostatic correlations in a self-consistently coupled manner. These correlations are manifest in the scattering properties of polyelectrolyte solutions using light, X-ray, and neutrons [Förster & Schmidt (1995), Muthukumar (2017)]. For salt-free polyelectrolyte solutions, the most characteristic feature of the dependence of scattering intensity $I(\mathbf{k})$ on scattering wave vector $\mathbf{k}$ is the presence of a scattering peak, known as the "polyelectrolyte peak," at k_m (Fig. 6.2a). The peak position depends on the polyelectrolyte concentration c and this dependence has been cast empirically as [Nierlich *et al.* (1979), Kaji *et al.* (1988), Johner *et al.* (1994), Nishida *et al.* (2001, 2002)]

$$k_m \sim c^\beta. \tag{6.1.3}$$

The value of β changes from 1/3 for extremely dilute solutions ($c < c^\star$) to 1/2 for dilute and semidilute solutions ($c > c^\star$). For extremely high polymer concentrations, $\beta \sim 1/4$ over a narrow range of c, close to the hydrated limit of the polyelectrolyte salt.

Another remarkable feature of the scattering intensity from salt-free polyelectrolyte solutions is that the intensity at near-zero scattering angles is enormous, which is indicative of the presence of very large scattering objects in the solution [Förster *et al.* (1990)]. The enhanced intensity at near-zero scattering angles always accompanies

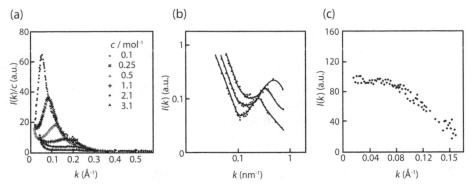

Figure 6.2 (a) Concentration dependence of the polyelectrolyte peak in salt-free solutions of sodium poly(styrene sulfonate). $I(k)$ is the X-ray scattering intensity. Polymer concentrations are denoted by various symbols in the figure [Nishida *et al.* (2001)]. (b) Enhanced zero-angle scattering intensity accompanying the polyelectrolyte peak. $I(k)$ is scattering intensity in SANS from aqueous solutions of quaternized poly(2-vinyl pyridine) in salt-free conditions. Polymer concentrations are 25 mg/L (upper curve), 10 g/L (middle curve), and 5 g/L (lower curve) [Förster *et al.* (1990)]. (c) Absence of polyelectrolyte peak and enhanced zero-angle scattering in the presence of added salt. $I(k)$ is the scattering intensity in SANS on aqueous solutions of sodium poly(styrene sulfonate) with 0.25 mol/L polymer concentration and 0.172 mol/L ionic strength [Nishida *et al.* (2002)].

the polyelectrolyte peak (Fig. 6.2b). In the presence of sufficient amounts of low molar mass salts, both the polyelectrolyte peak and the zero-angle enhanced intensity disappear, and the scattering behavior is analogous to that of a solution of uncharged polymers (Fig. 6.2c). It is surprising that similarly charged macromolecules clump together (responsible for the enhanced zero-angle scattering) in a homogeneous solution under conditions where electrostatic repulsion is maximal, and that such clumps become unstable when the repulsion is reduced by screening the electrostatic interaction with added salt.

In general, the collective behavior of charged macromolecules exhibits many puzzles and counterintuitive results unseen in uncharged systems [Förster & Schmidt (1995)]. Development of necessary concepts to understand the various phenomenological results of homogeneous solutions of charged macromolecules is the primary focus of this chapter. In addressing this goal, the rest of the chapter is organized into four sections:

(1) *Basic thermodynamic and structural aspects of solutions of uncharged macromolecules:* This premise is the starting point for considering polyelectrolyte solutions, and the results are pertinent to charged solutions at high concentrations of added salt. We will discuss a mean field theory (Flory–Huggins theory), concentration fluctuations, excluded volume screening (Edwards screening), and scaling laws.

(2) *Coupling between electrostatic interactions and topological correlations in polyelectrolyte solutions:* The Debye screening of electrostatic interactions and the Edwards screening of excluded volume interactions are intimately coupled. A

self-consistent treatment of these two screenings, called the double screening theory, and its predictions will be described in the context of extensive experimental results.

(3) *Five regimes of polymer concentration:* Various structural organizations in polyelectrolyte solutions are described in terms of five polymer concentration regimes.

(4) *Spontaneous formation of large aggregates from similarly charged macromolecules:* Assembly of such structures, which is responsible for the enhanced zero-angle scattering, will be addressed.

6.2   Solutions of Uncharged Macromolecules

In this section we summarize theoretical concepts and models to treat homogeneous solutions of uncharged macromolecules, which are also applicable to solutions of charged macromolecules in the high-salt limit. We begin with a description of the Flory–Huggins mean field theory, followed by the Edwards theory of excluded volume screening, and finally we present scaling laws for various structural and thermodynamic quantities. The primary purpose of this section is to derive expressions for the Helmholtz free energy and thermodynamic quantities such as the osmotic pressure, correlation functions for concentration fluctuations, and size of labeled macromolecules in semidilute and concentrated solutions as functions of polymer concentration.

6.2.1   Flory–Huggins Theory

A homogeneous solution of an uncharged polymer in a nonpolar solvent can become unstable under certain experimental conditions, exhibiting the critical phenomenon of liquid–liquid phase separation, as will be described in Chapter 8. The boundary delineating stable homogeneous phase behavior of polymer solutions is predicted by a simple lattice theory, introduced independently by Flory and Huggins [Flory (1942), Huggins (1942a, b, c)], which is equivalent to the Bragg–Williams theory of small molecular mixtures or metallic alloys [Bragg & Williams (1934, 1935)].The assumption and predictions of the Flory–Huggins theory are well described in textbooks on polymers [Flory (1953a), Boyd & Phillips (1993), Rubinstein & Colby (2003), Heimenz & Lodge (2007)]. Therefore, only a brief outline of the theory is presented below, with details relegated to Appendix 5.

Consider a randomly mixed solution consisting of n_1 solvent molecules and n_2 polymer chains, with each chain made of N_2 segments. Taking the whole space of the solution to be discretized into a lattice, let the volume of the mixture be divided into n_0 sites of equal volume. For simplicity, we assume that the volume of each site is the same as that of a solvent molecule and of a polymer segment, as sketched in Fig. 6.3. Assuming that the system is incompressible, the total number of sites n_0 is $(n_1 + n_2 N_2)$ and the volume fraction of the solvent and polymer are given, respectively, by ϕ_1 and ϕ_2,

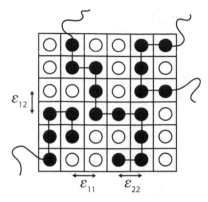

Figure 6.3 Lattice model of the Flory–Huggins theory. Filled circles denote polymer segments and open circles denote solvent molecules. The nearest neighbor interaction energies are $\epsilon_{11}, \epsilon_{12}$, and ϵ_{22}, respectively, for solvent–solvent, solvent–polymer, and polymer–polymer interactions.

$$\phi_1 = \frac{n_1}{(n_1 + n_2 N_2)}, \qquad \phi_2 = \frac{n_2 N_2}{(n_1 + n_2 N_2)}. \qquad (6.2.1)$$

In order to assess the feasibility of formation of such a mixture, we need to know the free energy of mixing ΔF_m associated with the formation of the mixture from its pure components. The necessary condition for the mixture to be homogeneous is $\Delta F_m < 0$. Otherwise, the mixing process is unfavorable and either instability or metastability will emerge in the solution. The free energy of mixing is given by

$$\Delta F_m = \Delta E_m - T \Delta S_m, \qquad (6.2.2)$$

where ΔE_m and ΔS_m are, respectively, the energy of mixing and entropy of mixing, and T is the absolute temperature.

Assuming a random distribution of polymer chains and solvent molecules in the lattice (Fig. 6.3), expressions for ΔS_m and ΔE_m can be easily derived. With a completely random distribution of polymer segments and solvent molecules, and without consideration of topological correlations due to chain connectivity, ΔS_m is given as (Appendix 5)

$$\frac{\Delta S_m}{k_B} = -n_1 \ln \phi_1 - n_2 \ln \phi_2, \qquad (6.2.3)$$

where ϕ_1 and ϕ_2 are given in Eq. (6.2.1). Since ϕ_1 and ϕ_2 are less than unity in a mixture, ΔS_m is always positive. Consequently, the entropic contribution to the free energy of mixing, $-T\Delta S_m$, is always negative favoring the mixing process.

The energy of mixing is calculated as follows: Let ϵ_{11} be the interaction energy between two solvent molecules in contact with each other by occupying two adjacent sites on the lattice, as denoted in Fig. 6.3. Similarly let ϵ_{12} and ϵ_{22} denote the nearest neighbor interaction energies for a pair of solvent molecule and polymer segment, and a pair of polymer segments, respectively. Assuming random mixing of solvent and polymer, the energy of mixing is given by (Appendix 5)

$$\frac{\Delta E_m}{k_B T} = n_0 \chi \phi_1 \phi_2, \tag{6.2.4}$$

where χ is known as the Flory–Huggins "chi" parameter defined as

$$\chi = \frac{z}{k_B T} \left[\epsilon_{12} - \frac{1}{2} \left(\epsilon_{11} + \epsilon_{22} \right) \right]. \tag{6.2.5}$$

Here, z is the coordination number of the lattice. Since the lattice is merely a representation of the continuous space of the system, z is an effective coordination number of order unity. The main feature of the Flory–Huggins parameter is that it is a unique combination of three parameters $\epsilon_{11}, \epsilon_{12}$, and ϵ_{22}, which in turn depend on the specific chemical details of the solvent and polymer. The χ parameter is therefore a measure of the "chemical mismatch" between the different components of the mixture. As we have already seen in Eq. (2.4.1), the Flory–Huggins χ parameter is related to the two-body excluded volume parameter v denoting the second virial coefficient among two polymer segments arising from polymer–polymer, polymer–solvent, and solvent–solvent interactions.

If polymer has a more favorable van der Waals attraction with its solvent in comparison to those for solvent–solvent and polymer–polymer contacts, that is ϵ_{12} is more negative than ϵ_{11} and ϵ_{22}, then χ is negative. Therefore, Eq. (6.2.5) shows that $\Delta E_m < 0$ and hence $\Delta F_m < 0$, since $-T\Delta S_m$ is always negative. As a result, the mixture is a completely miscible solution for $\chi < 0$. On the other hand, if the van der Waals interaction energies for polymer–polymer and solvent–solvent contacts are more attractive in comparison to polymer–solvent contacts, such that $(\epsilon_{11} + \epsilon_{22}) < 2\epsilon_{12}$, then χ is positive. Therefore, depending on the degree of chemical mismatch among the two components, ΔE_m can be either positive or negative. If $|\Delta E_m| < |-T\Delta S_m|$, then $\Delta F_m < 0$ and the mixture is homogeneous. On the other hand, if $|\Delta E_m| > |-T\Delta S_m|$, then $\Delta F_m > 0$ and the mixture is either metastable or unstable resulting in liquid–liquid phase separation.

Substituting Eqs. (6.2.3) and (6.2.4) in Eq. (6.2.2), we get the Flory–Huggins free energy of mixing for the polymer solution as

$$\frac{\Delta F_m}{k_B T} = n_1 \ln \phi_1 + n_2 \ln \phi_2 + \chi n_0 \phi_1 \phi_2. \tag{6.2.6}$$

When the number of segments N_2 per chain is one, the Flory–Huggins theory reduces to the Bragg–Williams theory for metallic alloys and small molecular mixtures. Since the Flory–Huggins theory assumes random mixing of the components in the system without any regard to chain connectivity, the contribution from topological correlations of polymer chains is ignored in this mean field theory.

The key thermodynamic function that relates to experimentally measurable quantities, such as the osmotic pressure of a solution and phase separation, is the chemical potential of the various components in the system. In general, the chemical potentials of the two components (μ_1 and μ_2) of a solution, containing n_1 molecules of component 1 with volume fraction ϕ_1 and n_2 molecules of component 2 with volume fraction ϕ_2, are given by the thermodynamic equations

$$\mu_1(\phi_2) - \mu_1(0) = \left(\frac{\partial \Delta F_m}{\partial n_1}\right)_{n_2, T}$$

$$\mu_2(\phi_2) - \mu_2(1) = \left(\frac{\partial \Delta F_m}{\partial n_2}\right)_{n_1, T}. \qquad (6.2.7)$$

Here $\mu_1(0)$ and $\mu_2(1)$ are $\mu_1(\phi_2 = 0)$ and $\mu_2(\phi_2 = 1)$ of pure components 1 and 2, respectively. Substituting Eq. (6.2.1) for a solution of n_2 polymer chains, each with N segments, and n_1 solvent molecules into Eq. (6.2.6), we get the chemical potentials of the solvent and polymer from Eq. (6.2.7) as

$$\frac{\mu_1(\phi_2) - \mu_1(0)}{k_B T} = \ln(1 - \phi_2) + \left(1 - \frac{1}{N}\right)\phi_2 + \chi\phi_2^2, \qquad (6.2.8)$$

and

$$\frac{\mu_2(\phi_2) - \mu_2(1)}{k_B T} = \ln\phi_2 + (1 - N)(1 - \phi_2) + \chi N(1 - \phi_2)^2. \qquad (6.2.9)$$

The chemical potential of the solvent is related to the osmotic pressure Π of the solution as

$$\frac{\Pi v_1}{k_B T} = -\frac{\mu_1(\phi_2) - \mu_1(0)}{k_B T}, \qquad (6.2.10)$$

where v_1 is the volume of a solvent molecule, namely, volume per site in the lattice model of the Flory–Huggins theory. From Eqs. (6.2.8) and (6.2.10), we get

$$\frac{\Pi v_1}{k_B T} = -\ln(1 - \phi_2) - \left(1 - \frac{1}{N}\right)\phi_2 - \chi\phi_2^2. \qquad (6.2.11)$$

For very dilute polymer solutions such that $\phi_2 \ll 1$, the osmotic pressure follows from Eq. (6.2.11) by expanding the logarithmic term as a virial series in ϕ_2, as

$$\frac{\Pi v_1}{k_B T} = \frac{\phi_2}{N} + \frac{1}{2}(1 - 2\chi)\phi_2^2 + O(\phi_2^3). \qquad (6.2.12)$$

Expressing the volume fraction $\phi_2 = nNv_1/V$ in terms of the polymer concentration $c = nN/V$, we get Eq. (5.10.6),

$$\frac{\Pi}{k_B T} = \frac{c}{N} + A_2 c^2 + O(c^3), \qquad (6.2.13)$$

where the second virial coefficient A_2 is

$$A_2 = \frac{1}{2}(1 - 2\chi)v_1. \qquad (6.2.14)$$

Note that the difference between M and N in Eqs. (5.10.6) and (6.2.13) is due to different definitions of the concentration c. According to the above equation, when $\chi = 1/2$, the second virial coefficient vanishes and the solution behaves like an ideal solution obeying the van't Hoff law. The combined condition of temperature and chemical mismatch at which $\chi = 1/2$ is called the "ideal" condition or the Flory condition, and the polymer solution is known as an "ideal solution." Since χ is inversely proportional to the temperature according to Eq. (6.2.5), $(1 - 2\chi)$ can be written as

$$1 - 2\chi = \left(1 - \frac{\Theta}{T}\right), \qquad (6.2.15)$$

where Θ depends on the chemical details of the two components, as evident from Eq. (6.2.5). At the temperature $T = \Theta$, A_2 is zero, and so Θ is known as the Flory temperature, or theta temperature, or simply the ideal temperature for a particular polymer–solvent system.

Although the polymer solution is ideal at $\chi = 1/2$, the functional form of A_2 in dilute solutions given in Eq. (6.2.14) is incorrect and disagrees with the rigorous result $A_2 \sim R_g^3/N^2$ described in Section 5.10. For example, in good solutions ($\chi < 1/2$, or equivalently the excluded volume parameter $v > 0$), $R_g \sim (1 - 2\chi)^{1/5}N^{3/5}$ so that

$$A_2 \sim (1 - 2\chi)^{3/5} N^{-1/5}, \qquad (6.2.16)$$

which is qualitatively different from the N-independent result $A_2 \sim (1 - 2\chi)$ given by the Flory–Huggins theory. This discrepancy is due to the assumption in the Flory–Huggins theory that the polymer segments and solvent molecules are randomly and uniformly distributed throughout the volume of the system, without any regard to chain connectivity. This assumption is unphysical for dilute polymer solutions, where polymer chains are sparse and well-separated in the solution, but the chain segments topologically correlated within each chain. Furthermore, fluctuations in the local polymer concentration are ignored in the Flory–Huggins mean field theory. In view of these features, deviations from the predictions of the Flory–Huggins theory can be considerable at very low polymer concentrations and semidilute conditions where concentration fluctuations are strong (Section 6.2.2).

6.2.2   Concentration Fluctuations

The topological correlations arising from chain connectivity and the inevitable fluctuations in the local polymer concentration are not addressed in the Flory–Huggins theory described above and in Appendix 5. We shall address these effects using the formalism of field theory first introduced by Edwards for polymers [Edwards (1966)]. Since the details of this theory are well documented in books [Doi & Edwards (1986), des Cloizeaux & Jannink (1990), Fujita (1990)], we present only the salient aspects of the theory.

Consider a collection of n chains, each of N segments, in volume V (Fig. 6.4a). Let the monomer number concentration $c = nN/V$ be higher than the overlap concentration $c^\star$ (defined in Eq. (6.1.1)), so that the chains fully intermingle throughout the volume of the solution. Any pair of segments, both intra-chain and interchain, undergo excluded volume interaction $v(\mathbf{r})$ parametrized as a short-ranged pseudo potential (Section 2.4),

$$v(\mathbf{r}) = v\ell^3\delta(\mathbf{r}), \qquad (6.2.17)$$

(a)   (b)   (c)

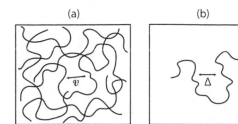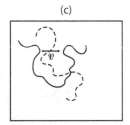

Figure 6.4 (a) Schematic of n coupled chains with intersegment interaction $v(\mathbf{r})$. (b) Generation of a field variable decouples the chains into n uncoupled effective chains with an effective interaction $\Delta(\mathbf{r})$ between any two segments. (c) Interpenetration of chains results in screening of the bare excluded volume interaction $v(\mathbf{r})$ into the screened interaction $\Delta(\mathbf{r})$.

where $\mathbf{r}$ is the distance vector between the interacting segments, and $\delta(\mathbf{r})$ is the Dirac delta function. By accounting for polymer–polymer, polymer–solvent, and solvent–solvent interactions, v is the excluded volume parameter defined in Eq. (2.4.1) and is related to the Flory–Huggins χ parameter as $v = (1 - 2\chi)$.

Using Eq. (2.4.6) for n chains, the contribution F_p from polymer chains to the Helmholtz free energy of the solution is given as

$$
e^{-\frac{F_p}{k_B T}} = \frac{1}{n!} \int \prod_{\alpha=1}^{n} \mathcal{D}[\mathbf{R}_\alpha(s_\alpha)] \exp\left\{ -\frac{3}{2\ell^2} \sum_{\alpha=1}^{n} \int_0^N ds_\alpha \left(\frac{\partial \mathbf{R}_\alpha(s_\alpha)}{\partial s_\alpha} \right)^2 \right.
$$
$$
\left. -\frac{v\ell^3}{2} \sum_{\alpha=1}^{n} \sum_{\beta=1}^{n} \int_0^N ds_\alpha \int_0^N ds_\beta \delta\left[\mathbf{R}_\alpha(s_\alpha) - \mathbf{R}_\beta(s_\beta) \right] \right\}, \quad (6.2.18)
$$

where $\mathbf{R}_\alpha(s_\alpha)$ is the position vector of the arc length variable s_α ($0 \le s_\alpha \le N$) of the αth chain, in the path integral representation of polymer chains.

In general, an interaction between two segments can be treated exactly as one segment present in the field created by the other segment. Performing this step enables the coupled interactions (v) between all chains in Eq. (6.2.18) to be written in terms of n uncoupled effective chains where the segments of each chain experience forces from the field Φ around them. In the presence of the field $\Phi[\mathbf{R}(s)]$ acting on the sth segment, the probability distribution function for each effective chain is written exactly, as a generalization of Eq. (2.1.7), as

$$
\left(\frac{\partial}{\partial N} - \frac{\ell^2}{6} \nabla^2 - \Phi(\mathbf{r}) \right) G(\mathbf{r}, \mathbf{r}'; N; [\Phi]) = \delta(N)\delta(\mathbf{r} - \mathbf{r}'), \quad (6.2.19)
$$

where $\mathbf{r}$ and $\mathbf{r}'$ are the position vectors of the two ends of the chain. The field variable $\Phi(\mathbf{r})$ in this equation is a function of the probability distribution function G and hence the above equation needs to be solved self-consistently to get G. This is the basis of the self-consistent field theory used in solving various polymer problems, which is usually performed numerically. In analytical calculations, G is expanded in Φ as a series and the various terms are represented as polymer diagrams which are integrals. Examples of polymer diagrams are given in Fig. 6.5. In the one-loop diagram (Fig. 6.5a), the segments s and s' of a chain interact with each other, mediated by the background field generated by all other chains. This interaction is further correlated by additional

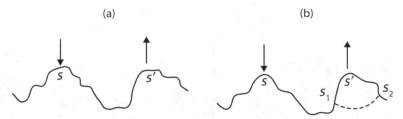

Figure 6.5 Polymer diagrams. (a) One-loop diagram used in the random phase approximation. The segments s and s' interact with each other, mediated by the potential field Φ (arrows) generated by all other chains. (b) Higher order correlations appear as vertex diagrams involving additional segments s_1 and s_2.

segments participating in the pairwise interactions, as depicted in the "vertex diagram" of Fig. 6.5b. After averaging such diagrams over chain conformations, the expanded series is summed back and then further averaged over the Φ-field.

The net effect of the field on the intersegment interactions in the effective chain is to transform $v(\mathbf{r})$ of Eq. (6.2.17) to $\Delta(\mathbf{r})$, as shown in Fig. 6.4. One of the key results of the Edwards field theory of polymer solutions is that the bare excluded volume interaction is screened as the polymer concentration is increased, due to intervention of other interpenetrating chains (Fig. 6.4c). In general, for solutions of uncharged polymers, $\Delta(\mathbf{r})$ is of the form

$$\Delta(\mathbf{r}) = v\ell^3 \left[\delta(\mathbf{r}) - \frac{1}{4\pi\xi^2} \frac{e^{-r/\xi}}{r} \right], \tag{6.2.20}$$

where ξ is the screening length for excluded volume interactions. The above equation shows that the bare excluded volume interaction $v\ell^3 \delta(\mathbf{r})$ between a pair of segments of every chain is screened by interpenetration by other chains. As we shall see below, ξ is also the correlation length for polymer concentration fluctuations. It is inversely related to the polymer concentration. At low polymer concentrations as in dilute solutions, ξ is very large so that $\Delta(\mathbf{r}) = v\ell^3 \delta(\mathbf{r})$, since the second term on the right-hand side of Eq. (6.2.20) vanishes. At very high polymer concentrations, ξ is small and the second term is one of the representations of the Dirac delta function, so that $\Delta(\mathbf{r})$ is essentially zero. Therefore, the excluded volume interaction is fully screened in concentrated solutions. As a result, chains obey Gaussian chain statistics ($R_g \sim \sqrt{N}$) at these concentrations, although the polymer solution is composed of a good solvent,

$$R_g = \begin{cases} \sqrt{N} & c \gg c^\star \\ N^\nu & c \ll c^\star. \end{cases} \tag{6.2.21}$$

This does not mean that a chain in concentrated solutions is self-intersecting as in the Kuhn chain model, but the result $R_g \sim \sqrt{N}$ is a consequence of excluded volume screening with the proportionality factor depending on c. Neutron scattering studies on labeled chains in concentrated solutions have verified the above result [Cotton *et al.* (1974), Daoud *et al.* (1975), Kirste *et al.* (1975), Richards *et al.* (1978)].

The extent of concentration fluctuations dictates the nature of the concentration dependence of the screening length. Generally speaking, two concentration regimes for $c > c^\star$, namely "semidilute" and "concentrated," can be identified to quantify concentration fluctuations. In the concentrated regime, there is less volume for chain conformations to fluctuate, thus local concentration fluctuations are weak. We expect this behavior for c higher than a certain threshold concentration $c^{\star\star}$, which is higher than the overlap concentration $c^\star$. On the other hand, for semidilute conditions, $c^\star < c < c^{\star\star}$, the concentration fluctuations are strong. The structural aspects and thermodynamic properties in these two concentration regimes are as follows.

6.2.2.1 Weak Fluctuations (Concentrated Solutions, $c > c^{\star\star} > c^\star$)

For weak concentration fluctuations, the field theory mentioned above can be adequately implemented by simply using the random phase approximation (RPA), where only the one-loop diagram (Fig. 6.5a) and its multiples as members of a geometric series are used. Higher order vertex diagrams as in Fig. 6.5b can be ignored in this concentration regime. With this RPA approach, Edwards derived the following results [Edwards (1966)].

(i) *Correlation function of concentration fluctuations:* Defining the monomer number concentration at location $\mathbf{r}$ as $c(\mathbf{r})$,

$$c(\mathbf{r}) = \sum_{\alpha=1}^{n} \int_0^N ds_\alpha \delta\left(\mathbf{r} - \mathbf{R}_\alpha(s_\alpha)\right), \qquad (6.2.22)$$

and the fluctuation $\delta c(\mathbf{r})$ as

$$c(\mathbf{r}) = c + \delta c(\mathbf{r}), \qquad (6.2.23)$$

the pair correlation function is given by

$$\langle \delta c(\mathbf{r}) \delta c(\mathbf{r}') \rangle = \frac{3c}{\pi \ell^2} \frac{e^{-|\mathbf{r}-\mathbf{r}'|/\xi}}{|\mathbf{r} - \mathbf{r}'|}. \qquad (6.2.24)$$

The correlation length ξ over which the concentration fluctuations are correlated, known as the Edwards length ξ_E, is given as

$$\xi = \xi_E = \left(\frac{1}{12cv\ell}\right)^{1/2} \sim c^{-1/2}. \qquad (6.2.25)$$

The correlation length ξ is the same as the excluded volume screening length appearing in Eq. (6.2.20).

(ii) *Scattering function:* In static scattering experiments using light, X-rays, and neutrons, the scattering intensity $I(k)$ at the scattering wave vector $\mathbf{k}$ is related to the Fourier transform of the pair correlation function of concentration fluctuations [Higgins & Benoit (1994), Doi & Edwards (1986)], as

$$I(\mathbf{k}) = V\langle c_\mathbf{k} c_{-\mathbf{k}} \rangle = cg(k), \qquad (6.2.26)$$

where $g(k)$ is the scattering function per segment,

$$g(k) = \frac{V}{c}\langle c_\mathbf{k} c_{-\mathbf{k}} \rangle, \qquad (6.2.27)$$

and

$$c_{\mathbf{k}} = \frac{1}{V} \int d\mathbf{r} e^{i\mathbf{k}\cdot\mathbf{r}} c(\mathbf{r}). \tag{6.2.28}$$

Using Eqs. (6.2.24), (6.2.27), and (6.2.28), the scattering function per segment follows as

$$g(k) = \frac{12}{\ell^2 \left(k^2 + \xi_E^{-2}\right)} = \frac{1}{cv\ell^3} \frac{1}{\left(1 + k^2 \xi_E^2\right)}. \qquad \text{(RPA)} \tag{6.2.29}$$

This equation is of the Ornstein–Zernike form. Using Eqs. (6.2.26) and (6.2.29), the correlation length can be determined by measuring the scattering intensity in the small $\mathbf{k}$ region and plotting its inverse against k^2, and measuring the slope of the linear plot,

$$\frac{I(k \to 0)}{I(k)} = \frac{g(0)}{g(k)} = 1 + k^2 \xi_E^2. \qquad (k \to 0) \tag{6.2.30}$$

(iii) *Size of a single chain:* Using the effective interaction $\Delta(\mathbf{r})$ of Eq. (6.2.20) for a pair of segments, the mean square end-to-end distance $\langle R^2 \rangle$ of a labeled chain in the concentrated solution can be calculated as

$$\langle R^2 \rangle = N\ell^2 \left[1 + \frac{12}{\pi} \frac{v\xi_E}{\ell}\right] = N\ell^2 \left[1 + \frac{\sqrt{12}}{\pi} \left(\frac{v}{c\ell^3}\right)^{1/2}\right], \qquad c \gg c^\star \tag{6.2.31}$$

where the terms inside the square brackets are independent of N. Thus, $\langle R^2 \rangle$ of a labeled chain in concentrated solutions is proportional to N, as envisaged in Eq. (6.2.21) based on the screening of excluded volume interactions. However, the chain is not a phantom Gaussian chain and the proportionality factor depends on the excluded volume parameter and polymer concentration.

(iv) *Free energy:* Calculation of F_p from Eq. (6.2.18) using RPA yields the polymer contribution to the Helmholtz free energy F of the polymer solution as a sum of the mean field free energy $F_{\mathrm{mf,p}}$ and contribution $F_{\mathrm{fl,p}}$ arising from fluctuations in the polymer concentration,

$$F = F_{\mathrm{mf,p}} + F_{\mathrm{fl,p}}. \tag{6.2.32}$$

The mean field contribution is

$$\frac{F_{\mathrm{mf,p}}}{k_B T} = -n \ln V + \ln n! + \frac{V}{2} v\ell^3 c^2. \tag{6.2.33}$$

Accounting for the translational entropy of n_1 solvent molecules and ignoring constant terms and terms of order c, the mean field contribution to F can be written as the Flory–Huggins free energy of mixing $F_{\mathrm{F-H}}$ given in Eq. (6.2.6) as,

$$\frac{F_{\mathrm{fl,p}}}{k_B T} = \frac{F_{\mathrm{F-H}}}{k_B T} = \frac{V}{\ell^3} \left[\phi_1 \ln \phi_1 + \frac{\phi_2}{N} \ln \phi_2 + \chi \phi_1 \phi_2\right], \tag{6.2.34}$$

where we have used $V = n_0 \ell^3$, $\phi_2 = c\ell^3$, and the solvent volume fraction ϕ_1 is $1 - \phi_2$. The fluctuation contribution $F_{\mathrm{fl,p}}$ obtained using RPA is

$$\frac{F_{\mathrm{fl,p}}}{k_B T} = -\frac{V}{12\pi} \frac{1}{\xi_E^3}. \qquad \text{(RPA)} \tag{6.2.35}$$

Substituting Eq. (6.2.25) into Eq. (6.2.35), we get

$$\frac{F_{\mathrm{fl,p}}}{k_B T} \sim -V c^{-3/2}. \qquad \text{(RPA)} \qquad (6.2.36)$$

The fluctuation contribution to the free energy is negative, and thus the concentration fluctuations lead to a reduction in the free energy of the solution in the high polymer concentration limit, in contrast with the behavior of semidilute solutions described in Section 6.2.2.2.

(v) *Osmotic pressure:* The osmotic pressure from the mean field contribution, Π_{mf}, is given in Eq. (6.2.11) as

$$\frac{\Pi_{\mathrm{mf}} v_1}{k_B T} = -\ln(1 - \phi_2) - \left(1 - \frac{1}{N}\right)\phi_2 - \chi \phi_2^2. \qquad (6.2.37)$$

The fluctuation contribution to the osmotic pressure, $\Pi_{\mathrm{fl}} = -\partial F_{\mathrm{fl,p}}/\partial V$, follows from Eq. (6.2.35) as

$$\frac{\Pi_{\mathrm{fl}}}{k_B T} = -\frac{1}{24\pi \xi_E^3} = -\frac{\sqrt{3}}{\pi}(vcl)^{3/2} \sim -c^{3/2}. \qquad \text{(RPA)} \qquad (6.2.38)$$

According to the RPA used in the Edwards theory, concentration fluctuations lead to a reduction in the osmotic pressure of polymer solutions. Also, the osmotic compressibility $\partial \Pi_{\mathrm{fl}}/\partial c$ from fluctuations is negative,

$$\frac{\partial}{\partial c}\left(\frac{\Pi_{\mathrm{fl}}}{k_B T}\right) = -\frac{3\sqrt{3}}{2\pi}(v\ell)^{3/2} c^{1/2} \sim -c^{1/2}. \qquad \text{(RPA)} \qquad (6.2.39)$$

The negative osmotic compressibility from fluctuations is unphysical, implying that the mean field contribution dominates over the fluctuation contribution. Combining Eqs. (6.2.37) and (6.2.38), the osmotic pressure follows as

$$\frac{\Pi}{k_B T} = \frac{c}{N} + \frac{1}{2}v\ell^3 c^2 - \frac{\sqrt{3}}{\pi}(vc\ell)^{3/2} + \cdots \qquad (6.2.40)$$

and the osmotic compressibility is given by

$$\frac{\partial}{\partial c}\frac{\Pi}{k_B T} = \frac{1}{N} + v\ell^3 c - \frac{3\sqrt{3}}{2\pi}(v\ell)^{3/2} c^{1/2} + \cdots \qquad (6.2.41)$$

where we have used $v_1 = \ell^3$ and $v = (1 - 2\chi)$. The condition that the osmotic compressibility must be positive gives a criterion for the concentration regime where RPA is valid. This criterion follows from Eq. (6.2.41) as

$$v\ell^3 c \geq \frac{3\sqrt{3}}{2\pi}(v\ell)^{3/2} c^{1/2}. \qquad (6.2.42)$$

In other words, the polymer concentration must be higher than the threshold concentration $c^{\star\star}$ given by

$$c \geq c^{\star\star} = \frac{27}{4\pi^2}\frac{v}{\ell^3}. \qquad (6.2.43)$$

This condition defines the regime of concentrated solutions ($c > c^{\star\star} > c^{\star}$), delineating it from the semidilute regime ($c^{\star} < c < c^{\star\star}$).

6.2.2.2   Strong Fluctuations (Semidilute Solutions, $c^\star < c < c^{\star\star}$)

In semidilute solutions, solvent content is substantial, which allows considerable conformational fluctuations, in contrast with concentrated solutions. When fluctuations are strong, the RPA is inadequate and it is necessary to include higher order correlations of polymer interactions involving vertex diagrams of the type given in Fig. 6.5b. Accounting for higher order correlations approximately, a general field theory was derived to address the strong concentration fluctuations in semidilute solutions and the crossover behavior to the weak fluctuations in the concentrated solutions. The main results of this theory for semidilute solutions are as follows [Muthukumar & Edwards (1982a)].

(i) *Correlation length ξ and scattering intensity:* The effective interaction $\Delta(\mathbf{r})$ can be approximated into the same screened form as in Eq. (6.2.20),

$$\Delta(\mathbf{r}) = v\ell^3 \left[\delta(\mathbf{r}) - \frac{1}{4\pi\xi^2} \frac{e^{-r/\xi}}{r} \right], \tag{6.2.44}$$

where the screening length in semidilute solutions is

$$\xi = 2^{-5/4} \left(\frac{3}{\pi} \right)^{1/2} v^{-1/4}\ell^{-5/4}c^{-3/4} \sim c^{-3/4}. \qquad \text{(semidilute)} \tag{6.2.45}$$

As noted in the preceding section, the screening length ξ is the correlation length for the pair correlation of concentration fluctuations given by Eq. (6.2.24).

The scattering function per segment $g(\mathbf{k})$ is given by the same formula as Eq. (6.2.29), except that ξ_E is replaced by ξ,

$$g(k) = \frac{1}{cv\ell^3} \frac{1}{(1 + k^2\xi^2)}. \tag{6.2.46}$$

Substitution of this result into Eq. (6.2.26) gives the scattering intensity as

$$I(\mathbf{k}) = \frac{1}{v\ell^3} \frac{1}{(1 + k^2\xi^2)}, \tag{6.2.47}$$

where $\xi \sim c^{-3/4}$ given in Eq. (6.2.45). Therefore, the correlation length can be determined experimentally using scattering techniques by plotting the inverse normalized scattering intensity versus the square of the scattering wave vector,

$$\frac{I(\mathbf{k} \to 0)}{I(\mathbf{k})} = 1 + k^2\xi^2. \tag{6.2.48}$$

While ξ is proportional to $c^{-3/4}$ in semidilute solutions with a good solvent, it is proportional to $c^{-1/2}$ in concentrated solutions. The theory of Muthukumar and Edwards provides an approximate crossover formula for ξ between Eq. (6.2.45) of semidilute solutions to Eq. (6.2.25) of concentrated solutions [Muthukumar & Edwards (1982a)]. This crossover behavior is sketched in Fig. 6.6a.

(ii) *Radius of gyration:* The concentration dependence of R_g in the semidilute region is given by the approximate formula [Muthukumar & Edwards (1982a)]

$$R_g = 2^{1/8} \left(\frac{1}{12\pi} \right)^{1/4} \ell^{5/8}v^{1/8}N^{1/2}c^{-1/8} \sim c^{-1/8}. \qquad (c^\star < c < c^{\star\star}) \tag{6.2.49}$$

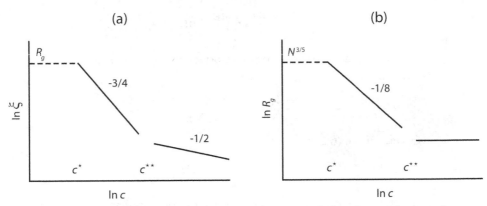

Figure 6.6 Regimes of dilute, semidilute, and concentrated solutions in a good solvent for (a) correlation length ξ for concentration fluctuations, and (b) radius of gyration.

Therefore, R_g of a labeled chain in a semidilute solution is proportional to $c^{-1/8}$, whereas it is essentially independent of c at higher concentrations (Eq. (6.2.31)),

$$R_g \to \frac{1}{\sqrt{6}} \sqrt{N}\ell. \qquad (c > c^{\star\star}) \qquad (6.2.50)$$

We have assumed the Gaussian chain result that R_g^2 is one-sixth of the mean square end-to-end distance. Note that due to excluded volume screening, the molecular weight dependence of R_g is the same ($R_g \sim \sqrt{N}$) in both the semidilute and concentrated solutions. The concentration-dependent crossover behavior of R_g between the semidilute and concentrated limits is sketched in Fig. 6.6b.

(iii) *Free energy and osmotic pressure:* The Helmholtz free energy F of good solutions of flexible polymers at concentration c above the overlap concentration is given by [Muthukumar & Edwards (1982a)]

$$\frac{F}{k_B T} = \frac{F_{FH}}{k_B T} + \frac{V}{24\pi} \frac{1}{\xi^3}, \qquad (6.2.51)$$

where ξ is given by Eqs. (6.2.45) and (6.2.25), respectively, for semidilute and concentrated solutions. The fluctuation contribution in this equation can be represented geometrically by imagining that the whole solution is a three-dimensional lattice of close-packed blobs, each of volume ξ^3. In this scaling picture, the number of blobs in the solution is proportional to V/ξ^3. Therefore, it follows from Eq. (6.2.51) that the solution equilibrates itself in such a manner that **the free energy per blob is the thermal energy** $k_B T$. This is a general result applicable to many situations in polymer systems [de Gennes (1979)].

Note the positive sign of the second term on the right-hand side showing that concentration fluctuations increase the free energy of the solution, in contrast to Eq. (6.2.35) derived using RPA. As shown in Section 6.2.2.1, the fluctuation contribution to the free energy in the weak fluctuation limit is smaller than the mean field contribution. On the other hand, this is not true for semidilute solutions with strong concentration fluctuations. The fluctuation contribution to the free energy for semidilute solutions follows from Eqs. (6.2.45) and (6.2.51) as

$$\frac{F_{fl}}{k_B T} = \frac{2^{3/4}}{9}\left(\frac{\pi}{3}\right)^{1/2} V\ell^{15/4}v^{3/4}c^{9/4} \sim v^{3/4}c^{9/4}. \tag{6.2.52}$$

The fluctuation contribution to the osmotic pressure, $\Pi_{fl} = -\partial F_{fl}/\partial V$, follows from Eq. (6.2.52) as

$$\frac{\Pi_{fl}}{k_B T} = -\frac{\partial}{\partial V}\left(\frac{1}{24\pi}\frac{V}{\xi^3}\right) = \frac{5}{96\pi}\frac{1}{\xi^3}. \tag{6.2.53}$$

In view of Eq. (6.2.45),

$$\frac{\Pi_{fl}}{k_B T} \sim v^{3/4}c^{9/4}. \tag{6.2.54}$$

Hence the fluctuation contribution to the osmotic pressure in semidilute solutions is positive and can dominate over the mean field term of c^2.

All conclusions predicted by the theory of semidilute solutions, given by Eqs. (6.2.45), (6.2.49), and (6.2.53), respectively, for correlation length, R_g, and osmotic pressure have been verified by experiments [Cotton *et al.* (1974), Daoud *et al.* (1975), Noda *et al.* (1981), Hamada *et al.* (1985)].

6.2.3   Scaling Laws

In this section, we present derivations of the scaling laws for the concentration dependence of the correlation length for concentration fluctuations (ξ), radius of gyration of a labeled chain (R_g), and the osmotic pressure (Π) of semidilute solutions, without resorting to the field theoretic calculations presented above. The fundamental concentration variable, which can make the polymer concentration c dimensionless, is the overlap concentration $c^\star$ (Eqs. (6.1.1) and (6.1.2)) [de Gennes (1979)]. We take the dimensionless polymer concentration as $c/c^\star \sim cN^{3\nu-1}$. Using $c/c^\star$ as the key variable in the following scaling arguments, and considering general solvent quality represented by the size exponent ν, we derive expressions for ξ, R_g, and Π.

6.2.3.1   Correlation Length

The correlation length ξ can in general be written as

$$\xi = R_g(c \to 0)f_1\left(\frac{c}{c^\star}\right), \tag{6.2.55}$$

where f_1 is some unknown function such that $f_1(x) \to 1$ as $x \to 0$, since the distance over which segment concentration is correlated is R_g in dilute solutions. The functional form of f_1 is guessed based on the following argument. For $c > c^\star$, there is no difference between a polymer solution containing n very long chains with N monomers and that with many x long chains each with X monomers such that $N = xX/n$. Therefore, only the polymer concentration, and not N, determines ξ. In other words, ξ is independent of N for $c > c^\star$. Since $R_g(c \to 0) \sim N^\nu$, Eq. (6.2.55) implies that $f_1(c/c^\star)$ *must* be a power of $(c/c^\star) \sim cN^{(3\nu-1)}$, for $c > c^\star$. Let this unknown power be denoted as y_1. Therefore,

$$\xi \sim N^\nu \left(\frac{c}{c^\star}\right)^{y_1} \sim N^\nu \left(cN^{3\nu-1}\right)^{y_1} \sim c^{y_1} N^{\nu+(3\nu-1)y_1}. \qquad (c > c^\star) \qquad (6.2.56)$$

Due to the N-independence of ξ for $c > c^\star$, the above result must scale as N^0. Hence we get

$$\nu + (3\nu - 1)y_1 = 0, \qquad (6.2.57)$$

so that $y_1 = -\nu/(3\nu - 1)$. Substituting this value for y_1 in Eq. (6.2.56), we obtain

$$\xi \sim c^{-\frac{\nu}{(3\nu-1)}}. \qquad (c > c^\star) \qquad (6.2.58)$$

This result is the same as Eq. (6.2.45) derived for $\nu = 3/5$, namely $\xi \sim c^{-3/4}$ for good solutions.

6.2.3.2   Radius of Gyration of a Labeled Chain

Given that the dimensionless concentration is $c/c^\star$ and $R_g \sim N^\nu$ in dilute solutions, the scaling form for R_g is written as

$$R_g \sim N^\nu f_2 \left(\frac{c}{c^\star}\right), \qquad (6.2.59)$$

where f_2 is some unknown function. However, we know from the field theory of polymer solutions that the bare excluded volume interactions are screened by interpenetrating chains such that $R_g \sim \sqrt{N}$ at higher polymer concentrations ($c > c^\star$). In order to satisfy this result, $f_2(c/c^\star)$ must be a power of $(c/c^\star)$ for $c > c^\star$. Letting this unknown power be y_2, we get

$$N^\nu \left(cN^{3\nu-1}\right)^{y_2} \sim N^{1/2}. \qquad (c > c^\star) \qquad (6.2.60)$$

Hence, by inspecting the exponents on N in this equation, we get

$$\nu + y_2(3\nu - 1) = \frac{1}{2}, \qquad (6.2.61)$$

so that

$$y_2 = -\frac{(2\nu - 1)}{2(3\nu - 1)}. \qquad (6.2.62)$$

Therefore, we get from Eq. (6.2.60)

$$R_g \sim \sqrt{N} c^{-\frac{(2\nu-1)}{2(3\nu-1)}}. \qquad (c > c^\star) \qquad (6.2.63)$$

This scaling result is the same as Eq. (6.2.49), $R_g \sim c^{-1/8}$, derived for good solutions ($\nu = 3/5$) under semidilute conditions.

6.2.3.3   Osmotic Pressure

Following the same arguments as above, the osmotic pressure can be written as

$$\frac{\Pi}{k_B T} = \frac{c}{N} f_3 \left(\frac{c}{c^\star}\right), \qquad (6.2.64)$$

where f_3 is some unknown function such that $f_3(x) \to 1$ as $x \to 0$, since the "ideal" van't Hoff law, $\Pi/k_B T = c/N$, must be recovered for very dilute solutions. The function f_3 is guessed as follows. As far as thermodynamic properties are concerned for

$c > c^\star$ (where chains interpenetrate), there is no difference between a polymer solution of n chains of N segments and that of x chains with $X = nN/x$ monomers. In other words, Π is independent of N for $c > c^\star$. Therefore, Eq. (6.2.64) implies that $f_3(c/c^\star)$ must be a power law of $c/c^\star \sim cN^{3\nu-1}$, for $c > c^\star$. Denoting this unknown power as y_3, we get

$$\frac{\Pi}{k_B T} = \frac{c}{N}\left(cN^{3\nu-1}\right)^{y_3}. \qquad (c > c^\star) \qquad (6.2.65)$$

Since the right-hand side of this equation must scale as N^0, we obtain

$$-1 + y_3(3\nu - 1) = 0, \qquad (6.2.66)$$

so that $y_3 = 1/(3\nu - 1)$. Substituting this value in Eq. (6.2.65), we get

$$\Pi \sim c^{\frac{3\nu}{3\nu-1}}. \qquad (6.2.67)$$

This is consistent with Eq. (6.2.54), $\Pi_{fl} \sim c^{9/4}$, for semidilute solutions in good solvents ($\nu = 3/5$).

6.2.3.4   Summary of Scaling Laws

We have presented the above scaling laws in semidilute solutions for arbitrary size exponent ν for uncharged polymers. For charged polymers, as shown in Chapter 5, $\nu \simeq 3/5$ in the high-salt limit and $\nu \simeq 1$ in salt-free conditions. Therefore, we expect from Eqs. (6.2.58), (6.2.63), and (6.2.67), the following scaling laws for semidilute solutions of polyelectrolytes,

$$\xi = \begin{cases} c^{-1/2} & \text{salt-free} \\ c^{-3/4} & \text{high salt,} \end{cases} \qquad (6.2.68)$$

$$R_g = \begin{cases} c^{-1/2} & \text{salt-free} \\ c^{-1/8} & \text{high salt,} \end{cases} \qquad (6.2.69)$$

$$\Pi = \begin{cases} c^{3/2} & \text{salt-free} \\ c^{9/4} & \text{high salt.} \end{cases} \qquad (6.2.70)$$

Summarizing this section, the derived scaling laws provide the proportionality relations for the concentration dependencies of ξ, R_g, and Π in semidilute solutions. Note that only the fluctuation contribution is captured by the scaling laws and the mean field contribution is not addressed. The latter can be quite significant as we shall see in Chapters 8 and 10. Additionally, note that the regime of weak fluctuations, where RPA is applicable, is not pertinent to the scaling laws derived above. The prefactors for the various scaling laws are required in calculations of phase diagrams and crossover behaviors between semidilute and concentrated solutions. These prefactors are provided in the theoretical results derived in Section 6.2.2.

6.3 Semidilute and Concentrated Polyelectrolyte Solutions

The nature of concentration fluctuations in polyelectrolyte solutions is significantly different from the results given in the preceding section for uncharged solutions. If the polyelectrolyte chains were to be broken into their charged monomers, then the solution of these ions and their counterions would behave like an electrolyte solution, with electrostatic screening (Debye screening) described in Chapter 3. On the other hand, if the chains are uncharged, then their intersegment excluded volume interaction is screened (topological screening), as detailed in Section 6.2.2. In polyelectrolyte solutions, both of these screenings occur simultaneously, with each coupled to the other in a self-consistent manner. We call this feature of coupled topological and electrostatic screenings as "double screening." In view of the self-consistent coupling, there is only one correlation length in the solution, which emerges as a confluence of both electrostatic and topological contributions. The field theory approach to a self-consistent treatment of topological correlations from chain conformations and electrostatic correlations arising from all charged species in the solution is the "double screening theory" [Muthukumar (1996a)]. This theory is a generalization of the field theory for uncharged polymer solutions described in Section 6.2.2. Due to the complex nature of the coupled long-ranged correlations in the system, the formulation of the double screening theory requires several field theoretic techniques. We relegate such technical details to the original reference [Muthukumar (1996a)]. However, we present below the essential features, such as the key assumptions and approximations behind the analytically tractable double screening theory. Following this, predictions from the theory, such as the effective intersegment interaction, correlation length for concentration fluctuations, scattering function, size of a labeled polyelectrolyte chain, free energy, and osmotic pressure will be described. One of the special predictions of the double screening theory is the emergence of intersegment attraction at intermediate distances as will be described in Section 6.3.2.

6.3.1 Essentials of Double Screening Theory

Consider a solution of n flexible polyelectrolyte chains each containing N segments, n_c counterions, n_γ ions of species γ from dissolved salt, and n_s solvent molecules in volume V. The polyelectrolyte concentration is above the overlap concentration. Note that $n_s = n_1$ in Section 6.2. Let α be the fixed degree of ionization per chain so that each of the N segments of the chain carries a charge $e\alpha z_p$ where e is the electronic charge. The total number of counterions is $n_c = \alpha z_p n N/z_c$ where z_c is the valency of the counterion. Let ez_i be the charge of the ith charged species. Using the path integral representation of polymer chains as continuous curves of length $L = N\ell$, where ℓ is the Kuhn step length, as in Eqs. (2.4.6) and (6.2.18), the Helmholtz free energy F of the system is given by

$$e^{-\frac{F}{k_B T}} = \frac{1}{n! n_c! n_s! \prod_\gamma n_\gamma!} \int \prod_{\alpha=1}^{n} \mathcal{D}[\mathbf{R}_\alpha] \int \prod_{i}^{n_c + n_s + \sum_\gamma n_\gamma} d\mathbf{r}_i$$

$$\times \exp\left\{-\frac{3}{2\ell^2}\sum_{\alpha=1}^{n}\int_0^N ds_\alpha\left(\frac{\partial \mathbf{R}_\alpha(s_\alpha)}{\partial s_\alpha}\right)^2 - U\left[\{\mathbf{R}_\alpha(s_\alpha)\},\{\mathbf{r}_i\}\right]\right\}. \quad (6.3.1)$$

Here $\mathbf{R}_\alpha(s_\alpha)$ is the position vector of the arc length variable s_α ($0 \le s_\alpha \le N$) of the αth chain and $\mathbf{r}_i$ is the position vector of the ith species. $U\left[\{\mathbf{R}_\alpha(s_\alpha)\},\{\mathbf{r}_i\}\right]$ is the total two-body potential interactions from polymer-polymer (U_{pp}), polymer-solvent (U_{ps}), solvent-solvent (U_{ss}), and polymer-ion (U_{pi}) contacts.

$U_{pp}(\mathbf{r})$ is the interaction energy between two segments of the chain separated by a distance $\mathbf{r}$,

$$U_{pp}(\mathbf{r}) = w\delta(\mathbf{r}) + \frac{\alpha^2 z_p^2 \ell_B}{r}, \quad (6.3.2)$$

where $w = v\ell^3$ is the excluded volume pseudopotential, which is related to the Flory–Huggins parameter χ according to $w = (1-2\chi)\ell^3$. $\delta(\mathbf{r})$ is the Dirac delta function and $r = |\mathbf{r}|$. The second term on the right-hand side of Eq. (6.3.2) represents the Coulomb interaction energy between the segments, where ℓ_B is the Bjerrum length (Eq. (1.3.2)). In writing this second term, we have assumed that the total charge $N\alpha z_p e$ of the chain is uniformly distributed along the chain skeleton. The short-ranged interactions between the polymer segments and solvent molecules and between solvent molecules are represented by

$$U_{ps}(\mathbf{r}) = w_{ps}\delta(\mathbf{r}) \qquad \text{and} \qquad U_{ss}(\mathbf{r}) = w_{ss}\delta(\mathbf{r}), \quad (6.3.3)$$

where w_{ps} and w_{ss} are the corresponding pseudopotential excluded volume parameters. The electrostatic interactions between charged segments and various ions are given by

$$U_{pi}(\mathbf{r}) = \frac{\alpha z_p z_i \ell_B}{r} \qquad \text{and} \qquad U_{ij}(\mathbf{r}) = \frac{z_i z_j \ell_B}{r}. \quad (6.3.4)$$

The above set of equations defines the model used in the double screening theory. Its theoretical formulation involves two major steps [Muthukumar (1996a)]. In the first step, all degrees of freedom associated with mobile (dissociated) counterions, electrolyte ions, and solvent molecules are integrated out, as cartooned as Step 1 in Fig. 6.7. This step is carried out with the Debye–Hückel theory of a charged plasma (here corresponding to the charged solution background which neutralizes the polyelectrolyte charges). As a result, the intersegment interaction between polymer segments is given by the screened Coulomb potential (in addition to the van der Waals contribution) as

$$v(\mathbf{r}) = w\delta(\mathbf{r}) + w_c \frac{e^{-\kappa r}}{4\pi r}, \quad (6.3.5)$$

where

$$w_c = 4\pi\alpha^2 z_p^2 \ell_B, \quad (6.3.6)$$

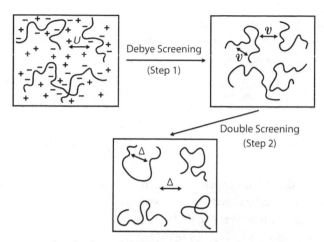

Figure 6.7 The computational scheme behind the double screening theory.

and

$$\kappa^2 = \frac{4\pi \ell_B}{V} \left(z_c^2 n_c + \sum_\gamma z_\gamma^2 n_\gamma \right). \tag{6.3.7}$$

After the first step, the system consists of only n polymer chains which are coupled through both intra-chain and inter-chain interaction potential $v(\mathbf{r})$ given by Eq. (6.3.5). The Helmholtz free energy F of the whole system is now given by the free energy F_p of these n chains and the free energy of the background F_b as

$$F = F_p + F_b. \tag{6.3.8}$$

The free energy of the background fluid consists of the entropy of mixing terms and the charge fluctuations in the neutralizing background,

$$F_b = F_{b0} + F_{fl,i}, \tag{6.3.9}$$

where

$$\frac{F_{b0}}{k_B T V} = c_s \ln c_s - c_s + c_c \ln c_c - c_c + \sum_\gamma \left[c_\gamma \ln c_\gamma - c_\gamma \right] + \frac{1}{2} w_{ss} c_s^2 + w_{ps} c c_s, \tag{6.3.10}$$

and

$$\frac{F_{fl,i}}{k_B T V} = -\frac{\kappa^3}{12\pi}. \tag{6.3.11}$$

It should be noted that the expression for $F_{fl,i}$ given by Eq. (6.3.11) is valid strictly within the Debye-Hückel regime, namely when the local electric potential is less than $k_B T$, and for point charges as discussed in Chapter 3. Extensions can be made to go beyond the linearized Poisson–Boltzmann formalism [Naji *et al.* (2013), Budkov *et al.* (2015)]. Including the finite size of the ions, Eq. (6.3.11) is generalized to (see Eq. (3.8.41))

$$\frac{F_{fl,i}}{k_B T} = -\frac{V}{4\pi \ell^3} \left[\ln(1 + \kappa \ell) - \kappa \ell + \frac{1}{2} \kappa^2 \ell^2 \right]. \tag{6.3.12}$$

The free energy F_p of n chains in the background where the interaction energy between any two segments separated by distance r is given by Eq. (6.3.5), follows from Eq. (6.3.1) as

$$e^{-\frac{F_p}{k_B T}} = \frac{1}{n!} \int \prod_{\alpha=1}^{n} \mathcal{D}[\mathbf{R}_\alpha(s_\alpha)] \exp \left\{ -\frac{3}{2\ell^2} \sum_{\alpha=1}^{n} \int_0^N ds_\alpha \left(\frac{\partial \mathbf{R}_\alpha(s_\alpha)}{\partial s_\alpha} \right)^2 \right.$$

$$\left. -\frac{1}{2} \sum_{\alpha=1}^{n} \sum_{\beta=1}^{n} \int_0^N ds_\alpha \int_0^N ds_\beta v \left[\mathbf{R}_\alpha(s_\alpha) - \mathbf{R}_\beta(s_\beta) \right] \right\}. \quad (6.3.13)$$

As described in Section 6.2.2, the situation of n coupled chains through the interaction $v(\mathbf{r})$ can be exactly mapped onto a collection of n uncoupled chains by introducing a potential field variable Φ. Now, the effective interaction between any two segments of a labeled chain is given by $\Delta(\mathbf{r})$,

$$v(\mathbf{r}) \to \Delta(\mathbf{r}), \quad (6.3.14)$$

where the arrow denotes the chain uncoupling with the introduction of the field Φ. Basically, the total pairwise potential interaction energy among all segments is replaced by the net energy of all segments such that each segment interacts with the field generated by all other segments in the system. $\Delta(\mathbf{r})$ is also the correlation function of the field variable separated by the distance r as $\Delta(r) = \langle \Phi(r)\Phi(0) \rangle$, with the angular brackets denoting the average over the field variable. The topological correlations of the chain connectivity and long-ranged electrostatic correlations are strongly coupled and these must be determined self-consistently in obtaining $\Delta(\mathbf{r})$ from $v(\mathbf{r})$ by the introduction of the field variable Φ. This is labeled as Step 2 in Fig. 6.7. These two steps constitute the technical aspects of the double screening theory. Based on this theory, expressions for the effective intersegment interaction in polyelectrolyte solutions and consequent results for the correlation length for concentration fluctuations, scattering function, size of a labeled chain, and the free energy of the solution are given below. The double screening theory covers both semidilute solutions where concentration fluctuations are strong and concentrated solutions with weak fluctuations. Although the random phase approximation (RPA) is sometimes used in describing semidilute polyelectrolyte solutions, this approximation is of limited validity as we shall see in Section 6.4.

6.3.2   Effective Interaction and Attraction at Intermediate Distances

The double screening theory provides full crossover expressions for the effective interaction $\Delta(r)$ between two monomers separated by distance r for polyelectrolyte concentrations ranging from semidilute to concentrated conditions. Here we consider only the limiting cases of salt-free and high-salt conditions. In the limit of low added salt concentration, $\Delta(r)$ is given by [Muthukumar (1996a)]

$$\Delta(r) = \frac{w_c}{4\pi r} e^{-\frac{1}{\sqrt{2}} \frac{r}{\xi_2}} \cos\left(\frac{1}{\sqrt{2}} \frac{r}{\xi_2} \right), \quad (6.3.15)$$

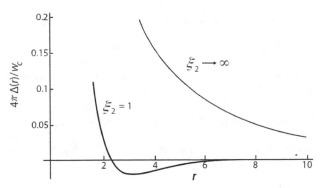

Figure 6.8 The normalized effective potential interaction $\Delta(r)$ is attractive at intermediate distances for finite polyelectrolyte concentrations in salt-free conditions (r and ξ_2 are in units of the Kuhn length ℓ) [Muthukumar (1996a)].

where ξ_2 is identified as the screening length for the simultaneous and self-consistent topological and electrostatic interactions. For distances larger than ξ_2, the intersegment interaction is screened. ξ_2 is also the correlation length for concentration fluctuations. As discussed later in Section 6.3.3, $\xi_2 \sim c^{-1/2}$ in salt-free semidilute solutions and $\xi_2 \sim c^{-1/4}$ in salt-free concentrated solutions. Therefore, as c increases, ξ_2 becomes shorter and hence the chains behave as Gaussian chains, although the interpenetrating chains are fully charged with the same sign.

One of the remarkable features of the effective interaction $\Delta(r)$ is that it is attractive at intermediate distances comparable to the correlation length ξ_2 in low salt concentration solutions. This attraction emerges despite the fact that all interacting segments have the same sign for the charge. An illustrative result for the emergence of attraction at intermediate distances is given in Fig. 6.8, where $4\pi\Delta(r)/w_c$ is plotted against the distance r. For the infinitely dilute limit such that $\xi_2 \to \infty$, the interaction is purely repulsive. As the polyelectrolyte concentration is increased, an attraction emerges due to topological correlations as shown by the bottom curve (for $\xi_2 = 1$) in Fig. 6.8. Here r and ξ_2 are in units of the Kuhn length ℓ. It must be emphasized that this attractive component to the effective pairwise interaction energy emerges solely due to chain connectivity. This effect is different from the van der Waals-type attraction in dense liquids.

The attraction between similarly charged polyelectrolytes at intermediate distances is present only if the polyelectrolyte concentration is sufficiently large and the salt concentration is sufficiently low. As the salt concentration increases, the long-ranged electrostatic interactions are screened and we are left with only the topological correlation, as described in Section 6.2.2 for uncharged solutions. Under these conditions, the double screening theory gives

$$\Delta(r) = \left(w + \frac{w_c}{\kappa^2}\right)\left[\delta(\mathbf{r}) - \frac{1}{4\pi r \xi_1^2}e^{-r/\xi_1}\right], \qquad (6.3.16)$$

where ξ_1 is the screening length. As given below, $\xi_1 \sim c^{-3/4}$ for semidilute conditions and $\xi_1 \sim c^{-1/2}$ for concentrated conditions. In the high salt concentration limit, the conclusions drawn in Section 6.2 are valid by replacing v by $v + w_c/(\kappa_2 \ell^3)$,

$$v \to v + \frac{w_c}{\kappa^2 \ell^3}, \tag{6.3.17}$$

where v and w_c are given, respectively, by Eqs. (2.4.1) and (6.3.6).

6.3.3   Correlation Length in Polyelectrolyte Solutions

For salt-free conditions, the double screening theory gives the correlation length for concentration fluctuations as

$$\xi_2 = \begin{cases} \left(\frac{6\sqrt{2}}{\pi} \right)^{1/3} \sqrt{\frac{3}{16}} w_c^{-1/6} \ell^{-1/3} c^{-1/2} & \text{semidilute } (c^\star < c < c^{\star\star}) \\ 6^{-1/4} w_c^{-1/4} \ell^{1/2} c^{-1/4} & \text{concentrated } (c^{\star\star} < c < c^{\star\star\star}), \end{cases} \tag{6.3.18}$$

where the electrostatic interaction is assumed to dominate over the van der Waals interaction represented by v.

On the other hand, for solutions with high enough salt, the screening length ξ_1 is given by the double screening theory as

$$\xi_1 = \begin{cases} \frac{3^{5/4}}{4\sqrt{\pi}} \left(w + \frac{w_c}{\kappa^2} \right)^{-1/4} \ell^{-1/2} c^{-3/4} & \text{semidilute } (c^\star < c < c^{\star\star}) \\ \frac{1}{\sqrt{6}} \left(w + \frac{w_c}{\kappa^2} \right)^{-1/2} \ell c^{-1/2} & \text{concentrated } (c^{\star\star} < c < c^{\star\star\star}). \end{cases} \tag{6.3.19}$$

The concentration dependencies of ξ_2 and ξ_1 in semidilute polyelectrolyte solutions are consistent with the scaling laws given in Eq. (6.2.68) assuming $\nu = 1$ in the salt-free limit and $\nu = 3/5$ in the high-salt limit. As we already know, the scaling laws do not capture the high polymer concentration behavior.

6.3.4   Scattering Function

The scattering function per segment $g(k)$ defined in Eq. (6.2.26), that is required to interpret static scattering experiments, is given by the general relation

$$\Delta(k) = v(k) - cg(k)v^2(k), \tag{6.3.20}$$

where $\Delta(k)$ is the Fourier transform of $\Delta(r)$, and

$$v(k) = w + \frac{w_c}{k^2 + \kappa^2}. \tag{6.3.21}$$

Substituting the result of $\Delta(k)$ from the double screening theory to get $g(k)$, and then using Eq. (6.2.26), the scattering intensity $I(k)$ in salt-free conditions is given by

$$I(k) = \frac{1}{w_c} \frac{k^2}{\left(1 + k^4 \xi_2^4 \right)}, \qquad (\kappa \to 0) \tag{6.3.22}$$

where ξ_2 is given by Eq. (6.3.18). In the high-salt limit, $I(k)$ follows from Eqs. (6.3.16) and (6.3.20) as

$$I(k) = \frac{1}{\left(w + \frac{w_c}{\kappa^2} \right)} \frac{1}{\left(1 + k^2 \xi_1^2 \right)}, \qquad \text{(high salt)} \tag{6.3.23}$$

where ξ_1 is given in Eq. (6.3.19). The predicted scattering intensity in Eq. (6.3.22) portrays the emergence of the polyelectrolyte peak in salt-free solutions (Fig. 6.2b). On the other hand, Eq. (6.3.23) predicts that there is no peak in the scattering function as seen in Fig. 6.2c.

6.3.5   Size of a Labeled Polyelectrolyte Chain

When the effective interaction between two segments is $\Delta(\mathbf{r})$, the probability distribution function for the end-to-end distance vector $\mathbf{R}$ of a labeled chain in a polyelectrolyte solution is given by

$$G(\mathbf{R}, N) = \int_0^{\mathbf{R}} \mathcal{D}[\mathbf{R}(s)] \exp \left\{ -\frac{3}{2\ell^2} \int_0^N ds \left(\frac{\partial \mathbf{R}(s)}{\partial s} \right)^2 \right.$$
$$\left. -\frac{1}{2} \int_0^N ds \int_0^N ds' \Delta[\mathbf{R}(s) - \mathbf{R}(s')] \right\}. \tag{6.3.24}$$

In view of the screened form of $\Delta(\mathbf{r})$ given in Eqs. (6.3.15) and (6.3.16), we anticipate that there is screening of both topological and electrostatic correlations, which eventually leads to Gaussian chain statistics as the concentration is increased.

According to the double screening theory, the radius of gyration R_g of a labeled polyelectrolyte chain is given as

$$R_g \sim \begin{cases} \sqrt{N} \left(w_c \ell^2 \right)^{1/12} c^{-1/4}, & \text{semidilute} \\ \sqrt{N} \ell, & \text{concentrated} \end{cases} \tag{6.3.25}$$

for salt-free conditions, and

$$R_g \sim \begin{cases} \sqrt{N} \left(w + \frac{w_c}{\kappa^2} \right)^{1/8} \ell^{1/4} c^{-1/8}, & \text{semidilute} \\ \sqrt{N} \ell, & \text{concentrated} \end{cases} \tag{6.3.26}$$

for salty solutions. The numerical prefactors and the extent of corrections to the apparent Gaussian chain statistics are available in the original publication [Muthukumar (1996a)].

These predictions, which are consistent with the scaling predictions of Eq. (6.2.69) for semidilute solutions (where screening is assumed *a priori*), have been validated experimentally [Prabhu *et al.* (2001, 2003)]. It is remarkable that even when all chains are charged similarly and uniformly, they obey the Flory theorem [Flory (1949)] of Gaussian statistics in the melt due to the screening of both excluded volume and electrostatic effects.

6.3.6   Free Energy and Osmotic Pressure

The evaluation of the effective interaction Δ and the consequent changes on the conformational fluctuations of chains lead to an expression for F_p defined in Eq. (6.3.13) as,

$$F_p = F_{p0} + F_{\text{fl,p}}(\Delta), \tag{6.3.27}$$

where F_{p0} is the contribution from the mean field part without concentration fluctuations (equivalent to the Flory–Huggins form for the polymer contribution) and $F_{fl,p}$ is the free energy contribution from fluctuations in the local polyelectrolyte concentration. In general, the polyelectrolyte fluctuation contribution to the free energy can be derived as [Muthukumar (1996a)]

$$\frac{F_{fl,p}}{k_B T} = \begin{cases} \frac{V}{24\sqrt{2}\pi} \frac{1}{\xi_2^3}, & \text{salt-free} \\ \frac{V}{24\pi} \frac{1}{\xi_1^3}, & \text{high salt} \end{cases} \qquad (6.3.28)$$

where the correlation lengths ξ_2 and ξ_1 are given in Eqs. (6.3.18) and (6.3.19). Due to the inherent nature of the coupled double screening (electrostatic and topological), ξ_2 and ξ_1 are crossover functions which reach the asymptotic results of Eqs. (6.3.18) and (6.3.19) at the various limits of polyelectrolyte concentration and added salt concentration.

For semidilute salt-free solutions,

$$\frac{F_{fl,p}}{k_B T} \sim V c^{3/2}, \qquad \text{(salt-free, semidilute)} \qquad (6.3.29)$$

where Eqs. (6.3.18) and (6.3.28) are used. The osmotic pressure from fluctuations in polymer conformations, $\Pi_{fl,p} = -\partial F_{fl,p}/\partial V$, is therefore given by

$$\Pi_{fl,p} \sim c^{3/2}. \qquad \text{(salt-free, semidilute)} \qquad (6.3.30)$$

For high-salt conditions, Eqs. (6.3.19) and (6.3.28) give

$$\frac{F_{fl,p}}{k_B T} \sim V c^{9/4}, \qquad \text{(high salt, semidilute)} \qquad (6.3.31)$$

so that $\Pi_{fl,p}$ follows as

$$\Pi_{fl,p} \sim c^{9/4}. \qquad \text{(high salt, semidilute)} \qquad (6.3.32)$$

The above equations for $\Pi_{fl,p}$ are consistent with the scaling laws of Eq. (6.2.70).

The Helmholtz free energy F of a polyelectrolyte solution is obtained by collecting all contributions to the free energy. For the sake of specificity, consider the model system introduced in Section 6.3.1. The Helmholtz free energy density f defined as $f \equiv F\ell^3/(Vk_B T)$, is given by

$$f = \frac{F\ell^3}{Vk_B T} = f_S + f_H + f_{fl,i} + f_{fl,p}. \qquad (6.3.33)$$

Here f_S represents the mixing entropy of ions, solvent molecules, and polymer chains,

$$f_S = \frac{\phi}{N} \ln \phi + \phi_c \ln \phi_c + \phi_+ \ln \phi_+ + \phi_- \ln \phi_- + \phi_s \ln \phi_s, \qquad (6.3.34)$$

where $\phi = nN\ell^3/V$ is the volume fraction of the polymer, $\phi_c = \alpha_1 \phi$ is the volume fraction of counterions from dissociated polyelectrolyte chains ($\alpha_1 = \alpha z_p/z_c$), $\phi_+ = n_+\ell^3/V$ and $\phi_- = n_-\ell^3/V$ are volume fractions of salt cations and anions, respectively, and $\phi_s = n_s\ell^3/V$ is the volume fraction of solvent.

The enthalpy part f_H in Eq. (6.3.33) is given by

$$f_H = \left(\frac{1}{2}w_{pp}\phi^2 + \frac{1}{2}w_{ss}\phi_s^2 + w_{ps}\phi\phi_s\right) + \phi^2\left(\frac{2\pi\alpha^2 z_p^2 \ell_B}{\kappa^2 \ell^3}\right) \qquad (6.3.35)$$

with

$$\kappa^2 \ell^2 = 4\pi\frac{\ell_B}{\ell}[z_c^2\phi_c + z_+^2\phi_+ + z_-^2\phi_-]. \qquad (6.3.36)$$

The enthalpy part of the free energy density represents the mean-field energy that includes the short-range interactions among solvent and neutral polyelectrolyte segments as well as the electrostatic interactions among polyelectrolyte segments. The short-range interactions can be represented via the Flory–Huggins parameter χ. The effective electrostatic energy between two charged polyelectrolyte segments is of the Yukawa form $v(r) = \alpha^2 z_p^2 \ell_B \exp(-\kappa r)/r$, with the inverse screening length κ defined by Eq. (6.3.7). When there are enough counterions and salt ions in the system, this becomes short-ranged as $(4\pi\alpha^2 z_p^2 \ell_B/\kappa^2)\delta(\mathbf{r})$. As shown in Eq. (5.3.15), this results in the modification of the χ parameter, as reflected in the second term of Eq. (6.3.35). For polyelectrolyte concentrations relevant to investigations of the phase behavior of solutions, this approximation is adequate even for "salt-free" solutions.

The free energy density due to ion fluctuations $f_{fl,i}$ is given by Eq. (6.3.12) as

$$f_{fl,i} = -\frac{1}{4\pi}\left[\ln(1 + \kappa\ell) - \kappa\ell + \frac{1}{2}(\kappa\ell)^2\right]. \qquad (6.3.37)$$

Finally, the free energy density due to polymer fluctuations $f_{fl,p}$, derived in the double screening theory for the high-salt and low-salt limits, can be written as a simple interpolation,

$$f_{fl,p} = \frac{\frac{2^{3/4}}{9}\sqrt{\frac{\pi}{3}}\left(\frac{3}{2}\right)^{-9/4}\left(4\pi z_p^2 \alpha^2 \ell_B/\ell\right)^{3/4}\phi^{9/4}}{(\kappa\ell)^{3/2} + 2^{5/4}\sqrt{\frac{\pi}{3}}\left(\frac{3}{2}\right)^{-3/4}\left(4\pi z_p^2 \alpha^2 \ell_B/\ell\right)^{1/4}\phi^{3/4}}. \qquad (6.3.38)$$

In Eq. (6.3.38), one can find that $f_{fl,p} \sim \phi^{3/2}$ in the low-salt limit and $f_{fl,p} \sim \phi^{9/4}$ in the high-salt limit. As it turns out, the contribution from polymer fluctuations to the free energy density is minor in comparison with the other terms so that elaborate crossover descriptions for $f_{fl,p}$ is unnecessary.

The set of Eqs. (6.3.33)-(6.3.38) form the basis for the calculation of theoretical phase diagrams described in Chapter 8.

6.4 Electrostatically Driven Structure in Salt-Free Polyelectrolyte Solutions

Electrostatic repulsion between polymer segments in salt-free polyelectrolyte solutions results in seemingly ordered structures, independent of whether the chains are interpenetrating or not, as evident from the polyelectrolyte peak in Fig. 6.2. In the presence of added salt, with the consequent screening of electrostatic repulsion, the

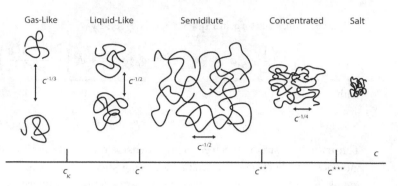

Figure 6.9 Five regimes of electrostatically driven structural organization in salt-free polyelectrolyte solutions.

polyelectrolyte peak is absent. Note that for polyelectrolyte solutions with higher concentrations of added salt, the various results of Section 6.2 for uncharged polymers are applicable with the modification of χ to χ_{eff} accounting for the presence of salt. The peak position k_m in salt-free solutions depends on polymer concentration as $c^{1/3}, c^{1/2}$, and $c^{1/4}$ in extremely dilute, dilute and semidilute, and concentrated solutions, respectively. The various aspects of the polyelectrolyte peak and thermodynamics of salt-free polyelectrolyte solutions are described in terms of five regimes of polymer concentration, as depicted in Fig. 6.9 [de Gennes *et al.* (1976), Muthukumar (2016a)]. For $c <$ $c^{\star}$, there are two regimes: (i) "Gas-like" (Regime I), and (ii) "Liquid-like" (Regime II). For $c > c^{\star}$, there are two regimes: (i) Semidilute $c^{\star} < c < c^{\star\star}$ (Regime III), and (ii) Concentrated $c^{\star\star} < c < c^{\star\star\star}$ (Regime IV), where $c^{\star\star\star}$ is a certain threshold concentration above which the solution is a hydrated melt (Regime V). The organization of molecules in these five regimes is given below.

Regime I: Electrostatically Uncorrelated Dilute "Gas-Like" Regime
$$(0 < c < c_\kappa)$$

When the polyelectrolyte concentration is extremely low, the average separation distance Λ between any two chains is so large that the strength of the electrostatic interaction between them is vanishingly small. Under these conditions,

$$c = \frac{nN}{V} \sim \frac{nN}{n\Lambda^3} \qquad (6.4.1)$$

so that

$$k_m \sim \frac{1}{\Lambda} \sim c^{1/3}. \qquad (6.4.2)$$

Defining the polyelectrolyte concentration c_κ as the concentration at which the average distance between any two chains is the Debye length,

$$c_\kappa \sim \frac{nN}{n\kappa^{-3}} \sim N\kappa^3, \qquad (6.4.3)$$

the "gas-like" regime corresponds to $0 < c < c_\kappa$.

Regime II: Electrostatically Correlated Dilute "Liquid-Like" Regime
$$(c_\kappa < c < c^\star)$$

In this regime, $c_\kappa < c < c^\star$, the average distance Λ between two chains is shorter than the Debye length and yet the chains have not overlapped, such that $2R_g < \Lambda < \kappa^{-1}$. Under these conditions, the structure factor $S(k)$, which is proportional to the scattering intensity $I(k)$, is approximately the product of the form factor $P(k)$ and the intermolecular structure factor given by Eq. (5.9.3) as

$$S(k) = P(k)[1 - \frac{c}{N}Q(k)], \tag{6.4.4}$$

where $Q(k)$ depends on the pair-potential $U(\mathbf{r})$ between two chains, with center of mass separation distance $\mathbf{r}$, given in Eq. (5.9.4). Substituting Eq. (5.9.11) into Eq. (5.9.4), and performing the usual high-temperature expansion, we get in the small angle scattering limit of $kR_g \ll 1$,

$$Q(\mathbf{k}) = \frac{N^2 w_c}{k^2}. \tag{6.4.5}$$

Combining this result with the form factor given in Eq. (5.9.18), the structure factor is obtained as

$$S(k) = \frac{1}{\left(1 + \frac{2v}{3}k^2 R_g^2\right)^{1/2v}} \left(1 - \frac{cNw_c}{k^2}\right). \tag{6.4.6}$$

This exhibits a peak at the maximum k_m given by

$$k_m^2 = cNw_c \left(\frac{1}{2} + v\right)\left[1 + \sqrt{1 + \frac{12}{(1 + 2v)^2 cNw_c R_g^2}}\right]. \tag{6.4.7}$$

For sufficiently large values of $cNw_c R_g^2 > 1$, this result gives

$$k_m \simeq \left(\frac{1}{2} + v\right)^{1/2} \sqrt{cNw_c}, \tag{6.4.8}$$

where v is the effective size exponent for the polyelectrolyte chain in dilute solutions. Therefore, the general result for the polyelectrolyte peak in dilute solutions with electrostatic correlations is

$$k_m \sim \sqrt{c}, \qquad \text{(dilute)} \tag{6.4.9}$$

although the polyelectrolyte concentration is below the overlap concentration $c^\star$.

Regime III: Semidilute Regime $(c^\star < c < c^{\star\star})$

In this regime, the polyelectrolyte concentration is above the overlap concentration $c^\star$ but below $c^{\star\star}$ at which concentration fluctuations become weak. For polyelectrolyte concentrations above $c^\star$, the double screening theory gives the scattering intensity $I(k)$ in the salt-free limit as (Eq. (6.3.22))

$$I(k) = \frac{1}{w_c} \frac{k^2}{(1 + k^4 \xi_2^4)}, \tag{6.4.10}$$

where w_c and ξ_2 are given in Eqs. (6.3.6) and (6.3.18), respectively. The dependence of the scattering intensity on the angularly averaged scattering wave vector, as given by Eq. (6.4.10), is presented in Fig. 10(a) for the choice of $w_c = 8.8$ nm and $\xi_2 = 10$ nm, exhibiting the polyelectrolyte peak. The peak position follows from Eq. (6.4.10), by finding the location of k at which $I(k)$ is a maximum, as ξ_2^{-1},

$$k_m \sim \frac{1}{\xi_2} \sim \sqrt{c}, \qquad \text{(semidilute)} \qquad (6.4.11)$$

where Eq. (6.3.18) is used. It must be noted that the peak position is directly related to the position of the attractive minimum in the effective interaction $\Delta(r)$. Although the exponent β for the concentration dependence of k_m, defined in Eq. (6.1.3), is the same for both the dilute correlated regime and the semidilute regime, their numerical prefactors are slightly different [Förster & Schmidt (1995)].

Regime IV: Concentrated Regime ($c^{\star\star} < c < c^{\star\star\star}$)

As the polyelectrolyte concentration becomes higher than a certain value $c^{\star\star}$, the concentration fluctuations become weak and RPA becomes applicable. According to the double screening theory, $I(k)$ is given by Eq. (6.3.22), with $\xi_2 \sim c^{-1/4}$ given in Eq. (6.3.18). Therefore, the position of the polyelectrolyte peak, $k_m \sim \xi_2^{-1}$, at very high polyelectrolyte concentrations, is

$$k_m \sim c^{1/4}. \qquad \text{(concentrated)} \qquad (6.4.12)$$

Regime V: Hydrated Melt ($c^{\star\star\star} < c$)

In this regime, the polymer concentration is higher than $c^{\star\star\star}$, beyond which only hydrated melt exists, without concentration dependence.

Before we compare the above predictions on the five regimes with experimental results, we note two additional important aspects. First, as already pointed out, the polyelectrolyte peak goes away upon addition of low molecular weight salt. According to the double screening theory, the scattering intensity for salty conditions is given by Eq. (6.3.23) as

$$I(k) = \frac{1}{(w + \frac{w_c}{\kappa^2})} \frac{1}{(1 + k^2\xi_1^2)}, \qquad (6.4.13)$$

where ξ_1 is proportional to $(w + w_c/\kappa^2)^{-1/4}c^{-3/4}$ and $(w + w_c/\kappa^2)^{-1/2}c^{-1/2}$, respectively, in semidilute and concentrated solutions. Now, the scattering intensity is of the Ornstein-Zernike form. As an example, $I(k)$ is plotted against $k\xi_1$ in Fig. 6.10(b) by choosing $(w + w_c/\kappa^2) = 1$ nm^3. The crossover behavior of $I(k)$ for intermediate salt concentrations is obtained from Eqs. (6.2.26) and (6.3.20) as

$$I(k) = \left(\frac{1}{v(k)}\right)\left[1 - \frac{\Delta(k)}{v(k)}\right], \qquad (6.4.14)$$

where $\Delta(k)$ is provided from the double screening theory and $v(k) = w + w_c/(k^2 + \kappa^2)$.

The second aspect is that, for polyelectrolyte concentrations higher than the added salt concentration, dynamic light scattering shows two modes of relaxation, known as

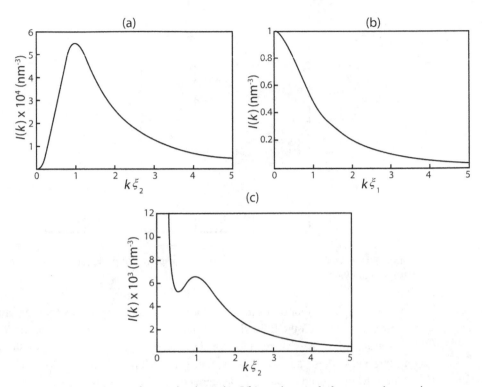

Figure 6.10 Dependence of scattering intensity $I(\mathbf{k})$ on the angularly averaged scattering wave vector $\mathbf{k}$. (a) salt-free ($w_c = 8.8$ nm, $\xi_2 = 10$ nm); (b) salty ($w + w_c/\kappa^2 = 1$ nm^3); (c) enhanced intensity at near zero angles and the polyelectrolyte peak in salt-free solutions ($w_c = 8.8$ nm, $\xi_2 = 3$ nm, $R_{g,\text{agg}} = 300$ nm) [Muthukumar (2016a)].

the "fast" and "slow" modes (Chapter 7). Concomitant to the emergence of the slow mode, the scattering intensity at very small angles is anomalously high. As addressed in Section 6.5, this is attributed to the formation of large aggregates of radius of gyration $R_{g,\text{agg}}$ in the order of hundreds of nanometers. The scattering intensity per segment follows from Eq. (5.9.2) for $kR_{g,\text{agg}} \ll 1$ as

$$I_{\text{agg}}(k) = \frac{1}{\left(1 + \frac{1}{3}k^2 R_{g,\text{agg}}^2\right)}. \tag{6.4.15}$$

Adding this contribution to the scattering intensity for the salt-free solution (Eq. (6.4.10)), we get

$$I(k) = \frac{1}{w_c} \frac{k^2}{\left(1 + k^4\xi_2^4\right)} + \frac{1}{\left(1 + \frac{1}{3}k^2 R_{g,\text{agg}}^2\right)}. \tag{6.4.16}$$

A typical plot of $I(k)$ versus k is given in Fig. 6.10c for the choice of $w_c = 8.8$ nm, $\xi_2 = 3$ nm, and $R_{g,\text{agg}} = 300$ nm. The occurrence of enhanced scattering intensity at near zero scattering angles and the presence of the polyelectrolyte peak are evident from this figure. Note that the intensity contribution in Eq. (6.4.15) is per segment and

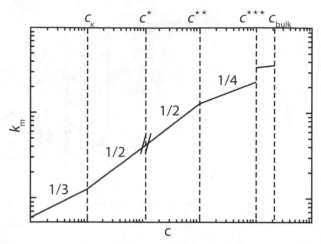

Figure 6.11 Concentration dependence of the polyelectrolyte peak in the five regimes [Adapted from Muthukumar (2016a)].

the intensity contribution from all segments in the aggregate is very high at near zero scattering angles. In the following section, we shall derive an estimate for $R_{g,\text{agg}}$.

Returning to the five regimes for the concentration dependence of the polyelectrolyte peak, these are sketched in Fig. 6.11 [Muthukumar (2016a)]. The extensive literature on the concentration dependence of k_m in salt-free aqueous solutions of sodium poly(styrene sulfonate) of different molecular weights, using light, X-ray, and neutron scattering is summarized in Fig. 6.12. In order to recognize the different concentration regimes and their crossover behaviors, it is necessary to estimate the overlap concentration for each molar mass of poly(styrene sulfonate). Using the formula of Eq. (6.1.1), we get

$$c^\star = \frac{M}{\frac{4}{3}\pi R_g^3 \mathcal{N}},\qquad (6.4.17)$$

where M is the molar mass and $\mathcal{N}$ is the Avogadro number. For a rigid rod of contour length L, R_g is $L/\sqrt{12}$. Since we know that a flexible polyelectrolyte chain is never fully extended in the experimental conditions of interest [Beer *et al.* (1997)], we take the span along the most extended direction as $L/2$, a value typically observed in computer simulations of short flexible polyelectrolyte chains [Liu & Muthukumar (2002)]. Therefore,

$$c^\star = \frac{144\sqrt{3}}{\pi}\frac{M}{\mathcal{N}L^3}.\qquad (6.4.18)$$

This estimate is the lowest bound for $c^\star$, because the radius of gyration is smaller than $L/(4\sqrt{3})$ for all molar masses considered in Fig. 6.12.

The specific details of the various data presented in Fig. 6.12 are the following. The molar masses of poly(styrene sulfonate) used in these investigations are 8, 18, 72, 100, 220, 252, 780, 1132, and 1200 kDa. The overlap concentration, estimated using Eq. (6.4.18), corresponding to these molar masses are 1218, 240, 15, 7.8, 1.61, 1.22, 0.13,

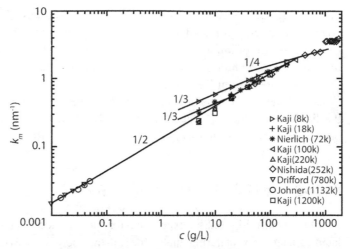

Figure 6.12 Dependence of the polyelectrolyte peak position k_m on polyelectrolyte concentration c in salt-free aqueous solutions of sodium poly(styrene sulfonate) at various molar masses and using different radiations. Different symbols denote the different molar masses as mentioned in the legend, where the key corresponds to the first author of the reference and the molar mass in Daltons. The nature of the radiation, molar mass, and the key in the legend are as follows: ▹ : X-ray, M = 8 kDa, Kaji (8k) [Kaji *et al.* (1988)]; + : X-ray, M = 18 kDa, Kaji (18k) [Kaji *et al.* (1988)]; ★ : neutrons, M = 72 kD, Nierlich (72k) [Nierlich *et al.* (1979)]; ◁ : X-ray, M = 100 kDa, Kaji (100k) [Kaji *et al.* (1988)]; △ : X-ray, M = 220 kDa, Kaji (220k) [Kaji *et al.* (1988)]; ◊ : X-ray, M = 252 kDa, Nishida (252k) [Nishida *et al.* (2001)]; ▽ : light, M = 780 kDa, Drifford (780k) [Drifford & Dalbiez (1984); o : light, M = 1132 kDa, Johner (1132k) [Johner *et al.* (1994)]; □ : X-ray, M = 1200 kDa, Kaji (1200k) [Kaji *et al.* (1988)]. The slopes of 1/3, 1/2, 1/2, and 1/4, expected respectively for infinitely dilute "gas-like," dilute "liquid-like," semidilute, and concentrated, are included as guides [Adapted from Muthukumar (2016a)].

0.06, and 0.054 g/L, respectively. Data collected over different combinations of molar mass and polymer concentration are able to cover the five concentration regimes providing a strong conceptual foundation on structural correlations in salt-free solutions. The small angle X-ray scattering (SAXS) data for M = 8 kDa show the "gas-like" regime, $k_m \sim c^{1/3}$, since $c^\star$ is much above the concentrations investigated for this molar mass [Kaji *et al.* (1988)]. For M=18 kDa ($c^\star$ = 240 g/L), SAXS data are in the "gas-like" regime for $c \ll c^\star$ and "liquid-like" ($k_m \sim \sqrt{c}$) for $c < c^\star$ [Kaji *et al.* (1988)]. The light scattering data on M = 780 kDa [Drifford & Dalbiez (1984)] and M = 1132 kDa [Johner *et al.* (1994)] are for $c < c^\star$, showing the "liquid-like" behavior, $k_m \sim \sqrt{c}$. The SAXS data of Kaji *et al.* (1988) for M = 100, 220, and 1200 kDa are for $c > c^\star$ showing the semidilute regime behavior, $k_m \sim \sqrt{c}$. Similarly, the neutron scattering data [Nierlich *et al.* (1979)] for 72 kDa exhibit $k_m \sim \sqrt{c}$ behavior for $c > c^\star$ and the crossover from the "gas-like" behavior for $c \ll c^\star$. For M = 252 kDa, SAXS data [Nishida *et al.* (2001)] at c = 200-700 g/L (much above $c^\star$ = 1.22 g/L) show the Regime IV law ($k_m \sim c^{1/4}$) expected for very high concentrated solutions. The SAXS data for M = 252 kDa at c = 1000 g/L represent essentially molten sodium poly(styrene sulfonate). These experimental observations summarized in Fig. 6.12 are in accord with the theoretical derivations given above for the five concentration regimes regarding structural correlations in salt-free conditions.

The presence of different concentration regimes and crossover behaviors between them are in full conformity with the above summary based on the double screening theory, as evident from Fig. 6.12.

Complementing the field theoretic double screening theory, approaches based on the liquid state theory [Yethiraj (1998, 2009), Shew & Yethiraj (1999, 2000)] have been implemented in addressing Regime I and Regime III.

6.5   Aggregation of Similarly Charged Polymers

As illustrated in Fig. 6.2(b), experimental results on the scattering intensity extrapolated to near zero scattering angles for salt-free polyelectrolyte solutions (Fig. 6.2b) indicate that there are large clusters consisting of many chains, although these chains have the same charge. We have already seen that when counterions adsorb on the charged monomers along chain backbones, dipoles form and the interaction energy u_0 between two randomly oriented dipoles is attractive. These dipole–dipole interactions can be quite strong compared to k_BT. For two freely rotating dipoles, the interaction energy is given in Eq. (1.3.8). Analogous to the two-body excluded volume parameter for uncharged polymers, the two-body dipole–dipole interaction parameter, v_{dd} can be defined using Eq. (1.3.8) (see Eq. (5.11.4)). At room temperature in water, $v_{dd} \simeq -10k_BT$ if the separation distance between the dipoles and the dipole length are 0.25 nm. Therefore, we expect that some segments of the chains will cling together due to formation of quadrupoles and the rest of the chains will repel each other. If several such quadrupoles can form between several intermingling chains, the resultant net attractive forces can effectively compete against the electrostatic repulsion between the chains. In addition, if the time required for the dissociation of the whole collection of quadrupoles is sufficiently long, an equilibrium description of such aggregates can be implemented to assess their relative stability.

Let us consider the equilibrium thermodynamics of the self-assembly of aggregates of polyelectrolyte chains, mediated by quadrupole formation [Muthukumar (2016b)]. When aggregation is feasible, the polyelectrolyte solution contains multichain aggregates with different numbers of chains, and unaggregated chains. As a concrete example, consider the formation of an aggregate of m polyelectrolyte chains (Fig. 6.13), where each quadrupole junction has an energy $-u_0$. Let the average number of segments between two adjacent quadrupole junctions be N. According to the classical equilibrium theory of self-assembly into micelles or aggregates [Tanford (1980), Israelachvili (2011)], the mole fraction of m-aggregates (m-mers) X_m is given as

$$X_m = m \left[X_1 e^{(F_1 - \frac{F_m}{m})/k_BT} \right]^m, \tag{6.5.1}$$

with the constraint of conservation of the total mole fraction X of the polymer in the solution,

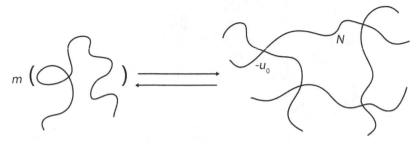

Figure 6.13 Cartoon of aggregation of m polyelectrolyte chains into a charged microgel with quadrupole energy of $-u_0$ and spacer length of N.

$$X = \sum_{m=1}^{\infty} X_m. \tag{6.5.2}$$

Here, the subscript 1 denotes the unaggregated chains. F_1 and F_m are the free energies of the unaggregated polyelectrolyte chain and the m-aggregate, respectively. After deriving expressions for F_1 and F_m in terms of variables such as the polymer concentration, salt concentration, and quadrupole energy, the distribution function of the mole fraction of m-mers can be obtained.

Since the enhanced scattering intensity occurs in salt-free solutions and disappears in salty solutions, we focus on the salt-free limits of F_1 and F_m. Substituting the equilibrium value of the end-to-end distance of a flexible polyelectrolyte chain given by Eq. (5.6.17) in Eq. (5.6.16), F_1 is given as

$$\frac{F_1}{k_B T} \simeq 1.11 \left(\frac{\alpha^2 z_p^2 \ell_B}{\ell} \right)^{2/3} N, \tag{6.5.3}$$

where the electrostatic interaction is recognized to overwhelm the van der Waals interactions among the chain segments. In order to obtain an expression for F_m, we use the approximate theory of polyelectrolyte gels developed in Chapter 10. For isotropically swollen salt-free polyelectrolyte gels containing m chains of N segments and $m/2$ crosslinks, the polymer volume fraction ϕ of the equilibrated gel is given by the swelling equilibrium, Eq. (10.3.4), as

$$\phi^{2/3} = \frac{\phi_0^{2/3}}{\alpha z_p N}, \tag{6.5.4}$$

where ϕ_0 is the volume fraction of the polymer in the hypothetical reference state where all strands between crosslinks obey Gaussian chain statistics. Substituting this result in Eq. (10.9.20), the free energy of the m-mer follows as

$$\frac{F_m}{k_B T} = m \left\{ -\frac{u_0}{2} + \frac{3\alpha z_p N}{2} \left[1 - \ln \left(\alpha z_p N^{4/3} \right) \right] \right\} + \text{constant}, \tag{6.5.5}$$

where the constant term is unspecified. In obtaining this expression, we have assumed that $\chi = \frac{1}{2}, \phi_0 = 1/\sqrt{N}, \alpha z_p \gg 1/(2N)$, and the polymer volume fraction is small ($\phi \ll 1$). As already noted, u_0 is the quadrupole energy.

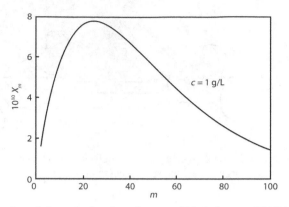

Figure 6.14 Distribution of the mole fraction of m-mers [Muthukumar (2016b)].

Using Eqs. (6.5.3) and (6.5.5), and choosing the constant term in Eq. (6.5.5) such that $F_m \to F_1$ for $m = 1$, we get

$$\frac{1}{k_B T}\left(F_1 - \frac{F_m}{m}\right) = \left(1 - \frac{1}{m}\right)\Theta, \qquad (6.5.6)$$

where

$$\Theta = 1.11\left(\frac{\alpha^2 z_p^2 \ell_B}{\ell}\right)^{2/3} N + \frac{u_0}{2} - \frac{3\alpha z_p N}{2}\left[1 - \ln\left(\alpha z_p N^{4/3}\right)\right]. \qquad (6.5.7)$$

Substituting Eq. (6.5.6) in Eq. (6.5.1), we obtain

$$X_m = m\left(X_1 e^{\Theta}\right)^n e^{-\Theta}, \qquad (6.5.8)$$

where Θ is given in Eq. (6.5.7). Combining Eqs. (6.5.4) and (6.5.8), X_1 is solved in terms of the total mole fraction X of the polyelectrolyte strands in the solution as [Israelachvili (2011)]

$$X_1 = \frac{\left(1 + 2Xe^{\Theta}\right) - \sqrt{1 + 4Xe^{\Theta}}}{2Xe^{2\Theta}}. \qquad (6.5.9)$$

The populations of aggregates with m chains and unaggregated chains for a prescribed total mole fraction of the polymer X are obtained using Eqs. (6.5.8) and (6.5.9) in terms of Θ given in Eq. (6.5.7). Above a certain critical polymer concentration, known as the critical aggregation concentration CAC at which $X_1 e^{\Theta} = 1$, stable aggregates form. As an example, for the choice of $\alpha z_p = 0.3, \ell_B/\ell = 3, u_0 = 10$, and $N = 16$, the mole fraction corresponding to CAC is $X_{\text{CAC}} = 8.2 \times 10^{-8}$. Note that the polymer concentration c in the unit of g/L is related to the mole fraction X according to $c = 55.5 M_w X$ in aqueous systems, where M_w is the molecular weight in g/mole. At $X = 5.15 \times 10^{-8}$, the calculated distribution function of m-mers is given in Fig. 6.14. On average, about 30 chains aggregate together with a radius of gyration of the aggregate $R_{g,\text{agg}}$ in the range of hundreds of nanometers. In general, for large m and $X > X_{\text{CAC}}$, the average number of chains in the aggregate is [Israelachvili (2011)]

$$\langle m \rangle = 2\sqrt{X}e^{\Theta/2}, \qquad (6.5.10)$$

where X is proportional to c. Taking the polymer volume fraction in the aggregate as

$$\phi = \frac{\langle m \rangle N \ell^3}{\left(\frac{4}{3}\pi R_{g,\text{agg}}^3\right)}, \tag{6.5.11}$$

and using Eq. (6.5.4) with $\phi_0 = 1/\sqrt{N}$, we obtain

$$R_{g,\text{agg}} = \left(\frac{3}{4\pi}\right)^{1/3} \left(\alpha z_p\right)^{1/2} \langle m \rangle^{1/3} N \ell, \tag{6.5.12}$$

in the salt-free limit addressed here. Since the average number of chains in the aggregates $\langle m \rangle$ is proportional to $\sqrt{c}$ according to Eq. (6.5.10), the concentration dependence of the radius of gyration of the aggregate $R_{g,\text{agg}}$ follows from Eq. (6.5.12) as

$$R_{g,\text{agg}} \sim N c^{1/6}. \tag{6.5.13}$$

Note that the above derivation is only at the mean field level. Conformational fluctuations and corrections to the approximations used in the theory of polyelectrolyte gels may contribute to additional details.

If the salt concentration is high, quadrupoles are unstable due to electrostatic screening and the aggregates cannot be sustained. As a result, aggregates are present only at low salt concentrations. Using the expressions for F_1 and F_m valid for moderate salt concentrations, it can be shown that aggregation can occur even at these salt concentrations, but only at higher polymer concentrations. As we shall see in the next chapter, the occurrence of the slow mode in salt-free solutions and its disappearance at higher salt concentrations is a manifestation of the spontaneous aggregation of similarly charged polyelectrolytes derived in this section.

7 Dynamics

7.1 Introduction

Communication from one part of the human body to another, and even between different locations inside a cell, requires movement of charged macromolecules and electrolyte ions in crowded environments. Without any motion of the various charged species constituting a cell, life as we know it cannot exist. This movement is orchestrated by several concomitant forces that emerge from the hydrodynamic interactions in the background fluid, electrostatic, and topological correlations among the macromolecules, molecular recognition and binding, and driving forces from gradients in chemical potentials of the species, pH, electrolyte concentration, electric potential, and temperature. In view of a plethora of dynamical properties a cell is required to exhibit in order to properly function, these forces are naturally coupled among themselves and work together either synergistically or antagonistically depending on the targeted purpose of transport of the macromolecules. In addition to the context of intracellular and intercellular communication through charged macromolecules, their designable mobility is of great significance in technologies that address, for example, separation science, controlled retention and release of macromolecular cargo, and new ways to characterize charged macromolecules.

Due to the aforementioned aspects, a comprehensive understanding of motion of charged macromolecules in *in vivo* biological environments is a difficult task. However, considerable progress toward this goal has been achieved with *in vitro* studies using folded proteins, polynucleotides, and synthetic polyelectrolytes in aqueous media. Even for these idealized situations, the phenomenology is quite complex in a manner completely unexpected from the knowledge on uncharged systems. Complexity of the dynamical behavior of solutions of charged systems arises from many sources. For example, the hydrodynamic interaction, which is long ranged, adds an additional dimension to the complexity of a system which already has two long-ranged correlations arising from chain connectivity and the electrostatic interactions. The necessity to treat these three long-ranged correlations in a self-consistent manner, along with effects from crowding and compartmentalization, poses a huge challenge in order to understand the rich phenomenology of the dynamics of charged macromolecules. In this chapter, we shall dissect the full challenge into several essential basic concepts and then assemble these together toward an understanding of the dynamics of solutions of charged macromolecules.

One of the fundamental laws of dynamics of a molecule or a suspended particle in a solution at nonzero temperatures is the law of **diffusion**. As a specific example, consider a rigid spherical uncharged particle of radius R immersed in a solvent made of relatively small molecules compared to the particle. The instantaneous movement of the particle is difficult to predict due to collisions by a large number of solvent molecules in random directions. As a result, the particle undergoes random motion, known as **Brownian motion** [Einstein (1956)]. If we track its center of mass position $\mathbf{R}_{cm}(t)$ at time t, the trajectory is random and the average displacement of the particle over a long period of time is zero. However, the mean square displacement of a particle undergoing Brownian motion over the time duration t is given by

$$\langle[\mathbf{R}_{cm}(t) - \mathbf{R}_{cm}(0)]^2\rangle = 6Dt, \qquad \text{(Einstein)} \qquad (7.1.1)$$

where the angular brackets denote averaging over thermal noise. Eq. (7.1.1) is the Einstein law of diffusion in three dimensions and D is the diffusion coefficient of the Brownian particle. The diffusion coefficient is given as

$$D = \frac{k_B T}{f_t}, \qquad (7.1.2)$$

where f_t is the translational friction coefficient of the particle. According to the Stokes law of friction, the translational friction coefficient depends on the particle radius and the viscosity η_0 of the solvent as,

$$f_t = 6\pi\eta_0 R. \qquad \text{(Stokes)} \qquad (7.1.3)$$

Substituting Eq. (7.1.3) into Eq. (7.1.2), we get the **Stokes–Einstein relation**,

$$D = \frac{k_B T}{6\pi\eta_0 R}. \qquad \text{(Stokes–Einstein)} \qquad (7.1.4)$$

Note that D is nonzero at nonzero temperatures and hence the particle undergoes diffusion at all temperatures except absolute zero. Furthermore, note that D depends on the nature of the particle only through its radius R. In contrast, if the particle follows Newtonian dynamics, the square of its displacement is proportional to the square of the elapsed time, t^2, with the proportionality factor depending on the mass of the particle. The diffusion coefficient can be experimentally measured using microscopy-based particle tracking methods or dynamic light scattering (DLS). The particle size is then inferred from the measured D using the Stokes–Einstein relation.

The Stokes–Einstein relation obtained above for a rigid spherical particle is also applicable for uncharged macromolecules and their self-assembled structures. However, there are additional features from topological correlations and excluded volume effects in these systems. As an example, consider an uncharged flexible polymer chain which is a topologically correlated structure. Every monomer of the polymer chain is subjected to stochastic Brownian forces arising from incessant collisions by solvent molecules. Therefore, each monomer has a tendency to undergo diffusion on its own right, but now this tendency is modified by a monomer's connectivity to its neighbors along the chain and excluded volume interactions among monomers. As a consequence of the non-independence of monomers, the Einstein law is not applicable for their mean square displacements for short durations. Instead, the monomers undergo an anomalous diffusion. However, for very long times and large distances compared to the polymer size, the center-of-mass of the chain obeys the Einstein law, where the diffusion coefficient now refers to the center-of-mass of the whole chain. Let τ be the duration of time after which the conformation of the chain is independent of its initial conformation. This characteristic time of the chain depends on the size and chemical details of the chain. The law of diffusion is valid for a polymer chain only if the time duration for its movement is longer than its characteristic time τ. This universal law of diffusion is validated by experiments on solutions containing uncharged macromolecules, independent of their chemical constitution or architecture. Furthermore, measurement of D at appropriately chosen time durations, in combination with use of the Stokes–Einstein law, is one convenient method to determine the size of a diffusing macromolecule.

The Stokes–Einstein law breaks down for charged macromolecules in drastic departure from the behavior of uncharged macromolecules. In general, when a macromolecule carries charges, its dynamical behavior exhibits many intriguing phenomena unseen in uncharged molecules (see Section 7.4). For example, the diffusion coefficient of a very large polyelectrolyte molecule, as measured with DLS, can be several orders of magnitude larger than expected from the Stokes–Einstein law, and it can be even as high as that of a metallic ion! We shall describe the resolution of these unexpected results in Sections 7.6 and 7.7.

In the case of flexible and semiflexible polyelectrolytes, chains can interpenetrate and entangle with each other upon an increase in polymer concentration. As shown in Chapter 6, three major concentration regimes of dilute, semidilute, and concentrated solutions can be identified and the dynamical properties in these regimes are distinctively different in terms of their dependencies on the characteristics of the macromolecules. The dynamical properties of interest include diffusion, mobility in the presence of an electric field, ionic current and viscosity of the solution, and relaxation spectra of internal chain dynamics as discerned from rheology and dielectric dispersion. In addition to incorporating the excluded volume and the electrostatic interactions which govern the equilibrium properties, we shall address the roles of hydrodynamic interactions in the solution, coupled dynamics of chains and their counterion clouds, and entanglement effects arising from the uncrossability of interpenetrating chains against each other.

Complementing the homogeneous solutions of charged macromolecules (Figs. 7.1a and 7.1b), there are many scenarios where charged macromolecules are compartmentalized and subsequently induced to move from one set of compartments to another set using various stimuli. An example is the translocation of a macromolecule such as DNA or RNA through a nanopore connecting two compartments (Fig. 7.1c) under the influence of an external stimulus. This setup is the basis of single molecule electrophoresis in the context of fast and accurate genome sequencing. In order for the macromolecule to translocate through a nanopore, it must squeeze itself into the nanopore and thus surrender a part of its conformational entropy. This is equivalent

(a)   (b)

(c)   (d)

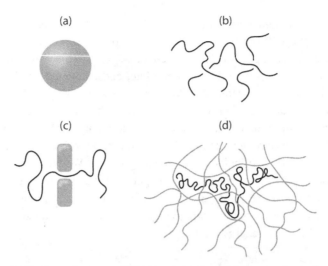

Figure 7.1 (a) A folded protein (or a colloidal particle) is modeled as a spherical particle carrying charges on its surface. (b) A collection of flexible polyelectrolyte chains. Three regimes of dilute, semidilute, and concentrated solutions can be identified depending on polymer concentration. (c) Sketch of navigation of a polymer chain through a single entropic barrier created by a membrane with a nanopore embedded in it. (d) Partitioning of a single polymer chain into multiple free energy traps results in non-diffusive topologically frustrated hierarchical dynamics for the chain.

to the creation of an entropic barrier for the translocation of the macromolecule, which arises from the topological correlation of the macromolecule through chain connectivity. The macromolecule must navigate through the entropic barrier in order to undergo successful translocation. Another example where topological correlations play a significant role is in the context of long charged macromolecules such as DNA trapped inside a hydrogel, where each chain is partitioned into multiple compartments (Fig. 7.1d). In order for the chain to diffuse and move around, it must simultaneously navigate through multiple entropic barriers. This is generally a difficult and time-consuming process. As a result, the chain is temporarily locked into a non-diffusive topologically frustrated dynamical state. Nevertheless, if the compartments are large enough to hold many monomers of the macromolecule, the internal dynamics of the trapped macromolecule are essentially unrestricted without compromising the functional properties of the macromolecule. Experimental handles such as the concentration of added salt can be designed to controllably release the trapped charged macromolecule from its dynamically frustrated states.

In this chapter, we shall present only the salient concepts that explain the various dynamical properties of charged macromolecules in the aforementioned contexts (Fig. 7.1). The technical details are relegated to the original publications. The organization of this chapter is as follows: (1) We shall begin with a brief introduction to the hydrodynamic interaction, which is one of the fundamental contributors to the dynamics of macromolecules in solutions. (2) Modeling macroions and folded proteins as spherical particles with surface charge, we shall describe their dynamics in dilute solutions and in the proximity of charged surfaces, and also inside nanopores. (3) We will present the basic models of dynamics of uncharged polymers: Zimm, Rouse, entropic barrier, and reptation. (4) Next, we will describe the dynamics of polyelectrolytes in dilute, semidilute, and entangled conditions. In this section, we shall address several important concepts such as hydrodynamic screening and electrostatically coupled dynamics between the macromolecule and its ion cloud, and phenomena such as the "ordinary–extraordinary" transition. We shall also describe diffusion, electrophoretic mobility, ionic current, relaxation spectra of macromolecules, and viscosity of polyelectrolyte solutions. (5) We will then describe the translocation of single macromolecules by navigation through a single entropic barrier. (6) Finally, we shall address the emergence of non-diffusive topologically frustrated hierarchical polymer dynamics, where single charged macromolecules are immobilized due to multiple free energy traps.

7.2   Hydrodynamic Interaction

When macromolecules are dispersed in a solvent, the characteristic size and time that describe the collective motion of the macromolecules are several orders of magnitude larger in comparison to those for individual solvent molecules. As a result, the solvent background can be assumed to be a hydrodynamic continuum. Adopting this useful and accurate proposition, let us consider a small volume element around the spatial location $\mathbf{r}$, which is large enough to consider the solvent as a continuum. The time dependence of the velocity field $\mathbf{v}(\mathbf{r},t)$ at $\mathbf{r}$ and time t is obtained from Newton's second law of motion, namely the rate of change of momentum of a fluid element is equal to the net force acting on it, as [Landau & Lifshitz (1959)]

$$\rho_0 \frac{\partial \mathbf{v}(\mathbf{r},t)}{\partial t} + \rho_0 \mathbf{v}(\mathbf{r},t) \cdot \nabla \mathbf{v}(\mathbf{r},t) - \eta_0 \nabla^2 \mathbf{v}(\mathbf{r},t) + \nabla p(\mathbf{r},t) = \mathbf{F}(\mathbf{r},t), \qquad (7.2.1)$$

where $p(\mathbf{r},t)$ is the local pressure, ρ_0 is the mass density of the fluid, and η_0 is the shear viscosity of the fluid. $\mathbf{F}(\mathbf{r},t)$ is an external force per unit volume, such as the gravitational or electric force.

Inertial forces are usually negligible at relatively long time scales relevant to macromolecules. In addition, the velocity fields are weak in most of the experimental situations of current interest, barring onset of turbulence. Therefore, the first two terms on the left-hand side of the above equation may be ignored resulting in the "zero-frequency" linearized Navier–Stokes equation [Landau & Lifshitz (1959)]

$$-\eta_0 \nabla^2 \mathbf{v}(\mathbf{r},t) + \nabla p(\mathbf{r},t) = \mathbf{F}(\mathbf{r},t). \qquad (7.2.2)$$

In addition, let us assume that the fluid is incompressible. This condition is given as

$$\nabla \cdot \mathbf{v}(\mathbf{r},t) = 0. \qquad (7.2.3)$$

The pressure term in Eq. (7.2.2) can be eliminated by using Eq. (7.2.3), so that the velocity field can be expressed in terms of any force field $\mathbf{F}$, independent of whether it is imposed externally or internally.

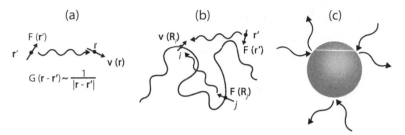

Figure 7.2 (a) Long-ranged hydrodynamic interaction between two fluid elements at $\mathbf{r}$ and $\mathbf{r}'$. The Oseen tensor $\mathbf{G}$ transmits a force $\mathbf{F}$ at $\mathbf{r}'$ into a velocity $\mathbf{v}$ at $\mathbf{r}$. (b) Each monomer is coupled hydrodynamically to all other monomers, both intra-chain and interchain. The velocity of the ith monomer depends on the force generated by the jth monomer as well as the force at $\mathbf{r}'$ in the solvent. (c) The role of monomers is replaced by surface elements in the case of rigid particles suspended in a solvent.

The fundamental property of the background fluid is how it transmits a force field $\mathbf{F}(\mathbf{r}',t)$ at the location $\mathbf{r}'$ to another location $\mathbf{r}$. Combining Eqs. (7.2.2) and (7.2.3), the velocity field at $\mathbf{r}$ can be written in terms of $\mathbf{F}$ at $\mathbf{r}'$ as

$$\mathbf{v}(\mathbf{r}) = \int d\mathbf{r}' \mathbf{G}(\mathbf{r} - \mathbf{r}') \cdot \mathbf{F}(\mathbf{r}'), \qquad (7.2.4)$$

where $\mathbf{G}$ is called the Oseen tensor,

$$\mathbf{G}(\mathbf{r} - \mathbf{r}') = \frac{1}{8\pi\eta_0 |\mathbf{r} - \mathbf{r}'|} \left[\mathbf{1} + \frac{(\mathbf{r} - \mathbf{r}')(\mathbf{r} - \mathbf{r}')}{|\mathbf{r} - \mathbf{r}'|^2} \right], \qquad (7.2.5)$$

where $\mathbf{1}$ is the unit tensor. The operator $\mathbf{G}$ may be taken as a pneumonic switch that converts a force at $\mathbf{r}'$ to a velocity at a different location $\mathbf{r}$ (Fig. 7.2a). If an angular average is performed on the Oseen tensor, the tensor inside the square brackets reduces to the factor 4/3 so that

$$G_{\text{ang.av}}(\mathbf{r} - \mathbf{r}') = \frac{1}{6\pi\eta_0 |\mathbf{r} - \mathbf{r}'|}. \qquad (7.2.6)$$

It is important to recognize that $\mathbf{G}$ is inversely proportional to the separation distance between the point source of a force and the point where velocity perturbation is monitored,

$$G(\mathbf{r} - \mathbf{r}') \sim \frac{1}{|\mathbf{r} - \mathbf{r}'|}. \qquad (7.2.7)$$

This long-ranged correlation between the force and velocity in the solvent continuum is called the **hydrodynamic interaction**.

The long-ranged nature of the hydrodynamic interaction, which is analogous to the long-ranged nature of the electrostatic interactions and topological correlations due to chain connectivity, leads to dynamical correlations among the various monomers of dispersed macromolecules, despite large separation distances between these monomers. An example is given in Fig. 7.2b, where the force $\mathbf{F}(\mathbf{R}_j)$ generated by the jth monomer at location $\mathbf{R}_j$ contributes to the velocity $\mathbf{v}(\mathbf{R}_i)$ of the ith monomer at location $\mathbf{R}_i$, in addition to the direct contribution from $\mathbf{F}(\mathbf{r}')$. The velocity of a specified monomer is affected by all sources of force, even those that are far away. These sources are other monomers of the same macromolecule or other macromolecules, charges from counterions and other electrolyte ions, and externally generated electric and flow fields. Note that the sources of force for rigid particles with surface charges are the surface points of the particles (Fig. 7.2c).

7.3   Dilute Solutions of Colloidal Particles, Macroions, and Folded Proteins

Let us model colloidal particles, macroions, and folded proteins as rigid spherical particles carrying charges only on their surfaces. In such systems, there are no internal structures and dynamics. As a general example, consider a collection of spherical particles of uniform radius R suspended in a very dilute electrolyte solution where the solvent viscosity is η_0. Each spherical particle carries a net charge Q which is uniformly distributed on its surface, as in Section 4.2. In very dilute solutions, the interparticle interactions are negligible, hence the Brownian motion of an isolated spherical particle is described by the results in Section 7.1. Without any consideration of its ionic environment, the diffusion coefficient D and the translational friction coefficient f_t of the particle are, respectively, given by the Einstein law (Eq. (7.1.2)) and the Stokes law (Eq. (7.1.3)) as

$$D = \frac{k_B T}{f_t}, \qquad f_t = 6\pi \eta_0 R. \tag{7.3.1}$$

In addition, if a uniform electric field $\mathbf{E}_0$ is imposed on the solution, the particle drifts toward the oppositely charged electrode. The movement of a charged particle or molecule relative to a stationary liquid by an applied electric force is the phenomenon of electrophoresis. For weak enough electric fields, and in steady state, the average velocity $\mathbf{v}_0$ of the particle is proportional to the electric field $\mathbf{E}_0$,

$$\mathbf{v}_0 = \mu \mathbf{E}_0, \tag{7.3.2}$$

where μ is called the electrophoretic mobility of the particle. Measurement of μ of a charged macromolecule or a particle is a commonly used experimental technique in characterizing the macromolecule or the particle in terms of its total charge and zeta potential (Fig. 4.6). In view of the importance of μ in experimental characterization of charged macromolecules and particles, we shall present the classical theories of electrophoretic mobility in the next three subsections. The role of coupling between an imposed electric field and flow fields on a charged particle undergoing drift and diffusion inside regions confined by charged surfaces will be treated subsequently.

7.3.1   Einstein's Electrophoretic Mobility

The electric force $\mathbf{F}_{\text{electric}}$ experienced by the particle of charge Q under the electric field $\mathbf{E}_0$, which is directed toward the oppositely charged electrode is

$$\mathbf{F}_{electric} = Q\mathbf{E}_0. \tag{7.3.3}$$

This force is opposed by the viscous resistance from the fluid, which is given by the frictional force,

$$\mathbf{F}_{friction} = -f_t \mathbf{v}_0, \tag{7.3.4}$$

where $\mathbf{v}_0$ is the velocity of the particle. From Newton's equation, the steady state velocity is attained when the net force from these contributions is zero (namely, there is no acceleration of the particle),

$$-f_t \mathbf{v}_0 + Q\mathbf{E}_0 = 0. \tag{7.3.5}$$

Therefore, the uniform velocity of the particle is given by

$$\mathbf{v}_0 = \frac{Q}{f_t} \mathbf{E}_0. \tag{7.3.6}$$

The electrophoretic mobility of the particle defined in Eq. (7.3.2) follows from Eq. (7.3.6) as

$$\mu = \frac{Q}{f_t}. \tag{7.3.7}$$

Substituting the Stokes result (Eq. (7.1.3)) for f_t in this equation, we get

$$\mu = \frac{Q}{6\pi\eta_0 R}. \tag{7.3.8}$$

In an equivalent manner, substitution of the Einstein result for the diffusion coefficient (Eq. (7.1.2)) in Eq. (7.3.7) gives

$$\mu = \frac{QD}{k_B T}. \tag{7.3.9}$$

Therefore, measurement of the electrophoretic mobility of a particle or a macromolecule enables the deduction of its net charge Q if either its radius or diffusion coefficient is known. Further, by defining $Z = Q/e$, where e is the electronic charge, Eq. (7.3.9) gives

$$\frac{\mu}{ZD} = \frac{e}{k_B T} = 39.6 \ \text{V}^{-1} \quad \text{at } 20°\text{C}. \tag{7.3.10}$$

This is a universal result given by Einstein's expression for electrophoretic mobility.

Although Eqs. (7.3.8)–(7.3.10) are sometimes used to determine the net charge of the particle or macromolecule, note that the Einstein result is derived only for an isolated particle without consideration of its ionic environment. In reality, the particle is never alone in the electrolyte solution: there are always oppositely charged ions due to the required electroneutrality condition. The particle is subjected to a force from its ion cloud in addition to $\mathbf{F}_{electric}$ and $\mathbf{F}_{friction}$ given above. The local charge density $\rho(\mathbf{r})$ at location $\mathbf{r}$ inside the ion cloud exerts a drag on the particle (Fig. 7.3). In fact, experiments on colloidal particles, proteins, and polyelectrolytes show significant deviations from Eqs. (7.3.8)–(7.3.10) in their electrophoretic mobility. It is necessary to account for the response of the ion cloud around a macroion when it moves under an electric field, as described below.

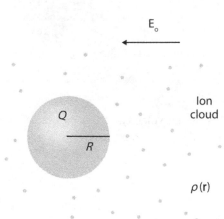

E_0

Ion cloud

Q

R

$\rho(r)$

Figure 7.3 Movement of a rigid sphere with net surface charge Q and radius R under an externally imposed electric field $\mathbf{E}_0$. Negatively charged sphere moves toward the positive electrode. The motion of the particle is resisted by the drag in the opposite direction due to the ion cloud around the particle. The local charge density at location $\mathbf{r}$ is $\rho(\mathbf{r})$.

7.3.2 Hückel Theory

The contribution of the ion cloud to the electrophoretic motion of a spherical charged particle is approximately accounted for in the Hückel theory of electrophoretic mobility [Hückel (1924)]. In general, the electrical force at location $\mathbf{r}$ due to the local electric field $\mathbf{E}(\mathbf{r})$ at that location is (Eq. (3.1.3))

$$\mathbf{F}(\mathbf{r}) = \rho(\mathbf{r})\mathbf{E}(\mathbf{r}), \tag{7.3.11}$$

where $\rho(\mathbf{r})$ is the local charge density. This force is transmitted through hydrodynamics to all surface elements of the particle. In the Hückel theory, the local electric field $\mathbf{E}(\mathbf{r})$ is assumed to adopt a uniform value $\mathbf{E}_0$. Using the Debye–Hückel theory to connect $\rho(\mathbf{r})$ and the electric potential around the spherical particle, and the Stokes method to describe hydrodynamics around the particle, Hückel's expression for the electric force on the particle is

$$\mathbf{F}_{\text{electric, Hückel}} = Q\mathbf{E}_0 - \frac{\kappa R}{(1 + \kappa R)}Q\mathbf{E}_0, \tag{7.3.12}$$

where κ is the inverse Debye length given in terms of the concentrations of counterions and dissociated electrolyte ions in solution. The first term on the right-hand side of Eq. (7.3.12) is the electric force acting directly on the charged particle, and the second term is the opposing force due to the ion cloud around the particle. Combining the two terms on the right-hand side of Eq. (7.3.12),

$$\mathbf{F}_{\text{electric, Hückel}} = \frac{Q}{(1 + \kappa R)}\mathbf{E}_0. \tag{7.3.13}$$

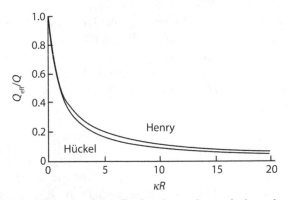

Figure 7.4 Dependence of the ratio of the effective charge Q_{eff} to the bare charge Q on the ratio κR of the particle radius to the Debye length, according to the theories of Hückel and Henry for electrophoretic mobility.

Therefore, the particle and its ion cloud move together as a collective entity (quasiparticle) with an effective charge Q_{eff} on the particle given as

$$Q_{eff} = \frac{Q}{(1 + \kappa R)}. \qquad (7.3.14)$$

The effective charge decreases with an increase in κ, and can be significantly smaller than the true charge at high salt concentrations (Fig. 7.4).

The net electric force acting on the particle given in Eq. (7.3.13) can be alternatively expressed in terms of the electric potential ψ_s at the surface of the particle. According to Eq. (4.2.9), ψ_s is given as

$$\psi_s = \frac{Q}{4\pi\epsilon_0\epsilon R}\frac{1}{(1 + \kappa R)}. \qquad (7.3.15)$$

Hence, Eq. (7.3.13) gives

$$\mathbf{F}_{\text{electric, Hückel}} = 4\pi\epsilon_0\epsilon R\psi_s\mathbf{E}_0. \qquad (7.3.16)$$

The total force acting on the particle is $\mathbf{F}_{\text{electric, Hückel}} + \mathbf{F}_{\text{friction}}$, where the latter is given in Eq. (7.3.4). In the steady state, the total force is zero so that

$$\mathbf{F}_{\text{electric, Hückel}} + \mathbf{F}_{\text{friction}} = 0. \qquad (7.3.17)$$

Substituting Eqs. (7.3.1), (7.3.4) and (7.3.16) in this equation, we get

$$(6\pi\eta_0 R)\mathbf{v}_0 = 4\pi\epsilon_0\epsilon R\psi_s\mathbf{E}_0. \qquad (7.3.18)$$

The electrophoretic mobility defined in Eq. (7.3.2) follows from this equation as

$$\mu = \frac{2}{3}\frac{\epsilon_0\epsilon\psi_s}{\eta_0}. \qquad (7.3.19)$$

As an alternative equivalent expression for the electrophoretic mobility, Eqs. (7.3.1), (7.3.13), and (7.3.17) give

$$\mu = \frac{Q}{6\pi\eta_0 R}\frac{1}{(1 + \kappa R)}. \qquad (7.3.20)$$

Eqs. (7.3.19) and (7.3.20) represent the **Hückel law of electrophoretic mobility**.

Note the modification of the Einstein law, Eq. (7.3.8) by the factor $(1+\kappa R)^{-1}$ arising from the counterions and electrolyte ions in the solution. If the Einstein law were to be empirically used to analyze experimental data, Eq. (7.3.20) can be interpreted in terms of the Einstein form by defining an effective charge Q_{eff} instead of the bare charge Q,

$$\mu = \frac{Q_{\text{eff}}D}{k_B T}. \tag{7.3.21}$$

The Hückel prediction for the ratio $Q_{\text{eff}}/Q = 1/(1 + \kappa R)$ is presented in Fig. 7.4 as a function of κR. The ratio Q_{eff}/Q decreases sharply from one to a small value as κR increases. For example, for a nanoparticle or a protein of radius 1.5 nm in 1 M KCl, $\kappa R = 5$, and the effective charge is merely one-fifth of the bare charge, as seen in Fig. 7.4. Therefore, the electrophoretic mobility of a charged particle can be significantly lower than the value given by the Einstein result, depending on the ionic strength of the solution. Although the Hückel theory is approximate, the conclusion drawn from Fig. 7.4 is generally valid. Among several theoretical attempts to improve the Hückel theory the most notable is the Henry theory [Henry (1931)], described next.

7.3.3   Henry Theory

In general, the electric field $\mathbf{E}(\mathbf{r})$ around a charged sphere is not uniform as assumed in the Hückel theory. Addressing the nonuniform nature of $\mathbf{E}(\mathbf{r})$, the Hückel theory was revised by Henry (1931) to obtain the electric force acting on a spherical particle as

$$\mathbf{F}_{\text{electric, Henry}} = \mathbf{F}_{\text{electric, Hückel}} \, f_{\text{Henry}}(\kappa R), \tag{7.3.22}$$

where $\mathbf{F}_{\text{electric, Hückel}}$ is given by Eq. (7.3.12) and $f_{\text{Henry}}(x)$ is

$$f_{\text{Henry}}(x) = 1 + \frac{x^2}{16} - \frac{5x^3}{48} - \frac{x^4}{96} + \frac{x^5}{96} + \frac{x^4}{8}\left(1 - \frac{x^2}{12}\right)e^x E_1(x). \tag{7.3.23}$$

Here $x = \kappa R$ and

$$E_1(x) = \int_x^\infty \frac{e^{-y}}{y}\,dy. \tag{7.3.24}$$

The Henry function $f_{\text{Henry}}(\kappa R)$ increases from unity to 3/2 as κR increases from zero to infinity. The limits of the Henry function for small and large values of κR are

$$f_{\text{Henry}}(\kappa R) = \begin{cases} 1 + \frac{\kappa^2 R^2}{16} + \cdots, & \kappa R \ll 1 \\ \frac{3}{2}\left(1 - \frac{3}{\kappa R} + \cdots\right), & \kappa R \gg 1. \end{cases} \tag{7.3.25}$$

By balancing the electric force from Eq. (7.3.22) with the frictional force of Eq. (7.3.4), the electrophoretic mobility follows as

$$\mu = \frac{2}{3}\frac{\epsilon_0 \epsilon \psi_s}{\eta_0} f_{\text{Henry}}(\kappa R). \tag{7.3.26}$$

Substituting the limiting values of $f_{Henry}(\kappa R)$ as given in Eq. (7.3.25), the electrophoretic mobility is given by the Hückel expression for $\kappa R \ll 1$,

$$\mu = \frac{2}{3} \frac{\epsilon_0 \epsilon \psi_s}{\eta_0}, \qquad \kappa R \ll 1 \qquad (7.3.27)$$

and for $\kappa R \gg 1$,

$$\mu = \frac{\epsilon_0 \epsilon \psi_s}{\eta_0}, \qquad \kappa R \gg 1. \qquad (7.3.28)$$

This expression is known as the **Helmholtz–Smoluchowski equation**.

Substituting the expression for the surface potential ψ_s from Eq. (7.3.15), the electrophoretic mobility in the small and large limits of κR follows as

$$\mu = \begin{cases} \frac{Q}{6\pi\eta_0 R} \frac{1}{(1+\kappa R)}, & \kappa R \ll 1 \\ \frac{Q}{4\pi R^2 \eta_0 \kappa} = \frac{\sigma_s}{\eta_0 \kappa}, & \kappa R \gg 1, \end{cases} \qquad (7.3.29)$$

where σ_s is the surface charge density $Q/(4\pi R^2)$. The dependence of the ratio of the effective charge Q_{eff} to the bare charge Q given by the Henry theory is included in Fig. 7.4 as a comparison with the prediction from the Hückel theory.

7.3.4 Helmholtz–Smoluchowski Theory

Let us consider the nature of flow of an electrolyte solution near a charged planar surface in the presence of a constant electric field $\mathbf{E}_0$, without any pressure gradient. This situation reduces to electrophoresis of a spherical particle of sufficiently large radius when we assume that the velocity of fluid at the interface is the same as the velocity of the particle (known as the stick or no-slip boundary condition). Let the velocity at location $\mathbf{r}$ be $\mathbf{v}(\mathbf{r})$ and the charge density be $\rho(\mathbf{r})$. According to the linearized Navier–Stokes equation pertinent to the current situation, the velocity field follows from Eq. (7.2.2) as

$$-\eta_0 \nabla^2 \mathbf{v}(\mathbf{r}) = \rho(\mathbf{r}) \mathbf{E}_0. \qquad (7.3.30)$$

Using the Poisson equation, Eq. (3.2.12), we get

$$-\eta_0 \nabla^2 \mathbf{v}(\mathbf{r}) = -\epsilon_0 \epsilon \nabla^2 \psi(\mathbf{r}) \mathbf{E}_0, \qquad (7.3.31)$$

where $\psi(\mathbf{r})$ is the local electric potential.

Let the interface be an infinite plane along x- and z-directions at $y = 0$ with a surface potential ψ_s, as sketched in Fig. 7.5. The externally imposed uniform electric field is parallel to the surface in the positive x-direction, with magnitude E_x. For infinitely long distances along x, the fluid velocity and the charge density are uniform so that the derivatives with respect to x are zero. The variations of the velocity and charge density are only along the y-direction given by Eq. (7.3.31) as

$$-\eta_0 \frac{\partial^2 v_x}{\partial y^2} = -\epsilon_0 \epsilon \frac{\partial^2 \psi}{\partial y^2} E_x. \qquad (7.3.32)$$

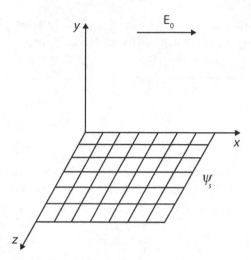

Figure 7.5 Fluid flow near a charged surface in the presence of an external uniform electric field E_0 directed along the x-direction. The surface is on the (x,z) plane at $y = 0$. ψ_s is the surface potential.

Integration of this equation gives

$$\eta_0 \frac{\partial v_x}{\partial y} = \epsilon_0 \epsilon \frac{\partial \psi}{\partial y} E_x + C_1, \tag{7.3.33}$$

where C_1 is an integration constant. At $y \to \infty$ far away from the charged surface, $\partial v_x / \partial y = 0$ and $\partial \psi / \partial y = 0$, so that $C_1 = 0$. Integration of Eq. (7.3.33) gives

$$\eta_0 v_x = \epsilon_0 \epsilon \psi E_x + C_2. \tag{7.3.34}$$

Here, the integration constant C_2 is zero, because both v_x and ψ are zero at distances much larger than the electric double layer thickness (see Section 4.1.2).

For the situation of electrophoresis of a spherical particle, the above equation is applicable if the particle radius R is so large that its surface is locally planar. In this scenario, ψ at $y = 0$ is the surface potential ψ_s, and the fluid velocity v_x at $y = 0$ is the velocity of the particle v_{0x} along the direction of the electric field E_x, so that

$$v_{0x} = \frac{\epsilon_0 \epsilon \psi_s}{\eta_0} E_x. \tag{7.3.35}$$

Therefore, the electrophoretic mobility given by the Helmholtz–Smoluchowski theory [Probstein (1989)] is

$$\mu = \frac{\epsilon_0 \epsilon \psi_s}{\eta_0}, \tag{7.3.36}$$

as given in Eq. (7.3.28) as the $\kappa R \gg 1$ limit in the Henry theory.

7.3.5   Electroosmotic Flow

Complementary to electrophoresis, the phenomenon of electroosmosis is the movement of a liquid relative to an immobile charged surface by an applied electric field

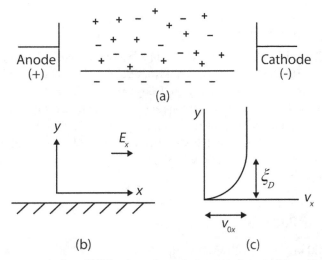

Figure 7.6 Electroosmotic flow (EOF) near a charged planar surface. (a) Experimental setup. (b) The applied electric field is uniform and along the x-direction. y is the direction normal to the planar surface. (c) Profile of fluid velocity along the x-direction with respect to the distance normal to the surface. v_{0x} is the terminal velocity and ξ_D is the Debye length.

[Probstein (1989)]. Due to the charge density of the surface, more counterions are present near the surface than in the bulk far from the surface (Section 4.1). These unadsorbed mobile counterions move collectively toward the oppositely charged electrode under an externally imposed electric field, while carrying solvent molecules with them. As a result, there is a net flow of the liquid, with its direction determined by the direction of the counterions. For example, negatively charged surfaces make the positively charged counterions move toward the negatively charged cathode, so that there is a net electroosmotic flow (EOF) toward the cathode, as illustrated in Fig. 7.6.

The EOF velocity is obtained by using the same setup as in Section 7.3.4 and Eq. (7.3.33), with the appropriate boundary conditions to solve Eq. (7.3.33). At the surface, the electric potential is ψ_s and the velocity of the liquid is zero with the assumption of the no-slip (or stick) boundary condition. Using these boundary conditions, the integration constant C_2 in Eq. (7.3.34) becomes

$$C_2 = -\epsilon_0 \epsilon \psi_s E_x. \tag{7.3.37}$$

Substituting this result in Eq. (7.3.34), we obtain

$$v_x(y) = \frac{\epsilon_0 \epsilon}{\eta_0} \left[\psi(y) - \psi_s \right] E_x. \tag{7.3.38}$$

The dependence of the electric potential on the distance from the planar surface is given in Section 4.1. For example, with the Debye–Hückel approximation, $\psi(y)$ falls exponentially with y and the characteristic decay length is the Debye length (Eq. (4.1.13) and Fig. 4.4). Therefore, at distances larger than the Debye length away from the surface, the electric potential vanishes and the velocity of the fluid acquires the terminal velocity v_{0x} given by

$$v_{0x} = -\frac{\epsilon_0 \epsilon \psi_s}{\eta_0} E_x. \tag{7.3.39}$$

This relation is the same Helmholtz–Smoluchowski relation given in Eqs. (7.3.28) and (7.3.36). A sketch of the result from Eq. (7.3.38) is included in Fig. 7.6. The thickness of the "skin" is essentially the Debye length. The EOF can be quite significant. As an example, for aqueous solutions at room temperature with the surface potential of 0.1 V and an electric field of 10^5 V/m, the terminal EOF velocity is about 10^7 nm/s, which is substantial in nanofluidics. In general, immobilization of large charged particles in the presence of externally applied electric fields can create substantial fluid velocities around the particles.

7.3.6   Zeta Potential

Although ψ_s in the above equations is defined as the surface potential at the particle radius R, it is in reality not at this location. As discussed in Section 4.1, many counterions spontaneously adsorb on a charged interface to form the Stern layer, thus reducing the charge density of the interface [Hiemenz & Rajagopalan (1997), Evans & Wennerström (1999), Israelachvilli (2011)]. Furthermore, the fluid exhibits relative motion with respect to the interface only at a distance from the interface further into the double layer. The location of the boundary between the immobile and mobile fluid regions is ambiguous and is referred to as the surface of shear (Fig. 4.6). Therefore, the value of ψ_s deduced from either electrophoretic mobility or EOF is appropriate only for the electric potential at the surface of shear and is known as the **zeta potential** ψ_ζ. This is not identical to the surface potential as sketched in Fig. 7.5. Therefore, for practical purposes, ψ_s in Sections 7.3.4 and 7.3.5 are replaced by ψ_ζ. For example, the electrophoretic mobility in the low-salt and high-salt limits of the Helmholtz–Smoluchowski equation are

$$\mu = \frac{2}{3}\frac{\epsilon_0 \epsilon}{\eta_0}\psi_\zeta, \qquad (\kappa R \ll 1) \tag{7.3.40}$$

and

$$\mu = \frac{\epsilon_0 \epsilon}{\eta_0}\psi_\zeta. \qquad (\kappa R \gg 1) \tag{7.3.41}$$

The crossover behavior between the above limits is given by the Henry formula in Eq. (7.3.23).

Summarizing the results so far in Section 7.3, the ion cloud around a charged particle or a macroion significantly modifies the Einstein law of electrophoretic mobility given in Eq. (7.3.9). The contribution from the ion cloud must be taken into account in understanding various experimental observations on electrophoretic mobility of charged particles. Charged macromolecules are no exception. The various experimentally observed deviations from the Einstein law of electrophoretic mobility in solutions of charged macromolecules arise from the correlations between the molecules and their ion clouds, as we shall see in Section 7.4.

(a)

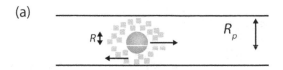

(b)

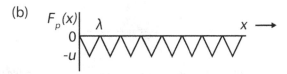

Figure 7.7 (a) A charged spherical particle of radius R undergoing electrophoresis inside a cylindrical pore of radius R_p. Counterion cloud drags the particle in the opposite direction and endows an apparent charge different from the particle's bare charge. Confinement of the particle inside the pore and the interactions between the particle and the pore modify the diffusion coefficient of the particle. (b) The attractive pore-particle interaction is modeled as a one-dimensional periodic free energy landscape $F_p(x)$ with depth (or equivalently height) u and period λ.

7.3.7   Mobility through a Nanopore

Measurement of the electrophoretic mobility of charged particles and macromolecules inside a nanopore is an important experimental technique in attempts to directly determine their charge and size. In addition to the role of ion clouds described in Sections 7.3.2 and 7.3.3, confinement of a charged particle inside a nanopore and its interaction with the interior wall of the pore contribute to the electrophoretic mobility. All of these contributions must be accounted for in order to correctly deduce the charge and size of the particle from experimental data on particle velocity under an electric field.

As a specific example, consider a uniform cylindrical pore of radius R_p of infinite length (to ignore end-effects), containing an electrolyte solution and a charged spherical particle of radius R, as depicted in Fig. 7.7a. The particle is electrophoretically driven by an externally applied constant electric field $\mathbf{E}_0$. The particle drifts toward its favorable electrode, opposed by its counterion cloud. At the same time, the drift and diffusion of the particle are hindered by the physical boundary of the pore and the interactions between the particle and the pore, such as adsorption. Generally speaking, the interaction between the particle and the internal wall of the pore can be quite complex. The particle can adsorb at the interface for a certain duration of time before it slips off the interface. For the sake of simplicity, let us model the particle-pore interaction as a saw-tooth periodic variation in the free energy of the particle $F_p(x)$ along the pore axis x, as sketched in Fig. 7.7b. Let the period of the free energy profile be λ and the depth be $-u$. The free energy profile is equivalent to a barrier of $+u$ in every period of length λ. An approximate generalization of Eq. (7.3.9), allowing for particle-counterion-cloud dynamics, confinement effects from the pore and particle-pore interactions, is as follows. As shown in Section 7.3.2, the effect of the counterion cloud can be represented by defining an effective charge for the particle (Eq. (7.3.14)). Theoretical treatment of the effects from confinement is difficult due to the nature of the long-ranged electrostatic and hydrodynamic interactions with finite boundaries and

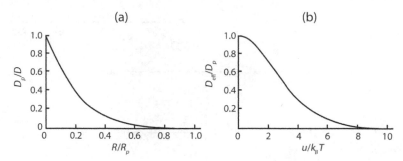

Figure 7.8 (a) Plot of the ratio of diffusion coefficient D_p inside the pore to the bulk value D as a function of the ratio of particle radius to pore radius, R/R_p. (b) Plot of the ratio of the diffusion coefficient D_{eff} in the presence of the free energy barrier to that without the barrier, versus the barrier height u/k_BT [Adapted from Muthukumar (2014)].

potentially complicated adsorption isotherms under flow. Nevertheless, approximate generalization of Eq. (7.3.9) is possible as given below [Muthukumar (2014)].

When a particle is confined inside a narrow pore, it moves along the pore axis due to the excluded volume effect with the pore that is responsible for the particle confinement. For an uncharged particle inside a cylindrical pore, the hard excluded volume effect and the hydrodynamic interactions between the particle and the pore wall lead to a significantly reduced diffusion coefficient of the particle in comparison to the Stokes–Einstein value in bulk solutions [Faxen (1922), Ferry (1936)]. Denoting the diffusion coefficient of the particle inside the cylindrical pore as D_p, the ratio of D_p to the bulk value D is given by (known as the Renkin equation) [Renkin (1954)]

$$\frac{D_p}{D} = \left(1 - \frac{R}{R_p}\right)^2 \left[1 - 2.104\left(\frac{R}{R_p}\right) + 2.09\left(\frac{R}{R_p}\right)^3 - 0.95\left(\frac{R}{R_p}\right)^5\right]. \qquad (7.3.42)$$

A plot of D_p/D versus R/R_p is given in Fig. 7.8a, showing substantially smaller values of D_p in comparison with D for strong confinements. A more rigorous consideration of restricted hydrodynamics inside the pore along with modifications to the contribution of counterion clouds (Section 7.3.2) is presently lacking, but the deviation from Eqs. (7.3.14) and (7.3.42) is expected to be only minor.

In order to assess the effect of the pore-particle interaction on the diffusion and drift of the particle under the influence of the externally applied electric field $\mathbf{E}_0$, we resort to the model given in Fig. 7.7b. In the presence of the saw-tooth periodic free energy profile, the diffusion coefficient of the particle is modified to a new value D_{eff}. An exact calculation of this modification gives [Muthukumar (2014)]

$$D_{\text{eff}} = D_p \frac{\left(\frac{u}{k_BT}\right)^2}{\left(1 - e^{-\frac{u}{k_BT}}\right)\left(e^{\frac{u}{k_BT}} - 1\right)}. \qquad (7.3.43)$$

This result is presented in Fig. 7.8b as a plot of D_{eff}/D_p versus u/k_BT. The diffusion coefficient decreases significantly as u/k_BT increases. Note that the above equation is symmetric in the sign of u. The lowering of D_p to D_{eff} is the same independent of whether the pore-particle interaction results in free energy traps or barriers. The asymptotic limits of Eq. (7.3.43) for weak and strong barriers are

$$D_{\text{eff}} = D_p \begin{cases} 1 - \frac{1}{12} \left(\frac{u}{k_B T} \right)^2 + \cdots, & u \ll k_B T, \\ \left(\frac{u}{k_B T} \right)^2 e^{-u/k_B T}, & u \gg k_B T. \end{cases} \tag{7.3.44}$$

The large barrier result is reminiscent of the Kramer's formula for barrier crossing, but with different prefactor $(u/k_B T)^2$. This asymptotic behavior is a good approximation for $u \geq 4k_B T$.

For weak electric fields ($\mathbf{E}_0$), the average velocity of the particle in the steady state is given by

$$\mathbf{v}_0 = \frac{Q_{\text{eff}} D_{\text{eff}}}{k_B T} \mathbf{E}_0, \tag{7.3.45}$$

where Q_{eff} is given in Eq. (7.3.14) and D_{eff} is given by Eqs. (7.3.42) and (7.3.43). Therefore, the electrophoretic mobility is given as

$$\mu = \frac{Q}{6\pi\eta_0 R} \frac{1}{(1 + \kappa R)} f_{\text{Henry}}(\kappa R) \left(1 - \frac{R}{R_p} \right)^2 \left[1 - 2.104 \left(\frac{R}{R_p} \right) + 2.09 \left(\frac{R}{R_p} \right)^3 \right.$$

$$\left. -0.95 \left(\frac{R}{R_p} \right)^5 \right] \frac{\left(\frac{u}{k_B T} \right)^2}{\left(1 - e^{-\frac{u}{k_B T}} \right) \left(e^{\frac{u}{k_B T}} - 1 \right)}. \tag{7.3.46}$$

Note that the length of the period λ does not enter in this result. The above formula enables the determination of the actual charge and size of the particle from experimentally measured electrophoretic mobility, instead of their apparent values obtained by using Eq. (7.3.9).

7.4   Phenomenology of Dynamics of Polyelectrolyte Solutions

The movement of charged flexible macromolecules in electrolyte solutions is a net effect of motion of all monomers constituting the molecules. The dynamics of the monomers are highly correlated due to chain connectivity, excluded volume and electrostatic interactions, hydrodynamic interactions, and inter-molecular correlations. As a manifestation of these correlations, the internal dynamics of a macromolecule exhibit rich behavior and constitute the earmarks of its specificity and its environment, such as polymer concentration, salt concentration, viscosity, and temperature. In contrast, internal dynamics are absent in rigid particles described in Section 7.3, and are relatively weak in well-folded globular proteins. Nevertheless, for flexible uncharged molecules at sufficiently long times and large length scales to be identified below, the whole molecule can move like a Brownian particle in quiescent solutions, in accordance with the Einstein theorem at nonzero temperatures. However, at both local and global levels, where the internal dynamics and large-scale motion are respectively explored, new laws of dynamics emerge for polyelectrolyte solutions that are qualitatively different from the behavior of solutions of uncharged macromolecules. These laws are described below by first reviewing general considerations and representative experimental results and then summarizing the advances made in the understanding of the dynamics of polyelectrolyte solutions.

The general features of the dynamical properties of polyelectrolyte solutions are qualitatively different from those of solutions of uncharged polymers. These differences are seen in the diffusion of polyelectrolytes, the viscosities of their solutions, and the occurrence of electrophoretic mobility. In dilute solutions (Fig. 7.1b), a polyelectrolyte chain is an expanded coil of radius of gyration R_g, which depends on its linear charge density, molecular weight, and the concentration of added salt. These molecules undergo diffusion and act like the Brownian particles discussed in Section 7.1. Although macromolecules are not rigid spheres, it is possible to treat them as roughly spherical coils with radii comparable to their R_g. Using this analogy, let us consider the diffusion of polyelectrolyte chains in dilute solutions, viscosity, and electrophoretic mobility, based on our knowledge on suspensions of rigid spherical particles. We shall then contrast the predictions from this general consideration with unexpected experimental facts.

7.4.1   Diffusion and Relaxation Time

In a solution without any externally imposed flow fields, the center-of-mass of a macromolecule $\mathbf{R}_{cm}(t)$ is expected to undergo diffusion just as a rigid spherical Brownian particle (Section 7.1),

$$\langle [\mathbf{R}_{cm}(t) - \mathbf{R}_{cm}(0)]^2 \rangle = 6Dt, \tag{7.4.1}$$

where t is a time duration that exceeds the relaxation time τ (defined below), and D is the center-of-mass diffusion coefficient,

$$D = \frac{k_B T}{f_t}. \tag{7.4.2}$$

The friction coefficient f_t of the whole macromolecule follows the same Stokes law (Eq. (7.1.3)) of proportionality to η_0 times the radius of the polymer coil, given as

$$f_t = 6\pi\eta_0 R_h, \tag{7.4.3}$$

where R_h is the hydrodynamic radius (Eq. (5.5.4)). Therefore, the Stokes–Einstein law for a macromolecule is given as

$$D = \frac{k_B T}{6\pi\eta_0 R_h}. \tag{7.4.4}$$

As seen in Section 5.5.1, R_h is proportional to R_g. Since R_g is related to the number of monomers in the chain N through the size exponent v as $R_g \sim N^v$, the diffusion coefficient follows as

$$D \sim N^{-v}. \tag{7.4.5}$$

Note that Eqs. (7.4.1) and (7.4.4) are valid only for time durations longer than the relaxation time τ of the whole molecule. We estimate τ using the following general argument. The Einstein law of diffusion (Eq. (7.1.1)) can be generally paraphrased as

$$\langle (\text{distance})^2 \rangle \sim (\text{diffusion coefficient}) \times (\text{time}). \tag{7.4.6}$$

Let us implement this result for a polymer chain of radius of gyration R_g. If the center-of-mass of the chain moves a distance comparable to R_g, the new polymer conformation is uncorrelated with its initial conformation. Therefore, the correlation time τ over which the conformations of a single chain are correlated is the time taken by the chain to explore a distance comparable to R_g by the diffusion process. Since the diffusion coefficient is inversely proportional to R_g, we get from the above equation,

$$R_g^2 \sim \frac{1}{R_g}\tau, \tag{7.4.7}$$

where numerical coefficients are not considered. Therefore, the scaling law for the correlation time is proportional to the molecular volume of the polymer chain, and in view of $R_g \sim N^\nu$,

$$\tau \sim R_g^3 \sim N^{3\nu}. \tag{7.4.8}$$

The correlation time is also the longest relaxation time taken by the chain to return to its equilibrium from a nonequilibrium conformation. In view of this, τ is also the relaxation time of the chain.

In non-dilute polymer solutions, two kinds of diffusion coefficients emerge, namely tracer (or self) diffusion coefficient D_t and cooperative (or mutual) diffusion coefficient D_c. While D_t denotes the diffusion coefficient of a labeled chain in solution, D_c corresponds to the diffusion coefficient for the collective diffusion of local concentration fluctuations which involve many chains. Of course, D_t and D_c are the same in dilute solutions. D_t is usually measured using techniques such as particle tracking and pulsed field gradient nuclear magnetic resonance, whereas D_c is measured using techniques such as dynamic light scattering. In dilute solutions, both D_t and D_c are the same as D given in Eq. (7.4.4).

The inverse relation between D_t and N given in Eq. (7.4.5) becomes stronger as the polymer concentration increases. As we shall see in Section 7.7, the hydrodynamic interactions between the monomers are screened as the polymer concentration is increased. When hydrodynamic interactions are absent, the friction coefficient of a chain is proportional to its length, $f_t \sim N$, so that the tracer diffusion coefficient is inversely proportional to N. Upon a further increase in polymer concentration, interchain entanglements set in. When entanglement effects are present, D_t is smaller with a more sensitive dependence on N, $D_t \sim N^{-\alpha}$, where α is in the range of 2.0 to 3.0 for neutral polymers. The most commonly used value of α is 2.0, based on the theory of reptation (Section 7.5.3). For solutions of neutral polymers, the three regimes of infinitely dilute, semidilute without entanglements, and semidilute and concentrated with entanglements are, respectively, referred to as the Zimm, Rouse, and reptation regimes,

$$D_t \sim \begin{cases} N^{-\nu}, & \text{Zimm} \\ N^{-1}, & \text{Rouse} \\ N^{-2}, & \text{reptation.} \end{cases} \tag{7.4.9}$$

Based on this result, the relaxation time for a single labeled chain follows from Eq. (7.4.6) as

$$\tau \sim \begin{cases} N^{3\nu}, & \text{Zimm} \\ N^2, & \text{Rouse} \\ N^3, & \text{reptation,} \end{cases} \tag{7.4.10}$$

since $R_g \sim \sqrt{N}$ in semidilute and concentrated solutions due to the screening of excluded volume and electrostatic interactions (see Section 6.3.5).

The cooperative diffusion coefficient in solutions with polymer concentrations higher than the overlap concentration is given in terms of the correlation length ξ for concentration fluctuations as

$$D_c = \frac{k_B T}{6\pi \eta_0 \xi}. \tag{7.4.11}$$

Although ξ in this equation is the hydrodynamic screening length (Section 7.7), it is proportional to the screening length described in Section 6.2.2 for equilibrium conditions. For convenience, let us take these screening lengths to be the same. Since ξ is inversely related to polymer concentration c as $c^{-1/2}$ and $c^{-3/4}$, respectively, for salt-free and high salt semidilute solutions, D_c is expected to be

$$D_c \sim \begin{cases} c^{1/2}, & \text{salt-free} \\ c^{3/4}, & \text{high salt.} \end{cases} \tag{7.4.12}$$

The above expected results on the diffusion coefficient in solutions of polyelectrolytes are not observed in experiments on solutions of a variety of charged macromolecules. An example of the dramatic deviation of experimental data from the above arguments is the **ordinary–extraordinary transition** described in the following section.

7.4.2   Ordinary–Extraordinary Transition

Experimental determination of the center-of-mass diffusion coefficient of flexible polyelectrolytes in dilute solutions at high salt concentrations shows that the Stokes–Einstein law is applicable. Since $R_g \sim N^{3/5}$ in high salt solutions (Eq. (5.6.6)), we get from Eqs. (7.4.4) and (7.4.5),

$$D \sim \frac{1}{R_g} \sim \frac{1}{N^{3/5}}, \tag{7.4.13}$$

which is observed in experiments using dynamic light scattering. As the concentration c_s of added salt is decreased, R_g is found to increase as described in Section 5.4. Therefore, the Stokes–Einstein prediction is that D should decrease with lowering c_s. However, DLS experiments reveal that D actually increases upon lowering c_s. In spite of the contradiction with the Stokes–Einstein law, the increase in D with a decrease in c_s is known as the **ordinary** behavior.

Even more remarkably, upon a further decrease in c_s, an additional diffusive mode with a very small diffusion coefficient emerges. This experimental observation suggests the presence of large aggregates at lower ionic strengths, even though the polyelectrolyte chains are similarly charged. It is surprising that similarly charged, and hence electrostatically repulsive, chains would aggregate at all, and that they disassemble when the electrostatic repulsion is screened by added salt. In view of this mysterious nature, the new diffusive mode is called **extraordinary**. The simultaneous emergence of the ordinary and extraordinary behaviors from the ordinary behavior upon reduction of c_s is known as the **ordinary–extraordinary transition** [Lin *et al.* (1978), Förster & Schmidt (1995)]. This phenomenon is commonly observed in DLS experiments on aqueous solutions containing a variety of charged macromolecules such as DNA, poly(styrene sulfonate), quaternized poly(2-vinyl pyridine), etc.

The diffusional modes corresponding to the ordinary and extraordinary behaviors are also called, respectively, the "fast" and "slow" modes with their respective diffusion coefficients D_f and D_s. Experimentally observed D_f and D_s in salt-free aqueous solutions are given in Fig. 7.9a as functions of polymer concentration c and degree of polymerization N. The data [Förster *et al.* (1990)] for quaternized poly(2-vinyl pyridine) with molecular weight $M = 1.09 \times 10^5$ g/mol, 7.8×10^5 g/mol, 5.8×10^5 g/mol, and 2.26×10^6 g/mol, with respective degree of quarternization $q = 0.65, 0.75, 0.4$, and 1.0 are given in this figure. The data [Sedlak & Amis (1992a,b)] for sodium poly(styrene sulfonate) with $M = 5 \times 10^3$ g/mol, 3.82×10^4 g/mol, 1.0×10^5 g/mol, and 1.2×10^6 g/mol are also included in this figure. It is evident from Fig. 7.9a that the occurrence of the fast and slow modes is a universal feature of polyelectrolytes independent of their chemical details. For comparison, the diffusion coefficient D_{SE} estimated from the Stokes–Einstein law and the diffusion coefficients of small electrolyte ions are also indicated in the figure.

Note several important features in Fig. 7.9(a). The fast diffusion coefficient D_f is several orders of magnitude higher than the expected Stokes–Einstein value D_{SE}. Remarkably, D_f for c above a threshold value is independent of c and N over several orders of magnitude. Even more remarkably, D_f of 10^6 g/mol sodium poly(styrene sulfonate) is only a factor of about 4 smaller than the diffusion coefficient of a metallic ion such as K^+, Na^+, etc.

The slow diffusion coefficient D_s is several orders of magnitude smaller than D_f and D_{SE} and it decreases strongly with an increase in either c or N. Implementation of the Stokes–Einstein law to such small values of D_s suggests presence of large aggregates made of many chains. Such clustering is also seen in suspensions of charged colloidal particles and worm-like micelles [Schmitz (1993)]. Concomitant to the onset of the slow mode, the static scattering intensity at very small scattering angles diverges as seen in Fig. 6.2b, consistent with presence of aggregates in the solution.

Upon addition of salt, the slow mode disappears, as illustrated in Fig. 7.9b for aqueous solutions of quaternized poly(2-vinyl pyridine) of molecular weight $M = 2 \times 10^5$ g/mol [Förster *et al.* (1990)]. Data for four c_s values (salt-free, 10^{-3} M, 10^{-2} M, and 10^{-1} M) are included in the figure. Consider the data at $c = 10^{-1}$ g/L. While both fast and slow modes are present in salt-free solutions, they are absent at $c_s = 10^{-3}$

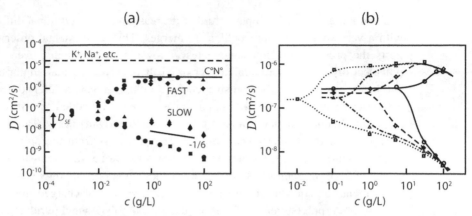

Figure 7.9 (a) Experimental data on diffusion coefficient for the fast (ordinary) mode, D_f, and the slow (extraordinary) mode, D_s, in salt-free polyelectrolyte solutions. The fast and slow modes emerge at polymer concentrations above a threshold value. In the fully developed two-mode regime, D_f is independent of polymer concentration c and degree of polymerization N over several orders of magnitude and is only a factor of about 4 smaller than the diffusion coefficient of a metallic ion such as Na$^+$ in water. D_s is smaller than D_f by three orders of magnitude and depends on c and N, suggesting formation of aggregates by similarly charged polymers. Different symbols denote data for quaternized poly(2-vinyl pyridine) with molecular weight $M = 1.09 \times 10^5$ g/mol, 7.8×10^5 g/mol, 5.8×10^5 g/mol, and 2.26×10^6 g/mol, with respective degree of quarternization $q = 0.65, 0.75, 0.4$, and 1.0 [Förster et al. (1990)] and data for sodium poly(styrene sulfonate) with $M = 5 \times 10^3$ g/mol, 3.82×10^4 g/mol, 1.0×10^5 g/mol, and 1.2×10^6 g/mol, respectively [Sedlak & Amis (1992a,b)]. (b) Dependence of D_f and D_s for quaternized poly(2-vinyl pyridine) of molecular weight $M = 2 \times 10^5$ g/mol on polymer concentration c at different KBr concentrations (square: salt-free, triangle: 10^{-3} M, diamond: 10^{-2} M, and circles: 10^{-1} M). Upon addition of the electrolyte, the slow mode disappears and D_f progressively becomes smaller, eventually approaching the diffusion coefficient D_{SE} expected from the Stokes–Einstein law. [Förster et al. (1990)].

M or higher. At higher concentrations of added salt, higher polymer concentrations are required for the occurrence of simultaneous fast and slow modes. The junction point at which one diffusion mode (fast) splits into two modes (fast and slow), either when c_s is decreased or c is increased, is identified as the ordinary–extraordinary transition point.

Note that one of the standard experimental methods to characterize uncharged macromolecules in solutions is DLS in combination with the Stokes–Einstein law. However, straightforward implementation of this procedure for charged macromolecules results in unreasonably low R_g values based on the fast mode and unreasonably high R_g values based on the slow mode. The occurrence of the ordinary and extraordinary behaviors exhibited in Fig. 7.9, as a spectacular deviation from the Stokes–Einstein expectation, undermines the utility of the Stokes–Einstein law for solutions of charged macromolecules at lower salt concentrations. In view of the lessons in the preceding chapters and Section 7.3.2, we can imagine the origin of the discrepancy between experimental observations and the Stokes–Einstein law to reside in the role of counterions, as we shall see in Sections 7.6 and 7.7.

7.4.3   Electrophoretic Mobility

For a charged particle with net charge Q and diffusion coefficient D, the Einstein result for the electrophoretic mobility in dilute solutions is given by Eq. (7.3.9). Since the total charge Q of a uniformly charged polyelectrolyte chain is proportional to the number of segments N in the chain, and since D is proportional to $1/R_g$ and hence to $N^{-\nu}$, we get from Eq. (7.3.9),

$$\mu = \frac{QD}{k_B T} \sim N^{1-\nu}. \tag{7.4.14}$$

For salt-free conditions, the size exponent ν is close to one, corresponding to rod-like conformations. In this limit, μ is independent of N. On the other hand, as the concentration of added salt increases, ν decreases toward 3/5. Therefore, Eq. (7.4.14) predicts that the electrophoretic mobility increases with molecular weight of the polyelectrolyte chain at high salt concentrations.

The above expectation based on the Einstein relation for electrophoretic mobility is not observed in experiments. It is well known [Hoagland *et al.* (1999), Stellwagen *et al.* (2003)] that the electrophoretic mobility of long polyelectrolyte molecules in solution electrophoresis is independent of molecular weight at all salt concentrations,

$$\mu \sim N^0. \tag{7.4.15}$$

This result is illustrated in Fig. 7.10a where the electrophoretic mobility of sodium poly(styrene sulfonate) in solutions containing sodium chloride salt in the range of 10^{-4} M to 3.0 M is plotted against the degree of polymerization N [Hoagland *et al.* (1999)]. As noted in the figure, the data sets are from capillary electrophoresis and electrophoretic light scattering. In another set of experiments, the diffusion coefficient and the electrophoretic mobility were measured independently for ssDNA and dsDNA at many chain lengths in order to check the validity of the Einstein relation given in Eqs. (7.3.9) and (7.3.10) [Stellwagen *et al.* (2003)]. The diffusion coefficient was found to be $D \sim N^{-\nu}$, with $\nu \simeq 0.6$, whereas μ was found to be independent of N in contradiction with Eq. (7.4.14). Writing the total charge Q as $Q = Ze$, where Z is the number of charged residues per polymer chain, the Einstein ratio $\mu/ZD = e/k_B T$ is given by Eq. (7.3.10) as 39.6 V^{-1} at 20°C. Plots of the experimentally determined values of the Einstein ratio against the number of repeat units are given in Fig. 7.10b for ssDNA and dsDNA at the ionic strength 20 mM [Stellwagen *et al.* (2003)]. The constant value 39.6 V^{-1} is not observed, demonstrating the inapplicability of the Einstein relation of electrophoretic mobility to polyelectrolyte molecules. The discrepancy between the results portrayed in Fig. 7.10 and Eq. (7.4.14) is due to the fact that a charged macromolecule is not alone but is surrounded by its electrostatically correlated ion cloud, as we shall describe in Section 7.6.

7.4.4   Viscosity

While macromolecules distributed in a solvent undergo diffusion, the shear viscosity η_0 of the solvent is modified by the presence of the macromolecules to the shear

(a)   (b)

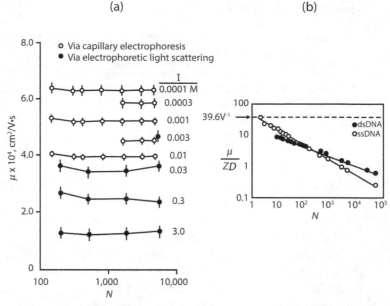

Figure 7.10 (a) Electrophoretic mobility of sodium poly(styrene sulfonate) in dilute solutions is independent of molecular weight at all salt concentrations of added NaCl salt. The ionic strength I is indicated for each data set. Data are from capillary electrophoresis and electrophoretic light scattering. [Hoagland *et al*. (1999)] (b) Plot of the Einstein coefficient μ/ZD versus the number of repeat units (N) of ssDNA and dsDNA, showing the inapplicability of the Einstein electrophoretic mobility to polyelectrolyte chains. Instead of the constant value (dashed horizontal line), μ/ZD can be orders of magnitude smaller. [Stellwagen *et al*. (2003)]

viscosity η of the whole solution. A scaling relation between the change in viscosity $(\eta - \eta_0)$ and the molecular weight of the macromolecule can be obtained from the well-known Einstein result for suspensions of spherical particles. According to the Einstein theory of viscosity [Einstein (1956)], the "specific viscosity" $(\eta - \eta_0)/\eta_0$ of a suspension containing n spherical particles of radius R in volume V is given by

$$\frac{\eta - \eta_0}{\eta_0} = \frac{5}{2}\frac{n}{V}\frac{4\pi R^3}{3}. \qquad \text{(Einstein)} \qquad (7.4.16)$$

In obtaining this result, interparticle interactions are ignored.

For a dilute solution containing n polymer chains of radius of gyration R_g, we implement the above Einstein law for the specific viscosity as

$$\frac{\eta - \eta_0}{\eta_0} \sim \frac{nR_g^3}{V}. \qquad (7.4.17)$$

Note that the porous nature of the internal structure of polymer chains, in contrast with a rigid particle, is reflected only in the numerical prefactor in Eq. (7.4.17), as we shall see in Section 7.6. Rewriting the above result in terms of the monomer concentration $c = nN/V$, we get

$$\frac{\eta - \eta_0}{\eta_0} \sim c\frac{R_g^3}{N}. \tag{7.4.18}$$

Since $R_g \sim N^\nu$, the "reduced viscosity" of the solution is given by

$$\frac{\eta - \eta_0}{\eta_0 c} \sim N^{3\nu-1}. \tag{7.4.19}$$

In the limit of $c \to 0$, the reduced viscosity is called the "intrinsic viscosity" $[\eta]$,

$$[\eta] = \lim_{c \to 0} \frac{\eta - \eta_0}{\eta_0 c} \sim N^{3\nu-1}. \tag{7.4.20}$$

As the polymer concentration is increased in dilute solutions, the specific viscosity $(\eta - \eta_0)/\eta_0$ can be written as a virial series in polymer concentration,

$$\frac{\eta - \eta_0}{\eta_0} = [\eta]c\left[1 + k_H[\eta]c + \cdots\right], \tag{7.4.21}$$

where the numerical factor k_H is known as the Huggins coefficient. Rewriting this equation in terms of the reduced viscosity, we get

$$\frac{\eta - \eta_0}{\eta_0 c} = [\eta]\left[1 + k_H[\eta]c + \cdots\right]. \tag{7.4.22}$$

Therefore, the intercept of a plot of the reduced viscosity versus polymer concentration c is the intrinsic viscosity, from which R_g and the size exponent ν can be determined using Eqs. (7.4.18) and (7.4.20).

Note that the specific viscosity is a monotonically increasing function of $[\eta]c$. Since $[\eta]$ is proportional to the inverse of the overlap concentration $c^\star$ (see Eqs. (6.1.2) and (7.4.20)), the specific viscosity is written as a monotonically increasing function of $c/c^\star$,

$$\frac{\eta - \eta_0}{\eta_0} = f_\eta\left(\frac{c}{c^\star}\right), \tag{7.4.23}$$

where f_η is some function to be addressed in Section 7.7. Equivalently, the reduced viscosity is expressed as

$$\frac{\eta - \eta_0}{\eta_0 c} = [\eta]f_r\left(\frac{c}{c^\star}\right), \tag{7.4.24}$$

where $f_r = f_\eta/[\eta]c$.

For solutions of uncharged macromolecules, a plot of the reduced viscosity versus polymer concentration is a monotonically increasing function. However, this is not true for polyelectrolyte solutions at low salt concentrations. Early experimental observations on dilute salt-free solutions showed that the reduced viscosity appears to diverge as the polymer concentration is reduced according to the Fuoss law,

$$\frac{\eta - \eta_0}{\eta_0 c} \sim \frac{1}{\sqrt{c}}. \qquad \text{(Fuoss law)} \tag{7.4.25}$$

This observation prohibited the use of viscometry and Eq. (7.4.20) in characterizing charged macromolecules. However, further measurements at infinitely dilute solutions showed that the reduced viscosity increases with polymer concentration before exhibiting the Fuoss behavior. This result is illustrated in Fig. 7.11 where the reduced

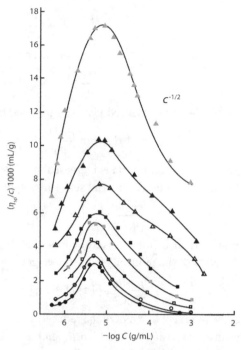

Figure 7.11 Plot of reduced viscosity of aqueous salt-free solutions of sodium poly(styrene sulfonate) versus logarithm of polymer concentration. The molecular weight from the top to bottom curve is 690 kDa, 345 kDa, 212 kDa, 177 kDa, 138 kDa, 88 kDa, 31 kDa, and 16 kDa [Cohen *et al*. (1988)].

viscosities of aqueous salt-free solutions of sodium poly(styrene sulfonate) are plotted versus the logarithm of polymer concentration c for different molecular weights in the range of 16 kDa to 690 kDa. The region where the reduced viscosity decreases with c corresponds to the Fuoss regime (Eq. (7.4.25)). The nonmonotonic dependence of the reduced viscosity on polyelectrolyte concentration is in marked contrast with the monotonic behavior for neutral polymers, or polyelectrolytes at the high salt limit.

The conceptual framework in addressing the rich phenomenology on polyelectrolyte solutions illustrated in Figs. 7.9, 7.10, and 7.11, will be presented in the following sections. In order to describe the various concepts with a progressive increase in polymer concentration, we shall consider three regimes in polymer concentration, namely, dilute, semidilute without entanglement effects, and concentrated solutions where entanglement effects dominate (Fig. 7.12). Before addressing these regimes in Sections 7.6, 7.7, and 7.8, respectively, we present basic models of polymer dynamics.

7.5   Models of Dynamics of Uncharged Polymers

The general theoretical setup to treat polymer dynamics in homogeneous solutions is as follows. Consider a collection of n uncharged chains distributed randomly in a

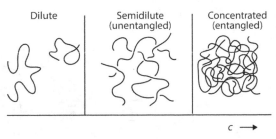

Figure 7.12 Three regimes of polyelectrolyte concentration in the treatment of dynamics of polyelectrolyte solutions: dilute, unentangled semidilute, and entangled semidilute and concentrated solutions.

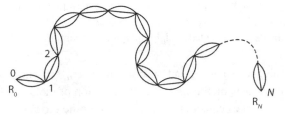

Figure 7.13 A conformation of a chain.

solvent. Each chain is modeled as a coarse-grained chain of N segments of segment length ℓ as illustrated in Fig. 7.13. Let us represent these segments as contiguously connected $N + 1$ beads. Let $\mathbf{R}_i(t)$ denote the position vector of the ith bead at time t. The set of vectors $\{\mathbf{R}_i\}(0 \leq i \leq N)$ represents a particular conformation of the chain. The ith bead of a chain in this solution is subjected to forces arising from chain connectivity to its adjacent segments and van der Waals interactions (and the electrostatic interactions, if charged) with all other segments of the same chain and other chains. In addition to these previously treated forces in equilibrium conditions, the ith bead experiences forces from its inertia, friction against the background fluid, and hydrodynamic interaction with all monomers in the system. The potential interaction V and hydrodynamic interaction $\mathbf{G}$ between a pair of beads are illustrated in Fig. 7.14a. Furthermore, the ith bead experiences a random force from incessant collisions with solvent molecules. For time scales relevant to typical situations associated with macromolecules, the contribution from inertia can be ignored. The equation of motion for the ith bead is the Newton equation,

$$0 = \mathbf{f}_{i,\text{connectivity}} + \mathbf{f}_{i,\text{interactions}} + \mathbf{f}_{i,\text{friction}} + \mathbf{f}_{i,\text{random}} + \mathbf{f}_{i,\text{ext}}. \qquad (7.5.1)$$

The first four terms on the right-hand side of Eq. (7.5.1) are, respectively, the aforementioned forces from chain connectivity, inter-segment potential interactions, friction, and random collisions by solvent molecules, which all act on the ith bead. $\mathbf{f}_{i,\text{ext}}$ is an external force, such as an electrical force acting on the ith bead if it is charged. The above equation in the presence of a random force, called Langevin equation, can be described only in terms of statistical averages in contrast to the deterministic Newton's equations. The frictional term actually arises from the coupling of the dynamics of the ith bead with the velocity field in the hydrodynamic continuum representing

(a)   (b)

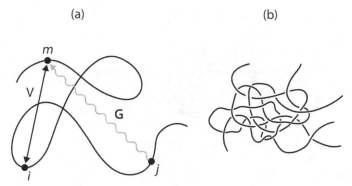

Figure 7.14 (a) Dynamics of the mth bead is controlled by connectivity to its neighbors $m-1$ and $m+1$, inter-segment potential interaction V and the hydrodynamic interaction $\mathbf{G}$. (b) Chains are entangled at higher polymer concentrations.

the solvent. An equal and opposite force $\sigma_{\alpha i} = -\mathbf{f}_{\alpha i,\text{friction}}$ from the ith bead of the αth chain acts on the solvent at the location of that bead $\mathbf{R}_{\alpha i}(t)$, so that the linearized Navier–Stokes equation Eq. (7.2.2) is modified to

$$-\eta_0 \nabla^2 \mathbf{v}(\mathbf{r}) + \nabla p(\mathbf{r}) = \mathbf{F}(\mathbf{r}) + \sum_{\alpha=1}^{n} \sum_{j=0}^{N} \delta(\mathbf{r} - \mathbf{R}_{\alpha i}) \sigma_{\alpha i}, \qquad (7.5.2)$$

The above two coupled equations are solved using the "no-slip" boundary condition between the velocity of the ith bead of the αth chain, $\partial \mathbf{R}_{\alpha i}(t)/\partial t$ and the fluid velocity $\mathbf{v}(\mathbf{R}_{\alpha i}(t))$ at its location,

$$\frac{\partial \mathbf{R}_{\alpha i}(t)}{\partial t} = \mathbf{v}(\mathbf{R}_{\alpha i}(t)). \qquad (7.5.3)$$

The solution is accomplished by eliminating $\sigma_{\alpha i}$ with the use of the above boundary condition, followed by averaging the results over all allowed polymer conformations in the system. Quantities such as the mean square displacement of a labeled monomer, mean square displacement of the center-of-mass of a labeled chain, tracer diffusion coefficient, cooperative diffusion coefficient, relaxation spectra of the internal chain dynamics, and dynamic structure factor are calculated from Eq. (7.5.1) in terms of molecular weight and concentration of the polymer and ionic strength in the medium. The large-scale properties such as the shear viscosity and elastic moduli of the whole solution are calculated from Eq. (7.5.2).

At very high polymer concentrations, entanglement effects arising from the uncrossability of polymer strands dominate the dynamics of the system (Fig. 7.14b). This situation belongs to a different class of dynamics where topological constraints play a major role.

The above outlined general approach to unravel the various aspects of the dynamics of polyelectrolyte solutions requires considerable technical effort. In an attempt to mainly focus on concepts, we relegate the technical details to the original references and Appendix 6 and present only the key models of polymer dynamics in the increasing level of complexity as follows.

(i) **Rouse model:** An isolated chain without inter-segment hydrodynamic interaction.

(ii) **Zimm model:** An isolated chain with hydrodynamic interaction.

(iii) **Reptation model:** A mean field description of entanglement effects in highly concentrated solutions and polymer melts.

(iv) **Entropic barrier model:** Movement of a polymer chain through an entropic barrier created by spatial confinements.

These models are addressed in the following subsections.

7.5.1   Rouse Dynamics

The simplest situation for treating the polymer dynamics is a Gaussian chain without any inter-segment potential interaction and hydrodynamic interaction [Rouse (1953)]. For this situation in the absence of any external force acting on the beads, only the terms from chain connectivity, friction, and random forces are present in Eq. (7.5.1). The equation of motion for the solvent is ignored, except to assume that its consequence is a phenomenological bead friction coefficient ζ_b for each of the beads in Eq. (7.5.1). As derived in Appendix 6, Eq. (7.5.1) for the mth bead is written in terms of these three terms as

$$\zeta_b \frac{\partial \mathbf{R}_m(t)}{\partial t} - \frac{3k_B T}{\ell^2} (\mathbf{R}_{m+1} - 2\mathbf{R}_m + \mathbf{R}_{m-1}) = \mathbf{f}_m(t), \qquad (7.5.4)$$

where the two terms on the left-hand side are due to friction against solvent and chain connectivity, respectively, and the term on the right-hand side is a random force. This equation is known as the **Rouse equation**.

Adopting the continuous representation of the chain, that is, m as a continuous variable, the Rouse equation becomes

$$\zeta_b \frac{\partial \mathbf{R}(m,t)}{\partial t} - \frac{3k_B T}{\ell^2} \frac{\partial^2 \mathbf{R}(m,t)}{\partial m^2} = \mathbf{f}(m,t). \qquad (7.5.5)$$

This equation is like a wave equation, and can be easily solved with Fourier transforms (normal modes of vibrations). Let us choose the Fourier series pair,

$$\mathbf{R}(m,t) = \sum_{p=-\infty}^{\infty} \hat{R}_p(t) \cos\left(\frac{\pi p m}{N}\right) \qquad (7.5.6)$$

$$\hat{R}_p(t) = \frac{1}{N} \int_0^N dm\, \mathbf{R}(m,t) \cos\left(\frac{\pi p m}{N}\right). \qquad (7.5.7)$$

The normal modes of the polymer $\hat{R}_p(t)$ are called **Rouse modes**. As can be seen from Eq. (7.5.7), $p = 0$ denotes the center-of-mass motion of the chain; $p = 1$ denotes an excitation over the whole contour of the chain; $p = 2$ denotes excitations over half of the distance along the string, etc., as illustrated in Fig. 7.15 for a polymer chain with its two ends fixed. When a chain relaxes from plucking, $p = 10$ represents the relaxation of one-tenth of the contour length of the chain. Expressing the Rouse equation, Eq. (7.5.5), in terms of the Rouse modes,

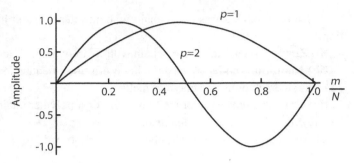

Figure 7.15 Rouse modes of a chain.

$$\zeta_b \frac{\partial \hat{R}_p(t)}{\partial t} + \frac{3\pi^2 k_B T}{\ell^2} \left(\frac{p}{N}\right)^2 \hat{R}_p(t) = \hat{f}_p(t). \tag{7.5.8}$$

The key results from this Rouse equation are well documented in books [Doi & Edwards (1986), Rubinstein & Colby (2003), Hiemenz & Lodge (2007)]. Here, we merely summarize these results without recourse to their derivations.

(1) Rouse relaxation times

The key time variable that sets the scale of time in monitoring the dynamics of chains is the characteristic relaxation time. For time durations shorter than the characteristic relaxation time, dynamics inside the polymer coil are explored. For longer times compared to the characteristic time, global properties of the polymer coil are explored. Inspection of the shape of Eq. (7.5.8), along with dimensional analysis, gives the characteristic relaxation time τ_p for the pth Rouse mode as

$$\tau_p = \frac{\zeta_b N^2 \ell^2}{3\pi^2 k_B T} \frac{1}{p^2}. \tag{7.5.9}$$

This equation constitutes the Rouse relaxation spectrum. The longest relaxation time (obviously for $p = 1$) is called the **Rouse time**,

$$\tau_R \equiv \frac{\zeta_b N^2 \ell^2}{3\pi^2 k_B T}, \tag{7.5.10}$$

and it is proportional to N^2 and the bead friction coefficient, and inversely proportional to temperature,

$$\tau_R \sim \frac{\zeta_b N^2}{T}. \tag{7.5.11}$$

(2) Center-of-mass diffusion

For durations of time longer than the Rouse time, the center-of-mass of the polymer undergoes diffusion, and the mean square displacement of the center-of-mass of the chain is given by

$$\langle [\mathbf{R}_{cm}(t) - \mathbf{R}_{cm}(0)]^2 \rangle = 6 D_{cm} t, \tag{7.5.12}$$

where the diffusion coefficient is

$$D_{cm} = \frac{k_B T}{\zeta_b N}. \tag{7.5.13}$$

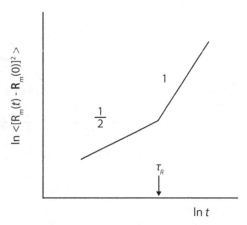

Figure 7.16 Different regimes of Rouse dynamics of a labeled monomer.

Note that Eqs. (7.5.11)-(7.5.13) are consistent with the scaling equation (Eq. (7.4.6)), where the characteristic time τ_R is the time taken by the center-of-mass of the Rouse chain to diffuse a distance comparable to its radius of gyration $R_g \sim \sqrt{N}$.

(3) Dynamics of a labeled monomer

If a labeled monomer is subjected to a random force and if its neighbors along the chain were absent, then it is expected to follow the usual Brownian motion with its mean square displacement proportional to the elapsed time. But, this is not the case when the monomer is connected to its neighbors along the chain. The chain connectivity modifies the monomer diffusion into a subdiffusion as long as the time duration is shorter than the Rouse time. For times longer than the Rouse time, labeled monomer moves along with the center-of-mass of the chain.

For times shorter than the Rouse time, $t < \tau_R$,

$$\langle [\mathbf{R}_m(t) - \mathbf{R}_m(0)]^2 \rangle \sim \sqrt{t}, \qquad t < \tau_R \qquad (7.5.14)$$

and for times longer than the Rouse time, $t > \tau_R$,

$$\langle [\mathbf{R}_m(t) - \mathbf{R}_m(0)]^2 \rangle \sim t, \qquad t > \tau_R. \qquad (7.5.15)$$

The subdiffusive behavior of a labeled monomer given in Eq. (7.5.14) is the earmark of the Rouse chain dynamics. These results are sketched in Fig. 7.16, where additional regimes occurring at extremely short times pertinent to monomer relaxation time are suppressed.

(4) Dynamic structure factor

The subdiffusion of a labeled monomer for short times and center-of-mass diffusion of the chain for long times can be measured using the dynamic structure factor $S(\mathbf{k}, t)$. This is the dynamical analog of the static structure factor defined in Eq. (5.9.1), defined as

$$S(\mathbf{k},t) = \frac{1}{N} \sum_m \sum_n \langle \exp(i\mathbf{k} \cdot [\mathbf{R}_m(t) - \mathbf{R}_n(0)]) \rangle = \frac{1}{N} \sum_{m,n} \exp\left(-\frac{k^2}{6} \langle [\mathbf{R}_m(t) - \mathbf{R}_n(0)]^2 \rangle \right),$$

$$(7.5.16)$$

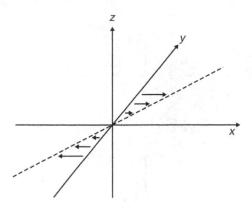

Figure 7.17 Shear flow.

where $\mathbf{k}$ is the scattering wave vector ($\frac{4\pi}{\lambda}\sin(\theta/2)$, where θ is the scattering angle and λ is the wavelength of the incident radiation). Using Eq. (A6.21) for the inter-segment dynamical correlation function $\langle[\mathbf{R}_m(t) - \mathbf{R}_n(0)]^2\rangle$ in Eq. (7.5.16), the dynamic structure factor measured in scattering experiments can be interpreted. Let us look at the following two limits of (i) very long times ($t > \tau_R$) and large lengths ($k^2 R_g^2 \ll 1$) and (ii) short times ($t < \tau_R$) and short lengths ($k^2 R_g^2 \gg 1$).

(i) $k^2 N \ell^2 \ll 1$

For long time durations where this limit is valid,

$$S(\mathbf{k},t) = N \exp\left(-k^2 D_{cm} t\right). \tag{7.5.17}$$

For long times and large length scales, the whole chain undergoes diffusion and the diffusion coefficient of the center-of-mass of the chain is given by Eq. (7.5.13).

(ii) $k^2 N \ell^2 \gg 1$

For short time durations where this limit is valid,

$$S(\mathbf{k},t) = S(\mathbf{k},0) \exp\left(-\frac{2}{\sqrt{\pi}}\sqrt{\Gamma_k t}\right), \tag{7.5.18}$$

which is a "stretched exponential," with the relaxation rate Γ_k given by

$$\Gamma_k = \frac{k_B T}{12\zeta_b} k^4 \ell^2. \tag{7.5.19}$$

The k^4 law for the dependence of the decay rate on the scattering wave vector is the earmark of Rouse dynamics and is a direct consequence of the subdiffusive dynamics of a labeled monomer given in Eq. (7.5.14).

(5) Viscoelasticity

Consider a shear flow with a shear rate $\dot{\gamma}(t)$ (Fig. 7.17), where the velocity components are $v_x = \dot{\gamma} y$, and $v_y = 0 = v_z$.

In general, the shear stress $\sigma_{xy}(t)$ is given by [Doi & Edwards (1986)]

$$\sigma_{xy}(t) = \int_{-\infty}^{t} dt' \, G(t - t') \, \dot{\gamma}(t'), \tag{7.5.20}$$

where G is the shear relaxation modulus. We can calculate $\sigma_{xy}(t)$ from the Rouse equation in the presence of flow and then match the result with the right-hand side of Eq. (7.5.20), resulting in

$$G(t) = \frac{cRT}{M} \sum_p \exp\left(-\frac{t}{\tau_p}\right). \tag{7.5.21}$$

Here, c is the monomer molar concentration, R is the ideal gas constant, M is the molar mass of the chain, and τ_p is now given by

$$\tau_p = \frac{\zeta_b N^2 \ell^2}{6\pi^2 k_B T} \frac{1}{p^2}. \tag{7.5.22}$$

Note that τ_p defined for $G(t)$ is half of the Rouse time defined in Eqs. (7.5.9) and (7.5.10).

For oscillatory shear, $\dot{\gamma}(t) = \dot{\gamma} e^{i\omega t}$, the frequency dependent modulus becomes

$$G(\omega) = i\omega \int_0^\infty dt\, e^{-i\omega t}\, G(t), \tag{7.5.23}$$

and the frequency dependent viscosity is $\eta(\omega) = G(\omega)/i\omega$. As shown in Appendix 6, the contribution of the chains to viscosity $\eta(\omega) = G(\omega)/i\omega$ follows from Eqs. (7.5.21) and (7.5.23) as

$$\eta(\omega) = \frac{cRT}{M} \sum_p \frac{\tau_p}{(1 + i\omega\tau_p)}. \tag{7.5.24}$$

The intrinsic viscosity of a polymer solution, defined as $[\eta] = \lim_{c \to 0}(\eta - \eta_0)/c\eta_0$, is

$$[\eta] = \frac{RT}{M\eta_0} \sum_p \frac{\tau_p}{(1 + i\omega\tau_p)}. \tag{7.5.25}$$

In the zero-frequency limit, Eqs. (7.5.22) and (7.5.25) give

$$[\eta] = \frac{N_A}{M\eta_0} \frac{\zeta_b N^2 \ell^2}{36}. \tag{7.5.26}$$

Therefore, according to the Rouse model, the change in viscosity due to the chains is given as

$$\frac{\eta - \eta_0}{\eta_0} \sim c\, \zeta_b\, N. \tag{7.5.27}$$

In the high frequency limit, $\omega\tau_R > 1$, Eqs. (7.5.22) and (7.5.25) give the frequency dependence of intrinsic viscosity as

$$[\eta] \sim \frac{1}{\sqrt{\omega}}. \tag{7.5.28}$$

Equivalently, the shear modulus at higher frequencies exploring the internal chain dynamics is given by

$$G \sim \sqrt{\omega}. \tag{7.5.29}$$

(6) Effect of size exponent on internal Rouse dynamics

If we consider a non-Gaussian chain with the size exponent ν ($R_g \sim N^\nu$) without any inter-segment hydrodynamic interaction, the Rouse relaxation spectra, the Rouse time, mean square displacement of a labeled monomer for times shorter than the Rouse time, and the rate of decay of dynamic structure factor for $kR_g \gg 1$ are given by (Appendix 6.4.1)

$$\tau_p \sim \frac{\zeta_b}{T}\left(\frac{N}{p}\right)^{2\nu+1},$$

$$\tau_R \sim \frac{\zeta_b}{T}N^{2\nu+1},$$

$$\langle[\mathbf{R}_m(t)-\mathbf{R}_m(0)]^2\rangle \sim t^{(2\nu/2\nu+1)},$$

$$\Gamma_k \sim k^{(2\nu+1)/\nu}. \tag{7.5.30}$$

The results given in Eq. (7.5.30) for $\nu \neq 1/2$ are of limited use, unless the chain is semiflexible, since it is difficult to realize experimental conditions where the potential interactions are not screened while hydrodynamic interaction is screened.

The above summarized results are the earmarks of the Rouse dynamics of polymer chains, and are summarized in Table 7.1 with $\nu = 1/2$.

7.5.2   Zimm Dynamics

Let us first consider a Gaussian chain and include hydrodynamic interactions among all segments. Recall from Section 7.2 that the long-ranged Oseen tensor $\mathbf{G}(\mathbf{r}-\mathbf{r}')$ transmits a force at $\mathbf{r}'$ to a velocity at $\mathbf{r}$ in the solvent. Therefore, in addition to the force from chain connectivity acting on the mth bead given by Eq. (7.5.4), there are multiple forces acting on the mth bead which emanate from the connectivity forces at all other segments (see Fig. 7.14a). For example, the force at the jth bead is transmitted to the mth bead modifying its velocity. The pairwise hydrodynamic interactions are then summed over all j beads except $j = m$. Also the mth bead affects the velocities of all other beads. The situation is like an orchestra where multiple bells ring simultaneously and the net result is a collective behavior of all beads constituting the macromolecule. There are no exact solutions for this nonlinear situation. However, analytical solutions are possible if we assume [Kirkwood & Riseman (1948), Zimm (1956)] that the hydrodynamic interaction is instantaneous and the inter-segment hydrodynamic couplings can be taken as averages over the chain conformations (known as the **preaveraging approximation**). Based on the resulting Kirkwood-Riseman-Zimm equation [Kirkwood & Riseman (1948), Zimm (1956), Yamakawa (1972), Appendix 6], the following results can be obtained.

(1) Zimm relaxation times

The relaxation time τ_p for the pth Rouse mode in the Zimm model for Gaussian chains is

$$\tau_p = \frac{1}{D_p}\frac{N^2\ell^2}{3\pi^2 k_B T}\frac{1}{p^2}, \tag{7.5.31}$$

where

$$D_p = \frac{1}{\zeta_b} + \frac{1}{\sqrt{3\pi^3}\eta_0\ell}\left(\frac{N}{p}\right)^{1/2}. \tag{7.5.32}$$

Note that, if hydrodynamic interactions are absent, we recover the Rouse result (the so-called "free draining" limit) from the first term on the right-hand side of this equation and plug back in Eq. (7.5.31), which now is the same as Eq. (7.5.9). If hydrodynamic interactions dominate, the second term on the right-hand side of Eq. (7.5.32) overwhelms the first term and leads to the Zimm results. This limit is referred to as the "non-free draining" limit.

When hydrodynamic interactions dominate, Eqs. (7.5.31) and (7.5.32) give

$$\tau_p = \frac{\eta_0 \ell^3}{\sqrt{3\pi} k_B T} \left(\frac{N}{p}\right)^{3/2}. \tag{7.5.33}$$

This equation constitutes the Zimm relaxation spectrum for Gaussian chains. The longest relaxation time (for $p = 1$) is called the **Zimm time**,

$$\tau_Z = \frac{1}{\sqrt{3\pi}} \frac{\eta_0 \ell^3}{k_B T} N^{3/2}. \tag{7.5.34}$$

Using the relation $R_g = \sqrt{N\ell^2/6}$ for Gaussian chains, the above equation is equivalently written as

$$\tau_Z \simeq 4.8 \frac{\eta_0}{k_B T} R_g^3 \sim \frac{\eta_0 R_g^3}{T}. \tag{7.5.35}$$

Thus the longest relaxation time for the Zimm model is proportional to R_g^3.

(2) Center-of-mass diffusion

$$\langle [\mathbf{R}_{cm}(t) - \mathbf{R}_{cm}(0)]^2 \rangle = 6 D_{cm} t, \tag{7.5.36}$$

where

$$D_{cm} = \frac{8 k_B T}{3 \sqrt{6\pi^3} \eta_0 \ell} \frac{1}{\sqrt{N}} = \frac{8}{3 \sqrt{\pi}} \frac{k_B T}{6\pi \eta_0 R_g}, \tag{7.5.37}$$

where $R_g = \sqrt{N\ell^2/6}$ is used.

Therefore,

$$D_{cm} \sim \frac{T}{\eta_0 R_g}. \tag{7.5.38}$$

This derived result from the Zimm model is in accordance with our earlier expectation based on the Stokes–Einstein law. Writing Eq. (7.5.37) in the form of the Stokes–Einstein law, we get

$$D_{cm} = \frac{k_B T}{6\pi \eta_0 R_h}, \tag{7.5.39}$$

where R_h is the **hydrodynamic radius**, as defined in Eq. (5.5.4). The hydrodynamic radius is related to R_g, for a Gaussian chain, by

$$\frac{R_g}{R_h} \simeq 1.5. \qquad \text{(Zimm, Gaussian chain)} \tag{7.5.40}$$

(3) Dynamics of a labeled monomer

For times shorter than the Zimm time, $t < \tau_Z$,

$$\langle [\mathbf{R}_m(t) - \mathbf{R}_m(0)]^2 \rangle \sim t^{2/3}, \qquad t < \tau_Z \tag{7.5.41}$$

and for times longer than the Zimm time, $t > \tau_Z$,

$$\langle [\mathbf{R}_m(t) - \mathbf{R}_m(0)]^2 \rangle \sim t, \qquad t > \tau_Z. \qquad (7.5.42)$$

(4) Dynamic structure factor

As in Section 7.5.1, the following two limits of experimental relevance may be identified.

(i) $k^2 N \ell^2 \ll 1$

$$S(\mathbf{k},t) \sim \exp\left(-k^2 D_{cm} t\right). \qquad (7.5.43)$$

(ii) $k^2 N \ell^2 \gg 1$

$$S(\mathbf{k},t) \sim \exp\left(-(\Gamma_k t)^{2/3}\right), \qquad (7.5.44)$$

which is a "stretched exponential," with the relaxation rate Γ_k given by

$$\Gamma_k \sim k^3. \qquad (7.5.45)$$

(5) Viscoelasticity

For oscillatory shear, $\dot{\gamma}(t) = \dot{\gamma} e^{i\omega t}$, the frequency dependent intrinsic viscosity, defined as $[\eta] = \lim_{c \to 0} (\eta - \eta_0)/c\eta_0$, becomes (analogous to the derivation for the Rouse model, see Section 7.5.1)

$$[\eta] = \frac{RT}{M\eta_0} \sum_p \frac{\tau_p}{(1 + i\omega\tau_p)}, \qquad (7.5.46)$$

where τ_p is one half of the expression given in Eq. (7.5.33),

$$\tau_p = \frac{\eta_0 \ell^3}{\sqrt{12\pi k_B T}} \left(\frac{N}{p}\right)^{3/2}. \qquad (7.5.47)$$

Combining Eqs. (7.5.46) and (7.5.47), we get

$$[\eta] \sim \begin{cases} \sqrt{N} & \text{(zero frequency)} \\ \omega^{-1/3}. & \text{(high frequency)} \end{cases} \qquad (7.5.48)$$

Note that the intrinsic viscosity $[\eta]$, in the zero-frequency limit, is proportional to $\sqrt{N}$. More explicitly, using Eqs. (7.5.46) and (7.5.47), we get

$$[\eta] = 6.25 N_A \frac{R_g^3}{M} \sim \sqrt{N}, \qquad (7.5.49)$$

where N_A is the Avogadro number.

This result is in accordance with our earlier prediction based on Einstein's result for the viscosity of suspensions,

$$\frac{\eta - \eta_0}{\eta_0} = \frac{5}{2}\phi \sim c\frac{R_g^3}{N}. \qquad (7.5.50)$$

Analogously, the frequency dependent modulus is given by the Zimm dynamics as

$$G(\omega) = \frac{cRT}{M} \sum_p \frac{i\omega\tau_p}{(1 + i\omega\tau_p)}, \qquad (7.5.51)$$

where τ_p is given by Eq. (7.5.47). The storage and loss moduli can be easily obtained from the above equation. As an example, the intrinsic modulus, in the higher frequency range ($\omega > 1/\tau_Z$) is

$$[G] \sim \omega^{2/3}. \tag{7.5.52}$$

(6) Effect of size exponent on internal Zimm dynamics

If we consider a non-Gaussian chain with the size exponent ν ($R_g \sim N^\nu$) with full inter-segment hydrodynamic interaction, the Zimm relaxation spectra, the Zimm time, mean square displacement of the center-of-mass of a chain and its diffusion coefficient, mean square displacement of a labeled monomer for times shorter than the Zimm time, the dynamic structure factor and its rate of decay for $kR_g \gg 1$, and the intrinsic viscosity and elastic modulus of the solution are given by (Appendix 6.4.2)

$$\tau_p \sim \frac{\eta_0}{T} \left(\frac{N}{p}\right)^{3\nu},$$

$$\tau_Z \sim \frac{\eta_0}{T} N^{3\nu} \sim \frac{\eta_0}{T} R_g^3,$$

$$\langle [\mathbf{R}_{cm}(t) - \mathbf{R}_{cm}(0)]^2 \rangle = 6D_{cm}t, \qquad (t > \tau_Z),$$

$$D_{cm} = \left(\frac{8}{3\sqrt{\pi}}\right) \frac{k_B T}{6\pi\eta_0 R_g} \sim \frac{T}{\eta_0 R_g},$$

$$\langle [\mathbf{R}_m(t) - \mathbf{R}_m(0)]^2 \rangle \sim t^{2/3}, \qquad (t < \tau_Z),$$

$$S(\mathbf{k},t) \sim \exp\left[-(\Gamma_k t)^{2/3}\right],$$

$$\Gamma_k \sim k^3,$$

$$[\eta] \sim \begin{cases} N^{3\nu-1} & \text{(zero frequency)} \\ \omega^{\frac{1}{3\nu}-1}, & \text{(high frequency)} \end{cases}$$

$$G(\omega) \sim \omega^{1/3\nu}. \quad \text{(high frequency)} \tag{7.5.53}$$

Note that the above Zimm results for the longest relaxation time, $\tau_Z \sim (\eta_0/T)R_g^3$, and the intrinsic viscosity are in accordance with the Einstein laws given in Eq. (7.4.8) and Eq. (7.4.19), respectively. Also, the k^3 dependence of the decay rate Γ_k is independent of the size exponent ν.

The predictions of the Zimm model for the various dynamical quantities in dilute solutions are summarized in Table 7.1, along with the corresponding results from the Rouse model for a Gaussian chain. Note that the Rouse model is relevant only at the experimental conditions where hydrodynamic interactions and excluded volume interactions for flexible polymers are screened.

7.5.3   Reptation Model

The Rouse and Zimm models described above address single chains in dilute solutions. However, in solutions with polymer concentration above a critical concentration and

Table 7.1 Dynamical properties from the Zimm model of a polymer chain with size exponent ν in dilute solutions, compared to the Rouse model of a Gaussian chain.

Property	Zimm ($R_g \sim N^\nu$)	Rouse ($\nu = 1/2$)
Longest relaxation time	$\tau_Z \sim \frac{\eta_0}{T} R_g^3 \sim \frac{\eta_0}{T} N^{3\nu}$	$\tau_R \sim \frac{\zeta_b}{T} N^2$
Center-of-mass diffusion coefficient	$D_{cm} \sim \frac{T}{\eta_0 R_g} \sim \frac{T}{\eta_0 N^\nu}$	$D_{cm} \sim \frac{T}{\zeta_b N}$
Mean square displacement of a monomer	$\langle [\mathbf{R}_m(t) - \mathbf{R}_m(0)]^2 \rangle \sim t^{2/3}$	$\langle [\mathbf{R}_m(t) - \mathbf{R}_m(0)]^2 \rangle \sim t^{1/2}$
Relaxation rate ($kR_g < 1$)	$\Gamma_k = D_{cm} k^2$	$\Gamma_k = D_{cm} k^2$
Relaxation rate ($kR_g > 1$)	$\Gamma_k \sim k^3$	$\Gamma_k \sim k^4$
Specific viscosity (zero frequency)	$\frac{\eta - \eta_0}{\eta_0} \sim cN^{3\nu-1}$	$\frac{\eta - \eta_0}{\eta_0} \sim cN$
Specific viscosity (high frequency)	$\frac{\eta - \eta_0}{\eta_0} \sim c\omega^{-(3\nu-1)/3\nu}$	$\frac{\eta - \eta_0}{\eta_0} \sim c\omega^{-1/2}$
Frequency dependence of shear modulus	$G(\omega) \sim c\omega^{1/3\nu}$	$G(\omega) \sim c\omega^{1/2}$

molecular weight above a critical value, entanglement constraints that two portions of either the same chain or different chains cannot intersect become more prevalent (Fig. 7.14b). These entanglement constraints lead to very long relaxation times for chains resulting in several universal features that are distinctly different from those of the Rouse-Zimm models. For example, the zero shear rate viscosity of a polymer melt varies with the number of segments per chain as $N^{3.4}$ for high molecular weights, instead of the Rouse behavior $\eta \sim N$ for low molecular weights,

$$\eta \sim \begin{cases} N & \text{(low molecular weight unentangled melt)} \\ N^{3.4}. & \text{(high molecular weight entangled melt)} \end{cases} \quad (7.5.54)$$

An elegant and simple theoretical model to describe the dynamics of a labeled chain in an entangled polymer system is the reptation model introduced by de Gennes (1971) and further developed by Doi & Edwards (1986).

When a labeled chain in the entangled state of Fig. 7.14b is projected onto a plane, the plane will look as shown in Fig. 7.18. The continuous curve is the projection of the labeled chain. The dots represent the various chains passing through the plane. Since the collective disappearance of all the dots surrounding the continuous curve will take a long time, let us assume that the labeled chain is essentially localized in a tube-like region [Edwards (1967)]. The contour of this tube can be approximately obtained by holding one end of the chain, pulling the other end, and finding the shortest path of the continuous curve which originally contains slacks. This contour is called **primitive path**, which is the shortest curve with the same topology as the real chain relative to other chains. The real chain is wriggling around the primitive path. Let the diameter of the tube-like region be d. Since the primitive path is only a mathematical construct, we further assume, for the sake of simplicity, that it is a random walk of Z steps, each of step length d. Since d represents the average distance between entanglements, we

Figure 7.18 Definition of the primitive path and imagination of a tube out of which a real chain undergoes reptation.

can write (from Gaussian chain statistics)

$$d^2 = N_e \, \ell^2, \tag{7.5.55}$$

where N_e is the number of Kuhn segments in the entanglement molecular weight M_e (molecular weight of the part of the chain between two adjacent entanglement constraints along the chain), and ℓ is the Kuhn segment length. Therefore, the number of primitive path steps Z is given by

$$Z = \frac{N}{N_e}, \tag{7.5.56}$$

where N is the number of Kuhn segments per chain.

Let us consider the motion of a chain confined by the tube enclosing the primitive path. Furthermore, as proposed by de Gennes (1971), let us assume that the motion of the polymer chain into the walls of the "tube" cannot be dominant in the zeroth-order picture, and that the chain changes its configuration only through the diffusion along the tube (primitive path). This one-dimensional random walk of the chain is called **reptation**. It is obvious from the model that in order for the chain to change its configuration, it must "disengage" from its original tube. The time required for this process is called the **disengagement time** τ_d. Based on the reptation model, quantitative aspects of the viscoelastic properties of entangled polymers are provided by the Doi–Edwards theory [Doi & Edwards (1986)]. Here, we will present only simple arguments for a few important quantities.

(1) Disengagement time τ_d

Since the chain is diffusing along the tube, the friction coefficient of the chain along the tube is $\zeta_{\mathrm{seg}} N$ (where ζ_{seg} is the friction coefficient of the segment and N is the number of Kuhn segments per chain). Therefore, the curvilinear diffusion coefficient along the tube D_{tube} follows from Einstein's law,

$$D_{\text{tube}} = \frac{k_B T}{\zeta_{\text{seg}} N}. \tag{7.5.57}$$

The mean square distance traveled by the chain is twice the product of D_{tube} and time duration t, for one-dimensional diffusion. When the time duration is the disengagement time τ_d, the mean square displacement is the mean square end-to-end distance of the primitive path, so that

$$\left[\left(\frac{N}{N_e}\right) d\right]^2 = 2\left(\frac{k_B T}{\zeta_{\text{seg}} N}\right) \tau_d. \tag{7.5.58}$$

In view of Eq. (7.5.55), we get a neat expression for the disengagement time,

$$\tau_d = \frac{\zeta_{\text{seg}} \ell^2}{2 N_e k_B T} N^3 \sim N^3. \tag{7.5.59}$$

(2) Center-of-mass diffusion, D_g

In time τ_d, a marked segment of the chain must have undergone a random walk of N steps since Gaussian statistics is assumed for the tube. Therefore, we expect the center-of-mass $\mathbf{R}_{cm}$ of the chain (which might lie on one of the segments of the chain) also to undergo a random walk of N steps in time τ_d. For the random walk of N steps of Kuhn step length ℓ, the mean square of the displacement is $N\ell^2$. Therefore, using the fundamental relation (Eq. (7.4.6)) between the characteristic time, diffusion coefficient, and mean square displacement, we have (for three dimensions)

$$N\ell^2 = 6 D_g \tau_d, \tag{7.5.60}$$

where D_g is the diffusion coefficient of the chain in the whole system. Using Eq. (7.5.59), we get

$$D_g = \frac{1}{3} \frac{N_e k_B T}{\zeta_{\text{seg}} N^2} \sim \frac{1}{N^2}. \tag{7.5.61}$$

(3) Mean square displacement of a labeled monomer, $r^2(t)$

Four regimes may be identified depending on time relative to the disengagement time and the Rouse time: (a) $t > \tau_d$, (b) $\tau_{\text{Rouse}} < t < \tau_d$, (c) $t < \tau_{\text{Rouse}}$, and (d) t very short so that the relaxation does not yet feel the presence of the tube. It can be derived that the respective exponents for the time dependence are $1, 1/2, 1/4$, and $1/2$. These four regimes are sketched in Fig. 7.19. The quarter power law of $r^2(t)$ is taken to be the earmark of reptation dynamics.

(4) Viscoelasticity

According to the Doi–Edwards theory of reptation, the stress relaxation in the terminal region is given by

$$G(t) = G_N^0 \frac{8}{\pi^2} \sum_{p=\text{odd}} \frac{1}{p^2} \exp\left(-\frac{p^2 t}{\tau_d}\right), \tag{7.5.62}$$

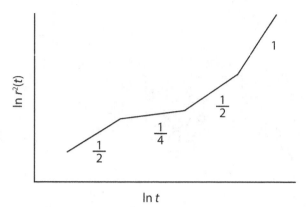

Figure 7.19 Double logarithmic plot of mean square displacement of a labeled monomer $r^2(t)$ versus time t.

where G_N^0 is known as the plateau modulus, given by

$$G_N^0 = \frac{\rho RT}{M_e},\qquad(7.5.63)$$

where ρ is the density of the melt, R is the ideal gas constant, and M_e is the molecular weight of the part of the chain between two adjacent entanglement constraints along the chain.

The dynamic moduli G' and G'' are obtained from the Fourier transform of the above equation (see Eq. (7.5.23)) as

$$G'(\omega) = G_N^0 \frac{8}{\pi^2} \sum_{p=\text{odd}} \frac{1}{p^2} \frac{\omega^2 \tau_d^2}{(1 + \omega^2 \tau_d^2)},\qquad(7.5.64)$$

$$G''(\omega) = G_N^0 \frac{8}{\pi^2} \sum_{p=\text{odd}} \frac{1}{p^2} \frac{\omega \tau_d}{(1 + \omega^2 \tau_d^2)}.\qquad(7.5.65)$$

The zero shear rate viscosity η is given by

$$\eta = \int_0^\infty dt\, G(t),\qquad(7.5.66)$$

where $G(t)$ is the stress relaxation at longer time durations given by Eq. (7.5.62). Performing the integral in Eq. (7.5.66) gives

$$\eta = \frac{\pi^2}{12} G_N^0 \tau_d.\qquad(7.5.67)$$

Note that η is proportional to the disengagement time τ_d. In view of Eqs. (7.5.59) and (7.5.63), the molecular weight dependence of the viscosity of entangled melts is given by

$$\eta \sim M^3.\qquad(7.5.68)$$

This is one of the important predictions of the reptation model.

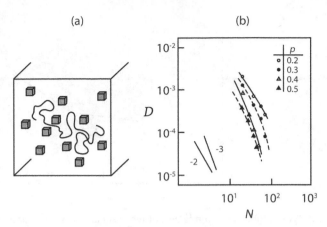

Figure 7.20 (a) Monte Carlo simulation of dynamics of a self-avoiding walk chain of N beads amidst randomly distributed fixed obstacles. (b) The center-of-mass diffusion coefficient D depends on N with a steeper power law than the reptation result, depending on the volume fraction p of the obstacles [Adapted from Muthukumar & Baumgärtner (1989a, 1989b)].

7.5.4   Entropic Barrier Model

The original context of the reptation model is the diffusion of a polymer chain in the presence of fixed obstacles in the system. Such a scenario can be experimentally realized by considering polymer chains trapped inside random porous media or gels. Although the reptation model is quite successful in describing polymer diffusion in melts, computer simulations of chain diffusion in a medium containing randomly distributed fixed rigid obstacles do not support the prediction $D \sim N^{-2}$ from the reptation model [Muthukumar & Baumgärtner (1989a, 1989b)]. As an example, consider the Monte Carlo simulations of the dynamics of an SAW chain ($\nu \simeq 3/5$) in a cubic lattice of lattice spacing a, made of 300^3 cells. Each of these cells of volume a^3 is occupied by a cubic solid particle with probability p, as shown in Fig. 7.20a. Thus, p is the volume fraction of the solid phase. A polymer chain is then introduced in the continuous space representing the fluid phase. Let the chain contain N hard spheres of diameter h, which are freely jointed together by $N - 1$ rigid bonds of length ℓ. The movement of the center-of-mass of the chain is then monitored for different values of p and N (without hydrodynamic interactions), from which the diffusion coefficient D is obtained in the long-time regime where the Einstein law of diffusion is applicable.

In the absence of obstacles, simulations show that $D \sim N^{-1}$ as expected from the Rouse dynamics. As p increases, D becomes smaller in an N-dependent fashion. The dependence of D on N at different values of p is presented in Fig. 7.20b. As seen in this figure, the reptation law, $D \sim N^{-2}$, is not supported. Instead, a more sensitive N-dependence, such as $D \sim N^{-3}$, is apparent for large values of N. This conclusion from the simulations is also observed in experiments on entangled polymer solutions and chains trapped inside gels, where $D \sim N^{-\alpha}$ with $\alpha \sim 3$ and sometimes even larger values [Kim *et al.* (1986), Calladine *et al.* (1991), Lodge & Rotstein (1991), Rotstein & Lodge (1992)]. Therefore, the reptation model is not applicable to polymer diffusion in restricted environments as addressed in the above mentioned simulations

(a)   (b)

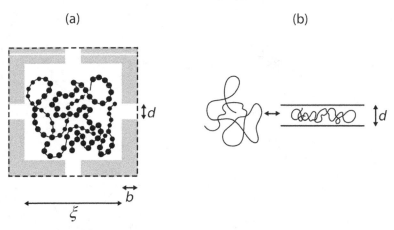

Figure 7.21 (a) Sketch of a chain captured initially inside a cubic cavity of linear dimension ξ. Each cavity is connected to its six neighbors through regular capillaries of length b and square cross section with side d. (b) Sketch of a chain being squeezed into a capillary which creates an entropic barrier.

and experiments. The details of the simulations reveal that the discrepancy between these observations and the reptation prediction is due to presence of bottlenecks in the random medium. The bottlenecks squeeze the chain and reduce its conformational entropy. As a result, the chain encounters several entropic barriers at random locations in its pathway.

In order to assess the nature of entropic barriers, let us consider the escape of a chain of radius of gyration R_g captured inside a cubic cavity of linear dimension ξ into one of the neighboring cavities in a large regular array of such cavities. Let R_g be comparable to ξ. A two-dimensional cross section of a cavity containing the chain is sketched in Fig. 7.21a. There are six capillaries connecting each cavity to its neighboring cavities. Let each capillary be of length b and the side of the square cross section d (Fig. 7.21a).

In general, when a chain in a solution is squeezed into a capillary, its conformations are substantially reduced in comparison to the situation in free solution (Fig. 7.21b). This results in an increase in the free energy of the chain, called **confinement free energy**, F_c. This can be estimated using the following scaling argument [de Gennes (1979)]. In the case of a chain with $R_g \sim N^\nu$ squeezed into a long capillary of cross section d, the key dimensionless variable is d/R_g, and hence the confinement free energy can be written in the scaling form

$$\frac{F_c}{k_B T} = f_c\left(\frac{d}{R_g}\right), \tag{7.5.69}$$

where f_c denotes an unknown function of the argument d/R_g. Since the confinement free energy is an extensive quantity, it is proportional to N for strong confinement of the chain. Therefore, under chain confinement, the function f_c must be such that it is proportional to N for $d/R_g < 1$,

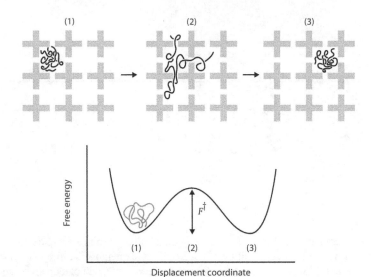

Figure 7.22 Top: A chain from one entropic trap (1) moves into another entropic trap (3) through a transitional stage (2) which creates an entropic barrier. Bottom: The corresponding free energy profile as a function of displacement of the center of mass of the chain.

$$f_c \left(\frac{d}{R_g} \right) \sim f_c \left(\frac{d}{N^\nu} \right) \sim N. \qquad (d < R_g) \qquad (7.5.70)$$

This condition is satisfied only if the function f_c is a power law of its argument d/N^ν, given as exponent x,

$$\left(\frac{d}{N^\nu} \right)^x \sim N. \qquad (d < R_g) \qquad (7.5.71)$$

Therefore, $x = -1/\nu$, giving the scaling expression for the confinement free energy as

$$\frac{F_c}{k_B T} \sim \frac{N}{d^{1/\nu}}. \qquad (7.5.72)$$

This is a general result for entropic confinement of a chain inside a restricted space.

Returning to the situation in Fig. 7.21a, each cavity acts as an entropic trap for the chain. Here, the confinement free energy of the chain is lower in comparison with other alternatives where the chain straddles through one or more bottlenecks separating the cavities, as illustrated in Fig. 7.22. As time passes by, the chain eventually gets out of one entropic trap and falls into the next entropic trap after negotiating the entropic barrier associated with restricted chain conformations inside the bottlenecks. The free energy landscape for the movement of a chain from one entropic trap to another is sketched in Fig. 7.22. The entropic barrier can be obtained by implementing Eq. (7.5.72) as

$$F^\dagger = F_{c,2} - F_{c,1}, \qquad (7.5.73)$$

where $F_{c,1}$ is the free energy of confinement of the chain fully inside one entropic trap, and $F_{c,2}$ is the free energy of the chain in its transitional state where the chain is in the process of passing through the bottleneck. Allowing for the chain to serially

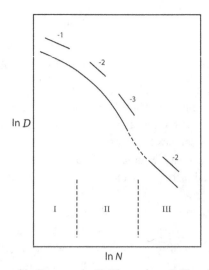

Figure 7.23 Sketch of a plot of $\ln D$ versus $\ln N$. The regimes I, II, and III denote, respectively, the Rouse, entropic barrier, and reptation regimes. In the Rouse regime, $D \sim N^{-1}$, and in the reptation regime, $D \sim N^{-2}$. In the intermediate crossover regime, D depends on N more sensitively according to $D \sim N^{-\alpha}$, where the apparent exponent α can be fitted to 2 or 3 (or even higher) depending on the details of the system. The dashed portion of the curve in the entropic barrier regime remains to be explored in experiments. [Muthukumar (1991)]

explore multiple bottlenecks in its pathway to move to the next cavity, $F^{\dagger}$ is obtained as [Muthukumar & Baumgärtner (1989a)]

$$\frac{F^{\dagger}}{k_B T} = A(\xi, b, d)N, \tag{7.5.74}$$

where A is a function of the cavity size ξ and the geometrical features of the capillaries separating the cavities.

The diffusion coefficient D of a chain to cross the free energy barrier $F^{\dagger}$ is given by the well-known Kramers formula [Hänggi et al. (1990)] as

$$D = D_0 \exp\left(-\frac{F^{\dagger}}{k_B T}\right), \tag{7.5.75}$$

where D_0 denotes the diffusion coefficient in the absence of the free energy barrier. In the present situation, D_0 is the diffusion coefficient of a Rouse chain, $D_0 \sim 1/N$. Therefore,

$$D \sim \frac{1}{N} e^{-AN}. \qquad \text{(entropic barrier)} \tag{7.5.76}$$

Note that if the concentration of fixed obstacles is very high such that there are no entropic traps, that is $\xi \sim \ell \ll R_g$, then the reptation model becomes applicable. If there are no entropic barriers at all at very low concentrations of the obstacles, then the Rouse model is valid (with no hydrodynamic interactions). Thus the entropic barrier model is applicable in the intermediate confinement regime between the Rouse and reptation regimes,

$$D \sim \begin{cases} \frac{1}{N} & \text{Rouse} \quad (\ell \ll R_g \ll \xi) \\ \frac{1}{N} e^{-AN} & \text{entropic barrier} \quad (\ell \ll R_g \sim \xi) \\ \frac{1}{N^2}, & \text{reptation} \quad (R_g \gg \xi \leq \ell) \end{cases} \tag{7.5.77}$$

as illustrated in Fig. 7.23. In the intermediate entropic barrier regime, D can be fitted with an apparent power law $N^{-\alpha}$, with a value of α around 3. In reality, the apparent value α depends on the range of N used in the data analysis. The value of α is not universal and denotes the transition behavior from Rouse to reptation behaviors.

The entropic barrier described above can be further modulated using the electrostatic interaction between the macromolecule and the solid phase, so that the free energy barrier $F^\dagger$ consists of both entropic and energy contributions. In the athermal limit, for example when both the macromolecule and the solid matrix are similarly charged, the intermediate regime is controlled by entropic barriers. In the case of oppositely charged macromolecules and the matrix, concentration of added salt can be used as an effective tuning variable to control the free energy barrier for diffusion of charged macromolecules.

7.6   Dilute Polyelectrolyte Solutions

Consider a dilute solution of polyelectrolyte chains containing added salt. Under this condition, the chains move around independently of one another. However, their movement is coupled to the movement of their counterions and the small ions from the added salt. Here, we are interested in the translational friction coefficient, electrophoretic mobility, and the diffusion coefficient of an isolated chain, as well as the modification of the solvent viscosity due to the presence of the polyelectrolyte chains. To address these considerations, we follow the general theoretical framework outlined in Section 7.5.

7.6.1   Translational Friction Coefficient

Using the preaveraging approximation mentioned in Section 7.5.2, the net frictional force acting on all monomers of a chain is $-f_t \dot{\mathbf{R}}^0$, where f_t is the translational friction coefficient of the chain and $\dot{\mathbf{R}}^0$ is the net drift velocity of the center-of-mass of the chain. f_t is derived in Appendix 6 (Eq. (A6.71)) as

$$f_t = \frac{\zeta_b N}{\left(1 + \frac{8}{3\sqrt{\pi}} \frac{\zeta_b N}{6\pi\eta_0 R_g}\right)}. \tag{7.6.1}$$

The second term in the denominator arises from the inter-segment hydrodynamic interactions, and the numerator denotes the friction coefficient of the chain as a sum of the frictional coefficients from the individual beads.

The translational friction coefficient in the two asymptotic limits of absence of the hydrodynamic interactions (free draining), and fully fledged hydrodynamic interactions (non-free draining), follows as

$$f_t = \begin{cases} \zeta_b N, & \text{free draining} \\ \left(\frac{3\sqrt{\pi}}{8}\right)\left(6\pi\eta_0 R_g\right), & \text{non-free draining.} \end{cases} \tag{7.6.2}$$

Using the Einstein relation between the diffusion coefficient and friction coefficient, we get the translation diffusion coefficient of the center-of-mass of the chain D in dilute solutions, where hydrodynamic interactions dominate over the bead friction, as $D = k_B T / f_t$ given by

$$D = \left(\frac{8}{3\sqrt{\pi}}\right)\left(\frac{k_B T}{6\pi\eta_0 R_g}\right). \tag{7.6.3}$$

Note that this result is identical to the result from the Zimm model, Eq. (7.5.53).

7.6.2 Electrophoretic Mobility

In the presence of an externally applied constant electric field, a polyelectrolyte chain undergoes electrophoretic drift, as a balance between the electric and frictional forces acting on it. Let the polyelectrolyte chain have a total charge $Q = N\alpha z_p e$ (where the charge sign is absorbed in z_p) and let the constant electric field be $\mathbf{E}_0$. The electric force on the chain is $Q\mathbf{E}_0$ and the net frictional force is $-f_t \dot{\mathbf{R}}^0$, as given in the preceding section. In the steady state where the chain does not accelerate, the net force acting on the chain is zero, so that the Einstein electrophoretic mobility follows as (Eq. (7.3.7))

$$\mu_{\text{Einstein}} = \frac{Q}{f_t}. \tag{7.6.4}$$

Since $f_t \sim R_g \sim N^\nu$ and $Q \sim N$, we get Eq. (7.4.14) as

$$\mu_{\text{Einstein}} \sim N^{1-\nu}. \tag{7.6.5}$$

This result is unphysical because this equation predicts an increasing electrophoretic mobility with an increase in N for $\nu < 1$. To resolve this issue, recall Section 7.3, where we noted that the polyelectrolyte chain does not move under the electric field as an isolated entity, but its motion is coupled to that of its ion cloud. Each polyelectrolyte chain is surrounded by $\left|N z_p / z_c\right|$ counterions, each with a charge $z_c e$. In addition, there are dissociated ions from the added salt. Let $\rho_j(\mathbf{r})$ be the charge density at $\mathbf{r}$ due to the counterion cloud surrounding the jth bead of the chain, as illustrated in Fig. 7.24. In the presence of the electric field $\mathbf{E}_0$, the charge density arising from counterion clouds of all beads exerts an electric force on the fluid element at $\mathbf{r}$ according to

$$\mathbf{F}_{\text{ext}}(\mathbf{r}) = \sum_{j=0}^{N} \rho_j(\mathbf{r})\mathbf{E}_0. \tag{7.6.6}$$

Accounting for this external force from the counterion cloud, the electrophoretic mobility is given as [Muthukumar (1997, 2005), Eq. (A6.78)]

$$\mu = Q\left(\frac{1}{f_t} - \hat{A}\right), \tag{7.6.7}$$

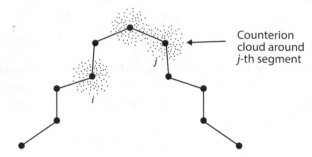

Figure 7.24 Presence of counterion cloud around every charged segment of the polymer. The counterion clouds contribute to the electrophoretic movement of the macromolecule.

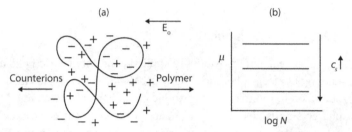

Figure 7.25 (a) Counterion cloud contributes to the electrophoresis of a charged macromolecule resulting in the molecular weight independence of its electrophoretic mobility at each salt concentration. (b) Presence of salt decreases the electrophoretic mobility, but the mobility is still independent of molecular weight.

where

$$\hat{A} = \frac{1}{N^2} \sum_{i=0}^{N} \sum_{j \neq i} \frac{1}{6\pi\eta_0} \left\langle \frac{1}{|\mathbf{R}_i - \mathbf{R}_j|} - \frac{e^{-\kappa|\mathbf{R}_i - \mathbf{R}_j|}}{|\mathbf{R}_i - \mathbf{R}_j|} \right\rangle. \tag{7.6.8}$$

Whereas the first term inside the brackets on the right-hand side of Eq. (7.6.7) gives the Einstein electrophoretic mobility, the second term gives the contribution from the counterion clouds around the charged segments of the polymer. Note the negative sign in the second term due to the motion of the counterion cloud in the opposite direction to that of the polymer, as illustrated in Fig. 7.25a. Furthermore, $\hat{A}$ is zero in the limit of $\kappa \to 0$. Therefore, μ can be written in the form

$$\mu = \frac{QD}{k_B T} \gamma_1(\kappa R_g), \tag{7.6.9}$$

where the multiplicative function γ_1 is a function of the ratio of the radius of gyration to the Debye length, and it reduces to 1 for $\kappa = 0$. Therefore, μ becomes the Einstein result only in the limit of $\kappa = 0$. Note that even in salt-free conditions, $\kappa \neq 0$ due to the presence of a finite number of dissociated counterions in solution.

Performing the conformational average in Eq. (7.6.8) and combining with Eq. (7.6.7) we get [Muthukumar (1997, 2005)]

$$\mu \sim Q \times \begin{cases} \dfrac{1}{\eta_0 R_g}, & \text{low salt} \\[2ex] \dfrac{1}{\eta_0 R_g^{1/\nu} \kappa^{(1-\nu)/\nu}}, & \text{high salt.} \end{cases} \tag{7.6.10}$$

Since $Q \sim N$ and $R_g \sim N^\nu$ ($\nu = 1$ in the no-salt limit), the above equation gives

$$\mu \sim N^0 f(c_s), \tag{7.6.11}$$

where $f(c_s)$ is a function of salt concentration c_s, and μ is independent of N. Including the dependence of R_g on κ, the electrophoretic mobility in the high salt limit is

$$\mu = \left(\frac{\alpha z_p e}{6\pi \eta_0 \ell}\right) \frac{3.3}{(\kappa \ell)^{2/3}} \frac{1}{\left(1 - 2\chi + \frac{4\pi \ell_B}{\kappa^2 \ell^3}\right)^{1/3}}, \tag{7.6.12}$$

where χ is the Flory–Huggins parameter. The above equation shows that μ is a decreasing function of c_s, as indicated in Fig. 7.25b. Detailed calculations show that μ decreases only by a factor of about five as the ionic strength is increased by four orders of magnitude [Hoagland et al. (1999)].

The major conclusion that μ is independent of N at each salt concentration is due to the significant role played by the counterion cloud. The above theoretical predictions are in qualitative agreement with the experimental data presented in Fig. 7.10a. The deviations from the Einstein ratio observed in solutions of ssDNA and dsDNA, as shown in Fig. 7.10b, are due to the term $\hat{A}$ in Eq. (7.6.7) arising from the counterion clouds around the charged repeat units. As a final note on the electrophoretic mobility of charged macromolecules, the above described theory is at the mean field level and consequences of the deformation of ion clouds and inhomogeneities of the electric field around the macromolecules are left for future consideration.

7.6.3   Coupled Diffusion Coefficient

The diffusion coefficient of the polyelectrolyte chain given in Eq. (7.6.3) can be valid only when the thickness of the ion cloud around the chain is infinitesimally small, namely, the concentration of the added salt is very high. At lower salt concentrations, or in the limit of the presence of only the counterions dissociated from the polyelectrolyte, the diffusion of the chains is strongly coupled to the dynamics of the counterions, analogous to the coupling in the presence of an external electric field treated in the preceding section. We first address this dynamical coupling in salt-free solutions, and then consider polyelectrolyte solutions containing added salt.

7.6.3.1   Salt-Free Dilute Polyelectrolyte Solutions
Let there be n polyelectrolyte chains of charge $Q = z_1 e$ and n_c counterions of charge $z_2 e$ in a solution of volume V. Furthermore, let the Stokes–Einstein diffusion coefficients of the polyelectrolyte chain and the counterion be D_1 and D_2, respectively. The average number concentration of the polyelectrolyte chains (c_1^0) and counterions (c_2^0) are n/V and n_c/V, respectively. Due to the required condition of electroneutrality,

$$z_1 c_1^0 + z_2 c_2^0 = 0, \qquad c_2^0 = \left|\frac{z_1}{z_2}\right| c_1^0. \tag{7.6.13}$$

In general, the local concentration of the polyelectrolyte chains and counterions at location $\mathbf{r}$ and time t are fluctuating quantities about their average values,

$$c_i(\mathbf{r}, t) = c_i^0 + \delta c_i(\mathbf{r}, t), \tag{7.6.14}$$

where $i = 1$ and 2 denote the polyelectrolyte and counterion, respectively, and $\delta c_i(\mathbf{r},t)$ is the fluctuation. The time evolution of the local concentration of the ith species is given by the continuity equation

$$\frac{\partial c_i(\mathbf{r},t)}{\partial t} = -\nabla \cdot \mathbf{J}_i, \qquad (7.6.15)$$

where $\mathbf{J}_i$ is the flux of the ith species into the volume element at $\mathbf{r}$. In the presence of the dynamical coupling between the polyelectrolyte and its ion cloud, the flux has two parts,

$$\mathbf{J}_i = -D_i \nabla c_i(\mathbf{r},t) + c_i(\mathbf{r},t)\mu_i \mathbf{E}_{\text{loc}}(\mathbf{r},t). \qquad (7.6.16)$$

The first term on the right-hand side is the diffusive flux with D_i the Stokes–Einstein diffusion coefficient. The second term on the right-hand side is the convective flux, where $\mu_i \mathbf{E}_{\text{loc}}(\mathbf{r})$ is the velocity due to the local electric field at $\mathbf{r}$ arising from the local ionic environment and μ_i is the electrophoretic mobility of the ith species. As derived in Appendix 6.5.3, the time dependence of $\delta c_i(\mathbf{r},t)$, to first order in fluctuations, becomes

$$\frac{\partial}{\partial t}\delta c_i(\mathbf{r},t) = D_i \nabla^2 \delta c_i(\mathbf{r},t) - c_i^0 \frac{\mu_i e}{\epsilon_0 \epsilon} \sum_{j=1}^{2} z_j \delta c_j(\mathbf{r},t). \qquad (7.6.17)$$

Thus the fluctuations in the local concentrations of the polyelectrolyte and the counterions, $\delta c_1(\mathbf{r},t)$ and $\delta c_2(\mathbf{r},t)$, are coupled. In the absence of coupling, the correlation function $\langle \delta\hat{c}_i(\mathbf{k},t)\delta\hat{c}_i(-\mathbf{k},0)\rangle$ (where $\delta\hat{c}_i(\mathbf{k},t)$ is the Fourier transform of $\delta c_i(\mathbf{r},t)$ with $\mathbf{k}$ as the scattering wave vector) decays exponentially with the decay rate $D_i k^2$. Due to the dynamical coupling between the fluctuations $\delta\hat{c}_1$ and $\delta\hat{c}_2$, two new modes with decay rates Γ_1 and Γ_2 emerge which replace the two individual uncoupled diffusion modes (with decay rates $D_1 k^2$ and $D_2 k^2$). Focusing on the polyelectrolyte, the time correlation function of fluctuations in the number concentration of polyelectrolyte chains is given as

$$\frac{\langle \delta\hat{c}_1(\mathbf{k},t)\delta\hat{c}_1(-\mathbf{k},0)\rangle}{\langle \delta\hat{c}_1(\mathbf{k},0)\delta\hat{c}_1(-\mathbf{k},0)\rangle} = A_1 e^{-\Gamma_1 t} + A_2 e^{-\Gamma_2 t}. \qquad (7.6.18)$$

The left-hand side is proportional to the normalized field correlation function measured in dynamic light scattering (DLS). The decay rates Γ_1 and Γ_2, and amplitudes A_1 and A_2, are given as (Appendix 6.5.3)

$$\Gamma_1 = D_1 \gamma_1 \kappa_1^2 + D_2 \kappa_2^2, \qquad \Gamma_2 = \frac{D_1 D_2 \left(\gamma_1 \kappa_1^2 + \kappa_2^2\right)}{\left(D_1 \gamma_1 \kappa_1^2 + D_2 \kappa_2^2\right)} k^2,$$

$$A_1 = \frac{D_1 \gamma_1 \kappa_1^2}{\left(D_1 \gamma_1 \kappa_1^2 + D_2 \kappa_2^2\right)}, \qquad A_2 = \frac{D_2 \kappa_2^2}{\left(D_1 \gamma_1 \kappa_1^2 + D_2 \kappa_2^2\right)}, \qquad (7.6.19)$$

where

$$\mu_1 = \frac{z_1 e D_1 \gamma_1}{k_B T}, \qquad \mu_2 = \frac{z_2 e D_2}{k_B T}, \qquad \kappa_1^2 = \frac{(z_1 e)^2 c_1^0}{\epsilon_0 \epsilon k_B T}, \qquad \kappa_2^2 = \frac{(z_2 e)^2 c_2^0}{\epsilon_0 \epsilon k_B T}. \qquad (7.6.20)$$

Recall that the factor γ_1 arises from the contribution from the counterion cloud of the polymer and it depends on concentration of added salt (Eq. (7.6.9)).

The decay rate Γ_1 is non-diffusive because it is not proportional to k^2. Instead, it is independent of k and is an effective relaxation rate of the ion cloud. This first mode with decay rate Γ_1 is called the **Debye mode**, or **plasmon mode**,

$$\Gamma_1 = \Gamma_{\text{plasmon}}. \tag{7.6.21}$$

This non-diffusive plasmon decay rate is too high ($> 10^6$ Hz) to be detected in conventional DLS measurements.

The second mode with decay rate $\Gamma_2 \sim k^2$ is diffusive, and the prefactor of k^2 in Eq. (7.6.19) is an effective diffusion coefficient referred to as the **fast** diffusion coefficient, $\Gamma_2 = D_{\text{fast}}k^2$, where

$$D_{\text{fast}} = \frac{D_1 D_2 \left(1 + \gamma_1 \left|\frac{z_1}{z_2}\right|\right)}{\left(D_2 + D_1\gamma_1 \left|\frac{z_1}{z_2}\right|\right)}, \tag{7.6.22}$$

where Eqs. (7.6.13) and (7.6.20) are used. In the case of a polyelectrolyte chain of charge $z_1 e = \alpha z_p N e$, Eq. (7.6.22) gives

$$D_{\text{fast}} = \frac{D_1 D_2 \left(1 + \gamma_1\alpha \left|\frac{z_p}{z_2}\right| N\right)}{\left(D_2 + D_1\gamma_1\alpha \left|\frac{z_p}{z_2}\right| N\right)}. \tag{7.6.23}$$

Since $D_1 \sim N^{-1}$ and $\gamma_1 = 1$ in salt-free solutions, the denominator of the above equation is comparable to D_2 within a numerical prefactor of about unity. Therefore, the above equation reduces to

$$D_{\text{fast}} \simeq D_1 \left(1 + \alpha \left|\frac{z_p}{z_2}\right| N\right). \tag{7.6.24}$$

Therefore, due to dynamical coupling, the effective diffusion coefficient for the fluctuations in the polyelectrolyte concentration approaches a higher value than D_1 expected for the large polyelectrolyte chain.

The result of Eq. (7.6.24) can be obtained in a simpler manner by assuming that the smaller counterions relax must faster than the larger polyelectrolyte chain. As shown in Appendix 6.5.3 (Eq. (A6.109)), D_{fast} is given as

$$D_{\text{fast}} = D_1 \left[1 + \frac{\gamma_1\kappa_1^2}{\kappa_2^2}\right]. \tag{7.6.25}$$

Note that Eq. (7.6.25) is independent of D_2 due to the assumption of the counterion fluctuation relaxing much faster than the polyelectrolyte concentration fluctuation. The general result without this assumption of separation of time scales for the two concentration fluctuations is Eq. (7.6.22).

7.6.3.2 Simple Electrolyte Solutions

Eq. (7.6.22) is general for $z_1 : z_2$ electrolytes dispersed in a solution, although it is derived above in the context of a polyelectrolyte and its counterions. In the case of a simple electrolyte, $z : z$ salt ($z_1 = z_2 = z, c_1^0 = c_2^0 = c_s^0, \gamma_1 = 1$), the above derivation is referred to as the **Nernst–Hartly theory**, and D_{fast} and the decay rate for the plasmon mode follow as [Berne & Pecora (1976)]

$$D_{\text{fast}} = \frac{2D_1 D_2}{(D_1 + D_2)} \tag{7.6.26}$$

$$\Gamma_1 = \Gamma_{\text{plasmon}} = \left(\frac{D_1 + D_2}{2}\right)\frac{1}{\xi_D^2}, \qquad \text{(plasmon)} \tag{7.6.27}$$

where the Debye length ξ_D in the present situation is given by

$$\xi_D^{-2} = \frac{2(ze)^2}{\epsilon \epsilon k_B T} c_s^0. \tag{7.6.28}$$

Therefore, the diffusion coefficient of the coupled mode is neither D_1 nor D_2, but a combination of these two individual diffusion coefficients. The characteristic time for the plasmon mode ($1/\Gamma_{\text{plasmon}}$) corresponds to the average time taken by the ion cloud to diffuse a distance comparable to the Debye length, with a diffusion coefficient given by the arithmetic average of D_1 and D_2.

7.6.3.3   Dilute Polyelectrolyte Solutions with Salt

The derivation of the coupled diffusion coefficient of the polyelectrolyte chain in salt-free conditions given in Section 7.6.3.1 can be extended to the presence of added salt. If there are a certain number of different kinds of charged species in the solution, then there are as many coupled equations for the fluctuations in the local concentrations of these species. This generalization of Eq. (7.6.17) leads to an equivalent number of coupled relaxation modes. For example, consider the situation of one kind of added salt in the solution, where one of the dissociated ions from the added salt is the same as the counterion of the polyelectrolyte. Now, there is an additional continuity equation for the fluctuation in the local concentration of the co-ions supplementing Eq. (7.6.17). This results in three modes of relaxation: one non-diffusive plasmon mode and two diffusive modes with diffusion coefficient $D_{\text{fast},1}$ as described above for the polymer and $D_{\text{fast},2}$ for the added salt. $D_{\text{fast},1}$ and $D_{\text{fast},2}$ are measurable using DLS and all these three modes are coupled [Jia & Muthukumar (2019), Muthukumar (2019)].

If we assume that the relaxation of the concentration fluctuations of the various small ions is rapid compared to the macromolecule, then the modification of the fast mode due to added salt can be easily derived. With this assumption, the formula for D_{fast} is the same as Eq. (7.6.25), except that κ_2^2 is replaced by $\kappa_2^2 + \kappa_s^2$ where κ_s^2 arises from the added salt. For $z : z$ electrolytes at number concentration c_s^0, κ_s^2 is given as

$$\kappa_s^2 = \frac{2(ze)^2}{\epsilon_0 \epsilon k_B T} c_s^0. \tag{7.6.29}$$

Combining this result with Eqs. (7.6.13), (7.6.20), and (7.6.25), we get

$$D_{\text{fast}} = D_1 \left[1 + \frac{\gamma_1 z_1^2 c_1^0}{\left(|z_1 z_2| c_1^0 + 2z^2 c_s^0\right)} \right]. \tag{7.6.30}$$

For monovalent counterions and salt ions, and $|z_p| = 1$ (so that the polyelectrolyte charge is αNe), we obtain

$$D_{\text{fast}} = D_1 \left[1 + \frac{\gamma_1 (\alpha N)^2 c_1^0}{\left(\alpha N c_1^0 + 2c_s^0 \right)} \right]. \qquad (7.6.31)$$

Note that γ_1 depends on the salt concentration c_s^0 as described in Section 7.6.2 and Eq. (7.6.9).

At higher concentrations of added salt, the second term on the right-hand side of the above equation becomes negligible compared to the first term, so that D_{fast} is given by D_1. Therefore, the dynamical coupling between the polyelectrolyte chain and its ionic environment is broken due to the small Debye length in this limit, and the diffusion coefficient of the polyelectrolyte chain approaches the value given by the Stokes–Einstein law. In the other limit of the salt-free condition ($c_s^0 = 0, \gamma_1 = 1$), where the Debye length is large, the dynamical coupling is strong and the effective diffusion coefficient is enhanced to a larger value by a factor of $(1 + \alpha N)$,

$$D_{\text{fast}} = \begin{cases} D_1 (1 + \alpha N), & c_s^0 = 0 \\ D_1, & c_s^0 \gg \alpha N c_1^0. \end{cases} \qquad (7.6.32)$$

Since the diffusion coefficient of a polyelectrolyte chain in salt-free conditions, without regard to its ion cloud, is inversely proportional to N, let D_1 be defined as D_{segment}/N so that D_{fast} follows from Eq. (7.6.32) as

$$D_{\text{fast}} = \left(D_{\text{segment}} \right) \alpha, \qquad (c_s^0 = 0) \qquad (7.6.33)$$

for large values of N. Since D_{segment} is comparable to that of an electrolyte ion and the typical value of the degree of ionization is about 1/4, the fast diffusion coefficient observed for the polyelectrolyte is only about a factor of 4 smaller than that of an electrolyte ion. Furthermore, note that D_{fast} is independent of both N and the number polymer concentration c_1^0. All of these conclusions offer an explanation of experimental results given in Fig. 7.9a.

7.6.4 Intrinsic Viscosity and Modulus

The frequency dependent intrinsic viscosity of a polyelectrolyte solution is given by Eq. (7.5.46) as

$$[\eta] = \lim_{c \to 0} \frac{\eta - \eta_0}{\eta_0 c} = [\eta'] - i[\eta''] = \frac{RT}{M\eta_0} \sum_{p=1}^{N} \frac{\tau_p}{(1 + i\omega\tau_p)}. \qquad (7.6.34)$$

Here, R is the gas constant, M is the molar mass, and c is the monomer concentration. $[\eta']$ and $[\eta'']$ are the real and imaginary parts of the intrinsic viscosity, respectively. τ_p is given by Eq. (7.5.53). In the zero frequency limit, the intrinsic viscosity is given by

$$[\eta] = \frac{RT}{M\eta_0} \sum_{p=1}^{N} \tau_p. \qquad (7.6.35)$$

Table 7.2 Dependence of intrinsic viscosity and intrinsic modulus on degree of polymerization of the polyelectrolyte and the frequency of oscillatory shear in dilute solutions.

Property	Salt level	$\omega \to 0$	$\omega \tau_Z \gg 1$
$\frac{\eta - \eta_0}{\eta_0 c}$	Low	N^2	$\omega^{-2/3}$
	High	$N^{4/5}$	$\omega^{-4/9}$
$\frac{G - G_0}{G_0 c}$	Low	ωN^2	$\omega^{1/3}$
	High	$\omega N^{4/5}$	$\omega^{5/9}$

Using the results given in Section 7.5.2, this equation yields

$$[\eta] \simeq 6.25 \frac{N_A}{M} R_g^3, \tag{7.6.36}$$

where N_A is the Avogadro number. Since $R_g \sim N$ and $R_g \sim N^{3/5}$, respectively, in the low salt and high salt limits, we get

$$[\eta] \sim \begin{cases} N^2, & \text{low salt} \\ N^{4/5}, & \text{high salt.} \end{cases} \tag{7.6.37}$$

The frequency dependence of $[\eta]$ follows from Eq. (7.5.53) as

$$[\eta] \sim \begin{cases} \omega^{-2/3}, & \text{low salt} \\ \omega^{-4/9}, & \text{high salt.} \end{cases} \tag{7.6.38}$$

The intrinsic modulus $[G]$ is related to the intrinsic viscosity as $[G] = i\omega[\eta]$. As shown in Section 7.5.2, the frequency dependence of the real and imaginary parts of the intrinsic modulus follows from Eq. (7.5.53) as

$$[G] \sim \begin{cases} \omega^{1/3}, & \text{low salt} \\ \omega^{5/9}, & \text{high salt.} \end{cases} \tag{7.6.39}$$

The above results on the viscoelastic properties of dilute polyelectrolyte solutions are summarized in Table 7.2.

7.7　Semidilute Polyelectrolyte Solutions

The primary goal of this section is to treat the various dynamical properties of poly-electrolyte solutions at polymer concentrations where the chains interact with each other (without entanglement constraints). At polymer concentrations above the overlap concentration $c^\star$, the chains intermingle and interact among themselves and behave collectively. Even at polymer concentrations below $c^\star$, the chains influence each other through hydrodynamic interactions as well as electrostatic and excluded volume interactions treated in Chapter 5. Before considering the collective behavior of chains in

(a)

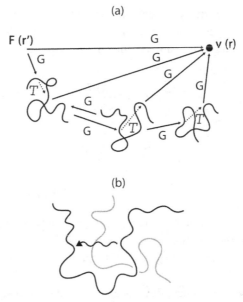

(b)

Figure 7.26 (a) Multiple scattering of force field among segments of all chains resulting in dynamical correlations. **G** denotes the Oseen tensor representing the hydrodynamic coupling among different chains. $\mathcal{T}$ denotes the hydrodynamic coupling within a chain. (b) Interpenetration of polymer strands results in hydrodynamic screening.

semidilute solutions with polymer concentrations above $c^\star$, let us address dilute solutions where the polymer solution is not infinitely dilute. Under such conditions, the inter-segment hydrodynamic interactions within a chain are perturbed by all other chains in the solution. For example, the velocity $\mathbf{v}(\mathbf{r})$ at location $\mathbf{r}$ is a confluence of hydrodynamic interactions arising from the force $\mathbf{F}(\mathbf{r}')$ and the forces exerted by all monomers of all chains, as sketched in Fig. 7.26a. Note that there are intra-chain hydrodynamic interactions denoted by the symbol $\mathcal{T}$. Furthermore, there are interchain hydrodynamic interactions as well. For these dilute conditions below the overlap concentration, the diffusion coefficient of the chains and viscosity of the solution can be expressed as virial series in polymer concentration.

When a specified polymer chain moves amidst other chains, which in turn influence the motion of the specified chain due to their own movements, two kinds of diffusion coefficients emerge (see Section 7.4.1). One is the **tracer diffusion** representing the diffusion of a labeled chain. This is sometimes also referred to as self-diffusion. The corresponding diffusion coefficient D_t is the tracer diffusion coefficient, which can be measured using techniques like labeling or pulsed field gradient nuclear magnetic resonance. The other diffusion is the **cooperative diffusion** representing collective average diffusion of fluctuations in the polymer concentration. The corresponding diffusion coefficient is the cooperative (or mutual) diffusion coefficient D_c. A common experimental technique that measures D_c is DLS. In infinitely dilute solutions, where interchain interactions are negligible, D_t and D_c are the same as the diffusion coefficient given by the Stokes–Einstein law.

Based on perturbative calculations to treat intermolecular interactions to the leading order in polymer concentration c, the virial series for the tracer diffusion coefficient D_t, cooperative diffusion coefficient D_c, and the specific viscosity $(\eta - \eta_0)/\eta_0$ of the solution are, respectively,

$$D_t = \frac{k_B T}{6\pi\eta_0 R_h} (1 - k_t[\eta]c + \cdots),$$ (7.7.1)

$$D_c = \frac{k_B T}{6\pi\eta_0 R_h} (1 + k_c[\eta]c + \cdots),$$ (7.7.2)

$$\frac{\eta - \eta_0}{\eta_0} = [\eta]c (1 + k_H[\eta]c + \cdots),$$ (7.7.3)

where the numerical coefficients k_t, k_c, and k_H are specific to the nature of the polymer and solution conditions. Recalling that $c[\eta]$ is proportional to $c/c^\star$, the above quantities are, in general, functions of the dimensionless polymer concentration variable $c/c^\star$.

For polymer concentrations above $c^\star$, the virial series are no longer applicable as seen for equilibrium properties described in Chapter 6. In this case, the chains interpenetrate into each other. For example, as sketched in Fig. 7.26b, the presence of a second chain alters the hydrodynamic coupling among the segments of the first chain and vice versa. In general, the various segments of all chains are coupled through hydrodynamic, electrostatic, and topological (chain connectivity) correlations. To describe the diffusion of chains and viscosity of the solution for $c > c^\star$, these three correlations among segments of all chains in the system must be addressed in a self-consistent manner. As derived in Appendix 6.1, the average velocity field $\langle \mathbf{v}(\mathbf{r}) \rangle$ is given in terms of a modified hydrodynamic interaction tensor $\mathcal{G}(\mathbf{r} - \mathbf{r}')$, instead of the bare Oseen tensor $\mathbf{G}(\mathbf{r} - \mathbf{r}')$ given in Eq. (7.2.5), as

$$\langle \mathbf{v}(\mathbf{r}) \rangle = \int d\mathbf{r}' \mathcal{G}(\mathbf{r} - \mathbf{r}') \cdot \mathbf{F}(\mathbf{r}'),$$ (7.7.4)

where the angular brackets denote averaging over chain conformations and angular average over the hydrodynamic interaction tensor. For length scales smaller than R_g but larger than the segment size ℓ, $\mathcal{G}$ follows as

$$\mathcal{G}(\mathbf{r} - \mathbf{r}') = \frac{e^{-|\mathbf{r}-\mathbf{r}'|/\xi_h}}{6\pi\eta_0 |\mathbf{r} - \mathbf{r}'|},$$ (7.7.5)

where ξ_h is a measure of the range of distance over which the hydrodynamic interaction is significant. Thus, interpenetration of chains leads to **hydrodynamic screening**. The long-ranged hydrodynamic interaction given by the Oseen tensor $G(\mathbf{r}-\mathbf{r}')$ given in Eq. (7.2.5) is screened at higher polymer concentrations and the hydrodynamic interaction becomes short-ranged (Fig. 7.27a). The range ξ_h over which the hydrodynamic interaction is significant is called the **hydrodynamic screening length**.

The origin of hydrodynamic screening is the coupling between chain dynamics and solvent dynamics. Note that its mathematical form, Eq. (7.7.5), is reminiscent of the electrostatic screening and excluded volume screening which are based on

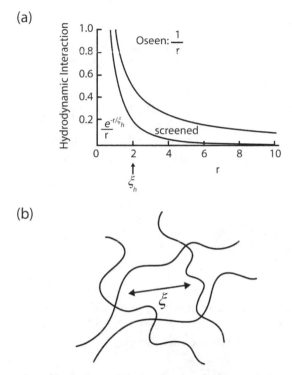

Figure 7.27 (a) Screening of hydrodynamic interaction in semidilute solutions. The hydrodynamic screening length ξ_h decreases as polymer concentration increases. (b) ξ_h is proportional to the static screening length ξ arising from the double screening of the excluded volume and electrostatic interactions.

potential interactions in equilibrium. However, theoretical calculations [Muthukumar & Edwards (1982b)] show that ξ_h is proportional to the correlation length ξ in equilibrium derived in Chapter 6, as sketched in Fig. 7.27b. Since the proportionality constant between ξ_h and ξ is of order unity, let us take ξ_h to be the same as ξ in interpreting experimental data,

$$\xi_h \simeq \xi. \tag{7.7.6}$$

The hydrodynamic interaction is progressively screened as the polymer concentration increases. If the polymer concentration is sufficiently high, then a labeled chain behaves as if there is no hydrodynamic interaction among its segments. Under such semidilute conditions (where the concentration is not too high to allow entanglement effects), Rouse dynamics, although developed for infinitely dilute solutions without hydrodynamic interactions, becomes a good description of single chain dynamics as far as molecular weight dependence is concerned.

Using $\mathcal{G}$, approximate closed-form expressions for tracer diffusion coefficient, electrophoretic mobility, cooperative diffusion coefficient, specific viscosity, and the ordinary–extraordinary transition can be derived for unentangled semidilute polyelectrolyte solutions. We present below the salient conclusions by following the theory of Muthukumar (1997) and leaving the details of the derivations to the original reference.

7.7.1   Tracer Diffusion Coefficient

Following the same procedure as in Section 7.6.1 and using $\mathcal{G}$ instead of $\mathbf{G}$, the translational friction coefficient f_t is calculated, which then gives the tracer diffusion coefficient D_t as

$$D_t = \frac{k_B T}{f_t} = \frac{k_B T}{3\pi\eta_0} \frac{\xi_h}{R_g^2(c)}, \tag{7.7.7}$$

where $R_g(c)$ is the radius of gyration of the polyelectrolyte chain at polymer concentration c given in Eqs. (6.3.25) and (6.3.26). The hydrodynamic screening length ξ_h is given as

$$\xi_h = \frac{12}{\pi} \frac{N}{c R_g^2(c)}. \tag{7.7.8}$$

Using Eq. (6.2.69), we get

$$\xi_h \sim \begin{cases} c^{-1/2}, & \text{low salt} \\ c^{-3/4}, & \text{high salt.} \end{cases} \tag{7.7.9}$$

Note that ξ_h is proportional to ξ given in Eq. (6.2.68), consistent with Eq. (7.7.6). Substituting Eqs. (7.7.8) and (6.2.69) in Eq. (7.7.7), we get the concentration dependence of the tracer diffusion coefficient as

$$D_t \sim \begin{cases} \frac{1}{N}, & \text{low salt} \\ \frac{1}{N\sqrt{c}}, & \text{high salt.} \end{cases} \tag{7.7.10}$$

The numerical prefactors in this equation can be obtained by directly using Eq. (7.7.7) and the results in Section 6.3. Note the lack of a concentration dependence in the low-salt limit. The scaling laws given in Eq. (7.7.10) can be obtained from scaling arguments as well, as given in Appendix 6.6.1, with the general scaling law

$$D_t \sim \frac{1}{N} c^{-(1-\nu)/(3\nu-1)}. \qquad c > c^{\star} \tag{7.7.11}$$

7.7.2   Electrophoretic Mobility

The derivation of the electrophoretic mobility of a labeled chain in semidilute solutions is the same as for dilute solutions given in Section 7.6.2, except that the screened hydrodynamic interaction $\mathcal{G}$ is used instead of the bare hydrodynamic interaction $\mathbf{G}$. The result is

$$\mu = \frac{Q}{6\pi\eta_0 R_g(c)} \mathcal{M}\left(\frac{\xi_h}{R_g(c)}\right), \tag{7.7.12}$$

where Q is the net charge on the chain, $R_g(c)$ is the concentration dependent radius of gyration in the semidilute solution, and $\mathcal{M}$ is a known function of the ratio of the hydrodynamic screening length to the radius of gyration of the chain. In the limit of $R_g \gg \xi_h, \mathcal{M}(\xi_h/R_g) \to 3\xi_h/R_g$ so that

$$\mu = \frac{Q\xi_h}{2\pi\eta_0 R_g^2(c)}. \tag{7.7.13}$$

In view of Eqs. (7.7.8) and (6.2.69), and since $Q \sim N$, the dependence of the electrophoretic mobility on N and c follows from the above equation as

$$\mu \sim \begin{cases} N^0 c^0, & \text{salt-free} \\ N^0 c^{-1/2}, & \text{high salt.} \end{cases} \tag{7.7.14}$$

Note that the electrophoretic mobility is independent of molecular weight in general and independent of concentration in salt-free limit. To date, no experimental investigation into the electrophoretic mobility of polyelectrolyte chains in semidilute solutions exists, and the above predictions remain to be validated.

7.7.3   Cooperative Diffusion Coefficient

Let the local segment concentration at location $\mathbf{r}$ and time t in a solution of volume V containing n polyelectrolyte chains each with N segments be denoted as $c(\mathbf{r},t)$,

$$c(\mathbf{r},t) = \sum_{\alpha=1}^{n} \sum_{i=0}^{N-1} \delta(\mathbf{r} - \mathbf{R}_{\alpha i}(t)), \tag{7.7.15}$$

where $\mathbf{R}_{\alpha i}$ is the position vector of the ith segment of the αth chain. Defining the Fourier transform of $c(\mathbf{r},t)$ as

$$\hat{c}(\mathbf{k},t) = \frac{1}{V} \int d\mathbf{r} e^{i\mathbf{k}\cdot\mathbf{r}} c(\mathbf{r},t), \tag{7.7.16}$$

the time evolution of $\hat{c}(\mathbf{k},t)$ is given by the standard theory of polymer dynamics as [Doi & Edwards (1986)]

$$\frac{\partial \hat{c}(\mathbf{k},t)}{\partial t} = -\Gamma_{\mathbf{k}} \hat{c}(\mathbf{k},t), \tag{7.7.17}$$

where the rate of concentration fluctuations resolved into the scattering wave vector $\mathbf{k}$ is given by

$$\Gamma_{\mathbf{k}} = k_B T \int \frac{d\mathbf{j}}{(2\pi)^3} \frac{\mathbf{k} \cdot \left(\mathbf{1} - \hat{\mathbf{j}}\hat{\mathbf{j}}\right) \cdot \mathbf{k}}{\eta_0 \left(j^2 + \xi_h^{-2}\right)} \frac{g\,(\mathbf{k}+\mathbf{j})}{g\,(\mathbf{k})}. \tag{7.7.18}$$

Here $\hat{\mathbf{j}}$ is the unit vector and $g(\mathbf{k})$ is the pair correlation function $\langle \hat{c}(\mathbf{k},t)\hat{c}(-\mathbf{k},t)\rangle$ described in Sections 6.2.2.1 and 6.3.4.

For salt-free conditions, $g(\mathbf{k})$ given by Eq. (6.3.22) exhibits the polyelectrolyte peak at $k = \xi_2^{-1}$ where ξ_2 is the screening length from the double screening theory given in Eq. (6.3.18) as

$$\xi_2 \sim c^{-1/2}. \quad \text{(semidilute)} \tag{7.7.19}$$

Using Eq. (6.3.22) for $g(\mathbf{k})$ in Eq. (7.7.18), the rate $\Gamma_{\mathbf{k}}$ is quadratic in k for small values of k pertinent to DLS experiments,

$$\Gamma_{\mathbf{k}} = D_c k^2 + O(k^4), \tag{7.7.20}$$

so that

$$\frac{\partial \hat{c}(\mathbf{k},t)}{\partial t} = -D_c k^2 \hat{c}(\mathbf{k},t), \qquad (7.7.21)$$

where D_c is the cooperative diffusion coefficient given by

$$D_c = \frac{k_B T}{6\pi\eta_0\xi_2} f\left(\frac{\xi_2}{\xi_h}\right). \qquad (7.7.22)$$

Here $f\left(\frac{\xi_2}{\xi_h}\right)$ is a numerical factor because ξ_h is proportional to ξ_2. Combining Eqs. (7.7.9) and (7.7.19) we get

$$D_c \sim \frac{T}{\eta_0} c^{1/2}. \qquad \text{(salt-free)} \qquad (7.7.23)$$

In the high-salt limit, the pair correlation function $g(\mathbf{k})$ is in the Ornstein-Zernike form (Eq. (6.3.23)) with the static correlation length $\xi_1 \sim c^{-3/4}$ as given in Eq. (6.3.19). Using Eq. (6.3.23) in Eq. (7.7.18), the rate of change in concentration fluctuations is given by the same Eq. (7.7.20), but now the cooperative diffusion coefficient is given by

$$D_c = \frac{k_B T}{6\pi\eta_0\xi_1} \frac{\xi_h}{(\xi_1 + \xi_h)}. \qquad (7.7.24)$$

Since $\xi_1 \sim \xi_h \sim c^{-3/4}$, we get

$$D_c \sim \frac{T}{\eta_0} c^{3/4}. \qquad \text{(high salt)} \qquad (7.7.25)$$

The results given in Eqs. (7.7.23) and (7.7.25) can be obtained by the scaling argument as well (Appendix 6.6.2), with the general scaling law,

$$D_c \sim c^{\frac{\nu}{3\nu-1}}. \qquad (c > c^\star) \qquad (7.7.26)$$

Note that D_c is an increasing function of c in contrast with the decreasing dependence of D_t.

7.7.4   Coupled Diffusion and Ordinary–Extraordinary Transition

In this section we shall address the onset of spontaneous "fast" and "slow" modes in polyelectrolyte solutions described in Section 7.4.2 and Fig. 7.9. As detailed below, the fast mode in semidilute solutions is due to the coupling between the cooperative diffusion of polymer segments and the dynamics of their counterion clouds. This is a generalization of the coupled diffusion in Section 7.6.3 to semidilute solutions. The slow mode arises from the aggregation of similarly charged polyelectrolyte chains described in Section 6.5 for polymer concentrations above a certain value and for concentrations of added salt below a certain value.

7.7.4.1   Fast Mode

Let us first consider a salt-free semidilute solution containing n interpenetrating uniformly charged polyelectrolyte chains each of N segments in volume V. The average polymer segment concentration is $c = nN/V$. Let each segment have charge $\alpha z_p e$. The number of dissociated counterions from the chains is $\alpha |z_p/z_c| nN$, where $z_c e$ is the charge of the counterion. The average concentration of counterions c_c is $\alpha |z_p/z_c| nN/V = \alpha |z_p/z_c| c$. Let the index $i = 1,2$ denote the polymer segment and counterion, respectively. Therefore, the average concentrations c and c_c, and the charges $\alpha z_p e$ and $z_c e$ are denoted as

$$c_{1,0} = \frac{nN}{V} = c, \quad z_1 = \alpha z_p, \quad z_2 = z_c, \quad c_{2,0} = c_c = \left|\frac{z_1}{z_2}\right| c_{1,0}. \quad (7.7.27)$$

Since the solution is overall electrically neutral,

$$z_1 c_{1,0} + z_2 c_{2,0} = 0, \quad (7.7.28)$$

so that

$$c_{2,0} = \left|\frac{z_1}{z_2}\right| c_{1,0} = \left|\frac{\alpha z_p}{z_c}\right| c. \quad (7.7.29)$$

Note that $c_{1,0}$ here in semidilute solutions is the average segment concentration, whereas c_1^0 is the chain number concentration in dilute solutions treated in Section 7.6.3. Let us take the local concentration of segments and counterions to fluctuate weakly around their average values as

$$c_i(\mathbf{r},t) = c_{i,0} + \delta c_i(\mathbf{r},t), \quad (i = 1,2) \quad (7.7.30)$$

where $\delta c_i(\mathbf{r},t)$ denotes the fluctuation.

The time evolution of the local concentration $c_i(\mathbf{r},t)$ is given by a combination of diffusive and convective fluxes (see Section 7.6.3) as

$$\frac{\partial c_i(\mathbf{r},t)}{\partial t} = D_i \nabla^2 c_i(\mathbf{r},t) - \nabla \cdot [c_i(\mathbf{r},t) \mu_i \mathbf{E}_{\mathrm{loc}}(\mathbf{r},t)]. \quad (7.7.31)$$

The first term on the right-hand side is the contribution from diffusion, where D_i is the cooperative diffusion coefficient of the ith species. The second term is the contribution from convection due to local electric field $\mathbf{E}_{\mathrm{loc}}$ arising from the local ionic environment. μ_i is the electrophoretic mobility of the ith species. For polymer segments D_1 is the cooperative diffusion coefficient $D_c \simeq k_B T/(6\pi\eta_0 \xi)$ described above, and μ_1 is the electrophoretic mobility of a segment $z_1 e D_{\mathrm{seg}}/k_B T$,

$$D_1 = D_c \simeq \frac{k_B T}{6\pi\eta_0 \xi}, \quad \mu_1 = \frac{z_1 e D_{\mathrm{seg}}}{k_B T} = \frac{z_1 e D_1 \gamma_1}{k_B T}, \quad \gamma_1 = \frac{D_{\mathrm{seg}}}{D_1}, \quad (7.7.32)$$

where γ_1 is defined as the ratio of the diffusion coefficient of a segment to the cooperative diffusion coefficient. Note that this γ_1 depends on polymer concentration, whereas the γ_1 defined in Section 7.6.3 for dilute solutions is different, even though the same symbol is used for dilute and semidilute solutions. For the counterion,

$$D_2 = \frac{k_B T}{6\pi\eta_0 a} \simeq D_{\mathrm{seg}}, \quad \mu_2 = \frac{z_2 e D_2}{k_B T}, \quad (7.7.33)$$

where a is the radius of a counterion. Note that D_2 is comparable to D_{seg} and is much larger than $D_1 = D_c$. Although D_1 is much smaller than D_2, μ_1 of a segment is comparable to μ_2.

As derived in Appendix 6.5.3, analogous to the situation in dilute solutions (Section 7.6.3.1), the fluctuations in the local monomer concentration δc_1 and the counterion concentration δc_2 are coupled. As a result of the two coupled equations, there are two coupled relaxation modes. The first is the non-diffusive superfast **plasmon mode** with a rate of relaxation given by

$$\Gamma_{\text{plasmon}} = D_1 \gamma_1 \kappa_1^2 + D_2 \kappa_2^2, \tag{7.7.34}$$

where

$$\kappa_1^2 = \frac{(z_1 e)^2 c_{1,0}}{\epsilon_0 \epsilon k_B T}, \qquad \kappa_2^2 = \frac{(z_2 e)^2 c_{2,0}}{\epsilon_0 \epsilon k_B T}. \tag{7.7.35}$$

Substituting Eqs. (7.7.32) and (7.7.33) in Eq. (7.7.34), and taking $|z_p| = 1 = |z_c|$, we get

$$\Gamma_{\text{plasmon}} = (1 + \alpha) D_{\text{seg}} \xi_{D,c}^{-2}, \tag{7.7.36}$$

where $\xi_{D,c}^{-2}$ is the square of the inverse Debye length for the counterion concentration $(= z_c^2 e^2 c_c / (\epsilon_0 \epsilon k_B T))$. Considering typical values of $D_{\text{seg}} \sim 10^{-5}$ cm^2/s and $\xi_{D,c} < 10$ nm, Γ_{plasmon} is too fast to measure with DLS.

The second mode of relaxation is the diffusive fast mode whose rate of relaxation is given by

$$\Gamma_{\text{fast}} = D_{\text{fast}} k^2, \tag{7.7.37}$$

where the fast diffusion coefficient D_{fast} follows from Eq. (7.6.22) as (see Appendix 6)

$$D_{\text{fast}} = \frac{D_1 \left(1 + \gamma_1 \left|\frac{z_1}{z_2}\right|\right)}{\left(1 + \frac{D_1 \gamma_1}{D_2} \left|\frac{z_1}{z_2}\right|\right)}. \tag{7.7.38}$$

Using $D_1 \gamma_1 = D_{\text{seg}} = D_2, z_1 = \alpha z_p$, and $z_2 = z_c$, we get

$$D_{\text{fast}} = \frac{D_1}{(1 + \alpha)} + \frac{\alpha}{(1 + \alpha)} D_{\text{seg}}. \tag{7.7.39}$$

Since $D_1 < D_{\text{seg}}$ for correlation lengths larger than the segment length ($\xi > \ell$), the second term dominates so that

$$D_{\text{fast}} \simeq \frac{\alpha}{(1 + \alpha)} D_{\text{seg}}. \tag{7.7.40}$$

Therefore the diffusion coefficient for the fast mode is close to that of a free segment. For example, if the degree of ionization is 1/3, D_{fast} is only one quarter of the segment diffusion coefficient,

$$D_{\text{fast}} \simeq \frac{1}{4} D_{\text{seg}}. \tag{7.7.41}$$

Furthermore, the fast diffusion coefficient is independent of polymer concentration and molecular weight in semidilute solutions,

$$D_{\text{fast}} \sim c^0 N^0. \tag{7.7.42}$$

All of the above main results on D_{fast} are observed in experiments, as indicated by the horizontal line marked for the fast mode in Fig. 7.9a.

7.7.4.2   Breakdown of Coupled Dynamics with Salt

The coupling between the diffusion of local polymer segments and counterion dynamics is broken when salt is added to the solution. Consider the presence of n_+ cations of charge $z_+ e$ and n_- anions of charge $z_- e$ from the added salt in addition to the polyelectrolyte chains and their counterions. Considering the general situation in which the counterions from the polymer are different from the dissociated ions from the added salt, let $c_{3,0}$ and $c_{4,0}$ be the average number concentrations of cations and anions from the salt, which is the same as the added salt concentration c_s. Generalizing Eq. (7.6.17) for the present situation of four charged species in terms of the Fourier transform of the concentration fluctuations, we obtain

$$\frac{\partial}{\partial t}\delta\hat{c}_i(\mathbf{k},t) = -D_i k^2 \delta\hat{c}_i(\mathbf{k},t) - c_{i,0}\frac{\mu_i e}{\epsilon_0 \epsilon}\sum_{j=1}^{4} z_j \delta\hat{c}_j(\mathbf{k},t), \qquad (7.7.43)$$

where $\mu_3 = z_3 e D_3/k_B T$ and $\mu_4 = z_4 e D_4/k_B T$, and where D_3 and D_4 are the corresponding diffusion coefficients of cations and anions of the added salt. As a result of the four coupled equations for the four species, there are four coupled relaxation modes in the system.

To simplify the calculations of the four coupled modes, let us assume that the fluctuations in the local concentrations of the small ions (namely, counterions and salt ions) relax to equilibrium more rapidly than the fluctuations in the polymer concentration. In view of this assumption, we approximate $\partial\delta c_i(\mathbf{k},t)/\partial t = 0$ for $i = 2, 3,$ and 4. With this approximation, we get (see Eq. (A6.126))

$$\frac{\partial}{\partial t}\delta\hat{c}_1(\mathbf{k},t) = -\left[D_c + \frac{c\alpha^2 D_{\text{seg}}}{(\alpha c + 2c_s)}\right]k^2 \delta\hat{c}_1(\mathbf{k},t), \qquad (7.7.44)$$

where we have taken the salt as monovalent at concentration $c_s = c_{3,0} = c_{4,0}, z_p = 1$, and $c_{2,0} = \alpha c$. The term inside the square brackets in the above equation is the coupled diffusion coefficient of the fast mode,

$$D_{\text{fast}} = D_c + \frac{c\alpha^2 D_{\text{seg}}}{(\alpha c + 2c_s)}. \qquad (7.7.45)$$

Whereas the first term on the right-hand side is the cooperative diffusion coefficient, the second term gives the contribution from the coupling between polymer segments and their ionic environment. For salt-free solutions, D_{fast} follows from the above equation as

$$D_{\text{fast}} = D_c + \alpha D_{\text{seg}}. \qquad (7.7.46)$$

As already noted in Eq. (7.7.40), $D_c < D_{\text{seg}}$ so that

$$D_{\text{fast}} \simeq \alpha D_{\text{seg}} \sim c^0 N^0, \qquad (7.7.47)$$

which is consistent with Eqs. (7.7.41) and (7.7.42), which were obtained without the approximation of rapid relaxation of counterion cloud compared to the relaxation of fluctuations. On the other hand, at higher salt concentrations such that $c_s > c$, the second term on the right-hand side of Eq. (7.7.45) becomes negligible so that D_{fast} is the cooperative diffusion coefficient D_c. In this limit, the dynamical coupling between polymer segments and small ions is broken.

The two limits of salt-free and high-salt conditions are summarized as

$$D_{fast} = \begin{cases} \alpha D_{seg} \sim c^0 N^0 & \text{salt-free } (c_s \ll c) \\ D_c \sim \frac{T}{\eta_0} c^{3/4}. & \text{high salt } (c_s > c) \end{cases} \tag{7.7.48}$$

The derivation of the above conclusions offers a molecular understanding of the various experimental data on the fast (ordinary) diffusion coefficient presented in Fig. 7.9a. In order to emphasize the independence of D_{fast} on c and N given by Eq. (7.7.47), a horizontal line is drawn in Fig. 7.9a for the fast mode in semidilute solutions. Furthermore, Eq. (7.7.45) shows that $D_{fast} \sim c^0 N^0$ emerges at increasingly higher polymer concentrations as the salt concentration is increased in agreement with the experimental data shown in Fig. 7.9b. The above results suggest that gradients in salt concentration can be used to modulate the diffusion of charged macromolecules.

7.7.4.3   Slow Mode

For polyelectrolyte concentrations higher than a certain threshold value, a second branch of diffusion emerges concurrent with the fast mode, as seen in Fig. 7.9b. This mode is known as the "slow mode" or "extraordinary mode" as described in Section 7.4.2 and Fig. 7.9. The corresponding diffusion coefficient D_{slow} is several orders of magnitude smaller than D_{fast} or the estimated Stokes–Einstein value D_{SE} based on single chain behavior. As seen in Fig. 7.9, D_{slow} decreases with an increase in either c or N. The emergence of the slow mode has been attributed to the spontaneous formation of aggregates of similarly charged polyelectrolytes described in Section 6.5. When a counterion adsorbs on a charged monomer of the chain, an ion pair with a dipole moment is formed at the adsorbed site. Due to the counterion adsorption, there are on average $(1 - \alpha)N$ dipoles per chain, where α is the degree of ionization. A pair of such dipoles can attract each other to form a quadrupole via dipole-dipole interactions (Eq. (1.3.8)). As a result, physical associations between distant monomers can occur leading to formation of aggregates from several chains, even though these chains seem to carry a net charge of the same sign. Assuming that the lifetime of the whole collection of quadrupoles in an aggregate is very long so that equilibrium thermodynamics is applicable to describe the aggregate, the radius of gyration of the aggregate $R_{g,agg}$ is given in Eq. (6.5.13) as

$$R_{g,agg} \sim N c^{1/6}. \tag{7.7.49}$$

Using the Stokes–Einstein law, the diffusion coefficient of such aggregates is given by [Muthukumar (2016b)]

$$D_{slow} = \frac{k_B T}{6\pi \eta_0 R_{g,agg}} \sim N^{-1} c^{-1/6}. \tag{7.7.50}$$

The concentration dependence of D_{slow} given in this equation is indicated in Fig. 7.9a with a straight line of slope -1/6, in qualitative agreement with experimental data.

Recall from Section 6.5 that aggregation can occur only if the polymer concentration is above a critical aggregation concentration. Hence, the slow mode can only be observed if the polymer concentration is above this critical polymer concentration. In other words, the polymer concentration required for the onset of the slow mode in Fig. 7.9 corresponds to the critical aggregation concentration. If the concentration of added salt increases, the quadrupole junctions become progressively unstable due to electrostatic screening, and as a result the aggregates are not sustainable. Consequently, the critical aggregation concentration required for the occurrence of the slow mode increases as the salt concentration is increased, as seen in Fig. 7.9b. Furthermore, the slow mode disappears at very high salt concentrations. At the same time, the fast diffusion coefficient becomes the cooperative diffusion coefficient. All of these conclusions are consistent with the compiled experimental data provided in Fig. 7.9. The apparently mysterious behavior of the ordinary–extraordinary transition observed in solutions of charged macromolecules arises due to the role of counterions. The fast mode arises due to the coupled dynamics of the chains and their counterions. The slow mode arises from aggregation of chains mediated by quadrupoles consisting of dipolar ion pairs which emerge from adsorption of counterions on the polymer.

Complementing the aggregation model for the slow mode, alternative arguments [Förster *et al.* (1990), Sedlak & Amis (1992b), Förster & Schmidt (1995), Zhou *et al.* (2009), Chremos & Douglas (2017)] have been suggested. Quantitative predictions from these arguments are yet to be obtained.

7.7.5   Viscosity

Using the multiple scattering theory [Muthukumar (1997)], the specific viscosity is given by

$$\frac{\eta - \eta_0}{\eta_0} = \frac{1}{\pi^2} \frac{R_g^2(c)}{\xi_h^2}, \tag{7.7.51}$$

where $R_g(c)$ is the concentration dependent radius of gyration (Section 6.3.5) and ξ_h is the hydrodynamic screening length given in Eq. (7.7.8). Substituting Eq. (7.7.8) into Eq. (7.7.51) we obtain

$$\frac{\eta - \eta_0}{\eta_0} = \frac{1}{144} \frac{c^2 R_g^6(c)}{N^2}. \tag{7.7.52}$$

Using the asymptotic expressions for $R_g(c)$ given in Eqs. (6.3.25) and (6.3.26) in the low salt and high salt limits, Eq. (7.7.52) gives

$$\frac{\eta - \eta_0}{\eta_0} \sim N \begin{cases} c^{1/2}, & \text{low salt} \\ c^{5/4}, & \text{high salt.} \end{cases} \tag{7.7.53}$$

If we are interested only in the scaling laws given in this equation, we can directly get them from the following scaling argument. As noted in Eq. (7.4.23), the specific viscosity is a function of the dimensionless concentration variable $c/c^\star$,

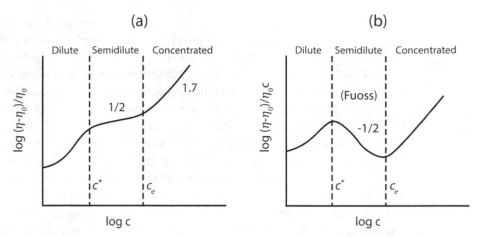

Figure 7.28 Sketch of concentration dependence of (a) specific viscosity and (b) reduced viscosity in salt-free solutions in dilute, semidilute, and entangled regimes.

$$\frac{\eta - \eta_0}{\eta_0} = f_\eta \left(cN^{3\nu-1} \right), \tag{7.7.54}$$

where f_η is addressed below. Since we expect the specific viscosity be proportional to N due to hydrodynamic screening and hence the applicability of the Rouse model, we get

$$\frac{\eta - \eta_0}{\eta_0} = f_\eta \left(cN^{3\nu-1} \right) \sim N, \qquad (c > c^\star) \tag{7.7.55}$$

so that $f_\eta \sim (cN^{3\nu-1})^{1/(3\nu-1)}$. Therefore, the concentration dependence of the specific viscosity follows the scaling law

$$\frac{\eta - \eta_0}{\eta_0} \sim Nc^{1/(3\nu-1)}. \tag{7.7.56}$$

This result is consistent with Eq. (7.7.53) for the low salt limit ($\nu = 1$) and the high salt limit ($\nu = 3/5$).

Let us now consider the experimental results on the viscosity of polyelectrolyte solutions presented in Fig. 7.11 for salt-free solutions. According to Eq. (7.7.53), the specific viscosity increases with polymer concentration as $\sqrt{c}$. This dependence is weaker than the linear dependence on c observed in very dilute solutions, as sketched in Fig. 7.28a. For $c < c^\star$, the specific viscosity increases linearly at very low concentrations and then more strongly due to the electrostatic interactions between the chains (analogous to the situation of colloidal dispersions). For concentrations above the overlap concentration, but below the entanglement concentration c_e, the specific viscosity increases less sensitively with polymer concentration as $\sqrt{c}$. Above the entanglement concentration, the viscosity of the solution increases with a stronger exponent on c, as we shall see in the next section.

When the above conclusions are depicted in terms of the reduced viscosity, by dividing the specific viscosity by a factor of concentration we get, for the dilute and semidilute regimes,

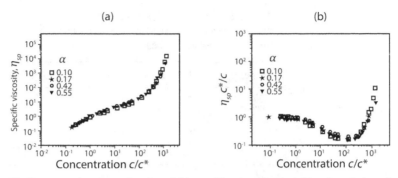

Figure 7.29 Concentration dependencies of (a) specific viscosity and (b) reduced viscosity of salt-free solutions of quaternized random copolymers of 2-vinyl pyridine and N-methyl-2-vinyl pyridinium chloride in ethylene glycol at different values of degree of quaternization (α = 0.10, 0.17, 0.42, and 0.55) [Adapted from Dou & Colby (2006)].

$$\frac{\eta - \eta_0}{\eta_0 c} \sim \begin{cases} [\eta]\,(1 + k_H[\eta]c + \cdots), & c < c^\star \\ Nc^{-1/2}, & \text{(Fuoss law)} \quad c > c^\star\,\text{(salt-free)}, \end{cases} \tag{7.7.57}$$

as presented in Fig. 7.28b. The concentration dependence of the reduced viscosity is nonmonotonic and the Fuoss law of $c^{-1/2}$ is apparent due to the screening of hydrodynamic interaction in semidilute solutions. Note that the viscosity of the solution is continuously increasing with polymer concentration, and that the $c^{-1/2}$ behavior is due to the division of the specific viscosity by a factor of c. The above analysis provides the explanation of the nonmonotonic behavior of the reduced viscosity with c seen in Fig. 7.11 for salt-free sodium poly(styrene sulfonate) solutions. In addition, consider the experimental data on salt-free solutions of quaternized random copolymers of 2-vinyl pyridine and N-methyl-2-vinyl pyridinium chloride in ethylene glycol, given in Fig. 7.29 [Dou & Colby (2006)]. The specific viscosity is given in Fig. 7.29a as a function $c/c^\star$ for different values of degree of quaternization (α = 0.10, 0.17, 0.42, and 0.55). The same data are presented in Fig. 7.29b as a plot of reduced viscosity versus $c/c^\star$. It is evident that the decrease in the reduced viscosity with c occurs for $c > c^\star$ and the slope in this region is about $-1/2$ as sketched in Fig. 7.28. Above the entanglement concentration c_e, the viscosity increases more sharply with c.

In contrast with the salt-free limit, the reduced viscosity of polyelectrolyte solutions in the presence of added salt monotonically increases with c,

$$\frac{\eta - \eta_0}{\eta_0 c} \sim \begin{cases} [\eta]\,(1 + k_H[\eta]c + \cdots), & c < c^\star \\ Nc^{1/4}, & c > c^\star\,\text{(high salt)}. \end{cases} \tag{7.7.58}$$

Note that the concentration dependent factor $c^{1/4}$ is due to chain interpenetration with the consequent hydrodynamic screening in semidilute solutions. Although the dependence on N is the same as that predicted by the Rouse model devised for infinitely dilute solutions, the concentration dependence reflects the collective dynamics of chains in semidilute solutions.

The frequency dependence of the reduced viscosity $(\eta - \eta_0)/\eta_0 c$ and the reduced shear modulus $(G - G_0)/G_0 c$ can be obtained [Muthukumar (2001)] using the same

Table 7.3 Dependence of reduced viscosity and reduced modulus of semidilute polyelectrolyte solutions on the frequency of oscillatory shear.

Property	Salt level	$\omega \to 0$	$\omega \tau_R \gg 1$
$\frac{\eta - \eta_0}{\eta_0 c}$	Low	$Nc^{-1/2}$	$c^{-1/4}\omega^{-1/2}$
	High	$Nc^{1/4}$	$c^{1/8}\omega^{-1/2}$
$\frac{G - G_0}{G_0 c}$	Low	$Nc^{-1/2}\omega$	$c^{-1/4}\omega^{1/2}$
	High	$Nc^{1/4}\omega$	$c^{1/8}\omega^{1/2}$

multiple scattering theory that is implemented in obtaining the above results. A summary of the results in the limits of $\omega \to 0$ and the high frequency range ($\omega \tau_R > 1$, where τ_R is the Rouse relaxation time) that explores chain dynamics is presented in Table 7.3.

7.8 Entangled Solutions

At polymer concentrations higher than the entanglement concentration, the various chains strongly interpenetrate and severely tangle among themselves. Development of an adequate theory to account for the electrostatic, excluded volume, and topological correlations among the entangled chains is a daunting challenge. In view of this we present only scaling arguments for the center-of-mass diffusion of a labeled chain and the viscosity of concentrated polyelectrolyte solutions.

Assuming that the reptation model (Section 7.5.3) is applicable to the movement of a labeled polyelectrolyte chain at very high polymer concentrations, the diffusion coefficient of the center-of-mass of a labeled chain D_{cm} is proportional to N^{-2}. The concentration dependence of D_{cm} can be obtained by writing it in scaling form as $R_g^{-1} f_e(c/c^\star)$ and requiring $D_{cm} \sim N^{-2}$ for $c > c^\star$, as

$$D_{cm} \sim \begin{cases} c^{-1/2}N^{-2}, & \text{low salt} \\ c^{-7/4}N^{-2}, & \text{high salt.} \end{cases} \tag{7.8.1}$$

Analogously, in view of Eq. (7.5.54), the scaling law for the viscosity of entangled polyelectrolyte solutions is given as

$$\eta \sim \begin{cases} c^{1.7}N^{3.4}, & \text{low salt} \\ c^{4.25}N^{3.4}, & \text{high salt.} \end{cases} \tag{7.8.2}$$

The low salt limit is indicated in Fig. 7.28a for $c > c_e$.

The above scaling results in the entangled regime are only suggestions without any consideration of dynamical consequences of entanglements. Several alternate forms have been suggested in the literature [Dobrynin *et al.* (1995), Lopez (2019a, 2019b)]. Experimental data on D_{cm} and η for entangled polyelectrolyte solutions are limited at present. Further investigations are required to fully understand polyelectrolyte

Table 7.4 Dependencies of tracer diffusion coefficient D_t, cooperative diffusion coefficient D_c, fast diffusion coefficient D_{fast}, slow diffusion coefficient D_{slow}, electrophoretic mobility μ, and the specific viscosity $(\eta - \eta_0)/\eta_0$ on c and N for various regimes of polyelectrolyte and salt concentrations.

Property	Salt level	Dilute (Zimm)	Semidilute (Rouse)	Concentrated (entangled)
D_t	Low	N^{-1}	$c^0 N^{-1}$	$c^{-1/2} N^{-2}$
	High	$N^{-3/5}$	$c^{-1/2} N^{-1}$	$c^{-7/4} N^{-2}$
D_c	Low	N^{-1}	$c^{1/2} N^0$	
	High	$N^{-3/5}$	$c^{3/4} N^0$	
D_{fast}	Low	N^0	$c^0 N^0$	
	High	$N^{-3/5}$	$c^{1/2} N^0$	
D_{slow}	Low		$c^{-1/6} N^{-1}$	
μ	Low	N^0	$c^0 N^0$	
	High	$N^0 c_s^{-2/3}$	$c^{-1/2} N^0$	
$\dfrac{\eta - \eta_0}{\eta_0}$	Low	$c N^2$	$c^{1/2} N$	$c^{1.7} N^{3.4}$
	High	$c N^{4/5}$	$c^{5/4} N$	$c^{4.25} N^{3.4}$

dynamics in the entangled state. In spite of the lack of full understanding of entangled polyelectrolytes in high polymer concentrations, the reptation model (Section 7.5.4) has been successfully implemented to understand the electrophoretic mobility of polyelectrolyte chains inside highly cross-linked gels in the context of gel electrophoresis. In this extension, the modification of reptation kinetics by uncorrelated entropic traps (Section 7.5.4) has been addressed [Zimm (1988, 1991, 1996), Slater & Wu (1995), Rousseau *et al.* (1997), Viovy (2000)]. Yet, the electrophoretic mobility of large macromolecules trapped inside hydrogels remains to be fully understood, in spite of the routine use of gel electrophoresis to separate and identify the various charged macromolecules (DNA, proteins, etc.) in a sample.

A summary of the above results for the tracer diffusion coefficient D_t, cooperative diffusion coefficient D_c, electrophoretic mobility μ, fast diffusion coefficient D_{fast}, slow diffusion coefficient D_{slow}, and the specific viscosity $(\eta - \eta_0)/\eta_0$ is given in Table 7.4 for various regimes of polyelectrolyte concentration and salt concentration.

7.9   Conductivity of Polyelectrolyte Solutions

Application of a constant electric field $\mathbf{E}_0$ to a polyelectrolyte solution generates an electric current density $\mathbf{I}$ due to the transport of polyelectrolyte chains and small ions through the solution toward their favored electrodes. For weak electric fields, Ohm's law is valid so that

$$\mathbf{I} = \sigma \mathbf{E}_0. \tag{7.9.1}$$

σ is the conductivity of the solution given by

$$\sigma = \sum_i z_i e \mu_i c_i, \tag{7.9.2}$$

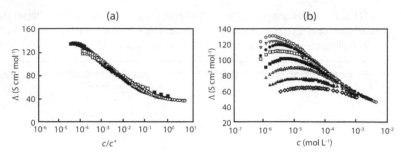

Figure 7.30 (a) Dependence of the equivalent conductivity Λ on $c/c^\star$ for salt-free aqueous solutions of poly(vinyl benzyl dimethyl n-butyl ammonium) chloride for various degrees of polymerization N (solid square: 27, open square: 56, solid triangle: 142, solid circle: 181, and open circle: 407 [Wandrey et al. (2000)]. (b) Concentration dependence of the equivalent conductivity for solutions of sodium poly(styrene sulfonate) (molar mass 8 kDa) at various NaCl concentrations. From top to bottom, respectively, salt-free, 1×10^{-6}, 2×10^{-6}, 4×10^{-6}, 1×10^{-5}, 2×10^{-5}, 5×10^{-5}, and 1×10^{-4} M. [Wandrey (1999)]

where $z_i e, \mu_i$, and c_i denote the charge, electrophoretic mobility, and the number concentration, respectively, of the ith charged species in the solution. Expressing σ as the specific conductivity (for a solution of 1 cm thickness and 1 cm^2 cross section), the equivalent conductivity Λ is defined according to

$$\Lambda = \frac{\sigma - \sigma_0}{c}, \tag{7.9.3}$$

where σ_0 is the conductivity of solvent and c is the molar concentration of the polymer.

Experimental data on Λ collected during the past several decades for a variety of charged macromolecules and added low molar mass electrolytes show rich phenomenology [Vink (1982), Colby et al. (1997), Wandrey (1999), Wandrey et al. (2000), Bordi et al. (2004)]. The control variables are the identity, molar mass, and concentration of the macromolecule and the identity and concentration of added low molar mass electrolyte. The experimental findings are quite broad and sometimes conflicting due to difficulties encountered in accurate measurement of Λ arising from experimental design and sample preparation. Nevertheless, there are several major features, which appear to be robust, as briefly mentioned here using a few examples. The equivalent conductivity in salt-free solutions of poly(vinyl benzyl dimethyl n-butyl ammonium) chloride at 20°C is given in Fig. 7.30a as a function of $c/c^\star$, where different symbols denote different degrees of polymerization (N). As the polymer concentration decreases below the overlap concentration, the equivalent conductivity increases independently of N. However Λ reaches a maximum at a certain lower polymer concentration, which depends on N. These features are observed for other polymers as well, demonstrating the universality of the behavior shown in Fig. 7.30. In the presence of added salt, the decrease of Λ with an increase in c becomes weaker. For example, the dependence of Λ on c for solutions of sodium poly(styrene sulfonate) of molar mass 8 kDa and NaCl at 20°C is given in Fig. 7.30b as a function of the salt concentration c_s. The polymer concentration corresponding to the maximum of Λ increases with c_s. Further inspection of the details in Fig. 7.30b reveals that (a) the maximum of Λ occurs at $c/c_s \sim 2$, (b) decrease in Λ with an increase in c occurs for $c > c_s$, (c) Λ increases with c for $c < c_s$, and (d) Λ is independent of N for $c > c^\star$.

An understanding of the features exhibited in Fig. 7.30 is presently lacking, although several attempts have been undertaken [Manning (1975, 1981), Vink (1982), Mandel (1988), Hoagland (2003), Bordi et al. (2004)]. Nevertheless, it is widely recognized that adsorption of counterions on the polyelectrolyte chains is a major factor and Λ depends on the free conducting ions. As c increases, the adsorption of counterions on polymer chains is enhanced (see Section 5.5) and hence the concentration of counterions in Eq. (7.9.2) decreases resulting in the decrease of Λ with an increase in c. Furthermore, the dynamics of the counterion cloud around a charged macromolecule is coupled to that of the molecule at low salt concentrations (Section 7.6.3). In addition, the mobility of small ions can occur by the sliding motion of adsorbed ions along chain backbone as well as by hopping from one adsorbed location to another [Muthukumar (2021)]. Treatment of the various contributing factors toward development of a satisfactory theory for the conductivity of solutions of charged macromolecules remains a challenge.

7.10   Dielectric Relaxation

Dielectric properties of aqueous polyelectrolyte solutions under weak periodically varying electric fields are significantly different from those of solutions of low molar mass electrolytes. The presence of macromolecules modifies the frequency dependence of the complex dielectric function in comparison to that given in Fig. 3.4 for water. The complex dielectric function $\epsilon^\star(\omega)$ is experimentally determined by measuring the current density $\langle \mathbf{I} \rangle$ through a solution as a linear response to an external oscillatory electric field $\mathbf{E}$ of angular frequency ω,

$$\langle \mathbf{I} \rangle = i\omega\epsilon_0\epsilon^\star(\omega)\mathbf{E}, \tag{7.10.1}$$

where ϵ_0 is the dielectric permittivity of vacuum. This technique of measurement of $\epsilon^\star(\omega)$ is known as impedance spectroscopy or dielectric spectroscopy [Kremer & Schönhals (2003)]. In general, the complex dielectric function $\epsilon^\star(\omega)$ is related to the correlation function $\phi(t)$ of the net dipole moment in a volume that is large enough to allow ensemble averaging, according to [Frohlich (1958)]

$$\frac{\epsilon^\star(\omega) - \epsilon_\infty}{\epsilon_s - \epsilon_\infty} = 1 - i\omega \int_0^\infty dt\, \phi(t)\, e^{-i\omega t}, \tag{7.10.2}$$

where ϵ_s and ϵ_∞ are the limiting low- and high-frequency permittivities, respectively.

If $\phi(t)$ is characterized by a single exponential relaxation with a characteristic relaxation time τ, Eq. (7.10.2) yields the well-known Debye relaxation function

$$\frac{\epsilon^\star(\omega) - \epsilon_\infty}{\epsilon_s - \epsilon_\infty} = \frac{1}{1 + i\omega\tau}. \tag{7.10.3}$$

On the other hand, if multiple relaxation processes occur in the system, $\phi(t)$ can be empirically represented by the Kohlrausch-Williams-Watts equation

$$\frac{\epsilon^{\star}(\omega) - \epsilon_{\infty}}{\epsilon_s - \epsilon_{\infty}} = \exp\left[-\left(\frac{t}{\tau}\right)^{\beta}\right], \qquad (7.10.4)$$

where $\beta < 1$. This result is referred to as "stretched exponential behavior." This behavior can equivalently be represented using a distribution of relaxation times $g(\tau)$ as

$$\frac{\epsilon^{\star}(\omega) - \epsilon_{\infty}}{\epsilon_s - \epsilon_{\infty}} = \frac{g(\tau)}{1 + i\omega\tau}. \qquad (7.10.5)$$

Using Eqs. (7.10.1) and (7.10.5), the number of relaxation processes in a polyelectrolyte solution and the corresponding relaxation times are determined by measuring the dielectric response of the solution to an externally imposed oscillatory electric field.

The standard theory of conductive dielectrics relates the complex dielectric function $\epsilon^{\star}(\omega)$ to the frequency dependent conductivity $\sigma(\omega)$ according to [Bordi *et al.* (2004)]

$$\epsilon^{\star}(\omega) = \epsilon'(\omega) - i\frac{\sigma(\omega)}{\epsilon_0\omega}, \qquad (7.10.6)$$

where $\epsilon'(\omega)$ is the real part of $\epsilon^{\star}(\omega)$. The real part denotes the storage of energy in the system. The imaginary part of $\epsilon^{\star}(\omega)$ is $\sigma(\omega)/\epsilon_0\omega$, denoting the dielectric loss. The net imaginary part of $\epsilon^{\star}(\omega)$ has two components, one from the dielectric process ϵ'', and another from the DC electrical conductivity σ_0, which is the low-frequency limit of $\sigma(\omega)$. As a result, the above equation is rewritten as

$$\epsilon^{\star}(\omega) = \epsilon'(\omega) - i\left[\epsilon''(\omega) + \frac{\sigma_0}{\epsilon_0\omega}\right]. \qquad (7.10.7)$$

Note that for polyelectrolyte solutions, which are conductive, interpretation of measured dielectric constants at low frequencies is challenging due to the dominance of the second term inside the square brackets on the right-hand side of the above equation. This difficulty precludes exploration of polyelectrolyte dynamics at very large time scales. On the other hand, intra-chain dynamics and dipoles formed by adsorbed counterions on chain backbone can be explored from measurements of $\epsilon^{\star}(\omega)$.

For example, the dielectric response of salt-free aqueous solutions of sodium poly(styrene sulfonate) of molar mass 177 kDa is presented in Fig. 7.31 at three different polymer concentrations ($c = 1.38, 0.45$, and 0.09 g/L) [Mandel (2000)] . There are several striking features evident in these data in comparison to the dielectric dispersion depicted in Fig. 3.4 for pure water. In the case of water, there is only one relaxation mode at about 10 GHz due to orientational relaxation of water molecules. In the case of polyelectrolyte solutions, additional relaxation modes appear that distinguish them from solutions of low molar mass salts. In the typical example portrayed in Fig. 7.31, there are two additional relaxation processes, one at $\sim 10^4$ Hz and another at about 1 MHz. The relaxation process associated with water molecules occurs at a much higher frequency of about 1 GHz (not shown in the figure). In addition, large increases in the dielectric constant with respect to the solvent are observed at very low frequencies. The dielectric dispersion at lower frequencies (of about 10^4 Hz) depends on

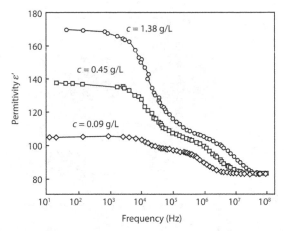

Figure 7.31 Permittivity of salt-free aqueous solutions of sodium poly(styrene sulfonate) of molar mass 177 kDa at various polymer concentrations (from top to bottom, respectively, 1.38 g/L, 0.45 g/L, and 0.09 g/L) over the frequency range from 10 Hz to 100 MHz. [Mandel (2000)]

the molar mass of the polyelectrolyte and can be associated with segmental dynamics mediated by counterions. The characteristic time corresponding to the dielectric region at higher frequencies of about 1 MHz is independent of molar mass of the polymer. This relaxation mode is associated with orientational relaxation of ion pairs constituted by adsorbed counterions and charged monomers, thus allowing investigation of the plasmon mode described in Section 7.6.3.

An adequate interpretation of the aforementioned dielectric results is presently unavailable. The concentration dependent dielectric increment of polyelectrolyte solutions at very low frequencies (Fig. 7.31) has been suggested to originate from polarization effects arising from the coupled charged macromolecule and its ion cloud under periodically varying electric fields [Mandel (1988)]. Due to such polarization effects, the induced dipole moment can become rather large resulting in significant dielectric increments. Investigation of dielectric relaxation in polyelectrolyte solutions containing added low molar mass electrolytes is not extensive, primarily due to difficulties arising from electrical conductivities that are too high. Development of a molecular theory of $\epsilon^\star(\omega)$ as well as frequency dependent conductivity $\sigma(\omega)$ remains a challenge, adding one more layer of complexity to the static electrical conductivity discussed in the preceding section.

7.11   Polymer Translocation

Controlled transport of charged macromolecules in restrictive aqueous environments is a ubiquitous way of life for biological cells. Central to these processes is the **translocation** of individual macromolecules across a single narrow constriction in space (for example, Fig. 7.22). One example is how a single charged macromolecule, such as ssRNA, dsDNA, or a protein, is translocated through a nanopore in salty aqueous solutions under an external electric field. Studies of single molecule

(a)

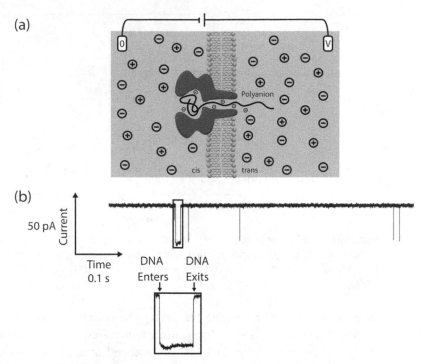

Figure 7.32 (a) Experimental setup to monitor translocation of a single charged macromolecule through a protein pore using ionic current measurements. A negatively charged polymer translocates toward the positive electrode. (b) Experimental current trace of ssDNA molecule stochastically captured into a α-hemolysin protein pore. A translocating molecule momentarily blocks the ionic current through the pore.

electrophoresis have enabled a molecular level understanding of the ubiquitous fundamental step of macromolecular transport in crowded aqueous environments [Kasianowicz *et al*. (1996), Meller (2003), Muthukumar (2011), Wanunu (2012), Muthukumar *et al*. (2015)]. This phenomenon can also be harnessed to develop separation technologies analogous to gel electrophoresis. In addition, these situations have opened a gateway for sensing polynucleotides and other biological macromolecules in the context of sequencing their chemical compositions. Furthermore, single molecule electrophoresis experiments hold promise to gain insight into the dynamics of charged macromolecules under nonequilibrium conditions involving several simultaneous synergistic and antagonistic forces [Sakaue (2007), Rowghanian & Grosberg (2012), Jeon & Muthukumar (2014), Chen *et al*. (2021)].

We shall describe below only the most common elements of single macromolecule translocation. A cartoon of the typical experimental setup to investigate translocation is given in Fig. 7.32a. A single protein pore (formed by the self-assembly of α-hemolysin, MspA, or CsgG proteins) is inserted into a thin membrane which separates the donor (*cis*) compartment and the receiver (*trans*) compartment. Both compartments contain an electrolyte solution. Under an externally applied voltage gradient, the electrolyte ions pass through the pore, resulting in an ionic current. The narrowest constriction in typical protein pores used in single molecule electrophoresis

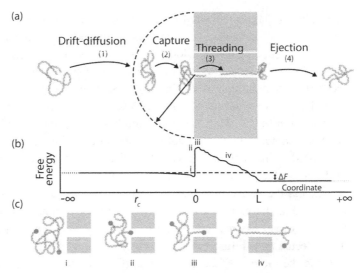

Figure 7.33 (a) Four stages of translocation: drift-diffusion, capture, threading, and ejection. (b) Sketch of the corresponding free energy landscape. The coordinate values 0 and L represent the points when the chain's head and tail enter and exit the pore, respectively. r_c is the capture radius. The states (i), (ii), (iii), and (iv) correspond to the typical chain conformations sketched in (c): (i) arrival of the chain at the pore with its ends unregistered with the pore, (ii) one end of the chain aligns with the pore entrance and enters the pore, (iii) the chain begins to unravel as the end moves through the pore, and (iv) the chain crosses the pore through a combination of electrophoretic drift and diffusion inside the pore. The transitions between the various steps are typically associated with a free energy barrier in the order of 10 $k_B T$. ΔF is the net free energy gain by the translocating molecule [Adapted from Muthukumar *et al.* (2015)].

is around 1.5 nm. When charged macromolecules with backbone thicknesses small enough to pass through the narrowest constriction of the protein pore are dispersed in one compartment, individual molecules can translocate in a single-file toward the other compartment due to the voltage gradient. As a specific example, consider the translocation of ssDNA from the *cis* compartment to the *trans* compartment (which has the positive electrode, Fig. 7.32a). The translocation events of single molecules cause temporal blockades of the ionic current flowing through the pore [Kasianowicz *et al.* (1996)], as illustrated in Fig. 7.32b. Durations, depths, and fine details of the ionic current blockades are collected from the recorded ionic current traces as a function of time. The identity of the macromolecule undergoing translocation and its mechanism of translocation are discerned from the collected data. For macromolecules with larger backbone thicknesses, such as dsDNA, which cannot pass through protein pores, solid-state nanopores with desirable diameters and lengths are used instead of protein pores. Development of technologies to fabricate solid-state nanopores and their analogs has enabled researchers to elicit the mechanism of translocation of a variety of charged macromolecules and to sense them at the single molecule level.

Independent of the particular type of nanopore, the translocation of macromolecules occurs via the same basic process, as illustrated in Fig. 7.33. First, a macromolecule must be captured at the entrance of the pore. Many biopolymers and synthetic polyelectrolytes are charged, so an applied electric field exerts an electrophoretic force that pulls the molecule toward the pore. The pull is strongest at the pore mouth and becomes progressively weaker at increasing distances from the pore. At the same time, the molecule undergoes diffusion, which can move it away from the pore. The region of space where drift due to electric force away from the pore is weak is designated as the "drift-diffusion regime" (1) in Fig. 7.33a. Near the pore, the electrophoretic pull overwhelms diffusion and this region of space is the "capture regime," as designated as regime (2) in Fig. 7.33a. The competition between diffusion and electrophoretic drift defines a capture radius r_c about the pore mouth [Wanunu *et al.* (2010), Muthuku-mar (2010)]. The capture radius can be orders of magnitude larger than the dimension of the nanopore. In addition, electro-osmotic flow and gradients in hydrostatic pressure across the pore can be manipulated to tune the capture radius. Once a molecule drifts within the capture radius, electrophoresis dominates over diffusion and the molecule eventually reaches the pore. At further distances outside the capture radius, the molecule may diffuse away.

Once the molecule reaches the pore, the next step is nucleation of translocation, namely, the insertion of one end of the polymer strand into the pore mouth, which is a requirement for single-file translocation. The key feature of macromolecules like polynucleotides and synthetic flexible polyelectrolytes is that they have enormous conformational entropy. At room temperature, the molecular conformation is continuously changing and the molecule typically approaches the pore mouth as a random coil, with neither of its two ends necessarily having the correct orientation for insertion into the pore. This state is depicted as (i) in Fig. 7.33c. For the molecule to enter the pore, one of its ends must escape the space within the polymer coil to find the pore entrance and adopt state (ii) in Fig. 7.33c. That step requires crossing a considerable entropic barrier on the order of $10k_BT$. The precise value of the barrier depends on the polymer length, electrolyte concentration, nature of interactions between the pore and the polymer end, and other details. We shall call **capture** of the chain as the event of chain attaining the state (ii) in Fig. 7.33. The free energy barrier associated with the transition from state (i) to state (ii) is sketched in Fig. 7.33b.

Subsequent to the capture of one chain end at the pore entrance, the chain undergoes **threading** under a net driving free energy change ΔF for translocation. This is referred to as the "threading regime" (3), as indicated in Fig. 7.33a. During the early stage of threading, there is an additional entropic barrier for inserting the chain into the pore, after the chain end finds the pore. When the chain is squeezed into state (iii) (Fig. 7.33c) along the narrow path of the pore, its range of allowed conformations is considerably reduced. As a result, the chain encounters a further conformational entropic barrier of roughly $2 - 3k_BT$ to proceed with the translocation process. The barrier connecting states (ii) and (iii) is sketched in Fig. 7.33b.

The combination of the above two entropic barriers constitutes the net barrier associated with nucleation of translocation of the chain. Once the nucleation barrier is crossed, sometimes with the aid of energetic interactions between the polymer and pore, the chain transits the pore through another drift-diffusion process. The drift arises from the electrophoretic pull of the single-file strand toward the *trans* side, while back-and-forth diffusion along the pore is due to thermal noise from the electrolyte solution

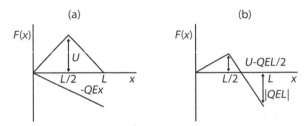

Figure 7.34 (a) Ramp-like profile for the free energy barrier across a distance L along the x-coordinate. The maximum of the barrier is U located at $x = L/2$. The electric field is uniform in the region $0 \leq x \leq L$; electric potential energy for the molecule at the location x is $-QEx$. (b) Free energy landscape by combining the barrier and electric potential energy. The maximum barrier is $U - |QEL/2|$ and the gain in free energy upon capture is $|QEL|$.

inside the pore. This state is denoted as (iv) in Fig. 7.33. This drift-diffusion process is further influenced by pore-polymer interactions specific to the chemistry of the pore and polymer. With a sufficiently strong free energy change that drives translocation, the chain exits the pore (region (4) in Fig. 7.33a). The transition of the chain from state (ii) to the completion of state (iv) until the chain exits the pore constitutes the threading process, whereas reaching state (ii) from the *cis* side is the capture process. Simple models to treat capture rate and translocation kinetics are considered next.

7.11.1   Polymer Capture through a Barrier

Consider a macromolecule of charge Q moving across a free energy barrier (which is primarily an entropic barrier arising from conformational changes of the macromolecule) in the presence of an externally imposed electric field of strength E along the x-direction. Let the net barrier for the movement of the center-of-mass of the molecule by distance L along the x-direction be U. For convenience of the calculation, let us assume that the barrier is triangular and symmetric with its maximum at $L/2$, as shown in Fig. 7.34a. As the center-of-mass of the molecule moves a distance x, the gain in electrical energy due to the external electric field is $-QEx$ (Fig. 7.34a). The resultant free energy landscape $F(x)$ for the capture of the macromolecule at $x = L$ from $x = 0$ is

$$F(x) = \begin{cases} 2U\frac{x}{L} - QEx, & 0 \leq x \leq L/2 \\ 2U\left(1 - \frac{x}{L}\right) - QEx, & L/2 \leq x \leq L, \end{cases} \quad (7.11.1)$$

as given in Fig. 7.34b, where the effective barrier height is $U - QEL/2$ and the net free energy gain due to capture is $|QEL|$.

As evident from Fig. 7.34, the capture of the macromolecule at $x = L$ requires that the molecule must cross the barrier. If the driving force is insufficient to surmount the barrier, the velocity of the molecule (and hence the capture rate) is essentially zero. For strong driving forces, the molecule undergoes electrophoretic drift. In the absence of the barrier and electric field, the molecule undergoes diffusion, and D is its diffusion

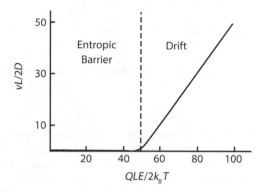

Figure 7.35 Plot of reduced flux $vL/2D$ against reduced voltage difference $|QLE|/2k_BT$ delineating the entropic barrier regime and drift-dominated regime. The curve is given by Eq. (7.11.2) for $U = 50k_BT$ [Adapted from Muthukumar (2010)].

coefficient. As derived in Appendix 6.7, the velocity v of the molecule in the steady state is given by

$$v = \frac{2D}{L}\left\{\frac{1}{\left(\frac{U}{k_BT} - \frac{QLE}{2k_BT}\right)}\left[e^{\left(\frac{U}{k_BT} - \frac{QLE}{2k_BT}\right)} - 1\right]\right.$$
$$\left. + \frac{1}{\left(\frac{U}{k_BT} + \frac{QLE}{2k_BT}\right)}\left[e^{\left(\frac{U}{k_BT} - \frac{QLE}{2k_BT}\right)} - e^{-\frac{QLE}{k_BT}}\right]\right\}^{-1}. \qquad (7.11.2)$$

This is a general formula for the capture velocity in terms of two dimensionless parameters U/k_BT and QEL/k_BT, applicable to all kinds of capture problems and nucleation process for movement of macromolecules trapped inside heterogeneous media. A typical consequence of the above equation is illustrated in Fig. 7.35 as a plot of $vL/2D$ versus $QLE/2k_BT$ for $U = 50k_BT$. For weak electric fields such that $QLE \ll U$, Eq. (7.11.2) leads to

$$v = \frac{D}{L}\frac{U/k_BT}{(e^{U/k_BT} - 1)}. \qquad (7.11.3)$$

For large barriers, this equation reduces to

$$v = \frac{D}{L}\frac{U}{k_BT}e^{-U/k_BT}. \qquad (7.11.4)$$

Therefore, if the applied energy $QLE/2$ on the molecule is weaker than large barrier heights U, then the molecule is essentially immobilized with exponentially small capture rates. On the other hand, in the opposite limit of $QLE \gg U$, the macromolecule undergoes drift according to

$$\frac{vL}{2D} = \frac{QLE}{2k_BT} - \frac{U}{k_BT}. \qquad (QLE > U) \qquad (7.11.5)$$

Therefore, as also seen in Fig. 7.35, the threshold electric field strength required for the electrophoretic velocity of the macromolecule to be practically nonzero is a measure

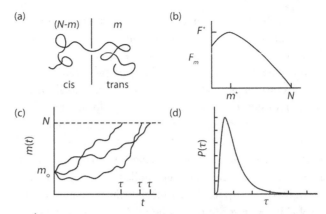

Figure 7.36 (a) A conformation of a chain undergoing translocation through a nanopore in a thin membrane with m monomers in the receiver compartment and $N - m$ monomers in the donor compartment. (b) Sketch of a generic free energy profile for translocation as a function of the number of monomers that are already translocated into the receiver compartment. (c) Sketch of three trajectories for the time evolution of the number of monomers translocated into the receiver compartment. The translocation time τ is the first passage time. (d) A typical distribution function $P(\tau)$ of the translocation time.

of the entropic barrier height U. Eq. (7.11.2) provides a convenient method to experimentally determine the free energy barriers for movement of charged macromolecules. The above model can be readily generalized to other shapes of the free energy barrier [Muthukumar (2011)].

7.11.2   Kinetics of Translocation

An example of the free energy profile for translocation (threading) of a chain through a nanopore is portrayed in Fig. 7.33b. Given a general form for the free energy profile, we now describe the time-dependent probability of realizing a particular state during translocation, followed by a derivation of the average time taken by a molecule for completing the translocation process.

Consider the simple situation of a single macromolecule undergoing single-file translocation through a nanopore embedded in a thin membrane, as sketched in Fig. 7.36a [Sung & Park (1996), Muthukumar (1999)]. At a particular time duration t from the initiation of translocation, let there be m monomers in the receiver compartment and $N - m$ monomers in the donor compartment. A sketch of the generic free energy profile for this situation is given in Fig. 7.36b, where F_m is the free energy of the chain of N monomers with m monomers having already translocated into the receiver compartment. As described in Fig. 7.33, there is an entropic barrier for nucleation of the translocation process, denoted by $F^\star$ in Fig. 7.36b, which occurs at $m^\star$. Only if the number of translocated monomers is larger than $m^\star$ will the translocation process proceed to completion.

Translocation is a stochastic process due to the omnipresent random thermal noise in the system. Each translocation event follows its own particular trajectory. As a result, we have an ensemble of translocation trajectories representing the various translocation events. In different realizations of the translocation process, the number of monomers that have translocated into the receiver compartment is different at a given time. Three examples of trajectories are illustrated in Fig. 7.36c, where the initial value of m at $t = 0$ is m_0. The time evolution of $m(t)$ is a random process due to diffusion of the monomers along the translocation coordinate concomitant with drift arising from the driving force. The translocation time τ for one translocation event is the first passage time for one trajectory as denoted in Fig. 7.36c. τ represents the time taken by the whole molecule to enter the receiver compartment for the first time. In this model, we assume that, once pulled into the receiver compartment, the molecules do not reenter the pore. Due to the stochastic nature of the translocation events, only a probabilistic description can be made for translocation. A typical result for the distribution of translocation time, $P(\tau)$ is given in Fig. 7.36d. The average translocation time $\langle \tau \rangle$ and the variance $\langle \tau \rangle - \langle \tau \rangle^2$ can then be determined from $P(\tau)$. The common theoretical approach for such a probabilistic description is the Fokker–Planck formalism [Risken (1989)]. In this formalism, the probability $P(x,t)$ of finding m monomers in the receiver compartment at time t is given by

$$\frac{\partial P(m,t)}{\partial t} = k_0 \left\{ \frac{\partial}{\partial m} \left[\frac{1}{k_B T} \frac{\partial F_m}{\partial m} P(m,t) \right] + \frac{\partial^2 P(m,t)}{\partial m^2} \right\}, \qquad (7.11.6)$$

where k_0 is the diffusion coefficient of the monomer during translocation, which is assumed to be independent of its chemical identity. The first term inside the curly brackets in the above equation represents the drift contribution while the second term denotes the diffusion contribution. Using F_m as the input and suitable boundary conditions that are appropriate for a particular experimental situation, the above equation is solved for $P(m,t)$, from which the distribution of translocation time and the average translocation time are obtained. Details of calculations of $P(m,t)$ and the average translocation time are available in the book on polymer translocation [Muthukumar (2011)].

The key assumption in the above commonly used Fokker–Planck method is that chain conformations equilibrate on a time scale shorter than the translocation time. This assumption is valid for short strands, but nonequilibrium effects come into play for longer ones, such as kilobase-length DNA. The role of nonequilibrium conformations during translocation, such as the propagation of tension along the chain contour [Sakaue (2007), Rowghanian & Grosberg (2012), Chen *et al.* (2021)], created by the electric field at the nanopore, is under active investigation. Nevertheless, the quasi-equilibrium approach has enabled molecular understanding of translocation processes in a variety of experimental situations through the intermediary of the free energy landscape which bridges the specifics of the experimental setup and the measured characteristics of the translocation events. In terms of a potential future direction, when the translocating polymer possesses ordered domains, as in multidomain proteins, features of translocation kinetics can be used to decode secondary structural details at a single molecule level.

7.12   Topologically Frustrated Non-diffusive Dynamics

The movement of a single macromolecule through a nanopore described above and diffusion of a chain out of one confining region into another (Fig. 7.21a) involve effectively a single entropic barrier. In these circumstances, the center-of-mass of a molecule undergoes diffusion at nonzero temperatures, coupled to drift, even though the diffusion coefficient is smaller in comparison to its value in free solution. The fact that the macromolecule undergoes diffusion in the presence of a single entropic barrier as well as in semidilute solutions and entangled melts is a manifestation of Einstein's fundamental theorem on diffusion.

The key results on the diffusion of macromolecules described in the preceding sections are summarized as follows. Consider a large macromolecule (guest) of N monomers and radius of gyration R_g embedded inside a host aqueous medium with spatial restrictions. As a specific example, let the host be a hydrogel, where the average mesh size (correlation length for local monomer concentration of the host gel) is ξ. As noted in Section 7.5.4 (Fig. 7.21), the movement of the guest molecule is dictated by the relative value of R_g to ξ, and the relative value of ξ to a length ℓ (comparable to that of a small number of monomers and not necessarily the Kuhn segment length [Chen & Muthukumar (2021)]). Based on experimental results and theories described in the preceding sections on polymer dynamics, four regimes (1,2,3, and 5 in Fig. 7.37a) can be identified for diffusion of the guest molecule inside the host. These are described, respectively, by the Zimm, Rouse, entropic barrier, and reptation models. In all these regimes, the guest molecule undergoes diffusion. The dependence of the diffusion coefficient D on N of the guest molecule is given by the scaling laws

$$
D \sim \begin{cases}
N^{-\nu}, & (R_g \ll \xi), \text{ Zimm} \\
N^{-1}, & (R_g < \xi), \text{ Rouse} \\
N^{-1}\exp(-AN), & (R_g \sim \xi), \text{ entropic barrier} \\
N^{-2}, & (R_g \gg \xi, \xi \sim \ell), \text{ reptation}
\end{cases}
\tag{7.12.1}
$$

where ν is the size exponent defined through $R_g \sim N^\nu$ and A is a nonuniversal numerical factor denoting the local structure of the host medium (Section 7.5.4).

Now, let us imagine a scenario where a single chain pervades multiple spatial domains, with each domain holding considerable number of monomers belonging to the same chain [Jia & Muthukumar (2018, 2021a)]. This corresponds to the condition $R_g \gg \xi \gg \ell$, labeled as regime (4) in Fig. 7.37a. Since the multiple domains holding a single molecule are connected only by narrower pathways (as in Fig. 7.22, for example), each spatial domain acts as an entropic trap, as sketched in Fig. 7.37b. In order for the whole chain to move, all of these multiple correlated barriers must be surmounted. Investigation of such scenarios reveals that a new universality class of polymer dynamics emerges as an apparent breakdown of the law of diffusion for practically relevant time scales. Here, the chains are locked into a dynamical state where they do not diffuse but are alive with their internal chain dynamics and functional properties. The multiple entropic traps pinning a single chain results in non-diffusive

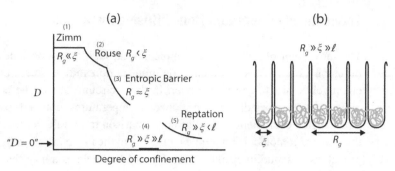

Figure 7.37 (a) Sketch of the dependence of the center-of-mass diffusion coefficient D of the guest molecule on degree of confinement (which increases with a decrease in the mesh size ξ). The four familiar regimes of polymer dynamics, namely, Zimm, Rouse, entropic barrier, and reptation, and their conditions of experimental relevance are denoted by 1, 2, 3, and 5, respectively. The corresponding scaling laws in these regimes connecting D and the degree of polymerization N of the guest chain are given in Eq. (7.12.1). Regime 4 is the topologically frustrated dynamical state (TFDS), where $D \simeq 0$ as denoted by the horizontal line at intermediate confinements. (b) Sketch of partitioning of a single chain into multiple deep entropic traps, which results in the emergence of TFDS.

hierarchical state of chain dynamics. We shall refer to this universality class of polymer dynamics as the **topologically frustrated non-diffusive dynamics**. As indicated in Fig. 7.37a, this non-diffusive **topologically frustrated dynamical state (TFDS)** occurs only at intermediate confinements, with the chain undergoing diffusion outside this intermediate window. Under the conditions for the emergence of TFDS, experiments using DLS and fluorescence studies on negatively charged hydrogels containing either similarly charged dsDNA or oppositely charged poly(L-lysine) show that the guest molecules do not diffuse. However, the DLS data show rich dynamical features at length scales comparable to the mesh size. Since $\xi \gg \ell$, the subchain dynamics inside each mesh is allowed [Jia & Muthukumar (2018, 2021a)]. As addressed in Sections 7.5.1–7.5.4, the dynamics of subchains are described by Zimm, Rouse, single entropic barrier, and reptation models depending on the polymer concentration inside each mesh, even though the center-of-mass of the guest molecule is localized. Performing the fluorescence tracking experiments for different mesh sizes of the gel and electric fields, the net barrier heights have been determined using Eq. (7.11.2). For λDNA as the guest molecule, the net free energy barrier that locks the molecule into immobility is in the range of 20 to 120 $k_B T$ depending on the mesh size [Chen & Muthukumar (2021)].

The emergence of TFDS adds a new facet to polymer dynamics in crowded environments that is complementary to the paradigm of diffusion. Realization of such immobile dynamical states that are still alive with functional properties enables several opportunities to manipulate at the single molecular level. For example, molecular engines can efficiently search their targets encoded on specific sequential domains of immobilized DNA, instead of chasing the constantly diffusing targets. Furthermore, the tunable capacity to elicit localized dynamics using electrostatic forces provides strategies to design scaffolds for controlled retention and release of charged macromolecular cargo in crowded aqueous media.

8 Self-Assembly and Phase Behaviors

8.1 General Premise

We know that oil and water do not mix. Similarly, when a concentrated solution of uncharged macromolecules is mixed with a poor solvent, the solution undergoes macroscopic phase separation where a dense phase of a polymer separates from its supernatant liquid. Analogously, homogeneous solutions of charged macromolecules in polar solvents can suddenly become thermodynamically unstable, when experimental variables such as the concentration of added salt or the temperature are changed. As a result, a parent single phase undergoes phase transitions into differently ordered single phases, or solutions with a coexistence of two or more phases. It is well-known that solutions of uncharged macromolecules undergo **liquid–liquid phase separation (LLPS)**, which is also called **macrophase separation**. Its equivalent in solutions of charged macromolecules exhibits more complex critical phenomena and phase transitions. The complexity in charged systems primarily emerges from two competing forces. As in the case of uncharged polymers, the van der Waals interaction between the backbone of charged macromolecules and the solvent can lead to thermodynamic instability of the solution. On the other hand, similarly charged macromolecules repel each other and hence promote their uniform distribution in the solution, leading to stability of the homogeneous phase. The relative strengths of the competing van der Waals and electrostatic interactions can be easily modulated by adding salt or by changing the temperature of the system. Furthermore, as we have seen in Chapter 6, even the so-called homogeneous phase is not entirely homogeneous at all length scales, as evident from the occurrence of the polyelectrolyte peak and aggregation of even similarly charged macromolecules. These features are amplified when the homogeneous phase is subjected to thermodynamic instability by varying the added salt concentration or temperature. The resulting phase behavior of solutions of charged macromolecules is very rich, and depends delicately on the various controlling experimental variables.

As an example of the rich phase behavior of polyelectrolyte solutions, Fig. 8.1 shows the phase diagrams of aqueous solutions of poly(vinylsulfonic acid) (PVSA) containing the salt XCl (X = K, Na, and Rb) or NaY (Y = I, Br, Cl, F, and NO_3) [Eisenberg & Mohan (1959), Eisenberg & Casassa (1960)]. For a given set of concentrations of the polymer and the particular salt, the phase-separation temperature T_p is determined by monitoring the first appearance of cloudiness as the solution is cooled.

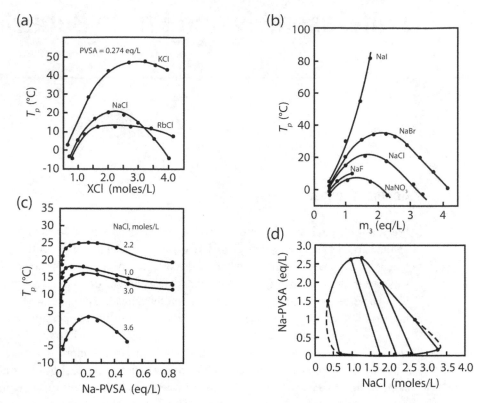

Figure 8.1 Phase diagrams of aqueous solutions of poly(vinylsulfonic acid). (a) Effect of identity of monovalent cation from the added salt XCl (X = K, Na, and Rb) at 0.274 monomer-molar polymer concentration. (b) Effect of identity of monovalent anion from the added salt NaY (Y = NO₃, F, Cl, Br, and I). (c) Effect of salt concentration. (d) Dependence of polymer concentration on salt concentration at a fixed temperature 0°C. (a), (c), and (d): Adapted from [Eisenberg & Mohan (1959)]; (b): Adapted from [Eisenberg & Casassa (1960)].

For temperatures above T_p, the solution is homogeneous, and for temperatures below T_p, the solution undergoes liquid–liquid phase separation. The dependence of T_p of 0.274 monomer-molar PVSA solution on the concentration of XCl is given in Fig. 8.1a for X = K, Na, and Rb, showing significant changes in T_p due to the identity of the cation and the amount of added salt. These results overwhelm the intuitive expectation that all monovalent ions behave more or less equivalently, emphasizing the importance of specificity of the ions. Analogously, Fig. 8.1b shows a plot of T_p against salt concentration at constant polymer concentration (0.2 monomer-molar) with Na as the counterion in each case. It is seen that T_p increases considerably with the salt anion in the order of NO₃, F, Cl, Br, I. Thus Figs. 8.1a and 8.1b demonstrate the substantial role played by the electrolyte ions in shifting the phase-transition temperature by even 80°C. It is to be noted that such a broad range in T_p occurs although the counterion and co-ion carry the same monovalent charge. Also, note that there is no LLPS in aqueous solutions if X (in XCl salt) is H, Li, Cs, or NH₄. Therefore, the phase behavior of solutions of charged macromolecules is extremely sensitive to the specificity of the electrolyte ions.

For fixed concentrations of NaCl, the dependence of T_p on the concentration of the polymeric salt NaPVSA is given in Fig. 8.1c. For each NaCl concentration, T_p exhibits an upper critical solution behavior. When NaCl concentration is increased, the phase-separation temperature increases first, as expected from enhanced electrostatic screening. However, surprisingly, T_p decreases upon further increase in NaCl concentration. Additionally, the phase diagram of NaPVSA concentration versus NaCl concentration at a fixed temperature (0°C) is given in Fig. 8.1d. It is evident that there exists a lower and an upper limit of polymer concentration between which phase separation occurs at a given temperature and salt concentration. Also, at a given temperature and polymer concentration, there exists a lower and an upper limit of added salt concentration between which phase separation occurs. Furthermore, as already mentioned in Chapter 1, there is the universal reentrant precipitation of charged macromolecules such as NaPSS and ssDNA from aqueous solutions containing trivalent salts (such as $LaCl_3$ and spermidine) (Fig. 1.10). For a fixed polyelectrolyte concentration, precipitation is observed as the salt concentration c_s is increased. If c_s is increased further, the precipitate becomes soluble again. The range of c_s for the precipitation of the polyelectrolyte decreases as the polymer concentration is increased. We shall later identify this reentrant precipitation phenomenon as liquid–liquid phase separation.

In general, the starting point to address phase behaviors of multicomponent solutions of charged macromolecules is the Helmholtz free energy F. As described in Chapter 6, F depends on the translational entropy of all species in the system, van der Waals and electrostatic interactions, conformational fluctuations of the macromolecules, and charge density fluctuations due to counterions and electrolyte ions. The various contributions to F are written as

$$F = F_S + F_H + F_{fl,p} + F_{fl,i}. \qquad (8.1.1)$$

F_S is the contribution from the translational entropy of the various species, and F_H is the contribution from all excluded volume interactions among all polymer segments and solvent molecules. These interactions include both the short-ranged van der Waals-type interactions and the electrostatic interactions. $F_{fl,p}$ is the contribution from fluctuations in polymer conformations described in Section 6.3.6. $F_{fl,i}$ is the contribution from the fluctuations in the charge distribution of the various small ions discussed in Sections 3.8.6 and 6.3.1. Adopting appropriate expressions for the various terms in Eq. (8.1.1) relevant to a particular system of interest, the free energy F is used to describe thermodynamic stability and phase separation of the system. We illustrate the methodology by first considering uncharged systems and then extending the same protocol for charged systems.

Although the various experimental results of the phase behaviors of solutions of charged macromolecules with added salt have been well established, an adequate understanding is presently lacking. Nevertheless, this chapter is devoted to a description of the current level of understanding and predictions based on simple models. First, we shall introduce the key concepts of the critical point, coexistence curves, and spinodal curves for uncharged polymer solutions, based on the Flory–Huggins theory described in Section 6.2.1. Next, we shall consider the role of electrostatic interactions on critical phenomena in electrolyte solutions without macromolecules, based on the restricted primitive model of Section 3.8. By combining the above two approaches, we will present a mean field theory of phase behavior for solutions of charged macromolecules. The roles of charge regularization due to counterion adsorption on the charged macromolecules and the concentration fluctuations near a critical point will be addressed next, followed by a brief discussion on kinetics of LLPS. Next, we shall describe micellization and microphase separation in systems containing polymers with blocky sequences of uncharged and charged repeat units. In the final sections, we will address phase transitions in solutions containing charged rods, folded proteins, and intrinsically disordered proteins. The phase behavior of solutions containing polycations and polyanions in an electrolyte solution, known as **coacervation**, will be described in Chapter 9.

8.2   Liquid–Liquid Phase Separation in Solutions of Uncharged Polymers

Consider a two-component solution consisting of a flexible uncharged polymer dissolved in a solvent. In thermodynamic equilibrium, the chemical potentials of the components are uniform in the solution, without any gradients. However, gradients in their chemical potentials can be generated when the system is forced out of equilibrium using external stimuli such as variations in temperature. Depending on the nature and extent of the stimuli, the gradients in chemical potentials can either relax into uniform chemical potentials inside the whole system, or result in phase separation. In the phase-separated state of a two-component system, two thermodynamically stable phases coexist, where the chemical potential of each component is the same in both phases.

The main features of the onset of thermodynamic instability, phase separation, and associated critical phenomena can be adequately captured using the Flory–Huggins theory described in Section 6.2.1. Incorporation of additional roles of fluctuations in polymer conformation and concentration does not modify the generic qualitative description of liquid–liquid phase separation in solutions of uncharged polymers. As derived in Section 6.2.1, the chemical potentials of the two components of a solution, where ϕ_2 is the volume fraction of the polymer (component 2) and $\phi_1 = 1 - \phi_2$ is the volume fraction of the solvent (component 1) are given by Eqs. (6.2.8) and (6.2.9). For the case of monodisperse polymer chains, each with N segments, the chemical potential of the solvent μ_1 and that of the polymer μ_2 are

$$\frac{\mu_1(\phi_2) - \mu_1(0)}{k_B T} = \ln(1 - \phi_2) + \left(1 - \frac{1}{N}\right)\phi_2 + \chi\phi_2^2, \qquad (8.2.1)$$

and

$$\frac{\mu_2(\phi_2) - \mu_2(1)}{k_B T} = \ln\phi_2 + (1 - N)(1 - \phi_2) + \chi N(1 - \phi_2)^2. \qquad (8.2.2)$$

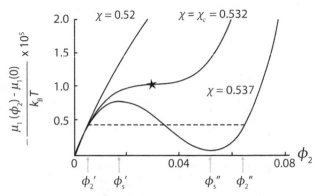

Figure 8.2 Plot of osmotic pressure, given by $\Pi v_1/k_BT = -(\mu_1(\phi_2) - \mu_1(0))/k_BT$, versus polymer volume fraction ϕ_2 for $N = 1000$ and $\chi = 0.52, 0.532$ and 0.537. For χ less than the critical value $\chi_c = 0.532$, the solution is homogeneous. The critical point is denoted by the star symbol. For $\chi > \chi_c$, the solution undergoes phase separation into two coexisting liquid phases with concentrations ϕ_2' and ϕ_2'' as denoted by the dashed line. The solution is unstable between the spinodal points ϕ_s' and ϕ_s''. For polymer concentrations between ϕ_2' and ϕ_s' and between ϕ_s'' and ϕ_2'', the solution is metastable.

Note that the osmotic pressure Π of the solution is related to the negative of the chemical potential of solvent given by Eq. (6.2.11) as

$$\frac{\Pi v_1}{k_BT} = -\ln(1 - \phi_2) - \left(1 - \frac{1}{N}\right)\phi_2 - \chi\phi_2^2, \qquad (8.2.3)$$

where v_1 is the volume of the solvent molecule.

The criterion for thermodynamic instability and the eventual phase behavior can be gleaned from the expression for the osmotic pressure of the solution. The dependence of $\Pi v_1/k_BT$ on ϕ_2, given by Eq. (8.2.3) is plotted in Fig. 8.2 for $N = 1000$ and different values of χ. For values of χ less than a certain critical value χ_c (which is 0.532 for $N = 1000$), Π increases with ϕ_2 monotonically, as illustrated for $\chi = 0.52$. For every value of ϕ_2, there is only one value of Π representing thermodynamically stable homogeneous solution. On the other hand, if χ is greater than χ_c, the $\Pi - \phi_2$ curve is nonmonotonic, as illustrated for $\chi = 0.537$. This curve exhibits a maximum at ϕ_s' and a minimum at ϕ_s''. For polymer volume fractions in between ϕ_s' and ϕ_s'', the osmotic pressure decreases with an increase in polymer concentration. Since this behavior is aphysical, the system is thermodynamically unstable at polymer concentrations between ϕ_s' and ϕ_s''. The boundary of total instability is given by the locations of the maximum at ϕ_s' and the minimum at ϕ_s'', which are given by the condition of extrema as

$$\frac{\partial \Pi}{\partial \phi_2} = 0. \qquad (8.2.4)$$

The solutions with polymer concentrations at ϕ_s' and ϕ_s'' are called spinodal points. The envelope of ϕ_s' and ϕ_s'', as χ is varied, is called the spinodal curve, as indicated in Fig. 8.3a. Since osmotic pressure is a first derivative of free energy with respect to

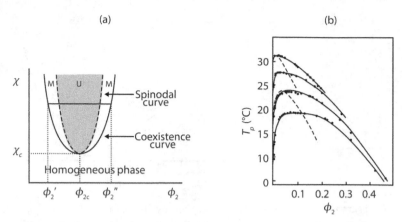

Figure 8.3 (a) Sketch of a phase diagram illustrating liquid–liquid phase separation for a solution of polymers with a fixed degree of polymerization N. For χ lower than the critical value χ_c, the solution is thermodynamically stable and exists as a single homogeneous phase. For $\chi > \chi_c$, liquid–liquid phase separation occurs resulting in two coexisting phases of compositions ϕ_2' and ϕ_2''. Inside the areas denoted as M between the coexistence curve and the spinodal curve, the solution is metastable. Inside the area U bounded by the spinodal curve, the solution is unstable. Outside the coexistence curve, the solution is homogeneous. The critical values χ_c and ϕ_{2c} are determined solely by the degree of polymerization N. (b) Coexistence curves of temperature (T_p) versus polymer concentration ϕ_2 for polystyrene solutions in cyclohexane. For this UCST system, χ is inversely related to T, and hence the orientation of the coexistence curves is opposite to that in (a). The molecular weights of polystyrene are 43.6, 89, 250, and 1270 kDa, respectively, from the bottom curve to the top curve. A comparison with theoretical predictions from the Flory–Huggins theory (dashed curves) is illustrated for two molecular weights [Adapted from Shultz & Flory (1952)].

concentration (Eqs. (6.2.7) and (6.2.10)), Eq. (8.2.4) gives the condition for obtaining the spinodal curve as

$$\frac{\partial^2(\Delta F_m/n_0 k_B T)}{\partial \phi_2^2} = 0, \qquad (8.2.5)$$

where ΔF_m is the free energy of mixing, and $n_0 v_1$ is the volume V of the solution. The spinodal curve gives the boundary of thermodynamic instability, and inside the spinodal curve, the system is unstable. The above condition is a general result, independent of which theory is used to obtain the free energy of mixing.

For polymer volume fractions outside the spinodal curve for $\chi > \chi_c$, the osmotic pressure increases with polymer concentration allowing two phases with the same value of osmotic pressure, meaning same value of chemical potential of the solvent. Among all pairs of values of ϕ_2 outside the range $\phi_s' \rightarrow \phi_s''$, only one unique pair of concentrations denoted by ϕ_2' and ϕ_2'' is allowed. Now the solution is said to undergo phase separation, with the two phases of concentrations ϕ_2' and ϕ_2'' coexisting in equilibrium. The unique choice of ϕ_2' and ϕ_2'' for a given $\chi > \chi_c$ arises from the criterion for thermodynamic equilibrium that chemical potentials of every component must be the same in all coexisting phases, given by

$$\mu_1(\phi_2') = \mu_1(\phi_2'')$$

$$\mu_2(\phi_2') = \mu_2(\phi_2''). \tag{8.2.6}$$

The first of these criteria is satisfied by any horizontal line in Fig. 8.2. But the horizontal line needs to be shifted vertically in order to satisfy the second criterion of Eq. (8.2.6). In solving Eq. (8.2.6) to get ϕ_2' and ϕ_2'', the expressions for μ_1 and μ_2 given by Eqs. (8.2.1) and (8.2.2) are employed. Instead of this procedure, ϕ_2' and ϕ_2'' can be alternatively determined by plotting $\Pi v_1/k_B T$ versus $1/\phi_2$ and using the equal-area theorem (Maxwell construction) [Maxwell (1875)] to locate the correct horizontal line connecting ϕ_2' and ϕ_2''. By repeating the calculation of ϕ_2' and ϕ_2'' at different values of χ, the coexistence curve is computed as sketched in Fig. 8.3a.

At the critical value $\chi_c = 0.532$, the dependence of osmotic pressure on the polymer concentration exhibits an inflection point denoted by the star symbol in Fig. 8.2. This point of inflection is the critical point for phase separation. At this point, both the first and second derivatives of Π with respect to ϕ_2 are zero. Therefore, the critical point for phase separation is given by

$$\frac{\partial \Pi}{\partial \phi_2} = 0 = \frac{\partial^2 \Pi}{\partial \phi_2^2}. \tag{8.2.7}$$

Since the osmotic pressure is related to the negative of the chemical potential μ_1 of the solvent and since μ_1 is already a first derivative of free energy of the solution (Eq. (6.2.7)), the general condition for the critical point is

$$\frac{\partial^2 (\Delta F_m/n_0 k_B T)}{\partial \phi_2^2} = 0 = \frac{\partial^3 (\Delta F_m/n_0 k_B T)}{\partial \phi_2^3}. \tag{8.2.8}$$

Using Eqs. (6.2.6) and (8.2.8), the critical values χ_c and ϕ_{2c} are given by

$$\frac{1}{1 - \phi_{2c}} + \frac{1}{N\phi_{2c}} - 2\chi_c = 0$$

$$\frac{1}{(1 - \phi_{2c})^2} - \frac{1}{N\phi_{2c}^2} = 0. \tag{8.2.9}$$

Solving these two equations for the two unknowns χ_c and ϕ_{2c}, we get

$$\chi_c = \frac{1}{2}\left(1 + \frac{1}{\sqrt{N}}\right)^2 \simeq \frac{1}{2} + \frac{1}{\sqrt{N}} + \cdots \tag{8.2.10}$$

and

$$\phi_{2c} = \left(\frac{1}{1 + \sqrt{N}}\right) \simeq \frac{1}{\sqrt{N}} + \cdots . \tag{8.2.11}$$

As seen in these equations, the critical values of the chemical mismatch parameter χ and the polymer concentration ϕ_2 are determined solely by the degree of polymerization of the polymer, according to the Flory–Huggins theory.

Following the above procedure to calculate the critical point, the spinodal curve, and the coexistence curve, phase diagrams for two-component polymer solutions are constructed using an empirically determined χ parameter which is specific to a particular polymer solution. A generic phase diagram is depicted in Fig. 8.3a, where χ is plotted against the polymer volume fraction ϕ_2. Above the critical point given by χ_c and ϕ_{2c}, liquid–liquid phase separation occurs when a solution is brought into the region bounded by the coexistence curve. The coexistence curve is the trace of ϕ_2' and ϕ_2'' corresponding to the polymer volume fractions of the two coexisting phases, as χ is varied. Outside the region of the coexistence curve, the solution exists as a homogeneous single phase. Inside the area bounded by the spinodal curve, the solution is thermodynamically unstable. When a solution is quenched into the spinodal region, phase separation occurs spontaneously through concentration fluctuations. The time evolution of such spontaneous fluctuations eventually leads to the two equilibrium phases of compositions ϕ_2' and ϕ_2''. This mechanism of phase separation is known as **spinodal decomposition**.

Inside the region between the coexistence and spinodal curves, the solution is metastable. If a solution is brought into this metastable region, phase separation occurs through the mechanism of **nucleation and growth**. Here, droplets (called nuclei) of the equilibrium phases, with concentration ϕ_2' or ϕ_2'', are nucleated in random locations in the "mother" solution. Once such droplets are stable enough against their surface tension, the droplets grow competitively among themselves until the two coexisting phases are fully formed. Independent of whether the spinodal decomposition or nucleation-and-growth mechanism is followed in terms of the kinetics of phase separation, the final state is the two coexisting phases with concentrations ϕ_2' and ϕ_2'' at any given value of χ above χ_c.

The general picture portrayed in Fig. 8.3a is valid for any two-component solution of uncharged polymers. The specificity of a polymer solution appears only through the chemical mismatch parameter χ. Once χ is determined experimentally for a particular polymer-solvent combination, its phase diagram can be predicted using the above described procedure. An example is given in Fig. 8.3b for an upper critical solution temperature (UCST) system, where χ is inversely related to temperature. In this case, the coexistence curve in a temperature versus ϕ_2 plot is upside down with respect to the χ versus ϕ_2 phase diagram. This is seen in Fig. 8.3b for solutions of polystyrene in cyclohexane, where the coexistence curves were determined by cloud point measurements. Data for four polystyrene molecular weights (43.6, 89, 250, and 1270 kDa) are shown in the figure. As the molecular weight increases, the coexistence curve moves toward higher temperatures for this UCST system. The dependence of the critical values of T_c and ϕ_{2c} on molecular weight are in qualitative agreement with the predictions from the Flory–Huggins theory given in Eqs. (8.2.10) and (8.2.11). On the other hand, if χ is an increasing function of temperature as in lower critical solution temperature (LCST) systems, then the $T - \phi_2$ coexistence curve maintains the same orientation as the $\chi - \phi_2$ coexistence curve. Now, the $T - \phi_2$ coexistence curve moves toward lower temperatures as the molecular weight of the polymer is increased. In general, experimental determination of the χ parameter shows the temperature dependence as

$$\chi_{\text{expt}} = \frac{a}{T} + b(T), \tag{8.2.12}$$

where a and b are empirical parameters specific to a particular system. The mapping of the Flory–Huggins $\chi - \phi_2$ phase diagrams into $T - \phi_2$ phase diagrams using χ_{expt} provides a variety of phase diagrams in qualitative agreement with experimentally determined phase diagrams, including the occurrence of both UCST and LCST behaviors. However, in simple polymer solutions, χ can be written as

$$\chi = \frac{\Theta}{2T}, \tag{8.2.13}$$

where Θ is the Flory temperature (Section 6.2.1) arising from the interaction parameters $\epsilon_{12}, \epsilon_{11}$, and ϵ_{22}.

The discussion in this section is only an illustration of how phase behavior in polymer solutions is addressed. Although we have focused only on the mean field theory of Flory and Huggins, the prescription for predicting phase diagrams for any system is general. To summarize, the protocol is as follows. First, we need to compose the free energy ΔF of the system based on the chosen models of the system. Next, the critical condition is determined from $\partial^2 \Delta F / \partial \phi_2^2 = 0 = \partial^3 \Delta F / \partial \phi_2^3$. The spinodal curve is given by $\partial^2 \Delta F / \partial \phi_2^2 = 0$, and the coexistence curve is given by the condition that the chemical potential of each component must be the same in all coexisting phases. The same protocol is adopted for multicomponent systems, which requires numerical computation. In the following sections, we shall only provide expressions for the free energy and the resulting phase diagrams without details of the computation.

8.2.1   Concentration Fluctuations and Landau–Ginzburg Theory

When the temperature of a homogeneous polymer solution with UCST behavior is lowered toward the condition of phase separation but still remaining in the homogeneous phase, concentration fluctuations become significant although they decay with time in order to maintain homogeneity of the solution over large distances. As we shall see below, these fluctuations diverge at the critical point and along the spinodal curve. The fluctuations in local polymer concentration lead to local inhomogeneities in the solution. In order to quantify the nature of these fluctuations, let us define an order parameter $\psi(\mathbf{r})$ as the deviation of the polymer concentration $\phi_2(\mathbf{r})$ at location $\mathbf{r}$ from the average value ϕ_2,

$$\psi(\mathbf{r}) = \phi_2(\mathbf{r}) - \phi_2. \tag{8.2.14}$$

The presence of any inhomogeneity, however weak it might be, is equivalent to formation of interfaces between adjacent regions in space. At the interface, there is a gradient in the local concentration. Therefore, the free energy of the solution depends on $\psi(\mathbf{r})$ and its local gradient $\nabla\psi(\mathbf{r})$. In general, the change in free energy ΔF to excite a fluctuation $\psi(\mathbf{r})$ is written in the Landau–Ginzburg form,

$$\frac{\Delta F}{k_B T} = \int \frac{d\mathbf{r}}{V} \left[\frac{A}{2}\psi(\mathbf{r})^2 + \frac{B}{4}\psi(\mathbf{r})^4 + \cdots + \frac{\kappa_0}{2}(\nabla\psi(\mathbf{r}))^2 + \cdots \right], \tag{8.2.15}$$

where higher order terms in $\psi(\mathbf{r})$ and its gradient are ignored. The coefficients A and B for a polymer solution with polymer concentration ϕ_2 and degree of polymerization N can be obtained by performing a Taylor series expansion on the free energy expression using the Flory–Huggins theory as

$$A = \frac{\partial^2 \Delta F_m / k_B T}{\partial \phi_2^2} = \frac{1}{(1 - \phi_2)} + \frac{1}{N\phi_2} - 2\chi, \qquad (8.2.16)$$

$$B = \frac{\partial^4 \Delta F_m / k_B T}{\partial \phi_2^4} = 2\left[\frac{1}{(1 - \phi_2)^3} + \frac{1}{N\phi_2^3}\right]. \qquad (8.2.17)$$

Expanding $\partial^2 \Delta F / \partial \phi_2^2$ further as a Taylor series expansion in $T - T_c$ or $T - T_s$ (where T_c and T_s are the critical temperature and spinodal temperature, respectively), we obtain from Eq. (8.2.16)

$$A = \left[\frac{1}{(1 - \phi_2)} + \frac{1}{N\phi_2} - 2\left(\frac{\partial}{\partial T}(T\chi)\right)_{T_s}\right]\left(\frac{T - T_s}{T_s}\right). \qquad (8.2.18)$$

Note that $T_s = T_c$ at the critical point. If χ is strictly proportional to $1/T$, then A reduces to

$$A = \left[\frac{1}{(1 - \phi_2)} + \frac{1}{N\phi_2}\right]\left(\frac{T - T_s}{T_s}\right), \qquad (8.2.19)$$

and can be rewritten as

$$A = a(T - T_s), \qquad (8.2.20)$$

where

$$a = \left[\frac{1}{(1 - \phi_2)} + \frac{1}{N\phi_2}\right]\frac{1}{T_s}. \qquad (8.2.21)$$

It is a common practice in polymer literature to define the "spinodal chi" χ_s from Eq. (8.2.16) as

$$2\chi_s = \left[\frac{1}{(1 - \phi_2)} + \frac{1}{N\phi_2}\right], \qquad (8.2.22)$$

so that when χ equals χ_s at the spinodal point, $A = 0$. If χ is of the form given in Eq. (8.2.13), then χ_s can be defined in terms of the spinodal temperature T_s (or T_c at the critical point) as

$$\frac{\Theta}{T_s} = \left[\frac{1}{(1 - \phi_2)} + \frac{1}{N\phi_2}\right]. \qquad (8.2.23)$$

Hence A follows from Eqs. (8.2.13) and (8.2.16) as

$$A = \Theta\left(\frac{1}{T_s} - \frac{1}{T}\right) = \Theta\frac{(T - T_s)}{T T_s}. \qquad (8.2.24)$$

Therefore A is proportional to the distance in T from the spinodal (critical) temperature T_s as we approach the spinodal point (onset of thermodynamic instability). The above equation is identical to Eq. (8.2.20) near the spinodal temperature. Note that the coefficient A changes sign from positive to negative as T decreases through T_s for UCST systems.

The square gradient term in Eq. (8.2.15) accounts for the penalty in free energy associated with the creation of interfaces. Since the interfaces are assumed to be very diffuse for weak concentration fluctuations, higher order terms in $\nabla \psi(\mathbf{r})$ are ignored. The square gradient term in the free energy expression is called van der Waals term, Ginzburg term, or Cahn–Hilliard term. The coefficient κ_0 is in general related to the range of interaction between different segments of the various molecules. For the case of polymers, it turns out that κ_0 is dominated by the entropic contributions arising from chain connectivity. The value of κ_0 dictates the interfacial tension between two coexisting phases. Based on the random phase approximation discussed in Chapter 6, κ_0 can be derived as [de Gennes (1979)]

$$\kappa_0 = \frac{\ell^2}{18} \frac{1}{\phi_2}. \tag{8.2.25}$$

Defining the Fourier transform of $\psi(\mathbf{r})$ according to

$$\psi(\mathbf{r}) = \sum_{\mathbf{k}} \psi_{\mathbf{k}} \exp(-i\mathbf{k} \cdot \mathbf{r}), \tag{8.2.26}$$

and

$$\psi_{\mathbf{k}} = \frac{1}{V} \int d\mathbf{r} \psi(\mathbf{r}) \exp(i\mathbf{k} \cdot \mathbf{r}), \tag{8.2.27}$$

we get

$$\frac{\Delta F}{k_B T} = \frac{1}{2} \sum_{\mathbf{k}} \left(A + \kappa_0 k^2 \right) |\psi_{\mathbf{k}}|^2. \tag{8.2.28}$$

Here we have kept only the leading terms in Eq. (8.2.15). Since the probability W to excite a fluctuation with the Fourier component $\mathbf{k}$ is given by the Boltzmann–Gibbs weight,

$$W \sim \exp \left[-\frac{1}{2} \left(A + \kappa_0 k^2 \right) |\psi_{\mathbf{k}}|^2 \right], \tag{8.2.29}$$

the mean square fluctuation with Fourier component $\mathbf{k}$, $\langle |\psi_{\mathbf{k}}|^2 \rangle$ is given as

$$\langle |\psi_{\mathbf{k}}|^2 \rangle = \frac{1}{A + \kappa_0 k^2}. \tag{8.2.30}$$

As mentioned in the preceding chapters, the left-hand side of this equation is the static structure factor $S(\mathbf{k})$ per unit volume with the scattering wave vector $\mathbf{k}$. The above form given by the Landau–Ginzburg theory is precisely of the Ornstein–Zernike form. Since the static structure factor is the Fourier transform of two-point correlation function $G(\mathbf{r})$ of the fluctuating order parameter, we obtain

$$G(\mathbf{r}) = \langle \psi(\mathbf{r}) \psi(\mathbf{0}) \rangle = \sum_{\mathbf{k}} \frac{\exp(-i\mathbf{k} \cdot \mathbf{r})}{A + \kappa_0 k^2}. \tag{8.2.31}$$

Converting the sum into an integral, we get

$$G(\mathbf{r}) = V \int \frac{d^3 k}{(2\pi)^3} \frac{\exp(-i\mathbf{k} \cdot \mathbf{r})}{A + \kappa_0 k^2}. \tag{8.2.32}$$

Performing the integral gives

$$G(\mathbf{r}) = \frac{V}{4\pi\kappa_0} \frac{e^{-r/\xi}}{r}, \qquad r \neq 0 \tag{8.2.33}$$

where ξ is given by

$$\xi = \sqrt{\frac{\kappa_0}{A}}. \tag{8.2.34}$$

ξ is the correlation length denoting the average distance over which the concentration fluctuations are significantly correlated. Using Eqs. (8.2.20), (8.2.21), and (8.2.25), we get

$$\xi^2 = \frac{\ell^2}{18} \left(\frac{1}{N} + \frac{\phi_2}{1 - \phi_2} \right)^{-1} \frac{T_s}{|T - T_s|}. \tag{8.2.35}$$

Therefore, the correlation length diverges as $T \to T_s$ according to

$$\xi \sim \frac{1}{|T - T_s|^{\nu}}, \tag{8.2.36}$$

where the critical exponent ν for the correlation length of concentration fluctuations based on the mean field Flory–Huggins theory is 1/2,

$$\nu = \frac{1}{2}. \qquad \text{(mean field Flory–Huggins theory)} \tag{8.2.37}$$

Substituting Eq. (8.2.34) in Eq. (8.2.30), the static structure factor follows as

$$S(\mathbf{k}) = \frac{1}{A} \frac{1}{(1 + k^2 \xi^2)}. \tag{8.2.38}$$

Since $A \to 0$ as T approaches the spinodal temperature, according to Eq. (8.2.20), the above equation predicts that the scattering intensity (which is proportional to $S(\mathbf{k})$) diverges at zero scattering wave vector. Using the relation between $S(\mathbf{k})$ and the isothermal compressibility per unit volume χ_T, Eq. (8.2.38) is rewritten as

$$S(\mathbf{k}) = \frac{k_B T \chi_T}{(1 + k^2 \xi^2)}. \tag{8.2.39}$$

The scattering intensity extrapolated at zero scattering wave vector $I(k \to 0)$ which is proportional to $1/A$ (and equivalently to χ_T) diverges as $T \to T_s$ according to

$$I(k \to 0) \sim \chi_T \sim \frac{1}{|T - T_s|^{\gamma}}, \tag{8.2.40}$$

where the critical exponent is given by the mean field Flory–Huggins theory as

$$\gamma = 1. \qquad \text{(mean field Flory–Huggins theory)} \tag{8.2.41}$$

The correlation length and the isothermal susceptibility are determined experimentally using the Ornstein–Zernike plot where the inverse of the scattering intensity is plotted versus the square of the scattering wave vector,

$$\frac{1}{I(\mathbf{k})} = \frac{1}{I(\mathbf{k} = 0)} \left(1 + k^2 \xi^2\right). \tag{8.2.42}$$

According to the predictions of the Flory–Huggins theory, Eqs. (8.2.20) and (8.2.36) show that $1/I(\mathbf{k} = 0) \sim |T - T_s|^\gamma$ and $\xi \sim |T - T_s|^{-\nu}$, with $\gamma = 1$ and $\nu = 1/2$. These results are observed for polymer solutions as long as the temperature is not too close to the spinodal temperature. However, deviations from these values of γ and ν are observed when the temperature is close to the spinodal or critical temperature. In fact, if higher order terms in Eq. (8.2.15) are taken into account, it can be rigorously shown that the critical phenomena of polymer solutions belong to the three-dimensional Ising universality class where the critical exponents are

$$\nu = 0.63 \quad \text{and} \quad \gamma = 1.26. \tag{8.2.43}$$

The experimental results for polymer solutions at temperatures close enough to criticality show the universal behavior given in Eq. (8.2.43). We shall see in Section 8.4.4 that polyelectrolyte solutions also belong to this universality class.

8.3 Coulomb Criticality in Electrolyte Solutions

Before discussing phase behavior of polyelectrolyte solutions, let us consider the nature of phase behavior if the solution contains only low molar mass salt ions without the presence of charged macromolecules. The phase behavior of simple electrolyte solutions is referred to as Coulomb criticality. This is addressed as follows: Consider a solution containing a certain amount of an electrolyte of type $C^{z_+}:A^{z_-}$, where z_+ and z_- are the valencies of the cation C and anion A, respectively. Our present goal is to derive an expression for the Helmholtz free energy F of the solution. We derive F by adopting the restricted primitive model of the Debye–Hückel theory to represent the electrolyte solution. Let ℓ be the diameter of the cation as well as the anion, and let there be n_+ cations in the solution. Since the solution is electrically neutral ($z_+ n_+ = z_- n_-$), the number of anions, n_-, is given as $n_- = z_+ n_+/z_-$. Let the total number of solvent molecules in the solution be n_0. We ignore any short-ranged interactions between the ions and solvent and among ions, except through the temperature dependence of the dielectric constant as discussed in Chapter 3. Adopting the lattice model for the solution, let every lattice site have volume ℓ^3. Assuming that the solution is incompressible, the total number of lattice sites is $n_t = n_+ + n_- + n_0$, and the volume of the system is $V = n_t \ell^3$. Therefore, volume fractions of the cation, anion, and solvent are given by

$$\phi_+ = \frac{n_+}{n_t} = \frac{n_+ \ell^3}{V}, \quad \phi_- = \frac{n_-}{n_t} = \frac{z_+}{z_-}\frac{n_+}{n_t} = \frac{z_+}{z_-}\phi_+, \quad \phi_0 = \frac{n_0}{n_t} = (1 - \phi_+ - \phi_-). \tag{8.3.1}$$

8.3.1   Free Energy

The free energy of the solution for this restricted primitive model of the Debye–Hückel theory is given by Eqs. (3.8.39)–(3.8.41) as

$$F = F_0 + F_{\text{fl,i}}, \tag{8.3.2}$$

where F_0 is the free energy contribution from the translational entropy of the ions and solvent molecules, and $F_{\text{fl,i}}$ is the Debye–Hückel free energy due to electrostatic correlations. F_0 is given by

$$\frac{F_0}{k_B T} = n_+ \ln \phi_+ + n_- \ln \phi_- + n_0 \ln \phi_0, \tag{8.3.3}$$

where the contribution from the solvent molecules is included to the result of Eq. (3.8.40) and constant terms are ignored. Using Eq. (8.3.1), the above equation becomes

$$\frac{F_0}{k_B T} = n_+ \ln \phi_+ + \frac{z_+}{z_-} n_+ \ln \phi_+ + \left(n_t - n_+ - \frac{z_+}{z_-} n_+ \right) \ln \left[1 - \left(1 + \frac{z_+}{z_-} \right) \phi_+ \right], \tag{8.3.4}$$

where the constant term $(z_+ n_+ / z_-) \ln(z_+ / z_-)$ is omitted.

The fluctuation contribution is given by Eq. (3.8.41) as

$$\frac{F_{\text{fl,i}}}{k_B T} = -\frac{V}{4\pi \ell^3} \left[\ln (1 + \kappa \ell) - \kappa \ell + \frac{1}{2} \kappa^2 \ell^2 \right], \tag{8.3.5}$$

where

$$\kappa^2 \ell^2 = \frac{e^2 \ell^2}{\epsilon_0 \epsilon k_B T} \left(z_+^2 \frac{n_+}{V} + z_-^2 \frac{n_-}{V} \right). \tag{8.3.6}$$

Using $\phi_+ = n_+ \ell^3 / V$ and $\phi_- = n_- \ell^3 / V$, and the definition of the Bjerrum length ℓ_B (Eq. (3.6.7)), we get

$$\kappa^2 \ell^2 = \frac{4\pi \ell_B}{\ell} z_+ (z_+ + z_-) \phi_+, \tag{8.3.7}$$

where we have used $\phi_- = z_+ \phi_+ / z_-$ (Eq. (8.3.1)).

The free energy density f defined as $F/(k_B T n_t)$ follows from Eqs. (8.3.4) and (8.3.5) as

$$f = \left(1 + \frac{z_+}{z_-} \right) \phi_+ \ln \phi_+ + \left[1 - \left(1 + \frac{z_+}{z_-} \right) \phi_+ \right] \ln \left[1 - \left(1 + \frac{z_+}{z_-} \right) \phi_+ \right]$$
$$- \frac{1}{4\pi} \left[\ln (1 + \kappa \ell) - \kappa \ell + \frac{1}{2} \kappa^2 \ell^2 \right], \tag{8.3.8}$$

where linear terms in ϕ_+ are ignored.

8.3.2   Instability and Criticality

The criterion for the onset of thermodynamic instability is obtained by calculating the spinodal curve given by (see Eq. (8.2.5))

$$\frac{\partial^2 f}{\partial \phi_+^2} = 0. \tag{8.3.9}$$

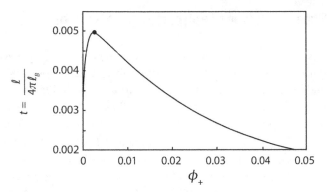

Figure 8.4 Spinodal curve for a solution of a monovalent electrolyte. The reduced temperature t is proportional to the temperature and temperature-dependent dielectric constant. ϕ_+ is the volume fraction of the cation.

The spinodal curve predicted by this equation is given in Fig. 8.4. The experimental variable T appears in Eq. (8.3.7) as its inverse through the Bjerrum length (Eq. (3.6.7)). Also, the dielectric constant ϵ appearing in ℓ_B is temperature dependent. Accounting for both of these features, let us define a reduced temperature t as

$$t \equiv \frac{\ell}{4\pi\ell_B}, \tag{8.3.10}$$

which is equal to $\epsilon_0\epsilon(T)k_BT\ell/e^2$. The calculated spinodal curve from Eqs. (8.3.8) and (8.3.9) for monovalent electrolytes is presented in Fig. 8.4 as a graph of t versus ϕ_+.

The critical point for the onset of instability follows from

$$\frac{\partial^2 f}{\partial\phi_+^2} = 0 = \frac{\partial^3 f}{\partial\phi_+^3}. \tag{8.3.11}$$

Substituting Eq. (8.3.8) in the above equations, the critical point denoted by ϕ_{+c} and t_c follows as [Stell *et al.* (1976), Fisher & Levin (1993)]

$$\phi_{+c} = \frac{1}{64\pi}\frac{z_-}{z_+ + z_-}$$

$$t_c = \frac{1}{64\pi}z_+z_-. \tag{8.3.12}$$

In obtaining the above results for the critical point, we have taken the limit of $\phi_+ \ll 1$, which is a justifiable approximation based on the spinodal curve given in Fig. 8.4. The critical temperature is given explicitly as

$$(\epsilon T)_c = \frac{z_+z_-}{64\pi}\frac{e^2}{\epsilon_0 k_B T\ell}. \tag{8.3.13}$$

Although the cations and anions are identical in charge valency and size, the spinodal curve is asymmetric in shape due to electrostatic correlations among ions. This asymmetry in phase diagram, characteristic of Coulomb criticality, is unlike the phase diagrams for uncharged small molecular mixtures.

8.4   Phase Behavior of Polyelectrolyte Solutions

The chemical backbones of most of synthetic polyelectrolytes, in the absence of any ionized monomers, are usually insoluble in polar solvents. As a result, the solution undergoes phase separation arising from unfavorable values of the χ parameter for miscibility. Let us refer to this macrophase separation phenomenon as "hydrophobicity criticality." On the other hand, when the polymer carries a significant number of ionized groups, the polymer can readily dissolve in a polar solvent due to electrostatic repulsion between the chains. If van der Waals type forces, collectively called the hydrophobic effect, can be neglected in comparison with the electrostatic effect, the solution can exhibit Coulomb criticality as described above. In general, both hydrophobic and electrostatic interactions are significant contributors to the phase behavior of most of polyelectrolyte solutions. As a result, the phase behavior of polyelectrolyte solutions, due to variations in composition and temperature, is theoretically addressed using the various χ parameters between the components (which are measures of hydrophobicity) and the Bjerrum length ℓ_B, which is a measure of the strength of the electrostatic interaction [Muthukumar (2002), Lee & Muthukumar (2009), Muthukumar *et al.* (2010)]. Therefore, we expect the phase diagrams of polyelectrolyte solutions to exhibit intermediate behaviors between the hydrophobicity criticality, as addressed using the Flory–Huggins theory, and the Coulomb criticality introduced in the preceding section.

The χ parameter and ℓ_B cannot be arbitrarily varied for any polyelectrolyte system because both of these quantities depend on temperature (see Eqs. (1.3.2) and (8.2.13)). Therefore, we shall write χ in terms of ℓ_B using a reduced temperature. As defined in Eq. (8.3.10), let t be the reduced temperature, $t = \ell/(4\pi\ell_B)$, since ℓ_B is inversely proportional to $T\epsilon$. The relation between the temperature T and the reduced temperature t depends on the choice of the value of the charge separation length ℓ along the backbone, and the temperature dependence of the dielectric constant of the solution. Choosing the χ parameter between the polymer and solvent as given by Eq. (8.2.13), χ is written in terms of the reduced temperature t as

$$\chi \equiv \frac{a_\chi}{20\pi t}. \tag{8.4.1}$$

The parameter $a_\chi (= 10\pi\epsilon_0\epsilon k_B\Theta/e^2)$ reflects the magnitude of the Flory Θ temperature and is taken as a measure of the hydrophobic interaction between the polyelectrolyte backbone and solvent. For the example of sodium poly(styrene sulfonate) in water, the charge separation along the backbone is ~ 0.25 nm and ℓ_B at room temperature is ~ 0.7 nm, so that $t \sim 1/(12\pi)$. Therefore, when $a_\chi = 1$, χ is about 3/5 indicating correctly that the polymer backbone is immiscible in water. The range of $0 \le a_\chi \le 1$ is reasonable for exploring the role of the hydrophobic effect on the phase behavior of aqueous polyelectrolyte solutions.

Since the translational entropy of counterions dominate the free energy of polyelectrolyte solutions over that of polymer chains, the molecular weight of the polyelectrolyte chains is not the key parameter in determining the phase diagram. This kind of Coulomb criticality is in sharp contrast to the hydrophobic criticality of phase diagrams of uncharged polymers (see Eq. (8.2.10)). Furthermore, these two kinds of criticality can interfere with each other and lead to multicritical phenomena. As already noted, presence of low molecular weight salt can dramatically complicate phase behavior. These issues are addressed below.

8.4.1 Salt-Free Polyelectrolyte Solutions

Consider a two-component solution of flexible polyelectrolyte chains in a polar solvent. Let the solution consist of n polyelectrolyte chains, each with N Kuhn segments dispersed in the solvent with n_0 molecules. Let α be the effective degree of ionization and z_p be the effective valency of each segment so that each segment has an effective charge $z_p\alpha$. To maintain the electroneutrality of the solution, the total number of unadsorbed counterions in the solution is $\alpha z_p nN/z_c$, with z_c being the valency of the counterion. Also, for simplicity, let us assume that all ions, polymer segments and solvent molecules have identical volume ℓ^3 (where ℓ is the Kuhn length).

As described in Section 8.1, the Helmholtz free energy density f defined as $f \equiv F\ell^3/(Vk_BT)$ is obtained by collecting contributions from translational entropy, energy, and fluctuations from ions and polymer concentration as

$$f = \frac{F\ell^3}{Vk_BT} = f_S + f_H + f_{\text{fl,i}} + f_{\text{fl,p}}. \tag{8.4.2}$$

f_S in Eq. (8.4.2) represents the mixing entropy of ions, solvent molecules, and polymer chains,

$$f_S = \frac{\phi}{N}\ln\phi + \phi_c\ln\phi_c + \phi_0\ln\phi_0, \tag{8.4.3}$$

where $\phi = nN\ell^3/V$ is the volume fraction of the polymer, $\phi_c = \alpha_1\phi$ is the volume fraction of counterions from dissociated polyelectrolyte chains ($\alpha_1 = \alpha z_p/z_c$), and ϕ_0 is the volume fraction of solvent. Using the electroneutrality condition, $\phi_c = \alpha_1\phi$ and the incompressibility constraint, $\phi + \phi_c + \phi_0 = 1$, we get

$$f_S = \left(\frac{1}{N} + \alpha_1\right)\phi\ln\phi + [1 - (1+\alpha_1)\,\phi]\ln\left[1 - (1+\alpha_1)\,\phi\right]. \tag{8.4.4}$$

The enthalpy part f_H in Eq. (8.4.2) is given by

$$f_H = \left(\frac{1}{2}w_{pp}\phi^2 + \frac{1}{2}w_{ss}\phi_0^2 + w_{ps}\phi\phi_0\right) + \phi^2\left(\frac{2\pi\alpha^2 z_p^2\ell_B}{\kappa^2\ell^3}\right) \tag{8.4.5}$$

with

$$\kappa^2\ell^2 = 4\pi\frac{\ell_B}{\ell}\left(z_c^2\phi_c\right). \tag{8.4.6}$$

The enthalpy part of the free energy density represents the mean-field energy that includes the short-range interactions among solvent and neutral polyelectrolyte segments as well as the electrostatic interactions among polyelectrolyte segments. The short-range interactions can be represented via the Flory–Huggins parameter χ. The effective electrostatic energy between two charged polyelectrolyte segments has the Yukawa form $v(r) = \alpha^2 z_p^2 \ell_B \exp(-\kappa r)/r$ with the inverse screening length κ defined by Eq. (8.4.6). Using the definition of χ (Eq. (6.2.5)), f_H becomes

$$f_H = -\chi \phi^2, \qquad (8.4.7)$$

where we have ignored terms linear in ϕ, which do not contribute to phase behavior.

The free energy density due to ion fluctuations $f_{\mathrm{fl,i}}$ is given by

$$f_{\mathrm{fl,i}} = -\frac{1}{4\pi} \left[\ln(1 + \kappa\ell) - \kappa\ell + \frac{1}{2}(\kappa\ell)^2 \right], \qquad (8.4.8)$$

where ℓ is the diameter of the ions. As in the Debye–Hückel theory, this equation is derived by solving a linearized Poisson–Boltzmann equation for a non-neutral charged plasma background [Muthukumar (2002)]. Note that the Debye–Hückel limiting law is recovered if the ion size approaches zero so that $f_{\mathrm{fl,i}} \to -(\kappa\ell)^3/(12\pi)$.

Finally, the free energy density due to polymer fluctuations in Eq. (8.4.2), $f_{\mathrm{fl,p}}$, as derived by Muthukumar (1996a) for both high salt and low salt limits, can be written as a simple interpolation (Section 6.3.6). It is given by Eq. (6.3.38) as

$$f_{\mathrm{fl,p}} = \frac{\dfrac{2^{3/4}}{9} \sqrt{\dfrac{\pi}{3}} \left(\dfrac{3}{2}\right)^{-9/4} (4\pi z_p^2 \alpha^2 \ell_B/\ell)^{3/4} \phi^{9/4}}{(\kappa\ell)^{3/2} + 2^{5/4} \sqrt{\dfrac{\pi}{3}} \left(\dfrac{3}{2}\right)^{-3/4} (4\pi z_p^2 \alpha^2 \ell_B/\ell)^{1/4} \phi^{3/4}}. \qquad (8.4.9)$$

8.4.1.1   Insignificance of Polymer Molecular Weight in Polyelectrolyte Phase Behavior

Using Eqs. (8.4.2)–(8.4.9), the critical point (ϕ_c, t_c) is determined from $\partial^2 f/\partial\phi^2 = 0 = \partial^3 f/\partial\phi^3$ as [Muthukumar (2002)]

$$\phi_c = \frac{1}{64\pi} \frac{1}{(\alpha_1 + \frac{1}{N})}; \qquad t_c = \frac{1}{64\pi} \frac{\alpha z_p z_c}{(\alpha_1 + \frac{1}{N})}. \qquad (8.4.10)$$

Here, the electrostatic contribution is taken to be dominant by assuming that $\chi = 0$. Since $\alpha_1 \ (= \alpha z_p/z_c)$ is of the order of 3 (by assuming that there are about 9 monomers per Kuhn segment, the average degree of ionization is 1/3, and the counterion is monovalent), $1/N$ in Eq. (8.4.10) is negligible in comparison with α_1. Therefore, the theory of phase diagrams of salt-free polyelectrolyte solutions predicts that the molecular weight of polyelectrolyte chains is an insignificant variable, in sharp contrast with the situation of uncharged polymers. This prediction is difficult to verify using experiments on aqueous solutions where salt-free solutions are homogeneous at large length scales. However, computer simulations [Orkoulas et al. (2003)] have validated the independence of phase behavior on molecular weight in salt-free solutions.

8.4.1.2   Interference between Coulomb Criticality and Hydrophobicity Criticality

If the hydrophobic term involving χ were to be dominant in Eq. (8.4.2), say in the limit of $\alpha \to 0$, the critical condition for phase separation is that of the Flory–Huggins theory (see Eqs. (8.2.10) and (8.2.11)),

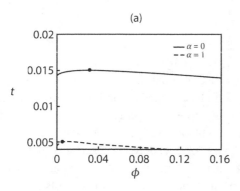
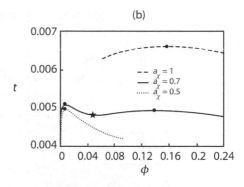

Figure 8.5 Coexistence curves for a salt-free polyelectrolyte solution in terms of the reduced temperature t versus polymer volume fraction ϕ. (a) $\alpha = 0$ and 1, for $a_\chi = 0.5$. The filled circles denote the critical points. (b) Crossover behavior between hydrophobicity criticality and Coulomb criticality and the emergence of a triple point ($\star$) for intermediate values of a_χ. The filled circles denote the critical points [Adapted from Lee & Muthukumar (2009)]

$$\phi_c = \frac{1}{\sqrt{N}+1}; \qquad \chi_c = \frac{1}{2}\left(1 + \frac{1}{\sqrt{N}}\right)^2. \qquad (8.4.11)$$

Defining $\tilde{\phi} = 64\pi\phi$ and $\tilde{t} = 64\pi t$, the hydrophobicity critical point (for $N = 1000$ and $a_\chi = 0.5$) is [Lee & Muthukumar (2009)]

$$\tilde{\phi}_c = 6.16 \qquad \text{and} \qquad \tilde{t}_c = 3.0, \qquad (\alpha = 0) \qquad (8.4.12)$$

where Eq. (8.4.1) is used in relating $\tilde{t}_c$ and χ_c. On the other hand, the Coulomb critical point for $\alpha = 1$ and $z_p = 1 = z_c$ is

$$\tilde{\phi}_c = 1, \qquad \text{and} \qquad \tilde{t}_c = 1. \qquad (\alpha = 1) \qquad (8.4.13)$$

These two limiting behaviors for $\alpha = 0$ and $\alpha = 1$ are shown in Fig. 8.5a, where the hydrophobicity parameter a_χ is 0.5. Therefore, the Coulomb critical temperature is reduced by a factor of three in comparison with the critical temperature for uncharged polymer solutions, which is usually around 300 K or above, but not 1800 K. As a result, it is impossible to observe macrophase separation in salt-free aqueous polyelectrolyte solutions.

The calculation [Lee & Muthukumar (2009)] of phase diagrams using Eqs. (8.4.2)–(8.4.9) shows that the two limits of critical points (corresponding to hydrophobicity criticality and Coulomb criticality) do not meet at an intermediate point. Instead they stop as critical end points with a narrow range of the hydrophobicity parameter a_χ, along with triple points, as shown in Fig. 8.5b for the degree of ionization $\alpha = 1$. For large values of a_χ, the phase diagram is dominated by the van der Waals interactions, and is close to that from the Flory–Huggins theory, as seen for $a_\chi = 1$. As a_χ is reduced, the electrostatic contribution becomes increasingly important. For $a_\chi = 0.5$, the phase behavior appears as Coulomb criticality. For the intermediate value $a_\chi = 0.7$, there are two critical points corresponding to hydrophobicity criticality and Coulomb criticality, and a triple point where three phases coexist. The theoretically predicted emergence of triple points for intermediate values of the χ parameter is yet to be verified by either simulations or experiments.

8.4.2   Polyelectrolyte Solutions with Salt

The presence of added salt to polyelectrolyte solutions raises the critical temperature by promoting hydrophobicity criticality and by screening electrostatic interactions. Rich phase behaviors emerge due to the constraint that the chemical potential of the salt must be the same in the coexisting phases, in addition to the same constraint for the polyelectrolyte and solvent. As an example, the calculated coexistence curves [Lee & Muthukumar (2009)] and critical points from Eqs. (6.3.33)–(6.3.38) with added monovalent salt are given in Fig. 8.6 for representative values of the hydrophobicity parameter a_χ (1/5, 4/9, and 1) and for $\alpha = 1$ and $N = 100$. Here $c_p = \phi + \phi_c, c_s = \phi_+ + \phi_-$ and the dissociated salt ions and the counterions of the polyelectrolyte are taken as distinct species. For relatively lower values of $a_\chi = 1/5$ (corresponding to χ of about 0.1 at room temperature, which is below χ_c for hydrophobicity criticality), the phase demixing region grows out from the left side of the plot and leans toward the bottom right side. For each temperature, the coexistence curve consists of two branches which are linked by tie lines and a critical point exists at the point where the two branches meet. For a relatively larger value of the hydrophobicity parameter, $a_\chi = 1$, the coexistence curves move up from a lower c_s and the region of homogeneity is reduced as the temperature is lowered. For intermediate values of a_χ (= 4/9), two sets of coexistence curves are predicted at higher temperatures (say $t = 0.004944$), each with its own critical point (Fig. 8.6b). As the temperature is lowered, the two demixing regions become broader and eventually merge into one demixing region. As in the salt-free case, triple points are predicted at lower temperatures.

If the added salt contains multivalent ions, the phase behavior becomes extremely rich. For example, when the added salt ion is trivalent, as in the case of $LaCl_3$, the predicted phase diagram from Eqs. (6.3.33) to (6.3.38) is given in Fig. 8.7 for $t = 0.03$ [Lee & Muthukumar (2009)]. In this figure, each dot represents an initial set of polymer and salt volume fractions that leads to liquid–liquid phase separation. The outer envelope of all points is the coexistence curve. The predicted liquid–liquid phase separation is analogous to the experimental observations at room temperature given in Fig. 1.10, although this effect was marked as reentrant precipitation [Olvera de la Cruz *et al.* (1995), Sabbagh & Delsanti (2000)].

Although the theoretically predicted phase diagrams for polyelectrolyte solutions with added salt show rich behaviors, and qualitatively match with those found experimentally, there is presently a gap in making quantitative comparisons between experiments and theory. For example, while experimental data given in Fig. 8.1 amply demonstrate the distinct role played by the specificity of electrolyte ions, the theoretical discussion given above treats all ions of a given valency as the same. In addition, experimental investigations have mainly focused on $c_s - c_p$ phase diagrams at room temperature, and studies of temperature dependence of phase behavior for polyelectrolyte solutions are rare. At present, only generic features of multicritical phenomena can be gleaned from theory. To enable fruitful predictions of experimental facts on polyelectrolyte phase behavior, the specificity of ions must be addressed in theoretical formulations, which continues to be a challenge. In addition, the change in

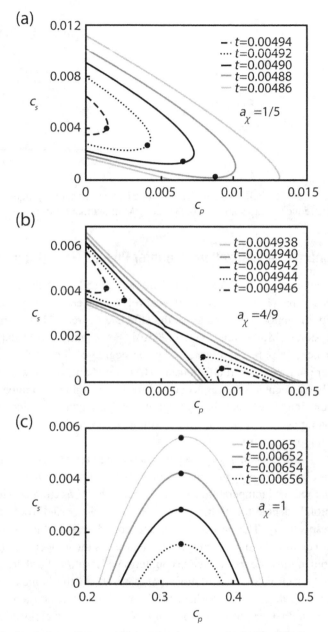

Figure 8.6 Calculated phase diagrams for a ternary system of polyelectrolyte, monovalent salt, and water ($\alpha = 1, N = 100$). The χ parameter increases from (a) to (c) [Adapted from Lee & Muthukumar (2009)].

degree of ionization of a polymer with changes in experimental conditions (which in turn depends on ion specificity) can significantly affect phase behavior. Adopting the parametrization of ion specificity as the parameter δ described in Section 5.2, let us now discuss the role of charge regularization in polyelectrolyte phase behavior.

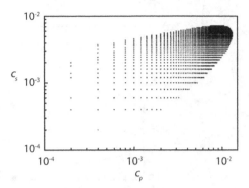

Figure 8.7 Calculated phase diagram for trivalent salt showing liquid–liquid phase separation at the reduced temperature $t = 0.03$ [Lee & Muthukumar (2009)].

8.4.3   Charge Regularization: Daughter Phases Have Different Charge Densities

The omnipresent charge regularization of the effective charge of polyelectrolyte chains accompanying changes in polyelectrolyte concentration, temperature, etc., leads to an additional order parameter as the effective degree of ionization, unseen in uncharged systems [Muthukumar *et al.* (2010)]. For example, consider a salt-free solution of a flexible polyelectrolyte and let the degree of ionization α change when temperature and polyelectrolyte concentration are changed in a self-consistent manner. This charge regularization is accounted for by changing the temperature through the Bjerrum length ℓ_B and the hydrophobicity parameter a_χ, and the dielectric mismatch parameter δ (to account for local ionization equilibria, Eq. (5.2.15), which depends on ion specificity).

The significant role played by charge regularization on phase behavior is illustrated in Fig. 8.8. In Fig. 8.8a, the calculated coexistence curve for $\delta = 2$ is given in terms of the reduced temperature t and polymer volume fraction ϕ. The filled circles and diamonds denote the higher concentration ϕ_b phase and lower concentration ϕ_a phase, respectively. The dashed, solid, and dotted vertical lines denote the polyelectrolyte concentration in the homogeneous phase. As the temperature is reduced, the degree of ionization α decreases due to counterion adsorption. Furthermore, α is higher at lower polyelectrolyte concentrations. As a result, in the homogeneous phase, the dependencies of α for the dashed, solid, and dotted lines on temperature are as shown in Fig. 8.8b. Once phase separation takes place, α increases as the polymer-poor phase becomes more and more diluted when the temperature is reduced. This is shown by the diamond symbol (α_a) in Fig. 8.8b. On the other hand, α decreases in the polymer-rich phase as the temperature is decreased as shown by the circle symbol (α_b) in Fig. 8.8b.

The difference in α between the daughter phases ($\Delta\alpha$) is plotted in Fig. 8.9 as the critical temperature is approached from below. The disparity in the charge of the daughter phases depends on the dielectric mismatch parameter δ (Fig. 8.9a) and the hydrophobicity parameter a_χ (Fig. 8.9b). If the dielectric mismatch parameter δ is decreased, there are increasingly more unabsorbed counterions stabilizing the homogeneous phase due to their increased translational entropy, as evident from

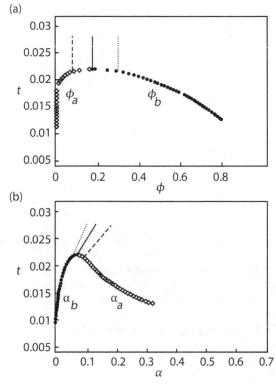

Figure 8.8 (a) Calculated phase diagram for a salt-free solution, with charge regularization. (b) Degree of ionization of the daughter phases are different [Adapted from M. Muthukumar *et al.* (2010)].

Fig. 8.9a. The corresponding difference in the polymer concentration $\Delta\phi$ between the polymer-poor and polymer-rich phases is also given in Fig. 8.9.

As expected from mean field theories, $\Delta\phi$ vanishes as

$$\Delta\phi \sim |t - t_c|^{1/2}, \tag{8.4.14}$$

when the critical point is approached. Analogously, $\Delta\alpha$ also vanishes as $t \rightarrow t_c$,

$$\Delta\alpha \sim |t - t_c|^{\beta_c}. \tag{8.4.15}$$

However, whether a universal critical exponent β_c can be identified with this additional order parameter α is yet to be established.

If charge regularization is omitted, then the coexistence curve is substantially lowered in t in comparison with the curve in Fig. 8.8a. Theoretical calculations show that there are significant modifications in phase behavior (substantially higher critical temperature and wider demixing region) by charge regularization, in comparison to computed phase behavior without charge regularization. This feature of additional cooperativity from charge regularization is true for all polyelectrolyte phase diagrams and must be addressed in quantitative comparisons between experiments and theory.

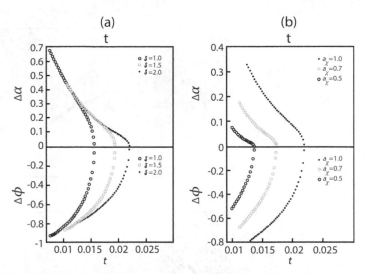

Figure 8.9 The differences in polymer concentration and degree of ionization approach zero differently as the critical point is approached. (a) Effect of dielectric mismatch parameter. (b) Effect of hydrophobicity parameter [M. Muthukumar *et al.* (2010)].

8.4.4   Concentration Fluctuations and Ginzburg Criterion

As seen in Section 8.2.1, when a solution of uncharged polymers is brought from a homogeneous phase to the proximity of the critical point or spinodal curve, composition fluctuations can be large and their correlation length ξ and zero-angle scattering intensity $I(0)$ diverge. Using the definition of χ in Eq. (8.2.13) and writing the spinodal χ_s as $\sim 1/T_s$, we recall from Section 8.2.1 that

$$\xi \sim \left(\frac{1}{T_s} - \frac{1}{T}\right)^{-\nu} ; \qquad I(0) \sim \left(\frac{1}{T_s} - \frac{1}{T}\right)^{-\gamma} . \qquad (8.4.16)$$

Very close to the critical point (or spinodal curve), the critical exponents ν and γ are 0.63 and 1.26, respectively. However, for conditions slightly farther from thermodynamic instability, mean field theories are adequate with $\nu = 1/2$ and $\gamma = 1$. The Ginzburg criterion [Robertson (1993)] for validity of the mean field theory is

$$A_G \ll \left|\frac{T - T_s}{T_s}\right| \ll 1, \qquad (8.4.17)$$

where A_G is a system-dependent constant.

This general picture of three-dimensional Ising universality class is valid for polyelectrolyte solutions as well. In the presence of modest amounts of divalent added salt (BaCl$_2$), aqueous solutions of sodium poly(styrene sulfonate) exhibit liquid–liquid phase separation. For this system, small angle neutron scattering experiments have been used to monitor concentration fluctuations in the homogeneous phase near the critical point (Fig. 8.10). As we have seen in Chapter 5, the presence of salt can be addressed by defining an effective χ parameter χ_{eff} as

$$\chi_{\text{eff}} = \chi - \frac{2\pi\alpha^2 z_p^2 \ell_B}{\kappa^2 \ell^3} . \qquad (8.4.18)$$

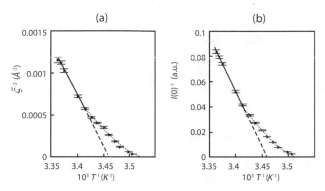

Figure 8.10 Concentration fluctuations in polyelectrolyte solutions near critical point obey 3-D Ising behavior. Away from the critical point by about 4°C, mean field behavior is observed for the correlation length ξ and scattered intensity extrapolated at zero angle [Adapted from Prabhu *et al.* (2003)].

If the added salt is monovalent, and if its number concentration c_s is large enough in comparison with the counterion concentration from the polyelectrolyte, then χ_{eff} can be written as

$$\chi_{\text{eff}} = \chi - \frac{\alpha^2 z_p^2}{4 c_s \ell^3}. \tag{8.4.19}$$

Substituting Eq. (8.4.18) into Eq. (8.2.16) and following the same steps as in Section 8.2.1, a straightforward expectation for polyelectrolyte solutions, with $\chi = \Theta/2T$ and $\chi_s = \Theta/2T_s$, is

$$\xi \sim \left(\frac{1}{T_s} - \frac{1}{T} + \frac{4\pi \alpha^2 z_p^2 \ell_B}{\Theta \kappa^2 \ell^3} \right)^{-\nu}; \quad I(0) \sim \left(\frac{1}{T_s} - \frac{1}{T} + \frac{4\pi \alpha^2 z_p^2 \ell_B}{\Theta \kappa^2 \ell^3} \right)^{-\gamma} \tag{8.4.20}$$

Small angle neutron scattering experiments on polyelectrolyte solutions provide plots of ξ and $I(0)$ versus $1/T$ as illustrated in Fig. 8.10 [Prabhu *et al.* (2003)]. Analysis of these data give $\nu = 0.63$ and $\gamma = 1.26$, as in the case of mixtures of uncharged polymers near their critical point for unmixing. Thus the critical phenomenon of liquid–liquid phase separation in polyelectrolyte solutions belongs to the same universality class as the three-dimensional Ising model. For temperatures sufficiently far from the critical temperature, the mean field exponents, $\nu = 1/2$, and $\gamma = 1$, are observed. The crossover region from Ising to mean field behaviors is about 4°C, as shown in Fig. 8.10.

8.5 Kinetics of Phase Separation

The study of kinetics of liquid–liquid phase separation in solutions of charged macromolecules remains largely an unchartered area of research. On the other hand, mechanisms of kinetics of macrophase separation in uncharged systems is well established [Gunton *et al.* (1983), Debenedetti (1996)]. When a homogeneous two-component solution is brought into thermodynamically unstable conditions, it undergoes macrophase separation via the spinodal decomposition mechanism (Section 8.2). If the solution is brought into a metastable region, the mechanism of phase separation is nucleation and growth. These general mechanisms of liquid–liquid phase separation are valid for polyelectrolyte systems as well, with modifications from electrostatic correlations.

8.5.1 Spinodal Decomposition

Consider a solution of polyelectrolytes of N segments and average polymer concentration ϕ_2, containing added salt. As described in Sections 8.2.1 and 8.4.4, concentration fluctuations in the homogeneous phase $\phi_2(\mathbf{r}) - \phi_2$ (where $\phi_2(\mathbf{r})$ is the local polymer concentration) become progressively stronger as the stability limit is approached. When the solution is quenched into the thermodynamically unstable state (Fig. 8.3a), it undergoes **spontaneous** phase separation, called spinodal decomposition. The local concentration $\phi_2(\mathbf{r})$ evolves with time t until it reaches the ultimate equilibrium values ϕ_2' and ϕ_2'' (Fig. 8.3a), as sketched in Fig. 8.11.

The time evolution of $\phi_2(\mathbf{r},t)$ is obtained by studying the continuity equation for $\phi_2(\mathbf{r},t)$,

$$\frac{\partial}{\partial t}\phi_2(\mathbf{r},t) = -\nabla \cdot \mathbf{J}, \tag{8.5.1}$$

where $\mathbf{J}$ is the flux of the molecules going into a volume element at $\mathbf{r}$, given by the Onsager law [Berne & Pecora (1976)]

$$\mathbf{J} = -\Lambda\nabla\left(\frac{\mu}{k_B T}\right). \tag{8.5.2}$$

$\nabla\mu$ is the chemical potential gradient at $\mathbf{r}$ (which drives the phase separation), and Λ (proportional to the diffusion coefficient of the polymer) is the Onsager coefficient. The chemical potential μ is the functional derivative $\delta\Delta F/\delta\phi_2(\mathbf{r})$, where ΔF is the free energy of the system. In general, ΔF can be expressed in terms of the order parameter $\psi(\mathbf{r}) = \phi_2(\mathbf{r}) - \phi_2$ as given in Eq. (8.2.15) with the coefficients A, B, and κ_0 assuming the appropriate values for a particular system.

For the situation of a polyelectrolyte solution with added salt, ΔF is given by Eq. (8.2.15) as

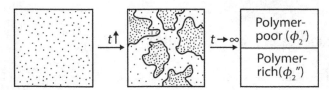

Figure 8.11 Sketch of spinodal decomposition where concentration fluctuations occur spontaneously. The domains of polymer-rich and polymer-poor fluctuations attain their eventual equilibrium concentrations of the daughter phases only at the very late stage of kinetics.

$$\frac{\Delta F}{k_B T} = \frac{1}{2} \int \frac{d\mathbf{r}}{V} \left[A\psi(\mathbf{r})^2 + \kappa_0 \left(\nabla \psi(\mathbf{r}) \right)^2 + \cdots \right], \tag{8.5.3}$$

where the higher order terms in $\psi(\mathbf{r})$ are ignored. This approximation is adequate to study the early stage of spinodal decomposition. κ_0 is given by Eq. (8.2.25) and A is given by

$$A = \frac{1}{(1 - \phi_2)} + \frac{1}{(N\phi_2)} - 2\chi_{\mathrm{eff}}, \tag{8.5.4}$$

where χ_{eff} is defined in Eqs. (8.4.18) and (8.4.19). Substituting the Fourier transform of $\psi(\mathbf{r})$, defined in Eq. (8.2.27) in the continuity equation for $\phi(\mathbf{r}, t)$, Eqs. (8.5.2) and (8.5.3) give

$$\frac{\partial}{\partial t} \psi_k(t) = -\Lambda k^2 \left(A + \kappa_0 k^2 \right) \psi_k(t) + \cdots. \tag{8.5.5}$$

The solution of this linear equation in ψ_k is

$$\psi_k(t) = \psi_k(0) e^{\Omega_k t}, \tag{8.5.6}$$

where the rate Ω_k is given as

$$\Omega_k = -\Lambda k^2 \left(A + \kappa_0 k^2 \right). \tag{8.5.7}$$

In the homogeneous phase, χ_{eff} is less than the spinodal value χ_s (see Eq. (8.2.22)) so that A is positive and the rate Ω_k is negative. Therefore, the polymer concentration fluctuations in the homogeneous phase decay with time. On the other hand, for $A < 0$, namely $\chi_{\mathrm{eff}} > \chi_s$, the fluctuations can grow with time. The dependence of the rate Ω_k on the wave vector k is sketched in Fig. 8.12. The concentration fluctuations grow with time only for wave vectors smaller than the critical value k_c and decay for $k > k_c$. Therefore, fluctuations with larger wavelengths (lower k values) grow at the cost of those with smaller wavelengths (higher k values). The growth rate is maximal at k_m. It follows from Eq. (8.5.7) that k_c and k_m are given by

$$k_c = \sqrt{\frac{|A|}{\kappa_0}} \quad \text{and} \quad k_m = \frac{1}{\sqrt{2}} \sqrt{\frac{|A|}{\kappa_0}}. \tag{8.5.8}$$

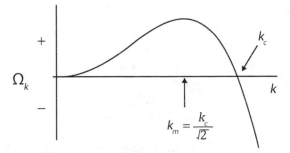

Figure 8.12 Wave vector dependence of the rate of change in concentration fluctuation in the linear regime of spinodal decomposition. Fluctuations grow for $k < k_c$ with a maximum rate at k_m. The spontaneously selected dominant domain size is $2\pi/k_m$.

Therefore, spontaneous selection of a characteristic length scale ($\sim 2\pi/k_m$) emerges for concentration fluctuations in the spinodal region.

Since the scattered intensity $I(\mathbf{k},t)$ is proportional to the mean square fluctuation in polymer concentration, we obtain

$$I(\mathbf{k},t) \sim \langle|\psi_{\mathbf{k}}(t)|^2\rangle = \langle|\psi_{\mathbf{k}}(0)|^2\rangle \exp(2\Omega_{\mathbf{k}}t), \tag{8.5.9}$$

where $\Omega_{\mathbf{k}}$ is given in Eq. (8.5.7). Therefore, in the early stage of spinodal decomposition where nonlinear terms in Eq. (8.5.5) are negligible, scattered intensity at wave vectors $k < k_c$ grow exponentially with time and the maximum intensity occurs at k_m.

For a solution of uniformly charged polyelectrolyte with monovalent salt, k_m at which the scattered intensity is maximal follows from Eqs. (8.2.25) and (8.5.4) as

$$k_m = \frac{3}{\ell}\left[\phi_2\left(\frac{1}{1-\phi_2} + \frac{1}{N\phi_2} - 2\chi + \frac{\alpha^2 z_p^2}{2c_s\ell^3}\right)\right]^{1/2}. \tag{8.5.10}$$

Therefore, as the electrostatic factor α^2/c_s increases, k_m increases. Hence, the spontaneously selected length scale in spinodal decomposition decreases with either an increase in the degree of ionization or a decrease in ionic strength.

In the early stage of spinodal decomposition, the growth rate is independent of time, while the intensity increases exponentially with time (Fig. 8.13a). As time progresses, the fluctuations become stronger and the nonlinear terms in Eq. (8.5.5) begin to contribute. As a consequence, k_m shifts to lower values with time, representing the growth of a spontaneously selected domain size (Fig. 8.13b). Generally speaking, the time dependence of k_m can be classified into three regimes. The first regime is the early regime of linear spinodal decomposition described above where the domain size is independent of time ($k_m \sim t^0$). In the second regime, different domains coarsen through a combination of the evaporation-condensation mechanism (called Ostwald ripening or Lifshitz–Slyozov mechanism, where molecules come out of one domain and merge with another domain) and coalesce [Gunton et al. (1983)]. In this regime,

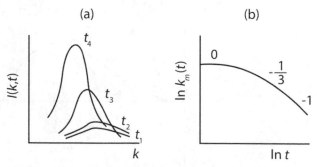

Figure 8.13 (a) Time dependence of the angularly averaged scattered intensity $I(k,t)$ at different times $t_1 < t_2 < t_3 < t_4$. (b) Time dependence of wave vector at maximum scattered intensity. The slopes 0, -1/3, and -1 correspond to the early, intermediate, and late stages, respectively, of spinodal decomposition.

$k_m \sim t^{-1/3}$. In the third regime, pertinent to very late stages of spinodal decomposition, k_m is proportional to t^{-1}. In this regime, hydrodynamics controls phase separation.

8.5.2 Nucleation and Growth

When a homogeneous solution is quenched into a metastable state, bounded by the coexistence and spinodal curves shown in Fig. 8.3a, phase separation occurs through the mechanism of nucleation and growth [Gunton *et al.* (1983), Debenedetti (1996)]. Thermal fluctuations in the metastable state lead to formation of "nuclei" with different sizes. Within each nucleus, the polymer concentration is the same as the equilibrium value, corresponding to either the phase with concentration ϕ_2' or the phase with concentration ϕ_2''. If the size of a nucleus is smaller than a certain critical size, it dissolves. On the other hand, if the size is larger than the critical value, then it provides a surface for further growth. The nuclei of sizes larger than the critical size grow competitively until the eventual completion of phase separation (Fig. 8.14).

In order to illustrate the approach to estimate the critical size of the nucleus in particular systems, consider the simplest case of the formation of a spherical droplet of radius r. The free energy of formation ΔF for such a droplet is given by the gain in free energy in its formation and the cost associated with the interface surrounding the droplet,

$$\Delta F = -\frac{4}{3}\pi r^3 (\Delta g) + 4\pi r^2 \sigma, \tag{8.5.11}$$

where Δg is the bulk free energy change per unit volume and σ is the interfacial tension per unit area. The driving force for phase separation given by Δg depends on temperature and polymer concentration, which can be expressed in terms of either quench depth in temperature (below the critical temperature for UCST) or quench depth in supersaturation (above the critical polymer concentration) or both. For example, $\Delta g \sim \Delta T$ for temperature quenches near the critical temperature T_c, where ΔT is $|T_c - T|$ at the temperature T of the nucleation process.

The dependence of ΔF on r is sketched in Fig. 8.15, which exhibits a free energy barrier $\Delta F^\dagger$ for formation of stable nuclei. The value of r at which ΔF is maximum is the critical radius r_c given by $\partial \Delta F / \partial r = 0$ as

$$r_c = \frac{2\sigma}{(\Delta g)}. \tag{8.5.12}$$

Substitution of r_c in Eq. (8.5.11) gives the nucleation barrier,

$$\Delta F^\dagger = \frac{16\pi}{3} \frac{\sigma^3}{(\Delta g)^2}. \tag{8.5.13}$$

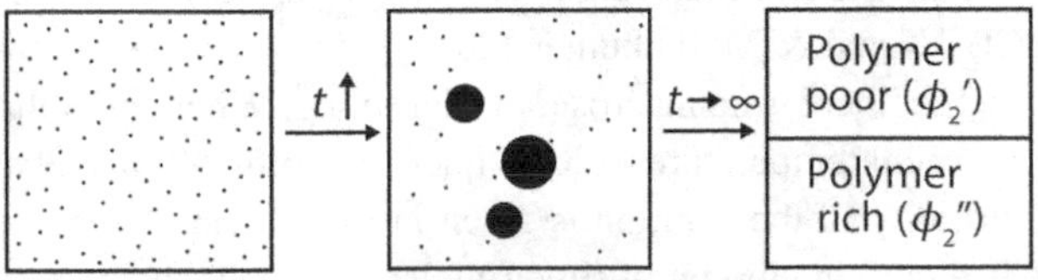

Figure 8.14 Nucleation and growth mechanism for liquid–liquid phase separation.

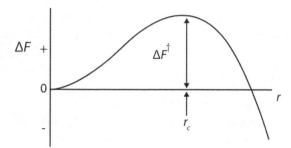

Figure 8.15 Nucleation barrier $\Delta F^\dagger$ for formation of a spherical droplet of critical radius r_c.

In order for stable nuclei to form, the nucleation barrier must be overcome. The time $\tau_{\text{nucleation}}$ required for nucleation to occur is proportional to the exponential of the nucleation barrier, $\tau \sim \exp(\Delta F^\dagger/k_BT)$ [Gunton $et\ al.$ (1983)].

$$\tau_{\text{nucleation}} \sim \exp\left[\frac{16\pi}{3k_BT}\frac{\sigma^2}{(\Delta g)^2}\right]. \qquad (8.5.14)$$

Since $\Delta g \sim \Delta T$ for temperature quenches, the critical nucleus size given in Eq. (8.5.12) diverges as $1/\Delta T$ for shallow quenches. The nucleation time also diverges as $\exp[1/(\Delta T)^2]$. As the quench depth increases, nucleation processes occur more readily. As noted above, for $r < r_c$ (ΔF of the nucleus being lower than the barrier), the nuclei dissolve. For $r > r_c$, nuclei can grow. The radius of a growing droplet depends on time t as $\sqrt{t}$ at early times and as $t^{1/3}$ at late times.

The above description is known as homogeneous nucleation and growth. If there are additional ingredients in the parent solution, such as dirt particles or externally imposed interfaces, then $\Delta F^\dagger$ given in Eq. (8.5.13) is appended by a multiplicative factor which can be substantially smaller than unity. This mechanism of phase separation is referred to as heterogeneous nucleation and growth [Debenedetti (1996)].

The key inputs in the above procedure to estimate critical sizes and nucleation barriers for the assembly of charged macromolecules are system-dependent expressions for Δg and σ. Results from Section 8.4 are used in obtaining these inputs.

8.5.3   Interlude of Aggregation in Phase Separation

While the study of kinetics of liquid–liquid phase separation in polyelectrolyte solutions remains scant, some recent experiments suggest that physical aggregation of chains interfere with the spinodal decomposition and nucleation and growth mechanisms of phase separation. As an example of this effect, consider an aqueous solution of NaPSS (molecular weight of 110,000 g/mol) at c_p=0.5 g/L, containing 0.015M BaCl$_2$ [Kanai & Muthukumar (2007)]. At 60°C, the solution is homogeneous in the sense that there is no macroscopic demixing. When this solution is quenched below the cloud point temperature 41°C to quench depths $\Delta T = 2, 3$, and 4°C, the time evolution of turbidity of the solution is given in Fig. 8.16a. At each quench depth, the solution takes a certain amount of time (nucleation time) to begin to build up turbidity, which

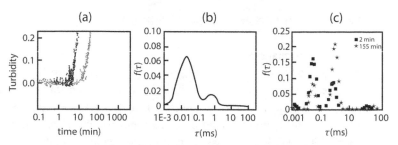

Figure 8.16   (a) Time-evolution of turbidity of aqueous solution of NaPSS (molar mass 110 kDa) at $c_p = 0.5$ g/L, containing 0.015 M $BaCl_2$. The quench depths ΔT are, from left to right, 4 and 3°C. (b) Distribution function of correlation time $f(\tau)$, measured in DLS at the scattering angle 90°, shows the fast and slow modes in the homogeneous phase. (c) The slow mode templates the phase separation kinetics upon quench into liquid–liquid phase separation region [Adapted from Kanai & Muthukumar (2007)].

is an indication of phase separation. After the nucleation time, turbidity increases linearly with time, demonstrating a linear growth rate during the early stages of growth. At later times, turbidity deviates from the linear growth regime and then saturates at very long times. Only the nucleation and initial growth of turbidity are presented in Fig. 8.16a. The data show that the nucleation time decreases with an increase in quench depth. Furthermore, the growth rate is steeper at larger quench depths. These results indicate the mechanism of nucleation and growth for phase separation in this system. However, details of the birth of new phases during the primordial stage of phase separation show more intricate behavior. These details can be gleaned from dynamic light scattering (DLS) studies which can explore shorter time scales.

The DLS data on this system exhibit two modes (fast and slow) as shown in Fig. 8.16b where the distribution function of correlation time $f(\tau)$ at the scattering angle 90° is given. As described in Section 6.5 and Chapter 7, the slow mode corresponds to aggregates and the fast mode corresponds to unaggregated chains. When this solution is quenched to 2°C below the cloud point temperature of 41°C, the time evolution of the DLS data is given in Fig. 8.16c. Two minutes after the quench, the histogram is similar to that of the homogeneous phase (Fig. 8.16b) with essentially the same fast and slow modes. Until 155 minutes, these two modes persist and the average decay rates of these modes remain unchanged. While the average decay rate remains unchanged, the population of the fast mode decreases continuously and the population of the slow mode increases continuously. Thus the population of aggregates increases by converting unaggregated chains into aggregates. After 155 minutes, the average decay time of the slow mode shifts to longer times and larger domains form.

These observations are illustrative of additional features of phase separation in charged solutions, where an enrichment of polymer aggregates of well-defined size distributions occurs in the very early stage of liquid–liquid phase separation. This first step is then followed by a growth process in the formation of the legitimate new phases dictated by the phase diagram. Polymer aggregates formed in the early stage act as templating nuclei for later stages of phase separation kinetics.

8.6   Micellization and Microphase Separation

So far in this chapter, we have considered phase behaviors of solutions of uniformly charged flexible polyelectrolytes, resulting in macrophase separation into coexisting daughter phases with macroscopic dimensions. When the charged macromolecules have blocky sequences of charged and uncharged repeat units, they can form domains with sizes of nanometers and even micrometers. In dilute solutions, these domains are **micelles** and their self-assembly process is called **micellization**. In highly concentrated solutions, the domains organize themselves into regular lattices with unit cells made of nanoscopic dimensions. This phenomenon is called **microphase separation**.

As the simplest example of how micellization and microphase separation emerge in solutions of charged macromolecules, consider diblock copolymers, where one contiguous block of N_A repeat units in a molecule is charged and the rest of the molecule (second block with N_B repeat units) is hydrophobic. The hydrophobic block of a molecule prefers to mingle with the hydrophobic blocks from other molecules to form globule-like domains, by pushing the charged hydrophilic blocks away from them and creating an interface between them. The size and shape of these domains are determined by N_A, N_B, polymer concentration, salt concentration, temperature, and the excluded volume parameters reflecting interactions among the various repeat units and solvent.

If $N_A \gg N_B$, spherical hydrophobic domains form in aqueous solutions, which are surrounded by water-soluble A block moieties. The junction between the A and B blocks is at the interface. In dilute solutions, the spherical domains are far from each other (Fig. 8.17a). On the other hand, at very high concentrations or in the melt state, these domains organize into a spherical morphology as in a body-centered-cubic lattice (Fig. 8.17b). If the relative length of the minority component N_B is increased with respect to the majority component N_A, cylinder-like micelles form in solutions and cylinder-shaped domains form in the melt state, organized into a cylindrical morphology (hexagonally close packed lattice) (Fig. 8.17b). When N_A is comparable to N_B, disc-like micelles form in solutions, whereas plate-like domains organize in the

(a)   (b)

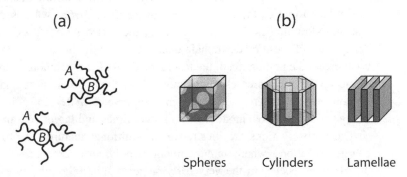

Spheres   Cylinders   Lamellae

Figure 8.17 Self-assembly of charged-hydrophobic diblock copolymers. N_A and N_B are the numbers of segments for the charged and the neutral blocks, respectively. (a) Micellization in aqueous dilute solutions into spherical micelles for $N_A \gg N_B$. (b) Microphase separation into spherical, cylindrical, and lamellar morphologies as N_A/N_B is progressively increased to unity. The microdomains are nanoscopic in size.

melt state into a lamellar morphology (Fig. 8.17b). Other morphologies such as gyroid phases can also form. The organization of these morphologies in highly concentrated solutions follows a series of phase transitions as experimental conditions are varied, such as a reduction in temperature [Leibler (1980), Marko & Rabin (1992), Kumar & Muthukumar (2007)]. These transitions include disorder–order transitions from a homogeneous system to spherical morphology, and order–order transitions from spherical morphology to cylindrical morphology, and order–order transitions from cylindrical morphology to lamellar morphology.

In the rest of this section, we shall first address the fundamentals of self-assembly of micelles from charged-hydrophobic diblock copolymers. Only spherical micelles are considered as specific examples. Next, we shall address the role of electrostatic interactions on the microphase separation of charged-hydrophobic diblock copolymers.

8.6.1 Self-Assembly of Charged Diblock Copolymes in Solutions

As a specific example, consider the self-assembly of charged-hydrophobic AB diblock copolymers in aqueous solutions containing salt into spherical micelles (sometimes called star-like aggregates) [Halperin (1987)] (Fig. 8.18). The A block is a flexible polyelectrolyte chain with charge $\alpha z_p e N_A$ uniformly distributed along its contour. The B block is hydrophobic and forms globule-like structures in aqueous media. When several such molecules interact, the B blocks mingle to form a large globule and this tendency is opposed by electrostatic repulsion between the polyelectrolyte blocks. As a result, a star-like micellar structure emerges for $N_A \gg N_B$, where the B blocks form the core, and the A blocks constitute the corona, with an interface between the core and corona. An optimal size distribution of the micelle emerges from the balance between globule formation inside the core and electrostatic repulsion inside the corona, mediated by the interfacial tension between the core and corona.

In order to determine the preferred size of the micelle, let us consider the assembly of m chains into a micelle of radius R with core radius R_0 and corona thickness L ($=R - R_0$, as shown in Fig. 8.18). As seen in Section 6.5, calculation of the free energy

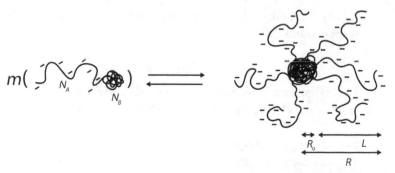

Figure 8.18 Self-assembly of m diblock copolymer chains into a spherical micelle. The A block is a flexible polyelectrolyte and the B block is hydrophobic forming a globule-like structure. N_A and N_B are the number of segments per chain in the A and B blocks, respectively. R_0 and R are the core radius and micelle radius, respectively. The thickness of corona is $L = R - R_0$.

F_m of the m-mer and that of the unaggregated unimer F_1 allows the determination of the critical aggregation concentration required for micellization, the distribution of micelle size, and the optimal value of m. In the rest of this section, we present only a scaling argument in which all numerical prefactors are ignored.

8.6.1.1   Free Energy of Single Diblock Chains

Recall from Section 2.6 that the radius of gyration R_g of a globule formed by one chain of N_B segments, with segment length ℓ, is

$$R_g \sim \ell \left(\frac{N_B}{|v|} \right)^{1/3}, \tag{8.6.1}$$

where v is the excluded volume parameter which is negative for globule formation. Writing $|v|$ as $\tau = (\Theta - T)/\Theta$, where Θ is the Flory theta temperature, the radius of the globule is given by

$$R_g \sim \ell \left(\frac{N_B}{\tau} \right)^{1/3}. \tag{8.6.2}$$

Using this result, the free energy of the globule formed by one chain follows from Eq. (2.6.1) as

$$\frac{F}{k_B T} \sim (-\tau^2 N_B). \tag{8.6.3}$$

Similarly, the R_g and free energy of a flexible polyelectrolyte chain of N_A segments in a solution with enough salt to qualify for the high-salt limit (Section 5.6.1) are given by (Eqs. (5.6.5) and (5.6.7)),

$$R_g \sim \ell v_{\text{eff}}^{1/5} N_A^{3/5}, \tag{8.6.4}$$

and

$$\frac{F}{k_B T} \sim v_{\text{eff}}^{2/5} N_A^{1/5}, \tag{8.6.5}$$

where

$$v_{\text{eff}} = v + \frac{\alpha^2 z_p^2}{2 c_s \ell^3}. \tag{8.6.6}$$

Adding the results of Eqs. (8.6.3) and (8.6.5), the free energy of a single diblock chain is approximated as

$$\frac{F_1}{k_B T} \sim v_{\text{eff}}^{2/5} N_A^{1/5} - \tau^2 N_B. \tag{8.6.7}$$

This expression is an approximation, and alternative expressions including surface terms of the globule can also be used [Khokhlov (1980), Joanny & Leibler (1990)].

8.6.1.2   Free Energy of a Spherical Micelle

The free energy F_m of the micelle depicted in Fig. 8.18, formed by m chains, is the sum of contributions from the core, interface, and corona,

$$F_m = F_{\text{core}} + F_{\text{interface}} + F_{\text{corona}}. \tag{8.6.8}$$

The core free energy F_{core} is the sum of the bulk free energy and the stretching free energy arising from the stretching of each B block chain with end-to-end distance R_0. It follows from Eq. (8.6.3) that the bulk contribution is $(-\tau^2 m N_B)$. The stretching free energy of a single chain of N_B segments with end-to-end distance R_0 is $3k_B T R_0^2/(2N_B \ell^2)$ (see Eq. (2.1.13)). Therefore, the free energy of the core per chain is given as

$$\frac{F_{\text{core}}}{mk_B T} \sim \frac{R_0^2}{N_B \ell^2} - \tau^2 N_B. \tag{8.6.9}$$

Since R_0 of the globule formed by m B-blocks follows from Eq. (2.6.3) as

$$R_0 \sim \ell \left(\frac{mN_B}{\tau}\right)^{1/3}, \tag{8.6.10}$$

we get

$$\frac{F_{\text{core}}}{mk_B T} \sim \frac{m^{2/3}}{\tau^{2/3} N_B^{1/3}} - \tau^2 N_B. \tag{8.6.11}$$

The free energy of the interface is the product of the interfacial tension γ (free energy per unit area) and the interfacial area proportional to R_0^2,

$$F_{\text{interface}} \sim \gamma R_0^2. \tag{8.6.12}$$

Using Eq. (8.6.10), we get the interfacial free energy per chain as

$$\frac{F_{\text{interface}}}{mk_B T} \sim \frac{\gamma \ell^2}{k_B T} \left(\frac{N_B}{\tau}\right)^{2/3} \frac{1}{m^{1/3}}. \tag{8.6.13}$$

The free energy of corona is obtained by the following scaling argument (for more details, see Appendix 7). As sketched in Figs. 8.18 and 8.19 for a spherical micelle, one end of the polyelectrolyte block is anchored at the interface and the m chains extend radially outward. The segment density of the polyelectrolyte blocks decreases from a higher value near the interface to lower values away from the interface until the micelle radius. As a consequence, the correlation length for monomer concentration ξ depends on the radial distance r from the center of the micelle. Generalizing the scaling laws developed in Section 6.2.3, we can imagine that there is radially a succession of blobs of size $\xi(r)$ as cartooned in Fig. 8.19. Based on the dependence of $\xi(r)$ on the local polymer concentration and geometrical considerations, F_{corona} is obtained as (see Appendix 7 for derivation)

$$\frac{F_{\text{corona}}}{mk_B T} \sim \sqrt{m} \ln\left(\frac{y}{m^{2/15}}\right), \tag{8.6.14}$$

where

$$y = v_{\text{eff}}^{1/5} \tau^{1/3} N_A^{3/5} N_B^{-1/3}. \tag{8.6.15}$$

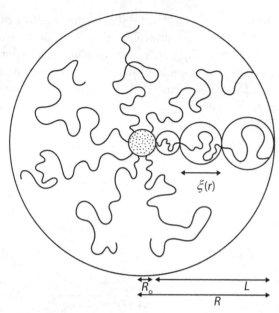

Figure 8.19 The correlation length ξ for the polyelectrolyte concentration increases with radial distance r from the center of the micelle. The construct of blobs of linear size $\xi(r)$ inside the corona leads to scaling laws for micelle radius R, corona thickness L, and free energy of corona.

8.6.1.3   Size and Free Energy of Equilibrated Micelle

Using the above derived contributions from the core, interface, and corona to the micelle free energy, which are all in terms of a specified number of chains m constituting the micelle, we need to minimize the total free energy with respect to m and determine its optimal value. Substituting this optimum value in the expressions for R_0, L, and F_m, we get the desired results as given below.

Combining Eqs. (8.6.11), (8.6.13), and (8.6.14), the free energy of the micelle per chain is given by,

$$\frac{F_m}{mk_BT} \sim \frac{m^{2/3}}{\tau^{2/3}N_B^{1/3}} - \tau^2 N_B + \frac{\gamma \ell^2}{k_BT}\left(\frac{N_B}{\tau}\right)^{2/3}\frac{1}{m^{1/3}} + \sqrt{m}\ln y - \sqrt{m}\ln m^{2/15}. \quad (8.6.16)$$

Since $y \gg m^{2/15}$ for large N_A and the first term on the right-hand side is negligible ($N_B \gg 1$) in comparison with the rest of the terms in the above equation, as can be checked by self-consistency based on the result below, the approximate scaling result for F_m is

$$\frac{F_m}{mk_BT} \sim \frac{\gamma \ell^2}{k_BT}\left(\frac{N_B}{\tau}\right)^{2/3}\frac{1}{m^{1/3}} + \sqrt{m}\ln y - \tau^2 N_B. \quad (8.6.17)$$

Minimizing this expression with respect to m, the number of chains in the micelle is given as

$$m \sim \left(\frac{\gamma \ell^2}{k_B T} \right)^{6/5} \left(\frac{N_B}{\tau} \right)^{4/5} (\ln y)^{-6/5}.$$ (8.6.18)

Note that the roles of the electrostatic interaction (through v_{eff} given in Eq. (8.6.6)) and the size of the polyelectrolyte block on the number of chains in the micelle is weak, since these factors appear merely as a logarithmic correction. The number of aggregated chains is controlled by the nature of the hydrophobic block and the interfacial tension.

Substituting the optimum value of m given by Eq. (8.6.18) into Eq. (8.6.10), we get the core radius as

$$R_0 \sim \ell \left(\frac{\gamma \ell^2}{k_B T} \right)^{2/5} \left(\frac{N_B}{\tau} \right)^{3/5} (\ln y)^{-2/5},$$ (8.6.19)

which depends only weakly on the properties of the polyelectrolyte block. Combining Eqs. (A7.9) and (8.6.18), the thickness of the corona is given by

$$L \sim \ell v_{\text{eff}}^{1/5} \left(\frac{\gamma \ell^2}{k_B T} \right)^{6/25} N_A^{3/5} \left(\frac{N_B}{\tau} \right)^{4/25} (\ln y)^{-6/25}.$$ (8.6.20)

The corona thickness is dictated by the properties of both the polyelectrolyte and hydrophobic blocks. In the high salt concentration limit treated in this section, L is essentially dictated by the radius of an isolated polyelectrolyte block, namely $v_{\text{eff}}^{1/5} N_A^{3/5}$, with a weak modification arising from the logarithmic term. The effect of monovalent salt concentration c_s on L is

$$L \sim \left(v + \frac{\alpha^2 z_p^2}{2 c_s \ell^3} \right)^{1/5} N^{3/5}.$$ (8.6.21)

If the bare excluded volume parameter v is close to the condition of the Flory theta temperature for the polyelectrolyte backbone, then the corona thickness is inversely related to the salt concentration as

$$L \sim \frac{1}{c_s^{1/5}}.$$ (8.6.22)

Therefore, the corona thickness decreases with an increase in the concentration of added salt. The predicted dependence of L on $N_A^{3/5} N_B^{4/25}$ is in agreement with experimental findings [Bluhm & Whitmore (1985), Halperin (1987)].

8.6.1.4   Distribution of Micelle Size and Critical Micelle Concentration

Following the same procedure outlined in Section 6.5, the mole fraction of m-mer aggregate, X_m, is given as

$$X_m = m \left[X_1 e^{\left(F_1 - \frac{F_m}{m} \right)/k_B T} \right]^m,$$ (8.6.23)

with the constraint,

$$X = \sum_{m=1}^{\infty} X_m. \tag{8.6.24}$$

Here, X is the total mole fraction of the diblock copolymer in the solution, and F_1 and F_m/m are given in Eqs. (8.6.7) and (8.6.17).

The critical micelle concentration ϕ_{cmc} follows from Eqs. (8.6.7), (8.6.17), and (8.6.23) as

$$\phi_{cmc} \sim e^{\left(\frac{F_m}{m} - F_1\right)/k_B T} \sim \exp\left[\left(\frac{\gamma \ell^2}{k_B T}\right)^{3/5} \left(\frac{N_B}{\tau}\right)^{2/5} (\ln y)^{2/5} - v_{eff}^{2/5} N_A^{1/5}\right]. \tag{8.6.25}$$

Therefore, the critical micelle concentration and the distribution of the size of micellar aggregates can be deduced from Eqs. (8.6.23)–(8.6.25) in terms of N_A, N_B, v, T, polymer concentration, and salt concentration.

The treatment given above for spherical aggregates can be readily implemented for other geometries such as cylindrical and plate-like assemblies as well. The above derivation is only at the bare minimum of essential features of micellization. More sophisticated treatments with additional details are available in the literature [Marko & Rabin (1992), Dan & Tirrell (1993), Borisov & Zhulina (2002)].

8.6.2   Microphase Separation

As a typical example of microphase separation in charged-neutral diblock copolymers in the melt state, consider the disorder–order transition of the formation of a lamellar morphology from a uniform phase. The melt state consists of n chains of an $A - B$ diblock copolymer, with the charged block A of N_A segments and the uncharged B block of N_B segments. The segment length for both blocks is ℓ. Let f be the fraction of charged segments, $f = N_A/N$, where $N = N_A + N_B$ is the total number of segments per chain. The number of charges per A block is $\alpha z_p N_A = \alpha z_p f N$. Therefore, the number of counterions n_c of valency z_c in the system is $\alpha z_p f n N / z_c$. In addition, there are n_γ ionic species of valency z_γ arising from the dissociation of added salt.

A brief outline of the theoretical procedure to address microphase separation is as follows: First, the free energy in terms of various intersegment interactions are presented through Eqs. (8.6.26)–(8.6.32). Identifying the density profile of one of the components through Eqs. (8.6.33)–(8.6.35), the free energy of the system is obtained in Eqs. (8.6.36)–(8.6.45). Finally, the consequences of the derived result in the high salt and low salt limits are discussed through Eqs. (8.6.46)–(8.6.49).

Generalizing Eqs. (2.1.8) and (6.3.1) for the present model system, the Helmholtz free energy is given in the Edwards path integral representation as

$$e^{-\frac{F}{k_B T}} = \frac{1}{n! n_c! \prod_\gamma n_\gamma!} \int \prod_{\alpha=1}^{n} \mathcal{D}[\mathbf{R}_\alpha] \int \prod_{i}^{n_c + \sum_\gamma n_\gamma} d\mathbf{r}_i \exp$$

$$\left[-\frac{3}{2\ell^2} \sum_{\alpha=1}^{n} \int_0^N ds_\alpha \left(\frac{\partial \mathbf{R}_\alpha(s_\alpha)}{\partial s_\alpha}\right)^2 - U\right], \tag{8.6.26}$$

where $\mathbf{R}_\alpha(s_\alpha)$ is the position vector of the arc length variable s_α ($0 \le s_\alpha \le N$) of the α-th chain. The symbol $\int \mathcal{D}[\mathbf{R}_\alpha]$ denotes the sum over all polymer conformations. The index i in the $\int d\mathbf{r}_i$ integral denotes all ions from the added salt and unadsorbed counterions. The first term inside the square brackets represents the entropic contribution from chain connectivity. The term U is the interaction energy among all polymer segments and ions.

Expressing U as the sum of contributions from polymer-polymer interaction U_{pp}, polymer-ion interaction U_{pi}, and ion-ion interaction U_{ii}, we get

$$U = U_{pp} + U_{pi} + U_{ii}. \tag{8.6.27}$$

U_{pp} is given by the sum of interactions within the individual blocks and between the inter-blocks as

$$U_{pp} = \frac{1}{2} \sum_{\alpha=1}^{n} \sum_{\beta=1}^{n} \left[\int_0^{N_A} ds_\alpha \int_0^{N_A} ds_\beta U_{AA}(\mathbf{r}) + \int_{N_A}^{N} ds_\alpha \int_{N_A}^{N} ds_\beta U_{BB}(\mathbf{r}) \right.$$
$$\left. +2 \int_0^{N_A} ds_\alpha \int_{N_A}^{N} ds_\beta U_{AB}(\mathbf{r}) \right] \tag{8.6.28}$$

where $\mathbf{r}$ is $[\mathbf{R}_\alpha(s_\alpha) - \mathbf{R}_\beta(s_\beta)]$ and $U_{AA}(\mathbf{r}), U_{BB}(\mathbf{r})$, and $U_{AB}(\mathbf{r})$ are the interaction energies between segments of different types separated by distance $\mathbf{r}$. For the charged-neutral $A - B$ diblock copolymer,

$$U_{AA}(\mathbf{r}) = w_{AA}\delta(\mathbf{r}) + \frac{\alpha^2 z_p^2 \ell_B}{r}; \quad U_{BB}(\mathbf{r}) = w_{BB}\delta(\mathbf{r}); \quad U_{AB}(\mathbf{r}) = w_{AB}\delta(\mathbf{r}), \tag{8.6.29}$$

where w_{AA}, w_{BB}, and w_{AB} are the intersegment excluded volume parameters and $r = |\mathbf{r}|$. U_{pi} in Eq. (8.6.27) is given by

$$U_{pi} = \sum_{\alpha=1}^{n} \sum_{i=1}^{n_c + \sum_\gamma n_\gamma} \left[\int_0^{N_A} ds_\alpha U_{Ai}[\mathbf{R}_\alpha(s_\alpha) - \mathbf{r}_i] + \int_{N_A}^{N} ds_\beta U_{Bi}[\mathbf{R}_\beta(s_\beta)) - \mathbf{r}_i] \right], \tag{8.6.30}$$

and U_{ii} is given by

$$U_{ii} = \frac{1}{2} \sum_{i=1}^{n_c + \sum_\gamma n_\gamma} \sum_{j=1}^{n_c + \sum_\gamma n_\gamma} U_{ij}(\mathbf{r}_i - \mathbf{r}_j) \Bigg\}. \tag{8.6.31}$$

In the above two equations, the interaction energies are

$$U_{Ai}(r) = \frac{\alpha z_p z_i \ell_B}{r}; \quad U_{Bi}(r) = 0; \quad U_{ij}(r) = \frac{z_i z_j \ell_B}{r}, \tag{8.6.32}$$

where i and j denote the small electrolyte ions and counterions, and the van der Waals-type excluded volume interactions involving counterions and salt ions are ignored.

As described in Chapter 6, field theoretic techniques are used on Eq. (8.6.26) to calculate the free energy of the system. Let $\phi_A(\mathbf{r})$ be the volume fraction of the charged A monomers at a location around $\mathbf{r}$ and $\psi(\mathbf{r})$ be the fluctuation in $\phi_A(\mathbf{r})$ from its value $\phi_A = f$ in the uniform phase,

$$\psi(\mathbf{r}) = \phi_A(\mathbf{r}) - \phi_A = \phi_A(\mathbf{r}) - f. \tag{8.6.33}$$

In order to simplify the calculations, we assume that the counterions and salt ions are point-like charges without volume and that the system is incompressible,

$$\phi_A(\mathbf{r}) + \phi_B(\mathbf{r}) = 1, \tag{8.6.34}$$

where $\phi_B(\mathbf{r})$ is the volume fraction of the uncharged B monomers at location around $\mathbf{r}$, which fluctuates around its average value $\phi_B = 1 - f$. Therefore, the order parameter can also be written in terms of $\phi_B(\mathbf{r})$ as

$$\psi(\mathbf{r}) = \phi_A(\mathbf{r}) - f = 1 - f - \phi_B(\mathbf{r}). \tag{8.6.35}$$

When the fluctuations $\psi(\mathbf{r})$ are weak, known as the weak segregation limit (WSL), the random phase approximation (RPA) can be fruitfully implemented to calculate the free energy (Section 6.2.2.1). This procedure can be accomplished analytically [Leibler (1980)]. On the other hand, if the fluctuations are strong (called the strong segregation limit), self-consistent field theory is implemented entailing numerical work, as described in Chapter 6. Since the details of the calculations are presented in the original references [Marko & Rabin (1992), Kumar & Muthukumar (2007)], only the key results on the role of electrostatics on microphase separation are given below.

In the weak segregation limit, where RPA is applicable, the free energy is calculated in terms of the order parameter $\psi(\mathbf{r})$ and keeping terms only up to the second order so that

$$F = F_0 + \Delta F, \tag{8.6.36}$$

where F_0 is independent of ψ and contains all mean field terms including the electrostatic correlations arising from counterions and salt ions. The leading fluctuation contribution is

$$\frac{\Delta F}{k_B T} = \frac{1}{2} \int \frac{d^3 k}{(2\pi)^3} S^{-1}(\mathbf{k}) \psi_{\mathbf{k}} \psi_{-\mathbf{k}}, \tag{8.6.37}$$

where $\psi_{\mathbf{k}}$ is the Fourier transform of $\psi(\mathbf{r})$ and $S(\mathbf{k})$ is the structure factor,

$$S(\mathbf{k}) = \langle \psi_{\mathbf{k}} \psi_{-\mathbf{k}} \rangle. \tag{8.6.38}$$

Here the angular brackets denote the averaging over all possible fluctuations. Using RPA, the inverse of the structure factor is given as

$$S^{-1}(\mathbf{k}) = S_0^{-1}(\mathbf{k}) + S_1^{-1}(\mathbf{k}), \tag{8.6.39}$$

where the first term on the right-hand side is the contribution if the A block (and B block) were uncharged, and the second term is the electrostatic contribution from the charges on the A block. $S_0^{-1}(\mathbf{k})$ is given as [Leibler (1980)]

$$S_0^{-1}(\mathbf{k}) = \frac{\ell^3}{N} \left\{ \frac{g(1,x)}{\{g(f,x)g(1-f,x) - [g(1,x) - g(f,x) - g(1-f,x)]^2/4\}} - 2\chi N \right\}, \tag{8.6.40}$$

where $g(f,x)$ is the Debye structure factor (see Eq. (2.1.10) for $f = 1$),

$$g(f,x) = \frac{1}{x^2}\left(fx + e^{-fx} - 1\right); \qquad x = \frac{k^2 N\ell^2}{6} = k^2 R_g^2, \qquad (8.6.41)$$

and R_g is the radius of gyration of a Gaussian chain of N segments.

The electrostatic contribution to the structure factor is [Marko & Rabin (1992), Kumar & Muthukumar (2007)]

$$S_1^{-1}(\mathbf{k}) = \frac{4\pi\alpha^2 z_p^2 \ell_B}{k^2 + \kappa^2}, \qquad (8.6.42)$$

where

$$\kappa^2 = 4\pi\ell_B\left(z_c^2 c_c + \sum_{\gamma} z_\gamma^2 c_\gamma\right). \qquad (8.6.43)$$

Here c_c and c_γ are the number concentrations of the counterion and the γ-type electrolyte ions. Since the number of counterions is $n_c = \alpha z_p f n N / z_c$ and the volume of the melt is $nN\ell^3$,

$$c_c = \frac{\alpha z_p f}{z_c \ell^3}. \qquad (8.6.44)$$

When the counterions and electrolyte ions are monovalent, Eq. (8.6.43) becomes

$$\kappa^2 = \frac{4\pi\alpha z_p \ell_B f}{\ell^3} + 8\pi\ell_B c_s, \qquad (8.6.45)$$

where c_s is the number concentration of added monovalent salt, and the first term on the right-hand side is entirely due to counterions.

In general, the structure factor for diblock copolymers exhibits a peak at a finite value of the wavevector k_m denoting the spontaneous self-selection of a characteristic length scale in the system. This result is illustrated in Fig. 8.20, where $S(k)$ is plotted against $k R_g$ for $\chi = 0.01049, N = 1000$, and $\alpha = 0$. As χ is increased further, the uniform phase reaches its stability limit and undergoes microphase separation. The stability limit is obtained from the spinodal condition, namely $\partial^2(\Delta F)/\partial\psi^2 = 0$, which occurs at a certain critical wavevector $k^\star$. Substitution of $k^\star$ in $S^{-1}(k^\star) = 0$ then gives the critical value $\chi^\star$ for the disorder–order transition. The domain spacing

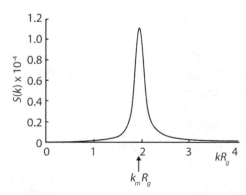

Figure 8.20 Spontaneous self-selection of a characteristic length scale $2\pi/k_m$ by diblock copolymers in the uniform phase. $N = 1000$, $\chi = 0.01049$, and $\alpha = 0$.

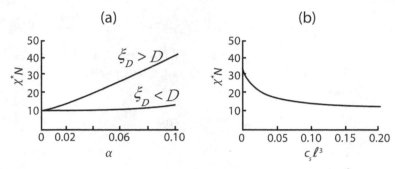

Figure 8.21 Dependence of $\chi^{\star}N$ on (a) α for $f = 1/2, N = 200, \ell_B/\ell = 3, c_s\ell^3 = 0.06022$ and (b) $c_s\ell^3$ for $f = 1/2, N = 200, \ell_B/\ell = 3, \alpha = 0.1$.

D in the ordered phase is given by $2\pi/k^{\star}$. Let us look at the high-salt and low-salt limits of Eq. (8.6.39).

8.6.2.1   High Salt Limit ($\xi_D = \kappa^{-1} < D$)

In the high salt limit, $\chi^{\star}$ follows from $S^{-1}(k^{\star}) = 0$ as

$$\left(\chi^{\star}N\right)_{\text{charged}} = \left(\chi^{\star}N\right)_{\text{uncharged}} + \frac{1}{\ell}\frac{2\pi\alpha^2 z_p^2 \ell_B N}{\left[\frac{4\pi\alpha z_p \ell_B f}{\ell} + 8\pi\ell_B c_s\ell^2\right]}, \tag{8.6.46}$$

where $\left(\chi^{\star}N\right)_{\text{uncharged}}$ is the stability limit if the A block does not carry any charges, and the salt is monovalent. The dependence of $\chi^{\star}N$ on α given by Eq. (8.6.46) is shown in Fig. 8.21a for $f = 1/2, \ell_B/\ell = 3, N = 200$, and $c_s\ell^3 = 0.06022$ for $\xi_D < D$. For $f = 1/2$, $\left(\chi^{\star}N\right)_{\text{uncharged}} = 10.495$. The dependence of the stability limit on the salt concentration is shown in Fig. 8.21b for $f = 1/2, N = 200$, and $\alpha = 0.1$. $\left(\chi^{\star}N\right)_{\text{charged}}$ decreases as c_s is increased. Hence electrostatic interactions enhance the stability of the uniform phase by elevating the stability criterion $\chi^{\star}N$ to higher values. On the other hand, the domain spacing is independent of electrostatics for $\xi_D < D$, as given by

$$D = \frac{\sqrt{2}\pi}{3^{3/4}} f^{1/4} (1 - f)^{1/4} \sqrt{N}\ell, \qquad \xi_D < D. \tag{8.6.47}$$

As illustrated in Fig. 8.22a, D is dictated only by the radius of gyration of the chain in the melt state and composition of the diblock, but not by electrostatics (due to electrostatic screening, $\xi_D < D$).

8.6.2.2   Low Salt Limit ($\xi_D > D$)

In the low salt limit, $\chi^{\star}N$ is given as [Kumar & Muthukumar (2007)]

$$\chi^{\star}N = \left(\chi^{\star}N\right)_{\text{uncharged}} + \sqrt{\frac{3}{4f^3(1 - f)^3}}\left[\left(1 + \frac{4\pi\alpha^2 z_p^2 N^2 f^2(1 - f)^2 \ell_B}{9\ell}\right)^{1/2} - 1\right]. \tag{8.6.48}$$

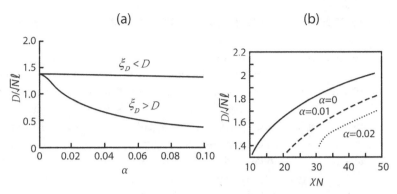

Figure 8.22 (a) Weak segregation limit: Dependence of domain spacing D on degree of ionization in the lamellar phase ($f = 1/2$). For $\xi_D > D$, D decreases with α, whereas D is independent of electrostatics for $\xi_D < D$. (b) Strong segregation limit: Dependence of D on χN and α for $f = 1/2$ and $N = 1000$. $\alpha = 0$ (solid line), 0.01 (dashed line), and 0.02 (dotted line). D decreases with α and increases with χN. For large values of χN, D approaches $N^{2/3}$ in the strong segregation limit, whereas $D \sim \sqrt{N}$ in the weak segregation limit at all values of χN [Adapted from Kumar & Muthukumar (2007)].

As the degree of ionization increases, $\chi^\star N$ increases more significantly in comparison with the system containing salt (Fig. 8.21a). This effect is due to the enhanced stabilization of the uniform phase by electrostatic repulsion between the charged segments. The domain spacing is given by

$$D = \frac{\sqrt{2}\pi}{3^{3/4}} \frac{f^{1/4}(1-f)^{1/4}\sqrt{N}\ell}{\left[1 + \frac{4\pi\alpha^2 z_p^2 N^2 f^2 (1-f)^2 \ell_B}{9\ell}\right]^{1/4}}. \tag{8.6.49}$$

As α increases, D decreases, as shown in Fig. 8.22a. This effect arises from a combination of entropy associated with redistribution of counterions upon microphase separation and electrostatic screening. Contrast this result for $\xi_D > D$ with that for $\xi_D < D$, as shown in Fig. 8.22a.

For χN sufficiently higher than $\chi^\star N$, the RPA used in the weak segregation limit is inadequate. It is necessary to account for higher order terms in ψ in calculating the free energy from Eq. (8.6.37). The inclusion of the higher order terms is accomplished using self-consistent field theory [Kumar & Muthukumar (2007)]. The numerically computed domain spacing for lamellar morphology in the strong segregation limit is given in Fig. 8.22b as a function of χN. In contrast with the weak segregation result $D \sim \sqrt{N}$, the domain spacing depends on N more sensitively upon an increase in χ, approaching the asymptotic limit, $D \sim N^{2/3}$, for $\chi N \gg 1$. Similar to the behavior in the weak segregation limit, D decreases with an increase in α due to redistribution of counterions in the microphase separated domains and electrostatic screening.

The computational procedures for the weak and strong segregation limits, illustrated above for microphase separation into lamellar phases from uniform phase can be generalized to the presence of solvent and formation of other morphologies. In particular, the above method can be implemented to study microphase separation in concentrated solutions of intrinsically disordered proteins, which is a complementary mechanism to the liquid–liquid separation.

8.7   Isotropic–Nematic Transition in Solutions of Charged Rods

In contrast with solutions of flexible charged macromolecules, where macrophase separated phases are isotropic, solutions of charged rod-like polymers exhibit a first order phase transition from an isotropic phase to an anisotropic (lyotropic liquid crystalline) phase as the polymer concentration is increased. The rods are randomly oriented in the isotropic phase, whereas they are collectively oriented in a preferred direction in the liquid crystalline phase. For charged rods, the liquid crystalline phase is usually the nematic phase, where there is only orientational order without any positional order. The extent of orientation is quantified using the orientational order parameter s, defined as the Legendre polynomial of the second order,

$$s = \frac{1}{2}\langle 3\cos^2\gamma - 1\rangle, \qquad (8.7.1)$$

where γ is the angle between two neighboring rods, and the angular brackets denote the averaging over all possible orientations. The order parameter undergoes a discontinuous transition from $s = 0$ in the isotropic phase to $s > 0$ in the nematic phase.

The isotropic–nematic transition was initially addressed by Onsager (1949) in the context of orientational ordering in solutions of tobacco mosaic virus (TMV) by considering translational and orientational entropy of hard rods in the athermal limit (that is, without considering temperature-dependent inter-rod potential interactions). Complementary theoretical approaches were later formulated by Flory (1956) and Maier & Saupe (1958). Whereas the Onsager theory is based on the second virial coefficient of rods so that it is mainly applicable to lower rod concentrations, the Flory theory, which is a generalization of the Flory–Huggins theory for flexible chains, is based on a lattice formulation applicable to concentrated solutions. The inclusion of the Flory–Huggins χ parameter in Flory's theory allows consideration of temperature effects on the isotropic–nematic transition. The Maier-Saupe theory is a molecular mean field theory, where the van der Waals inter-rod interactions are written in terms of the order parameter s, which is then self-consistently determined by accounting for the orientational distribution of rods and minimizing the free energy.

Since the theories of Onsager, Flory, and Maier & Saupe and their further modifications are described well in several textbooks [Chandrasekhar (1992), de Gennes & Prost (1993), Kleman & Lavrentovich (2003)], we only briefly mention the Onsager approach here. Consider a solution of n hard rods of diameter b and length L in volume V. The number concentration of rods is $c = n/V$ and the volume of each rod is $\pi b^2 L/4$ so that $\phi = c\pi b^2 L/4$ is the volume fraction of the rods in the solution.

While the free energy of solutions of hard spheres (and flexible chains) can be expressed only in terms of concentration c, the orientational distribution of hard rods must be specified for solutions of hard rods in addition to c. Let $f_{\mathbf{a}}$ be the probability for a given rod to point along the direction $\mathbf{a}$ with polar coordinates (θ, ϕ) within a small solid angle $d\Omega = \sin\theta d\theta d\phi$. By definition,

$$\int d\Omega\, f_{\mathbf{a}} = \int_0^\pi d\theta\, \sin\theta \int_0^{2\pi} d\phi\, f_{\mathbf{a}} = 1. \qquad (8.7.2)$$

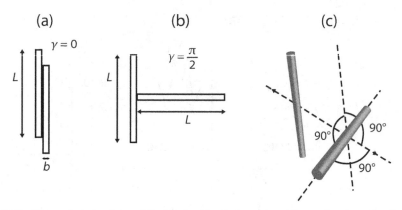

Figure 8.23 The excluded volume β for two hard rods depends on the angle γ between their orientations. (a) $\beta = 0$ for $\gamma = 0$. (b) $\beta = 2L^2b$ for $\gamma = \pi/2$. (c) Electrostatic repulsion between rods promotes twisting of rods with respect to one another.

Therefore, the number of rods per unit volume pointing in a small solid angle $d\Omega$ around a direction with unit vector $\mathbf{a}$ is $cf_\mathbf{a}d\Omega$. With this additional feature of rod orientations, the free energy in dilute solutions per rod is given as [de Gennes & Prost (1993)]

$$\frac{F}{k_BT} = \frac{F_0}{k_BT} + \int d\Omega f_\mathbf{a} \ln{(4\pi f_\mathbf{a}c)} + \frac{1}{2}c \int d\Omega \int d\Omega' f_\mathbf{a} f_{\mathbf{a}'} \beta(\mathbf{a},\mathbf{a}') + 0(c^2). \quad (8.7.3)$$

The first term is independent of $f_\mathbf{a}$. The second term is a representation of the reduction in entropy due to rod alignment. The third term accounts for the excluded volume effect. $\beta(\mathbf{a},\mathbf{a}')$ is related to the volume excluded by one rod in direction $\mathbf{a}$ with respect to another rod in direction $\mathbf{a}'$. For long rods, where end-effects can be ignored ($L \gg b$), β is given as [de Gennes & Prost (1993)]

$$\beta = 2L^2b|\sin\gamma|, \quad (8.7.4)$$

where γ is the angle between $\mathbf{a}$ and $\mathbf{a}'$. The excluded volume is minimal for parallel alignment ($\gamma = 0$) and maximal for perpendicular alignment ($\gamma = \pi/2$) as shown in Fig. 8.23a. Note that upon angular averaging, β is the second virial coefficient B_2 for hard rods given in Eq. (5.10.4) as

$$B_2 = \frac{\pi}{4}bL^2. \quad (8.7.5)$$

Minimization of F for all variations in $f_\mathbf{a}$, subject to the normalization constraint in Eq. (8.7.2), yields the self-consistent equation for $f_\mathbf{a}$ as

$$1 + \ln{(4\pi f_\mathbf{a})} - c \int d\Omega' \beta(\mathbf{a},\mathbf{a}') f_{\mathbf{a}'} = \lambda - 1, \quad (8.7.6)$$

where λ is determined using Eq. (8.7.2). Since it is difficult to solve the above nonlinear integral equation in $f_\mathbf{a}$ exactly, Onsager used a variational procedure with a trial function

$$f_\mathbf{a} = \frac{\alpha}{4\pi \sinh\alpha} \cosh(\alpha\mathbf{a} \cdot \mathbf{a}'), \quad (8.7.7)$$

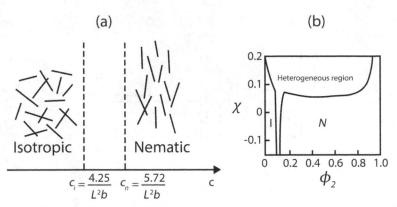

Figure 8.24 (a) Isotropic–nematic transition in solutions of hard rods. The solution is isotropic for $c < c_i$ and nematic for $c > c_n$. For $c_i < c < c_n$, the solution phase separates into coexisting isotropic phase of concentration c_i and nematic phase of concentration c_n. (b) Flory's phase diagram for a solution of rods, each of $N = 100$ segments, as a plot of the χ parameter versus volume fraction of rods. I and N denote the isotropic and nematic phases, respectively.

where α is a parameter. Note that $\alpha = 0$ corresponds to the isotropic phase. Substituting the trial function in Eq. (8.7.3), the free energy is expressed as a function $F(c, \alpha)$ of c and α. For small values of c, $F(c, \alpha)$ has only one minimum at $\alpha = 0$, which corresponds to the isotropic phase. If c exceeds a certain threshold value c_i, $F(c, \alpha)$ shows two minima, one at $\alpha = 0$ and another at a positive value of α corresponding to the nematic state. If c exceeds another threshold value c_n, the minimum at $\alpha = 0$ disappears and there is only one minimum at a larger positive value α. Therefore, Onsager's theory predicts the occurrence of the isotropic–nematic transition for solutions of hard rods. For $c < c_i$, the solution is isotropic; for $c_i < c < c_n$, phase separation takes place so that the isotropic phase of concentration c_i coexists with the nematic phase of concentration c_n; for $c > c_n$, the solution is nematic. For $L \gg b$, c_i and c_n are given by

$$c_i = \frac{4.25}{L^2 b}, \qquad c_n = \frac{5.72}{L^2 b}. \tag{8.7.8}$$

The predictions of Onsager's theory are summarized in Fig. 8.24a.

    The Onsager results are in the athermal limit, where temperature is not a variable. Using an adaptation of the lattice theory used in describing solutions of flexible chains (Section 6.2.1), Flory (1956) presented a theory for isotropic–nematic transition in solutions of rod-like molecules. In this theory, temperature and the chemical mismatch between the polymer and solvent are taken into account through the χ parameter (Eq. (6.2.5)). Consider a solution of n rod-like molecules and n_1 solvent molecules in volume V. Let each rod have N segments each of volume v_1, taken as the same as the volume of a solvent molecule. Accounting for the distribution of rods with their possible orientations, Flory presented the phase diagram of χ versus the volume fraction $\phi_2 = nNv_1/V$ of rods, as shown in Fig. 8.24b for $N = 100$. At lower values of χ, corresponding to higher temperatures for UCST systems, the solution is an isotropic phase at lower polymer concentrations and a nematic phase at higher polymer concentrations. The intervening "chimney" region corresponds to the phase separating heterogeneous region. As the temperature decreases, the heterogeneous region becomes wider as seen in Fig. 8.24b. The qualitative features of the isotropic–nematic transition in the athermal limit are the same for both the Onsager and Flory theories, although the values of c_i and c_n are different. This difference is expected, since the Onsager theory is pertinent to dilute solutions and Flory's lattice theory is applicable to concentrated solutions. The broadening of the heterogeneous region separating the isotropic and nematic phases is supported by experiments on rod-like molecules [Tohyama & Miller (1981)].

When the rod-like molecules or assemblies in a solution are charged, they also undergo isotropic–nematic transition. There are two additional features which contribute to their phase behavior. First, the repulsive electric potential between two identically charged rods results in an effective diameter b_{eff} which is larger than the bare diameter b. This effect can be expressed approximately as [Onsager (1949), Stroobants et al. (1986)]

$$b_{\text{eff}} = b + A_1 \xi_D, \tag{8.7.9}$$

where ξ_D is the Debye screening length and A_1 is a proportionality factor depending weakly on ξ_D. At lower salt concentrations (ξ_D is larger), the effective diameter can be much larger than the hard diameter. As the salt concentration is increased, b_{eff} becomes progressively smaller until it reaches the hard diameter b.

The second feature of the electrostatic interaction between identically charged rods is the twist of one rod against another in order to minimize the electrostatic repulsion between them [Stroobants et al. (1986), Carri & Muthukumar (1999)]. Due to twisting, the isotropic–nematic transition is expected to occur at higher rod concentrations than predicted without twist. Combining these two features, the Onsager result given in Eq. (8.7.8) can be approximately modified for charged rods as [Stroobants et al. (1986), Strzelecka & Rill (1990), Fraden et al. (1993)]

$$c_i = \frac{4.25}{L^2 b_{\text{eff}}} \left(1 - A_1' \frac{\xi_D}{b_{\text{eff}}} \right)^{-1}, \tag{8.7.10}$$

where A_1' is a proportionality factor for the twist effect and b_{eff} is given in Eq. (8.7.9). Since b_{eff} decreases with an increase in salt concentration, the general prediction from Eq. (8.7.10) is that c_i increases with salt concentration. Similarly c_n is predicted to increase with salt concentration. These predictions are supported by experiments on aqueous suspensions of tobacco mosaic virus ($L = 300$ nm, $b = 18$ nm), as shown in Fig. 8.25 [Fraden et al. (1989)]. Here, c_i and c_n are plotted against the ionic strength demonstrating the increase in c_i and c_n as the ionic strength is increased. Similar trends are observed in solutions of dsDNA as well, where other forms of liquid crystalline states, such as the cholesteric phase or the blue phase, can spontaneously form at higher solute concentrations. Although the qualitative aspects of the isotropic–nematic transition in solutions of charged rods are captured by the generalized Onsager theory, a rigorous theory valid for broader ranges of concentration, length, backbone diameter, and semiflexibility of charged macromolecules is yet to be fully developed.

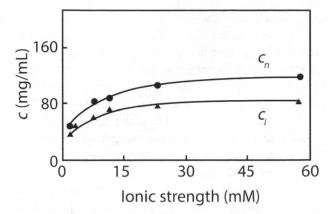

Figure 8.25 The concentrations of coexisting isotropic (triangles) and nematic (circles) phases of TMV plotted as a function of ionic strength [Adapted from Fraden *et al.* (1989)].

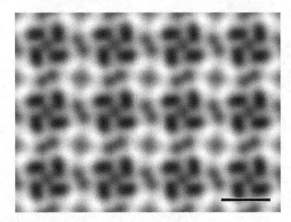

Figure 8.26 Transmission electron microscopy image of the structure of the S-layer protein SbpA (bar is 10 nm) [Pum & Sleytr (2014)].

8.8   Proteins, Fibrillization, and Membraneless Organelles

Solutions of proteins can also undergo phase transitions upon changes in experimental conditions such as protein concentration, salt concentration, temperature, and the presence of crowding agents, multivalent electrolyte ions, and polyions. If the protein sequences allow formation of folded structures, they undergo crystallization into well-ordered crystal lattices, with individual folded proteins constituting the unit cells with certain crystallographic symmetries. An example is given in Fig. 8.26, where a well-ordered two-dimensional lattice with crystallographic symmetry formed by S-layer proteins is illustrated [Pum & Sleytr (2014)].

8.8.1   Fibrillization

Many proteins self-assemble into linear structures in a polymerization process called **fibrillization**. The resultant fibrils are crucial to various biological functions. Examples include those formed by collagen (the most abundant protein in animals) and various amyloidogenic proteins such as α-synuclein. While the structure and concentration of collagen fibrils are essential for the stability of animal tissues such as the vitreous in the eye, amyloid fibrils are associated with a variety of detrimental neurodegenerative diseases such as Alzheimer's, Parkinson's, and Huntington's. The quest for an adequate understanding of the mechanism of this ubiquitous phenomenon of fibrillization continues to be actively pursued.

In general, fibrillization occurs by the nucleation and growth mechanism described in Section 8.5.2. When fibrils grow under suitable experimental conditions (pH, ionic strength $[I]$, protein concentration c, and temperature T), there are three stages: (i) nucleation, where the fibril grows and redissolves back and forth until its size exceeds a critical fibril length L_c; (ii) growth, where fibrils first grow linearly with time, followed by competitive growth among the various fibrils; and (iii) equilibrium, where the assembly has reached its saturation limit. These three regimes are sketched in Fig. 8.27, where a typical plot of turbidity τ (a measure of the extent of fibrillization) of the protein solution is plotted against time. The duration of the nucleation stage, called "lag time" t_{lag} is usually obtained as the intercept of the linear growth line with the abscissa. The growth rate G is estimated by fitting the growth kinetics at early times after nucleation with a straight line as shown in Fig. 8.27a. As time progresses, the various fibrils compete among themselves by continuous dissociation of monomers from the fibrils and their re-association with the most favorable fibrils. This mechanism depends on temperature and the diffusion constant of the monomers. In addition, the fibrils can merge during the coarsening stage. Overall, fibrillization can occur only

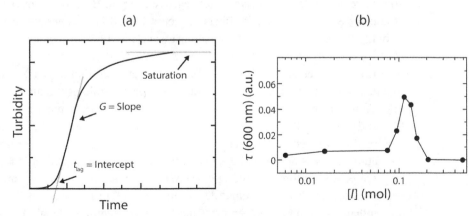

Figure 8.27   (a) Sketch of turbidity (a measure of the extent of fibrillization) versus time, showing the nucleation time, linear growth regime, and the saturation regime. (b) Fibril formation (turbidity τ) by type II collagen triple helices at pH 7.2 and 37°C after 48 hours of equilibration time, as a function of ionic strength $[I]$. Beyond $[I]$=250 mM, fibrillization does not occur [Adapted from Morozova & Muthukumar (2018)].

if the protein concentration is above a critical value and the temperature is below the disassembly temperature that is specific to a particular assembly process.

Specific interactions among the amino acids constituting the protein molecules and their sequences, as well as the electrostatic forces, play key roles in fibrillization. Focusing only on the electrostatic effects on fibrillization, we next consider only two examples, namely, collagen fibril growth and amyloid fibril growth. While the building units in collagen fibrillization are rod-like, they are roughly globule-like in amyloid fibrils.

8.8.1.1   Collagen Fibril Formation

Collagen is the most abundant protein in animals, and its structure and assembly characteristics have been extensively investigated. The literature on collagen fibril formation is vast [Shoulders & Raines (2009)]. The subunits of collagen that form fibrils are triple helices in which three polypeptide chains are hydrogen-bonded together. As a typical example, we illustrate the fibrillization of type II collagen and the role of electrostatics on its fibril formation. The building block of type II collagen is a triple helix with three identical left-handed helical chains that are hydrogen-bonded together in a right-handed manner. The triple helix is 300 nm in length and 1.5 nm in diameter. Approximately one-sixth of the overall amino acid sequence is ionizable and the isoelectric point of type II collagen is near pH 8.5. At the ends of each triple helix, there are globular domains which apparently are not involved in fibril formation. When fibrils are formed, the triple helices bundle together with both lateral and radial order [Hulmes *et al.* (1995), Holmes & Kadler (2006), Antipova & Orgel (2010)]. Laterally, the triple helices associate in a staggered manner with stagger spacing of about 67 nm (~234 amino acids).

Turbidity measurements on type II collagen solutions show that there is no fibrillization at pH < 2 (where the triple helices carry a net positive charge) or pH > 10 (where the triple helices carry a net negative charge). However, at intermediate pH around the isoelectric point, fibrillization occurs. Fig. (8.27b) shows the dependence of turbidity τ (measured using light of incident wavelength 600 nm) on the ionic strength [I] from sodium chloride at pH 7.2 and $T = 37°C$ after 48 hours of equilibration time [Morozova & Muthukumar (2018)]. Note that the turbidity is a non-monotonic function of ionic strength, with a peak at about the physiological ionic strength ~150 mM. At lower ionic strengths, τ is low but present. It reaches a maximum at about 0.1 M and decays rapidly to zero at higher ionic strengths. At about 0.2 M NaCl, fibrillization does not occur. In the narrow range of ionic strength where assembly occurs, the nature of fibrils delicately depends on the ionic strength as shown in Fig. 8.28. In this figure, atomic force microscopy (AFM) noncontact mode phase profiles of collagen samples with ionic strength [I] = 46, 96, and 156 mM at pH 7.4 and 37°C are plotted. At 46 mM ionic strength, the fibrils are semiflexible and tangled with a diameter of about 150 nm (compared to two orders of magnitude smaller diameter of the corresponding triple helix). As the ionic strength is increased to 96 mM, the fibril diameter increases to ~320 nm and the number density of the fibrils decreases. At the higher ionic strength of 156 mM, well-ordered thick fibers of diameter ~600

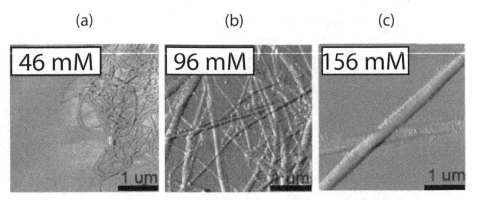

Figure 8.28 AFM images of type II collagen fibrils at pH 7.4 and 37°C at 46 mM, 96 mM, and 156 mM ionic strength. The scale bar is 1 μm [Adapted from Morozova & Muthukumar (2018)].

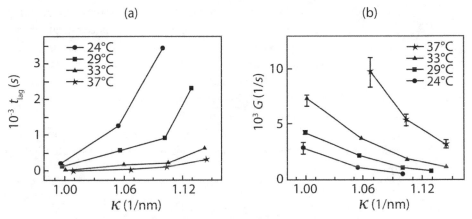

Figure 8.29 Kinetics of fibrillization of type II collagen as a function of ionic strength at different temperatures: (a) lag time t_{lag} and (b) initial growth rate G [Adapted from Morozova & Muthukumar (2018)].

nm form indicating the assembly structure at equilibrium. No fibrils form in solutions with ionic strength greater than 250 mM, consistent with Fig. 8.27b. Furthermore, the dependence of the lag time t_{lag} and the growth rate G (defined in Fig. 8.27a) on the ionic strength is given in Fig. 8.29 at pH 7.4 and different temperatures as plots of t_{tag} and G versus the inverse Debye length κ ($\kappa \sim \sqrt{[I]}$). As the ionic strength increases, the electrostatic interactions become progressively screened and the assembly takes longer. As seen in Fig. 8.29. the collagen assembly is slower at lower temperatures.

In summary, type II collagen fibrillization occurs only near isoelectric pH (~8.5) and if the ionic strength is below 250 mM. Furthermore, the assembly takes longer time if the ionic strength of the medium is increased. All of these features point to the important role of electrostatics on collagen assembly. Insight into this role can be gleaned by considering the electrostatic interaction energy between two triple helices.

Let us model each triple helix as a rigid cylinder [Wallace (1990), Morozova & Muthukumar (2018)]. The net surface charge on the triple helix can be estimated as a function of pH based on the pK_a values of the charged residues of the helix. This rough estimate of the surface charge is given in Fig. 8.30a as a function of pH. At pH < 4, the net charge is positive and the triple helices do not assemble due to inter-helix electrostatic repulsion. Similarly at pH > 10, fibrillization does not occur due to strong electrostatic repulsion between the negatively charged triple helices. At and around neutral pH, the net charge is roughly zero with the presence of both positively and negatively charged amino acids. In this situation, triple helices can pair using their complementary charges. Thus, the necessity of near neutral conditions for collagen fibrillization is a direct consequence of electrostatic interactions. It is remarkable that collagen assembly is exquisitely well-adapted to the tight constraints of physiological conditions.

The electrostatic contribution to the interaction energy between two triple helices of type II can be estimated by the following model and using the results in Section 4.3. Let each triple helix be a cylinder of length L and radius a. The cylinder axis is taken as a line charge with charges separated by a uniform distance ℓ (Fig. 8.30b). To specify the different charges of the various amino acids constituting the triple helix, each cylinder is imagined to be made of smaller cylinders each of length ℓ and radius a. For simplicity, we take each small cylinder to have a smeared charge of value 0, +3, or −3, depending on the amino acid sequence. Let another cylinder approach the first cylinder with parallel orientation with interaxis distance $d + 2a$ and a stagger length $n\ell$ (where n is an integer), as shown in Fig. 8.30b.

The electrostatic potential $\psi_j(r)$ due to a charge $z_j e$ inside a small cylinder j of the second cylinder at a radial distance r in the plane normal to the cylinder axis is given by Eq. (4.3.18) as

$$\psi_j(r) = \frac{z_j e}{2\pi\epsilon_0\epsilon\ell}\frac{1}{\kappa a}\frac{K_0(\kappa r)}{K_1(\kappa a)}, \tag{8.8.1}$$

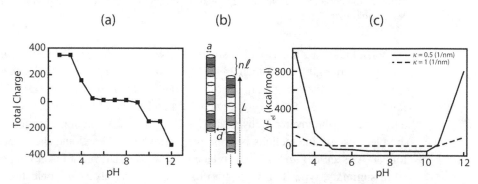

Figure 8.30 (a) Estimated net surface charge of a type II collagen triple helix, based on the pK_a values of charged amino acids. (b) Sketch of a model used in calculating pairwise interaction energy of cylinders of length L and radius a with interaxis distance d. The stagger length is $n\ell$ where ℓ is the length of a smaller cylinder described in the main text. (c) Plot of pairwise electrostatic interaction energy ΔF_{el} versus pH for two values of the inverse Debye length κ (0.5 nm^{-1} and 1.0 nm^{-1}) [Adapted from Morozova & Muthukumar (2018)].

where κ is the inverse Debye length, and K_0 and K_1 are the modified Bessel functions [Abramowitz & Stegun (1965)]. In getting the above equation from Eq. (4.3.18), the definition of the Bjerrum length is used. The electrostatic interaction energy $\Delta f_{\text{el}}(r_{ij})$ between two sub-cylinders i (with charge $z_i e$) and j (with charge $z_j e$) separated by distance r_{ij} is $z_i e \psi_j(r_{ij})$, such that

$$\Delta f_{\text{el}}(r_{ij}) = \frac{z_i z_j e^2}{2\pi\epsilon_0\epsilon\ell} \frac{1}{\kappa a} \frac{K_0(\kappa r_{ij})}{K_1(\kappa a)}. \tag{8.8.2}$$

For large values of κa and κr_{ij}, K_0 and K_1 can be approximated by [Abramowitz & Stegun (1965)]

$$K_\nu(x) \simeq \sqrt{\frac{\pi}{2x}} e^{-x}, \tag{8.8.3}$$

where the index ν is zero or one. Using this approximation, we get

$$\Delta f_{\text{el}}(r_{ij}) = \frac{z_i z_j e^2}{2\pi\epsilon_0\epsilon} \frac{1}{\kappa\ell} \frac{1}{\sqrt{ar_{ij}}} e^{-\kappa(r_{ij}-a)}. \tag{8.8.4}$$

The electrostatic interaction energy for two parallel cylinders separated by $d + 2a$ between the cylinder axes with stagger length $n\ell$ follows from the above equation as

$$\Delta F_{\text{el}} = \sum_{i=n+1}^{L/\ell} \sum_{j=1}^{(L-n\ell)/\ell} \frac{z_i z_j e^2}{4\pi\epsilon_0\epsilon} \frac{1}{\kappa\ell} \frac{1}{\sqrt{ar_{ij}}} e^{-\kappa(r_{ij}-a)}, \tag{8.8.5}$$

where L is the total length of the cylinder and

$$r_{ij} = \sqrt{(d+2a)^2 + (j-i)\ell^2}, \tag{8.8.6}$$

and a factor of $1/2$ is used in Eq. (8.8.5) to avoid double counting of pairwise interactions. Using representative values pertinent to type II collagen fibrils ($L = 300$ nm, $a = 0.75$ nm, $\ell = 0.29$ nm, $d = 1.3$ nm, $n = 234$, and $z_{i,j} = 0, +3$, or -3), an example of the calculated ΔF_{el} is given in Fig. 8.30c as a function of pH for two inverse Debye lengths ($\kappa = 0.5$ nm^{-1} and 1 nm^{-1}). The electrostatic contribution to the energy of formation of a pair of staggered triple helices is non-monotonic with respect to pH. For pH < 4 and pH > 10, ΔF_{el} is positive. In the intermediate pH range, ΔF_{el} is negative if the ionic strength is not high. When the Debye length κ^{-1} is 2 nm, ΔF_{el} is a minimum at pH 7. On the other hand, if κ^{-1} is 1 nm, ΔF_{el} is essentially zero in the intermediate pH range. The predictions in Fig. 8.30c based on an elementary electrostatic model for two interacting triple helices provide a qualitative rationale for the main experimental findings mentioned above for type II collagen fibrillization: at intermediate pH, collagen triple helices can assemble, and at extreme pH values, strong electrostatic repulsion prevents assembly; at lower ionic strengths (κ small), the attractive electrostatic energy associated with dipole (ion pair) formation between the triple helices is strong, leading to assemblies that dissolve at higher ionic strengths due to electrostatic screening. Analogously, electrostatic attraction at lower ionic strengths leads to faster nucleation and faster growth as exhibited in Fig. 8.29. Quantitative theories for nucleation and growth kinetics of fibrillization by accounting for electrostatic interactions remains to be fully developed.

8.8.1.2   Amyloid Fibril Formation

All amyloidogenic proteins exhibit fibrillar morphology. Under physiological conditions, the protein molecules can assemble into fibrils. Many *in vitro* experiments have revealed that the early stage of fibrillization occurs through several pathways involving fully unfolded, partially folded, and fully folded states of the single protein molecule [Bemporad *et al.* (2006)]. Generally speaking, there exists an intermediate oligomeric state of fibrils which is in equilibrium with unaggregated monomers and provides templates for formation of mature fibrils. For example, consider the fibrillization of the intrinsically disordered protein α-synuclein (described in Section 5.11.3). The typical experimental condition for fibrillization is incubation of monomeric α-synuclein at 37°C and pH 7.4. The rate of growth depends on other experimental variables such as ionic strength and mechanical agitation. The growth kinetics exhibits the three regimes of lag, growth, and saturation as shown in Fig. 8.27a. α-Synuclein monomers first form oligomers of 5 or 7 nm linear size in the late lag period. Subsequently, these oligomers form fibrils of length 500 nm to 3 μm and three distinct widths (3.8, 6.5, and 9.8 nm) [Fink (2006)], as shown in Fig. 8.31.

Even though mapping between this cascade of amyloid aggregation and the propensity of the onset of neurodegenerative diseases is still vigorously pursued, there is a preponderance of evidence suggesting that amyloid fibril formation is a critical step

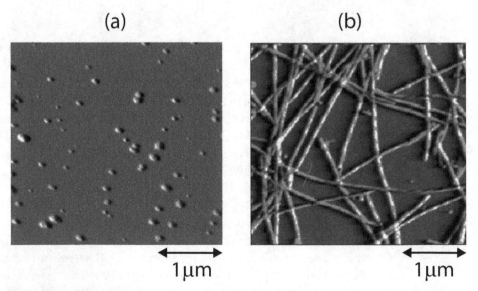

(a)   (b)

1 µm   1 µm

Figure 8.31 AFM images of (a) oligomers and (b) mature fibrils formed by α-synuclein. The heights of the oligomers are 5 or 7 nm and mature fibrils are 10 nm high [Adapted from Fink (2006)].

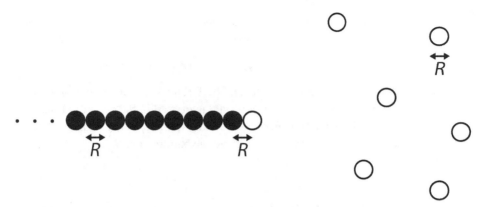

Figure 8.32 An elementary model to calculate the fibril elongation rate as a function of ionic strength in the bulk electrolyte solution. Each building unit in the fibril and solution is spherical with diameter R and charge ze. The electrostatic interaction in the solution is assumed to be given by the Debye–Hückel theory with the inverse screening length κ proportional to the square root of the ionic strength [I].

in the etiology of these diseases. In view of the significance of amyloid fibril formation in the context of averting neurodegenerative diseases, a molecular understanding of amyloid fibrillization is presently an active research area. The general theoretical paradigm to address amyloid fibrillization is the nucleation and growth mechanism described in the preceding sections. In this vast area of research addressing various biochemical and physical aspects, we focus here only on the electrostatic effects on the kinetics of fibril elongation.

Consider an idealized situation, cartooned in Fig. 8.32, where a one-dimensional fibril of building units (monomers or the oligomeric globule-like entities shown in Fig. 8.31a) grows by sequential capture of the building units dispersed in an electrolyte solution. Let each building unit in the assembly and solution be approximated as a spherical particle of diameter R and net charge ze. In the solution, the diffusion coefficient D of the building unit is $k_B T / 3\pi\eta_0 R$ (Stokes–Einstein) and c_0 is the concentration of the protein (strictly speaking, the building units) in the bulk solution. Furthermore, let us assume that the Debye–Hückel theory is applicable to the interactions among the building units, with the inverse Debye length κ proportional to the square root of the ionic strength in the solution ($\kappa \sim \sqrt{[I]}$).

The rate at which a building unit is captured at the growing end of a fibril is addressed as follows. Let us fix the origin of the coordinate system as the center of the building unit at the growing end of the fibril. The building unit that is being captured undergoes diffusion due to thermal forces and drift due to its potential interaction with the building unit at the end of the fibril (Fig. 8.32). As mentioned in Section 7.6.3.1, the time dependence of the local concentration $c(\mathbf{r}, t)$ of the building units at location $\mathbf{r}$ from the fibril end and time t is given by

$$\frac{\partial c(\mathbf{r}, t)}{\partial t} = -\nabla \cdot \mathbf{J}(\mathbf{r}, t), \qquad (8.8.7)$$

where $\mathbf{J}(\mathbf{r},t)$ is the flux of the building units into the volume element at $\mathbf{r}$. As mentioned above, $\mathbf{J}(\mathbf{r},t)$ arises from both diffusion and drift given by

$$\mathbf{J}(\mathbf{r},t) = -D\nabla c(\mathbf{r},t) + c(\mathbf{r},t)\mathbf{v}(\mathbf{r},t), \tag{8.8.8}$$

where D is the diffusion coefficient of the building unit and $\mathbf{v}(\mathbf{r},t)$ is the velocity of the building unit at $\mathbf{r}$ and t. Since we are interested in the capture of the building unit by growing fibril, we consider $\mathbf{v}(\mathbf{r},t)$ arising from the potential interaction $U(\mathbf{r})$ between the building unit at the fibril end and the building unit at $\mathbf{r}$,

$$\mathbf{v}(\mathbf{r},t) = \frac{D}{k_B T} [-\nabla U(\mathbf{r})], \tag{8.8.9}$$

where $D/k_B T$ is the mobility of the building unit and $(-\nabla U)$ is the force acting on it. In general, U consists of hydrophobic excluded volume component and the electrostatic component,

$$U(\mathbf{r}) = U_{\text{exc}}(\mathbf{r}) + U_{\text{el}}(\mathbf{r}). \tag{8.8.10}$$

For assembly involving similarly charged building units, the attractive $U_{\text{exc}}(\mathbf{r})$ drives the assembly against the electrostatic repulsion between the units. On the other hand, if the building units are oppositely charged as in the case of coacervation (Chapter 9), then capture can occur spontaneously. For the present goal of addressing the role of electrostatic effects on amyloid fibril growth, let us suppress the role of $U_{\text{exc}}(\mathbf{r})$ and take U, as given by the extended Debye–Hückel theory, as (Eq. (3.8.30))

$$\frac{U(\mathbf{r})}{k_B T} = \frac{z^2 \ell_B}{(1 + \kappa R)} \frac{e^{-\kappa(r-R)}}{r} = \gamma z^2 \ell_B \frac{e^{-\kappa r}}{r}, \tag{8.8.11}$$

where γ is defined as $\exp(\kappa R)/(1 + \kappa R)$. Ignoring all excluded volume interactions and using radial symmetry, Eqs. (8.8.7)–(8.8.9) yield

$$\frac{\partial c(r,t)}{\partial t} = \frac{1}{r^2} \frac{\partial}{\partial r} \left\{ r^2 \left[D \frac{\partial c(r,t)}{\partial r} + \frac{D c(r,t)}{k_B T} \frac{\partial U(r)}{\partial r} \right] \right\}. \tag{8.8.12}$$

When a building unit undergoing diffusion and drift reaches a distance R from the center of the building unit at the growing end of the fibril, we assume that the incoming building unit is absorbed. Furthermore, we assume that there is a constant supply of the building units to maintain the bulk concentration of building units at c_0 in the steady state. As derived in Appendix 8, the rate of growth I_{total} in the steady state is given as

$$I_{\text{total}} = (4\pi D R c_0) e^{\gamma z^2 \ell_B \kappa} \left(\frac{\gamma z^2 \ell_B}{R} \right) \left(e^{\gamma z^2 \ell_B / R} - 1 \right)^{-1}. \tag{8.8.13}$$

Note that, if there is no potential interaction ($z = 0$ in the present case), the above equation reduces to

$$I_{\text{total}}(z = 0) = 4\pi D R c_0, \tag{8.8.14}$$

which is the Smoluchowski diffusion limited capture rate [Chandrasekhar (1943)]. Instead of the factor 4π in Eq. (8.8.13) used for the spherical geometry of the building unit at the fibril end, a different prefactor has to be used depending on the patchiness of the surface of the building unit. Furthermore, a captured building unit can dissociate before the next unit is captured during the elongation process. In this case, there is an additional prefactor of the fraction of time $\tau_d/(\tau_d + \tau_r)$ when the capturing site is free [Buell et $al.$ (2010)]. Here, τ_r is the average residence time of the captured unit and τ_d is the characteristic diffusion time $\tau_d = (4\pi D R c_0)^{-1}$ given by Eq. (8.8.14). Hence, the elongation rate can be expressed as

$$I_{\text{total}} = \frac{4\pi D R c_0}{(1 + 4\pi D R c_0 \tau_r)} e^{\gamma z^2 \ell_B \kappa} \left(\frac{\gamma z^2 \ell_B}{R} \right) \left(e^{\gamma z^2 \ell_B / R} - 1 \right)^{-1}, \tag{8.8.15}$$

so that

$$I_{\text{total}} \sim e^{\gamma z^2 \ell_B \kappa}. \tag{8.8.16}$$

Therefore, the fibril elongation rate increases exponentially with κ. The fibril formation triggered by attractive hydrophobic interactions is amplified by screening the electrostatic repulsion between the assembling units at higher ionic strengths. Since $\kappa \sim \sqrt{[I]}$, a plot of log I_{total} versus square root of ionic strength is predicted to be linear based on the assumptions made in the above derivation. This behavior is observed in experiments [Buell et $al.$ (2013)]. For example, the effect of sodium chloride concentration on the insulin amyloid fibril elongation rate is given in Fig. 8.33a as a semilog plot of the rate versus $\sqrt{[I]}$, which is found to be in accordance with Eq. (8.8.16). However, as the ionic strength is increased, drastic deviation from the above prediction has been observed. This is illustrated in Fig. 8.33b, where the non-monotonic dependence of the fibril elongation rate with ionic strength is evident. Implementation of the full expression (Eq. (8.8.15)) and incorporation of hydrophobic contribution

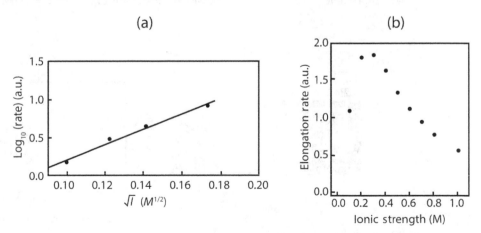

Figure 8.33 Dependence of insulin fibril elongation rate on ionic strength (measured using quartz crystal microbalance at 35°C). (a) Logarithm of the rate is linear with the square root of ionic strength, in accordance with Eq. (8.8.16). (b) Non-monotonic dependence of elongation rate on the ionic strength of the solution. At ionic strengths above 0.2 M NaCl, the fibril elongation rate decreases with an increase in the ionic strength [Adapted from Buell et $al.$ (2013)].

[Raman *et al.* (2005), Wetzel (2006), Eisenberg *et al.* (2006), Zhang & Muthukumar (2009), Buell *et al.* (2014)] in deriving the elongation rate are needed as a minimum effort to understand the origin of the non-monotonic feature portrayed in Fig. 8.33b.

8.8.2   Membraneless Organelles

In the case of intrinsically disordered proteins (IDPs), their solutions can form liquid-like biomolecular condensates upon certain changes in experimental conditions. There is a growing recognition that these condensates are membraneless organelles associated with several neurodegenerative diseases, ribosome assembly, gene regulation, and signal transduction [Uversky (2002), Kato *et al.* (2012), Pak *et al.* (2016), Banani *et al.* (2017)]. Such biomolecular condensates are postulated to arise from the liquid–liquid phase separation phenomenon described in the preceding sections, although there are several additional twists arising from protein sequence effects and specific electrostatic interactions.

As an example, solutions of hnRNPA1 (a mutant of heterogeneous nuclear ribonucleoprotein in stress granules, which are membraneless organelles) undergo liquid–liquid phase separation (LLPS) [Molliex *et al.* (2015)]. Turbidity of 500 μM hnRNPA1 solution increases upon reduction in temperature from 25°C to 4°C and differential interference contrast microscopy of the turbid solution at 10°C shows droplets containing protein-rich phase (Fig. 8.34a). These droplets are dynamical assemblies with continuous assembly and disassembly. In addition, the droplets show liquid properties with rapid exchange of molecules between the droplets and the surrounding solution. The coexistence curves for this LLPS are given in a plot of temperature versus protein concentration (Fig. 8.34b) at different concentrations of the crowding agent Ficoll, which promotes LLPS. The three curves from the top to bottom correspond to 150,

(a)   (b)

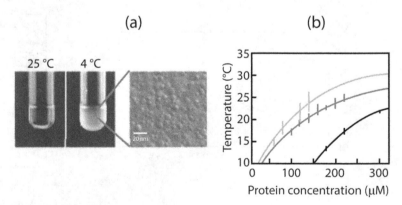

Figure 8.34   (a) Birth of droplets of hnRNPA1 protein upon reduction in temperature from 25°C to 4°C. Formation of droplets is seen in differential interference contrast microscopy on turbid hnRNPA1 solution at 10°C. (b) Coexistence curves for hnRNPA1 solutions at different concentrations of the crowding agent Ficoll (150, 100, and 75 mg/mL, from top to bottom). The solution is a single phase above the curves and two phases coexist below the curves [Molliex *et al.* (2015)].

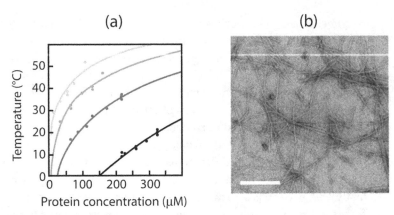

Figure 8.35 (a) Coexistence curves for solutions of Ddx4^{N1} at different NaCl concentrations (100, 150, 200, and 300 mM, from top to bottom). The solution is a single phase above the curves and two phases coexist below the curves [Nott *et al.* (2015)]. (b) Amyloid-like fibers constituting the hydrogels formed in solutions of the chimeric protein of fused-in-sarcoma RNA-binding protein (scale bar = 320 nm) [Kato *et al.* (2012)].

100, and 75 mg/mL Ficoll, respectively. The solution is a single homogeneous phase for conditions above the curve and consists of two coexisting phases for conditions below the coexistence curves.

The occurrence of LLPS upon reduction in temperature, increase in protein concentration, and increase in a crowding agent, are in qualitative agreement with the UCST behavior seen in homopolymer solutions. However, the salt concentration dependence is the opposite (see Section 8.1). For example, the cloud point temperature below which droplets form increases as ionic strength decreases, as shown in Fig. 8.35a for the assembly of the protein Ddx4^{N1} (primary constituent of germ or nuage granules) [Nott *et al.* (2015)]. The four curves from the top to bottom correspond to 100, 150, 200, and 300 mM NaCl concentration, respectively. Addition of salt suppresses LLPS for this system in contrast to the expectation based on the behavior of solutions of uniformly charged polyelectrolytes. This discrepancy suggests that identification of LLPS (that is observed in homopolymer solutions) as the mechanism of formation of membrane-less organelles is incomplete. Furthermore, different sequences of charges in Ddx4, with the same charge composition as the wild type, exhibit different phase separation propensity [Nott *et al.* (2015), Lin *et al.* (2020)]. These results emphasize that sequence-dependent ion-pair formation between charged repeat units belonging to several protein molecules plays a significant role.

In addition to such electrostatic correlations inside the droplets, there can be significant structural changes induced by the formation of droplets. For example, upon lowering the ionic strength via dialysis, hnRNPA1 assembles into hydrogels, instead of the organelles given in Fig. 8.34a. Similarly, chimeric protein of fused-in-sarcoma RNA-binding protein undergoes a reversible concentration-dependent phase transition into a hydrogel structure, composed of amyloid-like fibers (Fig. 8.35b) [Kato *et al.* (2012)]. As an additional example of hierarchical structural organization, when solutions of α-synuclein undergo LLPS, the liquid droplets containing concentrated

α-synuclein are embedded inside hydrogel matrices [Ray *et al*. (2020)]. Furthermore, the enhanced concentration of intrinsically disordered proteins in the liquid droplets can promote aggregation and fibrillization of proteins [Molliex *et al*. (2015), Kato *et al*. (2012), Ray *et al*. (2020)]. Thus, there is an evolution of hierarchical structures inside droplets of concentrated intrinsically disordered proteins.

Overall, the "LLPS" in solutions of intrinsically disordered proteins is much more complex than encountered in solutions of uniformly charged polyelectrolytes. Here, a hierarchy of structural organization accompanies the formation of membraneless organelles. The ability of different distant portions of the same molecule to be locally organized indicates that microphase separation (Section 8.6) is occurring concomitantly with macrophase separation. Association of oppositely charged domains either from the same molecule or from several interpenetrating molecules can readily occur by releasing their relevant counterions. Such associations formed by multiple ion pairs can spontaneously result in formation of networks involving many chains [Tanaka (2011), Muthukumar (2017), Das & Muthukumar (2022)]. Depending on the availability of such molecules in the system, these associations can lead to either clusters of finite size or a percolating gel. Even for the simplest situation of phase behavior of associating polymers, the phase diagram displays a homogeneous solution phase, percolated gel phase, percolation line separating solution and gel phase, and the coexistence of a dilute solution phase with a gel phase at lower temperatures. The percolation line meets the coexistence curve at a location that is specific to the system [Tanaka (2011)]. Inside the coexistence curve, the equilibrium situation is the coexistence of one solution phase and one gel phase at a given temperature. In reaching this equilibrium, it is not uncommon that the kinetic process of phase separation into the solution and gel phases can be slow due to barriers arising from the necessity of structural reorganization to reach equilibrium. Analogous to this well-known behavior of associating polymers, the role of percolation in concentrated solutions of biomolecules, modeled as sticker-spacer chains, has been recognized as an important factor to understand the phase behavior of biomolecular systems [Harmon *et al*. (2017), Choi *et al*. (2020a,b), Dar & Pappu (2020)]. The role of electrostatic interactions on the percolation line and phase behavior of biomolecular solutions remains to be addressed. Furthermore, if the sequence of the biomolecule contains domains of different kinds, then a collection of large numbers of such molecules can exhibit the phenomenon of microphase separation described in Section 8.6.2. At higher concentrations of biomolecules inside a droplet, microphase-separated structures can spontaneously form [Das & Muthukumar (2022)].

At present, a full understanding of the formation of membraneless organelles is only at its infancy. Nevertheless, the main theme behind the structural organization inside membraneless organelles is the electrostatic interactions among protein molecules working together with the formation of hydrophobic domains. This is evident from the significant role played by salt concentration in controlling conditions that promote the birth of condensates. In addition, the formation of condensates is enhanced by the presence of oppositely charged macromolecules in solution. We shall return to this effect in the context of coacervation described in Chapter 9.

9 Adsorption, Complexation, and Coacervation

9.1   Introduction

Charged macromolecules dispersed in a solution spontaneously adsorb to oppositely charged surfaces if the electrostatic attraction between them is strong enough (Fig. 9.1a). The electrostatic strength, controlled by the Bjerrum length and the Debye length, must be more than sufficient to overcome the loss of conformational entropy of the adsorbing macromolecules in the adsorbed state. Furthermore, adsorption in general is facilitated by the entropically favorable release of condensed counterions accompanying the adsorption process. This effect depends on the extent of condensed counterions before adsorption, which in turn depends on the strength of the electrostatic interactions in the system. Solvent reorganization also plays a role in establishing adsorption equilibria. Furthermore, the critical condition for adsorption depends on the geometry of the interface, whether it is planar, cylindrical, or spherical. Adsorption on curved surfaces is ubiquitous in colloid science and various biological situations. For example, complexation between a macromolecule and a protein or a vesicle can be addressed using the concepts of adsorption on a spherical surface (Fig. 9.1b). Complementing the formation of adsorbed layers, brush-like surfaces (Fig. 9.1c) carrying many charged macromolecules, with one end permanently anchored at the surface, are of common occurrence. Such charged brushes are of great significance in many biological contexts such as articular cartilage lubrication.

The same contributing factors to adsorption of macromolecules to oppositely charged surfaces are also operative when two oppositely charged macromolecules encounter each other to form intermolecular complexes (Fig. 9.1d). In addition, both partners must surrender their conformational degrees of freedom upon complexation in a self-consistent manner. The aqueous assembly of such intermolecular complexes is a confluence of inter-chain electrostatic attraction, intra-chain electrostatic repulsion, counterion release, the presence of other ions in the background, hydrophobic contribution, and rearrangement of water molecules. A delicate interplay between these simultaneously occurring forces contribute to the structure and dynamics of such assemblies. Intermolecular complexes proliferate in almost all biological processes as well as in the context of biomimetic assemblies. An illustrative example is genome packing in ssRNA viruses (Fig. 9.1e).

When oppositely charged macromolecules or molecules containing oppositely charged domains are abundant in an aqueous solution containing salt, the mixture

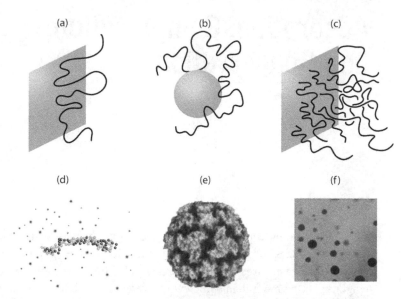

Figure 9.1 (a) Adsorption of a macromolecule at an oppositely charged planar surface. (b) Adsorption at a spherical surface. (c) Planar charged brush. (d) Complexation between two oppositely charged flexible chains [adapted from Ou & Muthukumar (2006)]. (e) The capsid of the rhinovirus with packaged ssRNA inside. The figure is obtained from VIPERdb (http://viperdb.scripps.edu) [Montiel-Garcia *et al*. (2021)]. (f) Confocal image of an equilibrated mixture of a coacervate phase with droplets of coexisting dilute phase [Spruijt *et al*. (2010)].

readily undergoes phase separation, resulting in coexistence of a polymer-rich phase and a polymer-poor phase. The polymer-rich phase is commonly called **coacervate phase** and the liquid–liquid phase separation in a mixture of oppositely charged macromolecules is called **coacervation**. An example is a mixture of poly(acrylic acid) and poly(N,N-dimethylaminoethyl methacrylate) in water containing potassium chloride, which results in droplets of polymer-poor phase amidst the background of the coacervate phase (Fig. 9.1f). Other examples of coacervates include membraneless dynamic biomolecular condensates, which are transient liquid-like droplets made of RNA and proteins.

In this chapter, we shall present the basic principles of assembly processes of charged macromolecules complexing with oppositely charged interfaces and macromolecules. Specifically, we shall address (i) adsorption at planar surfaces and spherical surfaces (both outside and inside), (ii) charged brushes, (iii) genome assembly inside RNA viruses, (iv) intermolecular complexation, (v) coacervation, and (vi) membraneless organelles.

9.2   Adsorption

Consider the adsorption of a single uniformly charged flexible chain on oppositely charged planar surfaces (Fig. 9.1a) or spherical surfaces (Fig. 9.1b). The electrostatic attraction between the polymer and a surface can result in adsorption. However, the adsorption is associated with a loss in conformational entropy of the polymer, which disfavors adsorption. Optimization between the electrostatic attraction and conformational entropy loss results in a critical condition as a necessary criterion for adsorption. If the attraction between the polymer and surface is weak, adsorption does not occur due to the more severe entropic penalty. For sufficiently strong attraction, adsorption occurs. Therefore, an evaluation of the electrostatic and entropic contributions enables derivations of adsorption criteria. As we shall show in the following text, the adsorption criterion depends on the Debye length and charge densities of the polymer and surface. By varying the experimental parameters that dictate the Debye length, such as salt concentration and pH, the capture and release of charged macromolecules by surfaces can be tuned using the derived adsorption criteria.

A general method to derive criteria for adsorption is based on the idea that the chain, taken as a guest molecule, is presented with an attractive potential from the host surface. This approach is also applicable to intermolecular complexation where the host is another macromolecule instead of a surface. The chain connectivity, excluded volume and the electrostatic interactions among polymer segments, and guest–host interactions are addressed by generalizing the path integral approach (Section 2.4) where a mapping is made with the time-dependent Schrödinger equation for a particle in a potential well (Eq. (2.4.6)). In this analogy, the kinetic energy of the particle is equivalent to conformational entropy of the guest and the potential energy of the particle is the guest–host electrostatic and excluded volume interaction energy. Accounting for chain connectivity and interactions of guest polymer segments among themselves and with the surface, Eq. (2.4.6) is generalized to get the probability distribution function $G(\mathbf{r}, \mathbf{r}'; N)$ for a flexible polyelectrolyte chain of contour length $N\ell$ with its ends at $\mathbf{r}$ and $\mathbf{r}'$ in the proximity of the surface, as

$$G(\mathbf{r}, \mathbf{r}'; N) = \int_{\mathbf{r}(0)=\mathbf{r}'}^{\mathbf{r}(N)=\mathbf{r}} \mathcal{D}[\mathbf{r}(s)] \exp\left[-\frac{3}{2\ell^2}\int_0^N \left(\frac{\partial \mathbf{r}(s)}{\partial s}\right)^2 - \frac{1}{k_B T}\int_0^N V_P[\mathbf{r}(s)]\right.$$
$$\left. -\frac{1}{k_B T}\int_0^N V_S[\mathbf{r}(s)]\right], \tag{9.2.1}$$

where $\mathbf{r}(s)$ is the position vector of the arc length variable s ($0 \le s \le N\ell$). V_P denotes the various intersegment interactions of the chain acting on the segment at $\mathbf{r}(s)$. V_S is the electrostatic potential from the charged surface acting on the polymer segment at $\mathbf{r}(s)$. The symbol $\int \mathcal{D}[\mathbf{r}(s)]$ denotes the functional integration representing the sum over all possible chain configurations under the constraints from all intra-chain interactions and interaction with the surface. The path integral representation of Eq. (9.2.1) can be equivalently written as

$$\left[\frac{\partial}{\partial N} - \frac{\ell^2}{6}\nabla_{\mathbf{r}}^2 + \frac{V_P(\mathbf{r})}{k_B T} + \frac{V_S(\mathbf{r})}{k_B T}\right] G(\mathbf{r}, \mathbf{r}'; N) = \delta(\mathbf{r} - \mathbf{r}')\delta(N). \tag{9.2.2}$$

As already noted, this equation is analogous to the time-dependent Schrödinger equation for a particle in a potential. The first two terms inside the square brackets represent the chain connectivity mapped into the kinetic energy of the particle and the third and fourth terms correspond to the energy from intra-chain interactions and the surface.

Using the bilinear expansion of $G(\mathbf{r},\mathbf{r}';N)$ in order to separate the variable N, Eq. (9.2.2) becomes

$$G(\mathbf{r},\mathbf{r}';N) = \sum_{m=0}^{\infty} \psi_m(\mathbf{r})\psi_m(\mathbf{r}')e^{-\lambda_m N}, \qquad (9.2.3)$$

where ψ_m and λ_m are the eigenfunctions and eigenvalues of the time-independent Schrödinger equation

$$\left[-\frac{\ell^2}{6}\nabla_{\mathbf{r}}^2 + \frac{V_P(\mathbf{r})}{k_BT} + \frac{V_S(\mathbf{r})}{k_BT} \right] \psi_m(\mathbf{r}) = \lambda_m \psi_m(\mathbf{r}). \qquad (9.2.4)$$

For a given set of potentials, V_P and V_S, and appropriate boundary conditions, the eigenvalue problem of Eq. (9.2.4) is solved for ψ_m and λ_m. The contribution from the first term inside the square brackets represents the entropic contribution and the rest of the terms denote the energy contribution to adsorption. Using the calculated ψ_m and λ_m, the probability distribution $G(\mathbf{r},\mathbf{r}';N)$ is obtained from Eq. (9.2.3). The adsorption criteria, and the density profile and free energy of an adsorbed chain follow from $G(\mathbf{r},\mathbf{r}';N)$.

If the combined potential $V(\mathbf{r})/k_BT = (V_P + V_S)/k_BT$ is attractive and sufficiently strong, the allowed solutions of the Schrödinger equation are bound states, as illustrated in Fig. 9.2. If the combined potential is weaker than a critical value, only scattering states are allowed. The occurrence of bound states corresponds to adsorbed states. Note that in the case of guest–host intermolecular interaction, the bound state corresponds to the formation of intermolecular complexes. The scattering states denote the absence of adsorption or complexation. The adsorption criterion corresponds to the condition at which at least one bound state is allowed by Eq. (9.2.4).

Once the adsorption criterion is established, the density profile and the free energy of the adsorbed state are calculated from the eigenfunctions ψ_m and eigenvalues λ_m.

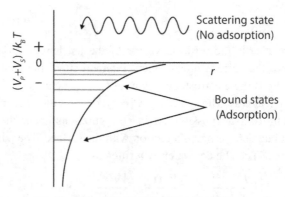

Figure 9.2 Long-ranged attractive potential $(V_P(\mathbf{r}) + V_S(\mathbf{r}))/k_BT$ emanating from a surface as a function of distance r from the surface. If the potential is attractive and stronger than a critical value, bound states are allowed denoting adsorbed states. If the potential is weakly attractive compared to the critical value or repulsive, only scattering states are allowed representing the unadsorbed state. The critical value corresponding to the appearance of at least one bound state provides the adsorption criterion.

For large polyelectrolyte chains where $N \gg 1$, and since $\lambda_m < 0$ in the adsorbed state, the leading term on the right-hand side of Eq. (9.2.3), corresponding to the ground state, dominates over the rest of the terms. Therefore, $G(\mathbf{r}, \mathbf{r}'; N)$ can be approximated by taking only the ground state contribution

$$G(\mathbf{r}, \mathbf{r}'; N) \simeq \psi_0(\mathbf{r})\psi_0(\mathbf{r}')e^{-\lambda_0 N}. \qquad (9.2.5)$$

This is known as the **ground state dominance (GSD) approximation**. The segment density profile in the adsorbed state $\rho(\mathbf{r})$ is generally given by

$$\rho(\mathbf{r}) = \frac{\int_0^N ds \int d\mathbf{r}_0 \int d\mathbf{r}_N G(\mathbf{r}, \mathbf{r}_0; s) G(\mathbf{r}_N, \mathbf{r}; N - s)}{\int d\mathbf{r}_0 \int d\mathbf{r}_N G(\mathbf{r}_N, \mathbf{r}_0; N)}, \qquad (9.2.6)$$

where $\mathbf{r}_0$ and $\mathbf{r}_N$ denote the positions of the chain ends, and $\mathbf{r}$ denotes the position of the sth segment. Substituting Eq. (9.2.5) from the GSD approximation, the segment density profile follows as

$$\rho(\mathbf{r}) = N\psi_0^2(\mathbf{r}). \qquad (9.2.7)$$

The free energy F of the adsorbed state is given by

$$e^{-F/k_B T} = \int d\mathbf{r} \int d\mathbf{r}' G(\mathbf{r}, \mathbf{r}'; N). \qquad (9.2.8)$$

Using the GSD approximation, the free energy is given as

$$\frac{F}{k_B T} = \lambda_0 N, \qquad (9.2.9)$$

where constant terms independent of N are left out. Since the "kinetic energy" part, $-\frac{\ell^2}{6}\nabla_{\mathbf{r}}^2$, in Eq. (9.2.4) corresponds to the conformational entropy of the chain and the rest of the terms on the left-hand side correspond to the energy of the system, the free energy per segment can be expressed as a sum of entropic contribution λ_s and energy contribution λ_u:

$$\frac{F}{N k_B T} = \lambda_s + \lambda_u. \qquad (9.2.10)$$

Using the above general procedure, we now consider the adsorption of a polyelectrolyte chain at oppositely charged planar surfaces, solid spherical surfaces, and for vesicles where adsorption can occur either in the exterior or interior.

9.2.1 Adsorption at a Planar Surface

Let $\sigma_s e$ be the charge density (e is the electronic charge and σ_s is the number charge density) on a thin and infinitely large planar surface and qe be the linear charge density of an oppositely charged flexible polyelectrolyte chain (Fig. 9.3). Note that $q\ell = \alpha z_p$, where ℓ is the Kuhn segment length, α is the average degree of ionization, and z_p is the number of ionizable repeat monomers per segment.

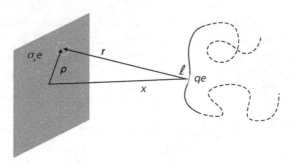

Figure 9.3 Attractive potential emanating from a large planar surface with charge density $\sigma_s e$ toward a segment of an oppositely charged polymer with linear charge density qe.

The potential $V_P[\mathbf{r}(s)]$ at the sth segment of the guest molecule arising from all intersegment interactions is given by Eq. (6.3.5) as

$$\frac{V_P[\mathbf{r}(s)]}{k_B T} = \frac{\ell^3}{2} \int_0^N ds' \, (1 - 2\chi) \, \delta \left[\mathbf{r}(s) - \mathbf{r}(s') \right] + \frac{q^2 \ell^2 \ell_B}{2} \int_0^N ds' \frac{e^{-\kappa |\mathbf{r}(s) - \mathbf{r}(s')|}}{|\mathbf{r}(s) - \mathbf{r}(s')|},$$

(9.2.11)

where all degrees of freedom of all dissociated small ions and solvent molecules are taken into account using the Debye–Hückel formalism (see Section 6.3.1).

The electrostatic potential energy $V_S[\mathbf{r}(s)]$ from the surface on the polymer segment at $\mathbf{r}(s)$ is obtained as follows. Consider a small surface area around the two-dimensional location $\vec{\rho}$ from the center of the surface, which is at the three-dimensional vectorial distance $\mathbf{r}$ from the location $\mathbf{r}(s)$ of the sth polymer segment (Fig. 9.3). Let x be the perpendicular distance of $\mathbf{r}(s)$ from the center of the surface. Using the Debye–Hückel potential (Eq. (3.8.16)) between two charges and averaging over all surface elements of the planar surface, we get

$$\frac{V_S[\mathbf{r}(s)]}{k_B T} = -|\sigma_s q| \ell_B \int d^2 \rho \frac{e^{-\kappa \sqrt{\rho^2 + x^2}}}{\sqrt{\rho^2 + x^2}}.$$

(9.2.12)

Performing the two-dimensional integration over $\vec{\rho}$, we obtain the electrical potential at x as

$$\frac{V_S[x]}{k_B T} = -2\pi |\sigma_s q| \frac{\ell \ell_B}{\kappa} e^{-\kappa x}.$$

(9.2.13)

Note that this potential is proportional to the one-dimensional result of the Gouy–Chapman theory given in Eqs. (4.1.13) and (4.1.14) within a factor of two.

The adsorption criterion and the properties of the adsorbed state are obtained by substituting Eqs. (9.2.11) and (9.2.13) in Eq. (9.2.4) and solving for the eigenfunctions ψ_m and eigenvalues λ_m. Even in the absence of any adsorbing surface, the chain statistics of an isolated polyelectrolyte chain is not exactly solvable. As described in Section 5.6, the square of the radius of gyration of a polyelectrolyte chain can be approximated in terms of the result for an effective Gaussian chain as

$$R_g^2 = \frac{N \ell \ell_{\text{eff}}}{6},$$

(9.2.14)

where ℓ_{eff} depends on N, ℓ_B, and κ and other excluded volume parameters pertinent to van der Waals-type interactions. Recall that the approximation of uniform chain expansion is used in writing the above expression for R_g. For Gaussian chains ℓ_{eff} is the Kuhn length ℓ. The limiting behaviors of ℓ_{eff} for the low-salt and high-salt limits are given in Eqs. (5.6.5) and (5.6.17) as (Appendix 2)

$$\ell_{\text{eff}} = \begin{cases} \frac{2}{3}\left(\frac{2}{5}\right)^{2/3}\frac{1}{\pi^{1/3}}\left(\frac{\alpha^2 z_p^2 \ell_B}{\ell}\right)^{2/3} N\ell, & \kappa R_g \ll 1, \\ \frac{6^{1/5}}{\pi^{3/5}}\left[(1-2\chi)+\frac{4\pi\alpha^2 z_p^2 \ell_B}{\kappa^2 \ell^3}\right]^{2/5} N^{1/5}\ell, & \kappa R_g \gg 1. \end{cases} \tag{9.2.15}$$

With this approximation of uniform swelling of the chain, the role of $V_P/k_B T$ in Eqs. (9.2.2) and (9.2.4) can be absorbed in the conformational entropic term by replacing one factor of Kuhn length by ℓ_{eff} [Muthukumar (1987)]. Therefore, Eq. (9.2.2) becomes

$$\left[\frac{\partial}{\partial N} - \frac{\ell\ell_{\text{eff}}}{6}\nabla_{\mathbf{r}}^2 + \frac{V_S(\mathbf{r})}{k_B T}\right] G(\mathbf{r},\mathbf{r}';N) = \delta(\mathbf{r}-\mathbf{r}')\delta(N), \tag{9.2.16}$$

where $\mathbf{r} = (x,y,z)$ and $\mathbf{r}_0 = (x_0,y_0,z_0)$ are the two ends of the chain. Since $V_S(\mathbf{r}) = V_S[x]$ depends only on x and not on y and z, the probability distribution function $G(\mathbf{r},\mathbf{r}';N)$ can be written as the product of the two-dimensional Gaussian chain distribution function $G_2((y,z),(y_0,z_0);N)$ and the one-dimensional distribution function $G_1(x,x_0;N)$:

$$G(\mathbf{r},\mathbf{r}';N) = G_2((y,z),(y_0,z_0);N) \times G_1(x,x_0;N), \tag{9.2.17}$$

where $G_2((y,z),(y_0,z_0);N)$ follows from Eqs. (2.1.6) and (2.1.7) as

$$G_2((y,z),(y_0,z_0);N) = \frac{3}{2\pi N\ell\ell_{\text{eff}}}\exp\left\{-\frac{3}{2}\frac{\left[(y-y_0)^2+(z-z_0)^2\right]}{N\ell\ell_{\text{eff}}}\right\} \tag{9.2.18}$$

and $G_1(x,x_0:N)$ is given by

$$\left[\frac{\partial}{\partial N} - \frac{\ell\ell_{\text{eff}}}{6}\frac{\partial^2}{\partial x^2} - 2\pi|\sigma_s q|\frac{\ell\ell_B}{\kappa}e^{-\kappa x}\right] G_1(x,x_0:N) = \delta(x-x_0)\delta(N), \tag{9.2.19}$$

where Eq. (9.2.13) is used. Analogous to the bilinear expansion in Eq. (9.2.3), $G_1(x,x_0:N)$ is expressed as

$$G_1(x,x_0;N) = \sum_{m=0}^{\infty}\psi_m(x)\psi_m(x_0)e^{-\lambda_m N}, \tag{9.2.20}$$

where ψ_m and λ_m are eigenfunctions and eigenvalues of

$$\left[-\frac{\ell\ell_{\text{eff}}}{6}\frac{\partial^2}{\partial x^2} - 2\pi|\sigma_s q|\frac{\ell\ell_B}{\kappa}e^{-\kappa x}\right]\psi_m(x) = \lambda_m\psi_m(x). \tag{9.2.21}$$

Using the boundary conditions that the segment density (proportional to $\psi_m^2(x)$) is zero at the surface ($x = 0$) and at distances far away from the surface ($x \to \infty$), the above equation can be solved exactly [Wiegel (1977), Muthukumar (1987)].

The criterion for adsorption is given by the occurrence of a bound state for the first time upon an increase in $|V_S(x)/k_BT|$ as

$$\left(\frac{48\pi\,|\sigma_s q|\,\ell_B}{\kappa^3\ell_{\text{eff}}}\right)_{\text{critical}} = 5.78. \qquad (9.2.22)$$

Adsorption occurs if the term on the left-hand side is greater than 5.78. Inserting the expression for the Bjerrum length ℓ_B (Eq. (1.3.2)), the condition for adsorption is

$$\frac{12\,|\sigma_s q|\,e^2}{\epsilon_0\epsilon\kappa^3\ell_{\text{eff}}k_BT} \geq 5.78. \qquad (9.2.23)$$

Note that the various experimental variables, namely, charge densities of the surface and the polymer, strength of the electrostatic attraction through temperature and dielectric constant of the solution, salt concentration through the inverse Debye length κ, and the electrostatically swollen size of the guest chain through the effective segment length ℓ_{eff} appear only as the unique combination $|\sigma_s q|\,/(\kappa^3\ell_{\text{eff}}\epsilon T)$. Any of these variables can be tuned to trigger adsorption or desorption. For example, if the surface charge density is the tuning variable, the condition for adsorption is

$$|\sigma_s| \geq 0.48\frac{k_B\epsilon_0\epsilon}{e^2}\frac{T\kappa^3\ell_{\text{eff}}}{|q|}. \qquad (9.2.24)$$

For fixed values of temperature, dielectric constant, and charge density of the guest polymer, the critical number charge density $|\sigma_s|_c$ on the surface depends on the inverse Debye length as

$$|\sigma_s|_c \sim \kappa^3\ell_{\text{eff}}. \qquad (9.2.25)$$

Substituting Eq. (9.2.15) into Eq. (9.2.25), we get

$$|\sigma_s|_c \sim \begin{cases} \kappa, & \kappa R_g \ll 1 \\ \kappa^{11/5}, & \kappa R_g \gg 1. \end{cases} \qquad (9.2.26)$$

Since the dependence of the inverse Debye length on the salt concentration c_s is $\kappa \sim \sqrt{c_s}$, higher surface charge density is required for adsorption at higher values of c_s. As a specific example, consider a system where $|\sigma_s|$ and κ are such that the inequality in Eq. (9.2.24) is satisfied so that adsorption occurs. If c_s (and hence κ) is increased to the level above which the inequality of Eq. (9.2.24) is no longer satisfied, then desorption occurs. In the limit of high concentrations of monovalent salt ($\kappa R_g \gg 1$), the theoretical prediction $|\sigma_s|_c \sim c_s^{1.1}$ given by Eq. (9.2.26) is consistent with experimental results [Fleer *et al.* (1993)].

For given values of the dimensionless parameter $|\sigma_s q|\,\ell_B/(\kappa^3\ell_{\text{eff}})$ above the critical condition for adsorption, the ground state eigenfunction $\psi_0(x)$ is calculated from Eq. (9.2.21). Substituting this calculated $\psi_0(x)$ in Eq. (9.2.7), the segment density profile is obtained. A typical example of the normalized segment density profile $\rho(x)$ is plotted in Fig. 9.4 versus κx for $(48\,|\sigma_s q|\,\ell_B)/(\kappa^3\ell_{\text{eff}}) = 26.38$. The density profile exhibits a depletion zone at the surface, a maximum at distances comparable to the Debye length, and an exponential decay with distance in the outer layer exposed to the solution.

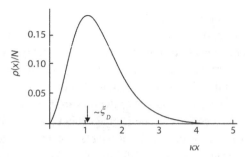

Figure 9.4 Plot of normalized segment density profile $\rho(x)/N$ from the GSD approximation versus κx for the dimensionless variable $(48\,|\sigma_s q|\,\ell_B)/(\kappa^3 \ell_{\mathrm{eff}}) = 26.38$.

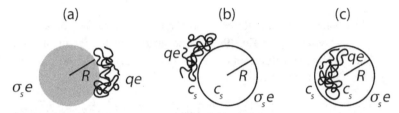

Figure 9.5 Adsorption of a polyelectrolyte chain of charge density qe on a sphere of radius R with surface charge density $\sigma_s e$. (a) Solid sphere. (b) Adsorption on the exterior surface of a spherical vesicle. (c) Adsorption from the interior of a spherical vesicle. The salt concentration c_s is the same inside and outside the vesicle in (b) and (c).

The thickness of the adsorbed layer is roughly proportional to the Debye length. This structure of the segment density profile is generic to polymer adsorption at interfaces.

9.2.2 Adsorption at Spherical Surfaces

Experiments on adsorption of synthetic polyelectrolytes on micelles, proteins, and dendrimers have shown an empirical law for the dependence of the critical surface charge density $|\sigma_c|$ on the inverse Debye length κ as [Dobrynin & Rubinstein (2005), Kizilay et al. (2011), Winkler & Cherstvy (2014)]

$$|\sigma_c| \sim \kappa^a, \qquad a \simeq 1 - 1.4. \qquad (9.2.27)$$

This situation can be addressed by considering a model system where the host is a solid sphere carrying charges on its surface and the guest is a polyelectrolyte chain, as sketched in Fig. 9.5a. This model is pertinent to adsorption on spherical colloidal particles as well. In addition, the situations of adsorption of a flexible polyelectrolyte chain on the outer surface of a vesicle (Fig. 9.5b) and on the interior surface of a vesicle (Fig. 9.5c) are of interest in several biological contexts [Alberts et al. (2015)]. Theoretical treatment of the above situations can be carried out by following the same procedure as outlined in Section 9.1, except that $V_S(\mathbf{r})$ is different for different situations, as described in the following text.

9.2.2.1   Adsorption on a Solid Sphere

Consider a solid spherical particle of radius R carrying charges on its surface with uniform charge density $\sigma_s e$ immersed in an electrolyte solution, as described in Section 4.2. There are no salt ions inside the sphere. An oppositely charged flexible chain dispersed in the electrolyte solution adsorbs at the surface of the sphere if the electrostatic attraction dominates over the loss in its conformational entropy accompanying adsorption. According to the Debye–Hückel theory, the electrostatic potential energy $V_S(\mathbf{r})/k_B T$ acting on the location $\mathbf{r}(s)$ of the sth segment, which is at the radial distance r from the center of the spherical particle, follows from Eq. (4.2.8) as

$$\frac{V_S(r)}{k_B T} = -4\pi \, |\sigma_s q| \, \ell \ell_B R^2 \frac{1}{(1 + \kappa R)} \frac{e^{-\kappa(r-R)}}{r}, \qquad r > R. \tag{9.2.28}$$

In obtaining the above result, integration over surface area is taken. In view of the radial symmetry in Eq. (9.2.4), and absorbing the effect of intra-chain electrostatic and excluded volume interactions into the renormalized effective Kuhn length ℓ_{eff}, Eq. (9.2.4) becomes

$$\left[-\frac{\ell \ell_{\text{eff}}}{6} \frac{1}{r^2} \frac{d}{dr} \left(r^2 \frac{d}{dr} \right) + \frac{V_S(r)}{k_B T} \right] \psi_m(r) = \lambda_m \psi_m(r), \tag{9.2.29}$$

where r is the radial polar coordinate and $V_S(r)$ is given in Eq. (9.2.28).

Since an exact solution of Eq. (9.2.29) is not possible, several techniques in quantum mechanics such as variational procedures and Wentzel–Kramers–Brillouin (WKB) approximation have been implemented to obtain the ground state eigenfunction $\psi_0(r)$ and its eigenvalue. Repeating the procedure outlined for the case of a planar surface (Section 9.2.1) and using a variational procedure [von Goeler & Muthukumar (1994), Shojaei & Muthukumar (2017)], the critical condition for adsorption is

$$|\sigma_s|_c \geq \frac{\kappa^2 \ell_{\text{eff}}(1 + \kappa R)}{24\pi \ell_B |q| R}. \tag{9.2.30}$$

In the limit of high salt concentrations ($\kappa R_g \gg 1$), which is usual in experiments, $\ell_{\text{eff}} \sim \kappa^{-4/5}$ as given in Eq. (9.2.15). Therefore, the dependence of the critical surface charge density for adsorption on the inverse Debye length κ follows from Eq. (9.2.30) as

$$|\sigma_s|_c \sim \begin{cases} \kappa^{6/5}, & \kappa R \ll 1 \quad \text{and} \quad \kappa R_g \gg 1 \\ \kappa^{11/5}, & \kappa R \gg 1 \quad \text{and} \quad \kappa R_g \gg 1. \end{cases} \tag{9.2.31}$$

The nature of the scaling relation between $|\sigma_s|_c$ and κ depends on the ratio of the particle radius to the Debye length κR. For radii of the adsorbing sphere shorter than the Debye length, $|\sigma_s|_c \sim \kappa^{6/5}$. This limit is pertinent to most of experiments on such systems, and the theoretical result is consistent with the empirical law given in Eq. (9.2.27). If the particle radius is larger than the Debye length, then $|\sigma_c| \sim \kappa^{11/5}$ is the theoretical prediction. Although the above results are obtained using a variational procedure, more rigorous WKB approximation gives the same scaling laws [Cherstvy & Winkler (2011), Winkler & Cherstvy (2014), Shojaei & Muthukumar (2017))].

9.2.2.2 Adsorption from the Exterior of a Vesicle

This situation is sketched in Fig. 9.5b, where a flexible polyelectrolyte chain of uniform linear charge density qe adsorbs from the exterior onto a uniformly charged spherical vesicle of radius R and charge density $\sigma_s e$. For the sake of simplicity, let the salt concentration c_s be the same in both the exterior and interior solutions of the vesicle. The electrostatic potential energy $V_S(r)$ at the location r of a polymer segment in the present situation is

$$\frac{V_S(r)}{k_B T} = -2\pi |\sigma_s q| \frac{\ell \ell_B}{\kappa} \left(1 - e^{-2\kappa R}\right) \frac{R}{r} e^{-\kappa(r-R)}, \qquad r > R. \tag{9.2.32}$$

Combining Eqs. (9.2.29) and (9.2.32), the critical condition for adsorption follows from a variational calculation as [von Goeler & Muthukumar (1994)]

$$|\sigma_c| \geq \frac{\kappa^3 \ell_{\text{eff}}}{12\pi \ell_B |q|} \frac{1}{\left(1 - e^{-2\kappa R}\right)}. \tag{9.2.33}$$

In the high-salt limit, the limiting behaviors for $\kappa R \ll 1$ and $\kappa R \gg 1$ are the same as in Eq. (9.2.31).

Note that the radius of the vesicle is an additional tuning variable for controlling adsorption, supplementing the roles from charge densities, salt concentration, and temperature. The segment density profiles along the radial distance from the surface are qualitatively similar to that given in Fig. 9.4 [Shojaei & Muthukumar (2017)].

9.2.2.3 Adsorption from the Interior of a Vesicle

For this situation (Fig. 9.5c), the electrostatic potential energy $V_S(r)$ is given by

$$\frac{V_S(r)}{k_B T} = -2\pi |\sigma_s q| \frac{\ell \ell_B}{\kappa} \left(1 - e^{-2\kappa R}\right) \frac{R}{r} \frac{\sinh(\kappa r)}{\sinh(\kappa R)}, \qquad r < R. \tag{9.2.34}$$

Combining Eqs. (9.2.29) and (9.2.34), calculations based on either a variational procedure or the WKB approximation yield the scaling law for the critical charge density as [Wang & Muthukumar (2011), Shojaei & Muthukumar (2017)]

$$|\sigma_c| \sim \kappa^{5/2}. \tag{9.2.35}$$

For all situations described in Section 9.2, the entropic and energy contributions to the free energy of the adsorbed state can be calculated using the ground state dominance approximation as detailed in the original publications [Muthukumar (2012b), Winkler & Cherstvy (2014), Shojaei & Muthukumar (2017)].

9.3 Charged Brushes

Charged brushes are ubiquitous in various lubrication systems in biological and aqueous media, such as knee joints and colloidal suspensions [Klein (1996)]. A charged brush is made of many polyelectrolyte chains with their one end permanently fixed at an interface. Such brushes are usually present in an electrolyte solution. The key variables that characterize a brush are the grafting density σ_g, which is the number of chains end-anchored per unit area of the interface, chain length $N\ell$, salt concentration, and curvature of the interface. The dependencies of the height, free energy, and monomer density profile of the brush on the various experimental variables are of fundamental interest. The force between two brushes brought together at certain distances is also of importance in the context of tribology and colloidal stability.

In this section, we focus on planar charged brushes and follow the scaling theory of Pincus (1991) to address the dependencies of the characteristic properties of brushes on the relevant experimental variables. It is convenient to separately consider the salt-free limit, where counterions play a dominant role, and the high-salt limit as follows.

9.3.1 No-Salt Limit

Consider a dense planar brush (Fig. 9.6a) of polyelectrolyte chains of uniform contour length $N\ell$ and charge $\alpha z_p e$ per segment length ℓ. As shown in Fig. 9.6a, L is the brush height or equivalently thickness. Let d be the spacing on the uncharged solid planar surface between two adjacent grafted chains. The grafting density is $\sigma_g = (\ell/d)^2$. Let us assume that the monomer density profile is a step function with uniform concentration c in the brush region of height L and no polymer concentration outside. This is known as the Alexander model of polymer brushes [Alexander (1977)] and the concentration profile is referred to as the Alexander-de Gennes profile. In view of this density profile, the segment concentration is $c = N/Ld^2$. The average polymer concentration can be equivalently written as the volume fraction ϕ by multiplying c by the segment volume ℓ^3. The variables characterizing the brush are summarized as

$$\sigma_g = \left(\frac{\ell}{d}\right)^2, \qquad c = \frac{N}{Ld^2}, \qquad \phi = \frac{N\ell^3}{Ld^2}. \tag{9.3.1}$$

In the salt-free limit, the equilibration of the brush is attained by optimizing the osmotic pressure from the counterions and the force working against the stretching of the grafted chains. It is convenient to separately consider the two limits of strongly charged and densely grafted chains and weakly charged and sparsely grafted chains. In

(a)   (b)

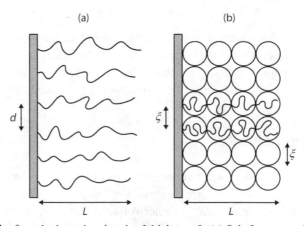

Figure 9.6 Sketch of a polyelectrolyte brush of thickness L. (a) Salt-free case with grafting density $\sigma_g = (\ell/d)^2$. (b) In the high-salt limit, $\sigma_g = (\ell/\xi)^2$, where ξ is the linear size of a blob used in the scaling argument.

the first case, counterions are abundant and the Debye length ξ_D arising from counterions is short in comparison with the brush height L. In the second case, ξ_D can be much longer than L, since the concentration of counterions is very low. This situation is analogous to the planar charged surface treated in Section 4.1 using the Gouy–Chapman theory. These two cases are described as follows.

9.3.1.1 Strongly Charged and Densely Grafted Brushes

For the case of strongly charged and densely grafted brush, the ideal osmotic pressure Π_{os} from counterions is

$$\frac{\Pi_{os}}{k_B T} = \alpha z_p c, \tag{9.3.2}$$

since there are $\alpha z_p N$ counterions per chain, which are translationally free. This pressure favors stretching of the brush. As given in Eq. (2.1.13), the free energy associated with stretching the end-to-end distance of a chain to L in one dimension is

$$\frac{F}{k_B T} = \frac{1}{2} \frac{L^2}{N \ell^2}. \tag{9.3.3}$$

The elastic pressure P_{el} is the force against stretching per unit area, given by $(-\partial F / \partial L)/d^2$, which follows from Eq. (9.3.3) as

$$\frac{P_{el}}{k_B T} = -\frac{1}{d^2} \frac{\partial F / k_B T}{\partial L} = -\frac{L}{N \ell^2 d^2}. \tag{9.3.4}$$

The elastic force disfavors stretching of the bristles. Combining the elastic pressure and the counterion pressure, the net pressure is

$$\frac{P_{el}}{k_B T} + \frac{\Pi_{os}}{k_B T} = -\frac{L}{N \ell^2 d^2} + \alpha z_p c. \tag{9.3.5}$$

At equilibrium, the net force must be zero. Therefore, equating the right-hand side of Eq. (9.3.5) to zero, we get

$$L = \sqrt{\alpha z_p} N \ell, \tag{9.3.6}$$

where the expression for c in Eq. (9.3.1) is used. Therefore, in the strongly charged and densely grafted limit, the brush thickness is independent of grafting density and is linear in chain length with the prefactor depending on the square root of degree of ionization of the chain.

9.3.1.2 Weakly Charged and Sparsely Grafted Brushes

For the case of weakly charged brushes ($\alpha z_p \ll 1$), counterion concentration is very low so that $\xi_D \gg L$. As noted above, this situation is similar to planar charged surfaces immersed in a solution containing counterions (Section 4.1). Since the screening length ξ_D is much outside the brush thickness, the brush can be taken as a planar surface with an effective number charge density given by

$$\sigma_s = \frac{\alpha z_p N}{d^2}. \tag{9.3.7}$$

In this analogy, the screening length is given by the Gouy–Chapman length (Eq. (4.1.5)) as

$$\xi_D = \frac{d^2}{2\pi \ell_B \alpha z_p N},$$ (9.3.8)

where Eq. (9.3.7) is used. Note that the counterion pressure inside the brush is due to only the fraction of the counterions inside the brush, which is approximately L/ξ_D. Therefore, the osmotic pressure from the counterions in the weakly charged limit is

$$\frac{\Pi_{os}}{k_B T} = \alpha z_p \left(\frac{L}{\xi_D} \right) c.$$ (9.3.9)

Combining this result with P_{el} of Eq. (9.3.4), the condition of equilibrium is

$$\frac{L}{N \ell^2 d^2} = \alpha z_p \left(\frac{L}{\xi_D} \right) c.$$ (9.3.10)

Using Eqs. (9.3.1) and (9.3.8), we get the expression for the brush height in terms of chain length, grafting density, and degree of ionization as

$$L = 2\pi \ell_B \left(\alpha z_p \right)^2 \sigma_g N^3.$$ (9.3.11)

Thus, the brush thickness is very sensitive to the bristle length ($L \sim N^3$) in the weakly charged limit, although L is much smaller than the fully stretched limit of $N\ell$. The brush thickness crosses over from this behavior to the stretched behavior (Eq. (9.3.6)) for strongly charged dense brushes.

9.3.2   High-Salt Limit

In the limit of high salt concentrations inside the brush, the brush can be treated as a semidilute good solution made of blobs, as sketched in Fig. 9.6b. As discussed in Section 6.2.2.2, the linear size of each blob is the correlation length ξ and the free energy per blob is $k_B T$. In view of the construction of the brush in terms of blobs of size ξ, the grafting density and the polymer concentration are given by

$$\sigma_g = \left(\frac{\ell}{\xi} \right)^2, \quad c = \frac{N}{L \xi^2}, \quad \phi = \frac{N \ell^3}{L \xi^2}.$$ (9.3.12)

As derived in Appendix 9, the free energy F and thickness L of the brush, based on scaling arguments, are given as

$$L \sim \ell v_{eff}^{1/3} \sigma_g^{1/3} N, \qquad \frac{F}{k_B T} \sim N v_{eff}^{1/3} \sigma_g^{5/6},$$ (9.3.13)

where v_{eff} is given by Eq. (8.6.6) for monovalent salt as

$$v_{eff} = v + \frac{\alpha^2 z_p^2}{2 c_s \ell^3}.$$ (9.3.14)

If electrostatic interactions dominate over the van der Waals excluded volume interactions, $v_{\text{eff}} \sim 1/c_s$. Therefore, the dependencies of the brush thickness L and the free energy F on salt concentration are given as

$$\frac{L}{\ell} \sim \frac{\sigma_g^{1/3} N}{c_s^{1/3}}, \qquad \frac{F}{k_B T} \sim \frac{\sigma_g^{5/6} N}{c_s^{1/3}}. \tag{9.3.15}$$

Thus, both the brush thickness and free energy decrease with an increase in salt concentration.

9.3.3   Disjoining Pressure

Consider the force per unit area when two opposing identical brushes are brought at a separation distance h between them. This is sometimes called the disjoining pressure. For separations larger than the brush thickness ($h > L$), the disjoining pressure P is exponentially small [Evans & Wennerström (1999), Israelachvili (2011)]. When the brushes are pushed together, P is given by the osmotic pressure Π inside the gap. In the present situation where one end of the grafted chains is permanently fixed, results derived in Section 10.3 for gels and networks are pertinent. Accounting for the free energy of mixing and the electrostatic interactions, Π is given at the mean field level by Eqs. (10.2.8) and (10.2.40) as

$$\frac{\Pi \ell^3}{k_B T} = \frac{1}{2} v_{\text{eff}} \phi^2. \tag{9.3.16}$$

Using $\phi = N\ell^3/(h\xi^2)$, the disjoining pressure ($P = \Pi$), follows from σ_g given in Eq. (9.3.12) and Eq. (9.3.14) as

$$P = \frac{k_B T}{\ell^3} \left(\frac{1}{2} - \chi + \frac{\alpha^2 z_p^2}{4 c_s \ell^3} \right) N^2 \sigma_g^2 \left(\frac{\ell}{h} \right)^2. \tag{9.3.17}$$

Thus, the pressure arising from compression of brushes is of longer range ($\sim 1/h^2$) than the attractive van der Waals force. Therefore, for sufficiently large grafting density, charged brushes provide a barrier against adhesion.

9.3.4   Density Profile

The monomer density of the brush in the above treatment is assumed to be a step function in terms of density versus the orthogonal distance from the brush surface. However, in reality, this assumption is inadequate due to topological correlations of segments in layers parallel to the brush surface. Deviations from the assumed uniform segment density and the constraint that chain ends are at the same distance from the grafting surface can readily occur. The contribution from these effects can be accounted for by using the self-consistent field theory (SCFT) [Miklavic & Marčelja (1988), Misra et al. (1989)]. The calculations show that the segment density profile is parabolic instead of a step function, analogous to the case of uncharged brushes

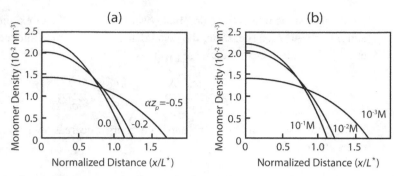

Figure 9.7 Parabolic monomer density profile for a polyelectrolyte brush based on self-consistent field theory. $N = 100, \ell = 1$ nm, $\sigma_g = 0.01$ nm^{-2}, and the height of the neutral brush $L^\star = 0.64N\ell$. (a) Role of segment charge αz_p at monovalent salt concentration $c_s = 10^{-3}$ M. (b) Role of c_s for $\alpha z_p = -0.5$ [Adapted from Miklavic & Marčelja (1988)].

[Milner *et al.* (1988)]. In fact the parabolic density profile for brushes is universal independent of the details of segmental interactions, as derived analytically in Appendix 10. Typical calculated results based on SCFT for the present planar charged brushes are given in Fig. 9.7. The role of segment charge αz_p on the monomer density profile is given in Fig. 9.7a where the distance x normal to the surface is in units of $L^\star$, the height of the neutral brush. In this example, $N = 100, \ell = 1$ nm, $L^\star = 0.64N\ell$, monovalent salt concentration $c_s = 1$ mM, and grafting density $\sigma_g = 0.01$ nm^{-2}. For the same system as in Fig. 9.7a, the role of salt concentration on the monomer density profile is shown in Fig. 9.7b for $\alpha z_p = -0.5$.

Although the parabolic segment density distribution function is distinctly different from the step function profile, the above derived results on the brush thickness remain valid in the SCFT calculations as well. The parabolic nature of the density profile is generic for all kinds of brushes as we shall show in the next section.

9.4 Genome Packaging in Viruses

Viruses are astronomically abundant in our world [Knipe & Howley (2001), Acheson (2007)]. The generic composition of viruses is an enclosure called capsid, which is made of a finite number of similar proteins, and one or a few RNA or DNA molecules constituting the genome present inside the capsid. The genome can be either ssRNA, dsRNA, ssDNA, or dsDNA. The shape of the capsid can be an icosahedron as in cowpea chlorotic mottle virus (Fig. 1.11a), rod-like as in tobacco mosaic virus, or polyhedron–rod composite as in a bacteriophage. Furthermore, the icosahedral viruses can have an additional envelope surrounding the capsid. The interior of the virus is an electrolyte solution containing additional charged macromolecules. The outer surface of the viruses are compatible with aqueous media. In addition, the capsid enclosure is porous allowing exchange of small electrolyte ions between its interior and exterior regions.

Among the myriads of viruses, let us consider only the class of icosahedral viruses containing ssRNA. Even in this class of RNA viruses, the number of different viruses is very large [Baker *et al.* (1999), Knipe & Howley (2001)]. The number of vertices

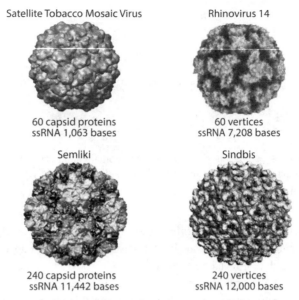

Satellite Tobacco Mosaic Virus

60 capsid proteins
ssRNA 1,063 bases

Rhinovirus 14

60 vertices
ssRNA 7,208 bases

Semliki

240 capsid proteins
ssRNA 11,442 bases

Sindbis

240 vertices
ssRNA 12,000 bases

Figure 9.8 Representative ssRNA viruses. The figures are from VIPERdb. (http://viperdb.scripps.edu) [Montiel-Garcia *et al*. (2021)].

in these icosahedra is not arbitrary but is quantized as $60 \times T$, where T is called the triangulation number. The triangulation number is expressed as $T = h^2 + hk + k^2$, where h and k can take any value of the positive integers including zero. A few examples of RNA viruses are shown in Fig. 9.8. The satellite tobacco mosaic virus consists of 60 identical copies of a single protein making up the virus capsid and 1063-nucleotide ssRNA genome. The human rhinovirus 14 (common cold virus) has 60 copies of three large capsid proteins and 7,208 nucleotides in the ssRNA genome. Both semliki and sindbis are $T = 4$ viruses with 240 capsid proteins and 11,442 and about 12,000 nucleotides, respectively, in their ssRNA genomes. In semliki and sindbis, the capsid is enveloped by a lipid bilayer membrane with an additional layer of envelope protein [Acheson (2007)]. As illustrated in Fig. 9.8, capsids of the same number of vertices do not package the same number of ssRNA nucleotides. In addition, although similar lengths of ssRNA are packaged inside capsids of the same physical size, the chemical details of the capsid proteins are quite different. These facts point to a general principle governing genome packaging in RNA viruses which goes beyond the various local details.

The structure of the proteins that constitute the capsid is quite intricate involving various forms of secondary structures and partially folded moieties [Hendrix (1999)]. In isolation, the structure of ssRNA is equally rich involving internal loops, pseudo knots, kissing hairpins, etc., arising from hydrogen bonding among the nucleotides [Palmenberg & Sgro (1997)]. Therefore, a detailed treatment of these complexities to understand genome packaging inside a capsid is a daunting task. Nevertheless, structural elucidation of assembled capsid proteins reveals their generic composition as having one hydrophobic domain and one hydrophilic domain, as illustrated in Fig. 9.9a. The hydrophobic domains form the capsid wall and the hydrophilic domains protrude into the interior of the capsid as tails. These tails generally carry

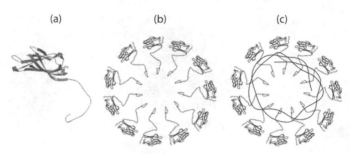

Figure 9.9 (a) Sketch of the hydrophobic domain and positively charged tail in a capsid protein. (b) Charged brush-like model for the capsid. (c) ssRNA lives inside the oppositely charged brush.

a net positive charge, while the ssRNA genome is negatively charged. Therefore, there are substantial electrostatic interactions between the capsid tails and the genome.

We shall now address the role of the electrostatic interactions in genome packaging in RNA viruses. Toward this goal, let us consider the following zeroth order model, which builds on the preceding sections in this chapter. In view of the chimeric nature of the assembled capsid protein shown in Fig. 9.9a, we imagine that an icosahedral capsid is a spherical polyelectrolyte brush where the hydrophilic tails, each carrying a net positive charge, are protruding toward the center of the virus. One end of each of these tails is covalently connected to its hydrophobic block which constitutes the capsid shell. Thus, in the absence of the genome, the interior of the capsid adjacent to the capsid wall appears as a positively charged polyelectrolyte brush (Fig. 9.9b) as considered in Section 9.3. In RNA viruses, the negatively charged genome lives inside the brush-like region (Fig. 9.9c). The optimal length of ssRNA packaged inside an RNA virus and the density profile of the RNA genome inside the virus are derived below based on this model and using the concepts developed in the previous sections.

Let us assume in the above described zeroth-order model of icosahedral RNA viruses that there are n_h bristles, each with N_h segments of segment length ℓ_h and charge ez_h per segment. The subscript h denotes that the capsid is taken as the host and z_h includes the degree of ionization. Let the total positive charge on the bristle be uniformly distributed along its backbone. We parametrize the ssRNA genome as a flexible polyelectrolyte chain of N segments, each of segment length ℓ and charge $z_p e$, where z_p includes the degree of ionization. In addition, electrolyte ions are present in the aqueous region inside the virus. Excluded volume interactions among the segments from the protein tails and the genome are present in addition to the electrostatic interactions.

The free energy of the present model system is derived by generalizing Eq. (6.3.1) to the presence of the negatively charged guest genome chain with the constraint that one end of the bristles is permanently anchored at the internal surface of the capsid wall. For the sake of simplicity, assume that the capsid surface is planar around the anchoring location of a protein tail. This assumption enables the description of the probability distribution function for a tail as a one-dimensional problem in the x-coordinate normal to the surface as in Section 9.3. However, the effective potential acting on every segment of the tail emerges from all species, including small ions and the genome in the three-dimensional brush.

Repeating the field theoretic method described in Section 6.3, as derived in Appendix 10, the self-consistently determined potential $V(x)$ at distance x from the capsid wall is parabolic in x,

$$V(x) = \frac{3\pi^2}{8\ell_h^2 N_h^2}\left(H^2 - x^2\right).$$ (9.4.1)

Here, H is the brush height at which $V(x)$ is zero and $V(x)$ is expressed in dimensionless units. The above result is universal, independent of details of the intersegment interactions between bristles and brush–genome interactions. The genome lives inside the brush where its segment at the location x is subjected to the potential $V(x)$, in addition to its chain connectivity. Repeating the procedure outlined in Section 9.2 and using the ground state dominance approximation, the segment density profile $\rho(x)$ of the genome follows from Eq. (9.2.7) as (Appendix 10)

$$\rho(x) = N\psi_0^2(x) = \frac{2N}{\sqrt{\pi}}\omega^{3/2}x^2 \exp\left[-\omega x^2\right],$$ (9.4.2)

where

$$\omega = \frac{3}{2}\pi\left|\frac{z_p}{z_h}\right|^{1/2}\frac{1}{\ell\ell_h N_h}.$$ (9.4.3)

Therefore, the dependence of the density profile of the genome on the distance from the capsid wall follows as

$$\rho(x) \sim x^2 e^{-(x/x_{max})^2},$$ (9.4.4)

where the location of the maximum genome density x_{max} is given by

$$x_{max} = \frac{1}{\sqrt{\omega}},$$ (9.4.5)

which depends on N_h, z_h, and z_p through Eq. (9.4.3).

The main features of the density profile given in Eq. (9.4.4) are that there is a depletion zone at the surface, followed by a single peak near the surface, which is then followed by a region with exponential decrease. Therefore, the nucleotides are predicted by the present model to pack in a single spherical shell with a gap between the nucleotides and capsid wall. The predicted density profile is in agreement with experimental results on many RNA viruses. Two examples are given in Fig. 9.10 [Zlotnick *et al.* (1997), Tihova *et al.* (2004)]. Cryoelectron microscopy data on hepatitis B virus are shown in Fig. 9.10a where packing of nucleotides in a spherical shell and the presence of a gap between the capsid and packed genome are evident. A quantitative comparison between the profile given in Eq. (9.4.4) and the experimental result for the flock house virus is presented in Fig. 9.10b. In this figure, the squares are experimental data and the curve is the best fit using Eq. (9.4.4) where x_{max} is taken as an adjustable parameter. The experimental points at distances beyond 6 nm from the capsid are believed to be due to instrumental error of cryoelectron microscopy and not nucleotide density [Conway & Steven (1999)].

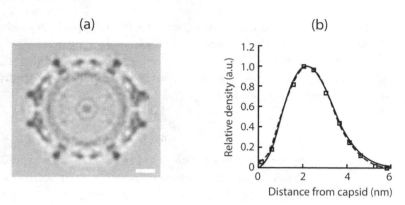

(a)   (b)

Figure 9.10 (a) Cryoelectron microscopy image of a cross section of hepatitis B virus, showing a gap between the capsid and densely packed shell of the genome [Zlotnick *et al.* (1997)]. (b) Nucleotide density inside the virus capsid of $\Delta 31$Bac mutant of flock house virus. Squares are experimental data based on cryoelectron microscopy [Tihova *et al.* (2004)]. Solid line is the best fit with Eq. (9.4.4) [Belyi & Muthukumar (2006)].

The relation between the packaged genome length and the net charge on all protein tails emerges from the Poisson equation (Eq. (3.8.2)) as an effective electroneutrality equation (Appendix 10),

$$|z_p| N \simeq n_h z_h N_h. \tag{9.4.6}$$

Therefore, the optimal condition for genome packaging is that the net negative charge of the packaged genome balances the net positive charge on all protein tails inside the virus. Note however that in this zeroth-order electrostatic model of genome packing, the role of chemical sequences of the protein tails and the genome are not accounted for, in addition to approximations such as local planarity of the capsid surface and the ground state dominance approximation.

The prediction from the electrostatic model that the length of packaged genome is proportional to the total charge of the polypeptide tails inside the virus is generally consistent with experimental facts. A representative collection of experimental data is presented in Fig. 9.11, where various known wild-type viruses from significantly different virus families infecting both animals and plants are selected [Belyi & Muthukumar (2006)]. This figure shows that the ratio of the genome length to the net charge on the capsid polypeptide arms is indeed conserved. The best fit for this ratio is 1.61 ± 0.03, which appears universal for all virus families included in Fig. 9.11. More detailed considerations of the excluded volume interactions, Donnan equilibrium, and charge distribution on the protein tails show that there can be variations from the slope of 1.61 shown in Fig. 9.11, although it is of order unity with an upper bound of 2.0 [Siber & Podgornik (2008), Ting *et al.* (2011)].

One of the main assumptions of the above electrostatic model of genome packing is that the effects from the sequences of RNA and protein tails are ignored. In other words, the main conclusion of electroneutrality between the genome and protein tails, which dictates the length of the packaged genome, is independent of the specificity of the genome sequence. The predicted sequence-independent genome packaging in

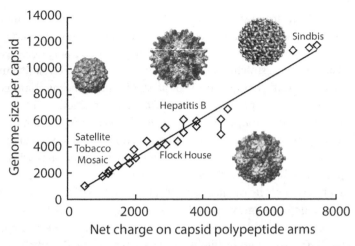

Figure 9.11 Universal dependence of genome size on the net charge of protein tails in ssRNA viruses. The experimental data for a variety of wild-type viruses, with four viruses shown explicitly. Connected points correspond to viruses that package genomes of varying lengths. The line is the best fit to the data with 1.61 slope. The images are from VIPERdb. (http://viperdb.scripps.edu) [Montiel-Garcia *et al.* (2021)]. [Adapted from Belyi & Muthukumar (2006)].

RNA viruses is observed in experiments as well [Tihova *et al.* (2004), Hu *et al.* (2008), Routh *et al.* (2012), Cadena-Nava *et al.* (2012), Stockley *et al.* (2013)].

Furthermore, the electroneutrality criterion has a significant implication in the context of molecular evolution. According to the paradigm of the central dogma of molecular biology, once genetic information has gotten into a protein, it cannot get out again, for example, never to RNA or DNA. Although violations of this dogma are known [Baltimore (1970), Temin & Mizutani (1970), Crick (1970)], the present scenario where genome length is self-consistently controlled by protein charge suggests co-evolution of proteins and RNA dominantly driven by electrostatic interactions.

The prediction of the universality of electroneutrality between the genome and proteins in ssRNA viruses, obtained without any regard to RNA sequence or invocation of the central dogma, and validated by experimental data, signals the importance of the electrostatic interactions in assembly of genomes in RNA viruses and virus-like particles. This result for flexible ssRNA is in contrast to the packaging and ejection of semiflexible dsNDA in a bacteriophage, where excluded volume effects dominate over the electrostatic interactions [Zandi *et al.* (2003), Siber *et al.* (2012), Mahalik *et al.* (2013)].

9.5   Intermolecular Complexation

Oppositely charged polyions interact in dilute aqueous solutions to spontaneously form a variety of water-soluble macromolecular complexes. These intermolecular complexes are called complex coacervates, and sometimes referred to as symplexes. The structures of such complexes are quite different from those of complexes formed by polyelectrolytes and oppositely charged low molecular inorganic counterions or surfactants. For example, consider the complexation between a large polyelectrolyte chain (labeled as host) and an oppositely charged polyelectrolyte (labeled as guest). The emergent structure of this system depends on the relative molecular weights and charge densities of the host and guest, pH, salt concentration, and temperature. From a conceptual point of view, the electrostatic interactions play a dominant role in the structure formation, along with van der Waals interactions, the hydrophobic interaction and hydrogen bonding. Unlike the situation of metallic counterions (see Fig. 5.11c), the topological correlations of the guest macromolecule, arising from chain connectivity, is self-consistently coupled with those of the host macromolecule, mediated by all relevant electrostatic and hydrophobic interactions. Due to the inevitable strongly interacting proximity of the host and guest molecules inside the complex, numerous structures without sequential registry can form. A few examples are shown in Fig. 9.12. The structure in Fig. 9.12a is known as "ladder-like" structure. In this ladder, oppositely charged monomers of the host and guest are paired sequentially to form a dipolar chain which looks like a twisted zipper. The dipolar chain is significantly less hydrophilic compared to its individual members. This feature renders new properties to the system unseen in solutions of polyelectrolytes with only one charge sign.

The ladder-like organization is the limiting situation with full electrostatic registry, where dipole formation dominates over conformational degrees of freedom of the member molecules. On the other hand, if the conformational entropic contribution is significant, then "scrambled-egg" structure can emerge as the other limiting situation, as shown in Fig. 9.12b. In between these two limiting situations, numerous

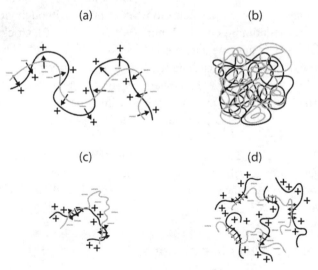

Figure 9.12 (a) A ladder-like dipolar chain forms when two oppositely charged polyelectrolyte chains fully complex together. (b) Scrambled-egg structure of a complex coacervate. (c) Sketch of a branched structure formed by two oppositely charged polyelectrolyte chains without full registry. (d) Sketch of microgel-like branched structure formed by several polycation and polyanion chains.

branched structures can form, where various dipolar stems are connected with loops of polycations and polyanions (Fig. 9.12c). When multiple chains participate in the complex formation, the aggregated complexes are microgels composed of dipolar domains, polycation strands, and polyanion strands, as sketched in Fig. 9.12d. As we shall see in Section 9.7, such microgel-like structures are prevalent in solutions of intrinsically disordered proteins and model systems of membraneless organelles.

The most commonly observed structures in dilute solutions are of the type sketched in Fig. 9.12c. An example of such a scenario is portrayed in Fig. 9.13a (Fig. 1.11b), where pUC19-supercoiled DNA of 2686 base pairs is complexed with a positively charged cylindrical brush polymer made of quaternized 2-vinyl pyridinium side chains [Störkle *et al.* (2007)]. Note the multiple loops of DNA emanating from the core. As another example [Borgia *et al.* (2018)], consider the complexation from two intrinsically disordered proteins: Linker histone H1, which carries a net +53 charge and is involved in chromatin condensation, and its nuclear chaperone prothymosin-α (ProTα) which carries a net -44 charge. These two IDPs spontaneously associate into a complex with picomolar affinity. Representative snapshots from the simulation of the complex are shown in Fig. 9.13b. Note that the complexes fully retain their structural disorder, long-range flexibility, and highly dynamical characteristics (as discerned from fluorescence studies) [Borgia *et al.* (2018)]. In spite of high affinity, these complexes exhibit extreme disorder. Furthermore, the dissociation constant of the complex increases by six orders of magnitude with a mere doubling of ionic strength from a physiological 165 mM to 340 mM. These facts demonstrate the important role of electrostatic control in macromolecular recognition and formation of intermolecular complexes.

In addition to understanding how intermolecular complexes form, an understanding of how such complexes break up upon imposed stimuli is also of considerable interest.

(a)   (b)

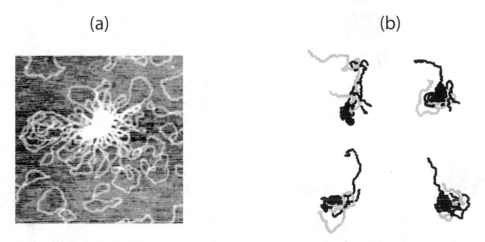

Figure 9.13 (a) AFM image of a complex of DNA and a positively charged polymer brush [Störkle *et al.* (2007)]. (b) Snapshots of simulated complex from two oppositely charged intrinsically disordered proteins, H1 (+53 charges) and ProTα (−44 charge). [Adapted from Borgia *et al.* (2018)].

One of the basic properties of polyelectrolyte complexes is their capacity to participate in competitive inter-polyelectrolyte substitution processes with other charged macro-molecules. Such processes are ubiquitous in various regulatory mechanisms and the functioning of the cell as a whole. Therapeutic procedures such as gene therapy also involve controlling the stability of complex coacervates constituted from DNA and release of DNA with suitable substitution processes. Although complex coacervates are the basis of a large number of applications in daily life as well as in industry, an adequate understanding of their structure and properties still remains to be reached. We give below only elementary models to address the mechanisms of formation of complex coacervates and release of guest macromolecules from a complex by another competitive guest molecule.

9.5.1   Counterion Release Drives Complex Coacervation

Let us address the driving force behind the spontaneous formation of a polymeric pair from two oppositely charged polymers. It turns out that entropic changes aris-ing from the release of adsorbed counterions from the participating polyelectrolyte chains contribute dominantly to the complex formation [Ou & Muthukumar (2006)]. The mutual electrostatic attraction among the oppositely charged monomers plays relatively a minor role in aqueous assemblies at ambient temperatures. The entropic contribution from counterion release is further augmented by reorganization of water molecules during complexation [Wang & Schlenoff (2014)].

The mechanism of counterion release can be readily seen in computer simula-tions of complexation. For example, consider the complexation of two uniformly and oppositely charged flexible polyelectrolyte chains in salt-free condition. Each poly-electrolyte chain is modeled as a coarse-grained chain of N united atoms (beads). Each bead has $+1$ charge for the polycation and -1 charge for the polyanion. Let the counterions of both polymers be monovalent. The snapshots of chain configurations at different stages during complexation, as obtained by Langevin dynamics simulations, are presented in Fig. 9.14. In this figure, $N = 60$ and the Coulomb strength parameter $\Gamma = \ell_B/\ell_0$ is 2.0 (where ℓ_B is the Bjerrum length and ℓ_0 is the charge separation dis-tance along a chain). The snapshots correspond progressively to the simulation time units of 71, 501, 811, and 966. As visible in the figure, each chain in the beginning of the complexation has a finite number of adsorbed counterions and is expanded due to intra-chain electrostatic repulsion. Once the two chains meet, there is a progressive release of adsorbed counterions during complexation. In the final state, the complex is essentially neutral without any adsorbed counterions on its participating chains. Therefore, the translational entropy of released counterions in the solution is much larger than that in the initial state where only the dissociated counterions can explore the whole solution. This entropic gain during complexation can be significant. In the final state, the size of the complex is much smaller in comparison to that of the individ-ual chains before complexation. The resultant complex is equivalent to a double chain where the two chains are paired by many dipoles formed by the oppositely charged monomers (Fig. 9.12a). Due to dipole–dipole attraction, the dipole chain is collapsed.

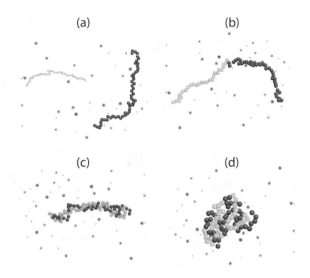

Figure 9.14 Snapshots of conformations of two oppositely charged polyelectrolytes and their counterions during complexation in salt-free conditions ($N = 60, \Gamma = 2.0$). Counterions are released progressively with time. (a), (b), (c), and (d) correspond to the simulation time units of 71, 501, 811, and 966, respectively. Two individual chains with expanded size become one dipolar chain of globule-like size. Entropy of counterion release drives aqueous assembly of complex coacervates [Ou & Muthukumar (2006)].

Therefore, upon complexation, two polyelectrolyte chains (each carrying a net charge) become one dipolar chain with essentially no net charge.

The computed free energy of simulated complexation $\Delta F/T$, and its energy contribution $\Delta E/T$ and entropic contribution ΔS, are given in Fig. 9.15 as functions of the Coulomb strength parameter Γ. At lower values of Γ, where the degree of counterion adsorption is low, the contribution from the entropy of released counterions is minor and hence, ΔF is governed by the electrostatic attraction given by ΔE. On the other hand, for larger values of Γ, where the extent of adsorbed counterions before complexation is high (see Fig. 5.13), $T\Delta S$ contributes to ΔF more than ΔE. The transition from dominance by electrostatic attraction to dominance by counterion release occurs at the Coulomb strength parameter $\Gamma^{\star} = 1.5$ for the parameters used in the simulation. For highly charged polyelectrolytes, for example poly(styrene sulfonate) and dsDNA in aqueous solutions, Γ is in the range of 2.0–4.0. Therefore, the aqueous assembly of complex coacervates is mainly driven by entropy gain associated with counterion release.

9.5.2   Competitive Substitution in Intermolecular Complexes

As an example of the substitution process in complex coacervates, consider the following reaction. A complex, made of poly(methacrylate) (PMA) polyanion and poly(N-ethyl-4-vinyl-pyridinium) (PEVP) polycation, is reacted with DNA in the presence of KCl salt. Let PMA_{50}, PMA_{1000}, and PMA_{2100} denote the average degree

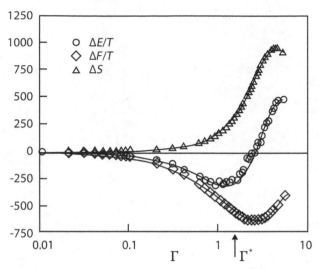

Figure 9.15 Relative contributions of energy (circles) and entropy (triangles) to the free energy change (diamonds) of complexation as functions of the Coulomb strength parameter Γ. Entropy dominates for $\Gamma > \Gamma^{\star} = 1.5$, pertinent to aqueous assembly of complex coacervates [Adapted from Ou & Muthukumar (2006)].

of polymerization of PMA as 50, 1000, and 2100, respectively. Similarly let $PEVP_{30}$ and $PEVP_{130}$ denote the average degree of polymerization of PEVP as 30 and 130, respectively. If the conditions are favorable, the competitive polyanion DNA replaces PMA. The ensuing chemical equilibrium is

$$PMA/PEVP + DNA \xrightleftharpoons{K^+Cl^-} DNA/PEVP + PMA \qquad (9.5.1)$$

The equilibrium degree of substitution q_e, namely the extent of the above reaction toward right, can be determined by fluorescence techniques. The dependence of q_e on KCl concentration is given in Fig. 9.16 for different samples of PMA and PEVP (at 25°C, 0.01 M Tris buffer, and pH 9) [Izumrudov $et\ al.$ (1995)]. A comparison of curves 1 and 2, respectively, for $PMA_{2100}/PEVP_{30}/DNA$ and $PMA_{50}/PEVP_{30}/DNA$ systems shows that the substitution reaction is facilitated by lowering the molecular weight of the polyanion which is being substituted by DNA. Similarly, a comparison of curve 3 corresponding to $PMA_{50}/PEVP_{130}/DNA$ system with curve 2 shows that the substitution of PMA by DNA is favored as the size of PEVP is increased relative to that of PMA. All of these features point to the conclusion that the substitution process is controlled by entropic changes accompanying the process. If several small guest molecules are involved in a complex, an invading larger guest molecule can successfully substitute the smaller ones due to their release into the solution whereby they gain translational entropy.

More insight into the underlying mechanism of competitive substitution in poly-electrolyte complexes can be gleaned from simulations [Peng & Muthukumar (2015)]. For example, consider an A–B complex (made from a flexible cation of $N_A = 60$ beads and a flexible polyanion of $N_B = 60$ beads, analogous to Fig. 9.14). Let this complex

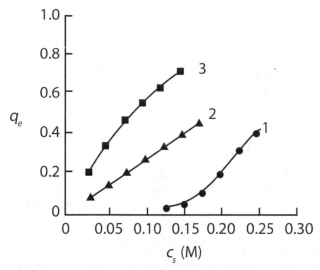

Figure 9.16 Dependence of the equilibrium degree of substitution q_e on KCl concentration c_s in the systems $PMA_{2100}/PEVP_{30}/DNA$ (curve 1), $PMA_{50}/PEVP_{30}/DNA$ (curve 2), and $PMA_{50}/PEVP_{130}/DNA$ (curve 3). Data support the entropic mechanism of substitution [Adapted from Izumrudov *et al.* (1995)].

be invaded by a polycation C of N_C beads under salt-free conditions. Langevin dynamics simulations of this competitive substitution process at the Coulomb strength $\Gamma = 2.8$ reveals that the substitution of the A chain by the C chain can occur if N_C is sufficiently larger than N_A. Snapshots of the substitution reaction for $N_C = 120$ at several time instances during the process are given in Fig. 9.17. Beginning from the instant at which the invading chain makes contact with the A–B complex, a long-lived temporary complex incorporating three chains (A–B–C) is formed. In this temporary complex, there is a continuous swapping between segments of the A–B pair and the B–C pair. If N_C is comparable to N_A, this swapping continues endlessly without the completion of the substitution process. On the other hand, if N_C is sufficiently larger than N_A, the swappings among the various segments yield to the C chain, which then dominates over the A chain resulting in completion of the substitution reaction. The time required for complete substitution depends on the ratio N_C/N_A of the length of the invading chain to that of the resident chain. When $N_C/N_A \sim 1$, the substitution time is essentially infinite. As this ratio increases, the substitution time decreases precipitously until a threshold value of about 1.5. Beyond this threshold value, the substitution time is insensitive to the length of the invading chain. Presence of salt facilitates the substitution reaction by weakening the electrostatic interactions, and reducing the substitution time.

Inspection of the electric potential difference during swapping of A–B and B–C segments and translational entropy of counterions released from the C chain inside the three-chain complex domain leads to the following conclusions. The electrostatic interactions between the C chain and the A–B complex is sufficient to initiate the substitution reaction, but not enough to complete the process. The gain in entropy from

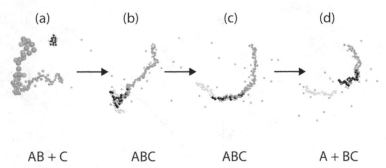

(a)   (b)   (c)   (d)

AB + C   ABC   ABC   A + BC

Figure 9.17 Snapshots during substitution reaction, where a C chain (squares) substitutes an A chain (light circles) in the A–B complex (B is dark circles) at the simulation times 1 (a), 110 (b), 2200 (c), and 3200 (d). [Adapted from Peng & Muthukumar (2015)].

the released counterions is required for completing the substitution reaction. Furthermore, the occurrence of the threshold length of the invading chain, beyond which the substitution time is insensitive to N_C, originates from counterion release. Once the monomers of A complexed with B are loosened by an adequate number of monomers of the C chain, the rest of the monomers of C do not release their adsorbed counterions, thus not contributing anymore to the driving force for the complexation process.

The formation of relatively long-lived intermediate metastable ternary complex (state ABC in Fig. 9.17) has important implications for the kinetic mechanisms of disordered polyelectrolyte complexes pertinent to biological contexts [Schuler *et al.* (2020)].

9.6   Coacervation

Non-dilute aqueous solutions of polycations and polyanions containing low molecular salt readily exhibit liquid–liquid phase separation (LLPS), resulting in coexistence of a polymer-rich phase and a polymer-poor phase. Each of these phases contains both polycations and polyanions. This kind of LLPS in solutions of oppositely charged polymers spontaneously occurs at ambient conditions below a critical salt concentration, in contrast to the behavior in solutions with only polycations or only polyanions. The polymer-dense phases so formed from polycations and polyanions are called coacervate phases or simply coacervates. The corresponding LLPS is known as the coacervation. Starting from the initial experimental observation of coacervation [Bungenberg de Jong & Kruyt (1929)] and subsequent industrial applications of coacervates, this phenomenon has a rich history and continues to attract considerable attention [Srivastava & Tirrell (2016)].

As a typical example of coacervation, consider an aqueous solution of poly(acrylic acid) (PAA) and poly(N,N-dimethylaminoethyl methacrylate) (PDMAEMA) containing KCl at pH 6.5. Upon mixing the components of the solution, coacervation occurs immediately if the salt concentration is low enough [Spruijt *et al.* (2010)]. An increase in salt concentration in a coacervate increases its solubility. The polymer concentration in the coexisting dilute phase increases at the expense of the polymer concentration inside the coacervate with an increase in salt concentration. At high enough salt concentrations, there is no LLPS. A quantitative summary of these features for phase separation in PAA-PDMAEMA solutions, at room temperature and pH 6.5, is presented in Fig. 9.18a. The salt-polymer phase diagrams given in this figure denote the binodal curves of coexisting phases for PAA-PDMAEMA mixtures at the 1:1 stoichiometric ratio. Data for four different degrees of polymerization of PAA (N = 20, 50, 150, and 510) are included in the figure. The filled symbols denote the polymer-dense coacervate phase and the open symbols denote polymer-poor dilute phase. The experimental critical salt concentration beyond which no LLPS is observed is indicated by the crossed symbols. The binodal data are the boundaries of the two-phase region, and outside this region, the mixture is a single phase. The difference in polymer concentrations in the two coexisting phases is larger for longer polymer chains. Furthermore, the dilute phase is more dilute for longer polymer chains at a given salt concentration and the corresponding coacervate phase is more concentrated. Usage of fluorescently labeled PAA in the PAA–PDMAEMA mixtures allows visualization of the coexisting phases. For example, confocal image of an equilibrated mixture of a coacervate phase for N = 150 and c_s = 1.0 M is given in Fig. 9.18b [Spruijt *et al.* (2010)]. Here the droplets of about 10 μm are the coexisting dilute phase in the continuous coacervate phase.

Note that, in marked contrast to solutions of polyelectrolytes with only one charge sign where lowering salt concentration suppresses phase separation, coacervation is instead promoted. This feature indicates that enhanced dissociation of intermolecular complexes with an increase in salt concentration plays a key role in determining the coacervation phenomenon. Therefore, the elements of complex coacervation described

(a)   (b)

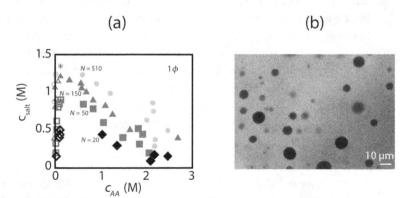

Figure 9.18 (a) Salt-polymer phase diagram for LLPS in 1:1 stoichiometric mixtures of PAA and PDMAEMA at pH 6.5, as a plot of salt concentration c_s versus monomer concentration of PAA, c_{AA}. N is the degree of polymerization of PAA. Solid symbols denote polymer-rich coacervate phase and open symbols denote polymer-poor dilute phase. Crossed symbols denote the lowest salt concentrations above which there is no phase separation. (b) Confocal image of coexisting dilute phase (droplets) and the coacervate (background) for the PAA-PDMAEMA system (N = 150, c_s = 1.0 M). Scale bar represents 10 μm. [Adapted from Spruijt *et al.* (2010)].

in Section 9.5 are required in order to fully understand coacervation. We present two different kinds of theoretical approaches to address coacervation in the next two subsections.

9.6.1   Voorn–Overbeek Theory

The earliest theory of coacervation, which still continues to be used extensively, is the Voorn–Overbeek theory (VOT). It combines the Flory–Huggins theory of polymer solutions and the Debye–Hückel theory of electrolytes. Consider the general situation of a solution containing n_1 polyanions of N_1 monomers, n_2 polycations of N_2 monomers, n_+ cations and n_- anions from fully dissociated salt, and n_0 solvent molecules. Let α be the degree of ionization for each monomer in both the polycation and polyanion. Let the volume fraction of the polyanion, polycation, cations from the salt, anions from the salt, and solvent be denoted as $\phi_1, \phi_2, \phi_+, \phi_-,$ and ϕ_0, respectively. Taking the counterions of the polyanions and polycations as monovalent, the volume fractions of the counterions from polyanions (ϕ_{c-}) and polycations (ϕ_{c+}) are given as $\phi_{c-} = \alpha\phi_1$ and $\phi_{c+} = \alpha\phi_2$, respectively.

It is straightforward to write the free energy of the above system using the Flory–Huggins theory of polymer solutions (Section 6.2.1) and the Debye–Hückel theory of electrolytes (Section 3.8.2) as described in the following text (Eq. (9.6.9)). However, the Voorn–Overbeek theory invokes additional drastic assumptions, although it was cast in the additive framework of Flory–Huggins theory and Debye–Hückel theory, as detailed in the following text.

In order to illustrate the original derivation of VOT, consider a symmetric salt-free solution of polycations and polyanions, namely the polycations and polyanions are equal in number and length. Let N be the number of monomers per chain and the linear charge density be σ (which is the degree of ionization per monomer, α). For this symmetric case, $\phi_1 = \phi_2$. Let the total polymer volume fraction be $\phi = \phi_1 + \phi_2$, so that $\phi_1 = \phi_2 = \phi/2$. The VOT invokes two key assumptions: (i) there are no counterions, and (ii) for treating the electrostatic correlations, polycations and polyanions are broken into their monomeric charged species as if they are simple electrolyte ions without any topological correlations due to chain connectivity. With these assumptions, the free energy density is given as

$$f_{\text{VO}} = f_{\text{S,VO}} + f_{\text{DH,VO}} \qquad (9.6.1)$$

where $f_{\text{S,VO}}$ is the free energy density from the entropy of mixing and $f_{\text{DH,VO}}$ is the Debye–Hückel contribution from the hypothetically broken-into monomers. $f_{\text{S,VO}}$ follows from the Flory–Huggins theory as

$$f_{\text{S,VO}} = \frac{\phi}{N} \ln\left(\frac{\phi}{2}\right) + (1 - \phi) \ln(1 - \phi), \qquad (9.6.2)$$

where the first term on the right-hand side corresponds to the translational entropy of polycations and polyanions (with irrelevant linear terms in ϕ ignored) and the second term corresponds to the translational entropy of solvent molecules (assuming that the solution is incompressible). Using Eqs. (3.6.7) and (3.8.11), the Debye–Hückel theory gives the inverse Debye length κ for this model according to

$$\kappa^2 \ell^2 = \frac{4\pi \ell_B}{\ell} (\alpha\phi),$$ (9.6.3)

where ℓ is the monomer diameter. Note that the degree of ionization α is the same as the linear charge density σ in the present model. $f_{\text{DH,VO}}$ follows from Eq. (3.8.42) and (9.6.3) as

$$f_{\text{DH,VO}} = -\frac{(4\pi \ell_B/\ell)^{3/2}}{12\pi}(\alpha\phi)^{3/2}.$$ (9.6.4)

Combining Eqs. (9.6.1)–(9.6.3), the free energy density of VOT is given as

$$f_{\text{VO}} = \frac{\phi}{N} \ln\left(\frac{\phi}{2}\right) + (1-\phi)\ln(1-\phi) - \frac{(4\pi \ell_B/\ell)^{3/2}}{12\pi}(\alpha\phi)^{3/2}.$$ (9.6.5)

This original expression of VOT can be generalized to the presence of monovalent salt of volume fraction ϕ_s $(= \phi_+ + \phi_- = 2\phi_+ = 2\phi_-)$ and the van der Waals contribution to the polymer-solvent interactions parametrized through the χ parameter as

$$f_{\text{VO, modified}} = \frac{\phi}{N} \ln\left(\frac{\phi}{2}\right) + \phi_s \ln\left(\frac{\phi_s}{2}\right) + \phi_0 \ln \phi_0 + \chi\phi\phi_0 - \frac{1}{12\pi}\left(\frac{4\pi \ell_B}{\ell}\right)^{3/2}(\alpha\phi + \phi_s)^{3/2},$$ (9.6.6)

where ϕ_0 is the volume fraction of the solvent. Since the first term on the right-hand side of this equation is negligible for large values of N, the coacervate phase diagram is dominated by the χ parameter, translational entropy of small ions and solvent, and the electrostatic correlations among small ions and the hypothetically fragmented charged monomers.

It is a common practice to use the modified VOT expression given by Eq. (9.6.6) to compute the coexistence curves for coacervation and fit with experimental data. However, the VOT suffers from a serious deficiency of ignoring the counterions. As seen in Section 8.4.1, counterion entropy plays a significant role in determining phase behavior in polyelectrolyte systems. A straightforward extension of the Flory–Huggins theory to the present situation gives the Helmholtz free energy density as

$$f_{\text{FH}} = \left(\frac{1}{N} + \alpha\right)\phi \ln\left(\frac{\phi}{2}\right) + \phi_s \ln\left(\frac{\phi_s}{2}\right) + \phi_0 \ln \phi_0 + \chi\phi\phi_0.$$ (9.6.7)

Since there are αN counterions from each of the polycations and polyanions and there are equal number of monovalent cations and anions from the salt, the Debye–Hückel theory gives

$$f_{\text{DH}} = -\frac{1}{12\pi}\left(\frac{4\pi \ell_B}{\ell}\right)^{3/2}(\alpha\phi + \phi_s)^{3/2}.$$ (9.6.8)

Note that this equation is serendipitously the same as Eq. (9.6.4) in the VOT, although the origins are different. In Eq. (9.6.8), the term $\alpha\phi$ arises from the counterions, whereas in the VOT it arises from the broken-up monomers of polymer chains. Adding the terms given in Eqs. (9.6.7) and (9.6.8), the free energy density is given as

$$f = \left(\frac{1}{N} + \alpha\right) \phi \ln\left(\frac{\phi}{2}\right) + \phi_s \ln\left(\frac{\phi_s}{2}\right) + \phi_0 \ln \phi_0 + \chi \phi \phi_0 - \frac{1}{12\pi} \left(\frac{4\pi \ell_B}{\ell}\right)^{3/2} (\alpha\phi + \phi_s)^{3/2}.$$
(9.6.9)

The key deficiency of the VOT is the neglect of the translational entropy of counterions. The prefactor of the $\phi \ln \phi$ term should be $(\alpha + 1/N)$, dominated by the degree of ionization of order one in comparison with the negligible term $1/N$. Therefore, the VOT artificially suppresses the entropic contribution from the polymer with a factor of $1/N$ instead of $(\alpha + 1/N)$. Hence the LLPS is artificially promoted in the VOT due only to solvent entropy and the Debye–Hückel term. Therefore, the correct combination of the Flory–Huggins theory and the Debye–Hückel theory, given in Eq. (9.6.9), must be used in predicting the coacervate phase behavior. However, this correct mean field theory leads to unphysical results. For example, the upper critical temperature for LLPS in aqueous coacervate systems is tens of degrees below 0°C in contradiction with reality [Adhikari et al. (2018)]. Furthermore, Eq. (9.6.9) in the salt-free limit is identical to Eqs. (8.4.2-8.4.7) derived for salt-free solutions containing polyelectrolyte chains of only one charge sign. The only difference is that ϕ in Eqs. (8.4.2-8.4.7) needs to be identified as the total polymer volume fraction from both polycations and polyanions. With this mapping, the critical point for the coacervate phase behavior follows from Eq. (8.4.10) as

$$\phi_c = \frac{1}{64\pi} \frac{1}{\left(\alpha + \frac{1}{N}\right)} \quad \text{and} \quad t_c = \frac{1}{64\pi} \frac{\alpha}{\left(\alpha + \frac{1}{N}\right)},$$
(9.6.10)

where t is the reduced temperature $\ell/(4\pi\ell_B)$ defined in Eq. (8.3.10). As in the case of salt-free polyelectrolyte solutions, the chain length does not play a significant role in determining the critical point for coacervate phase behavior and the result is approximately close to that of the restricted primitive model of solutions of simple electrolytes. Taking the dielectric constant as 80 and $\ell = 0.25$ nm, the predicted critical temperature for salt-free symmetric mixture of polycations and polyanions (with $N = 100$) is 52 K. Clearly this result, based on correct mean field theory, is unphysical and in contradiction with the experimental fact that the coacervates form rather readily at room temperatures in many mixtures of polycations and polyanions. Analogously, in the presence of salt, the predicted phase diagrams from the above mean field theory are not in satisfactory agreement with experiments, although rich behavior is predicted similar to those in Fig. 8.6 for polyelectrolyte solutions.

9.6.2   Dipolar Theory of Coacervation

The origin of the discrepancy between the predictions of the above mean field theory and experimental results lies in the neglect of formation of ion pairs among the monomers from polycations and polyanions. Ion-pair formation in coacervate systems has been addressed in several theoretical treatments [Kudlay et al. (2004), Salehi & Larson (2016), Muthukumar (2017)]. As described in Section 9.5, even in infinitely

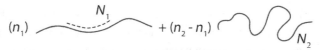

Figure 9.19 An elementary model for dipolar theory of coacervation. There are n_1 chains each with N_1 ion pairs taken as dipoles and $(N_2 - N_1)$ charged monomers, and $(n_2 - n_1)$ chains with N_2 charged monomers in the solution [Muthukumar (2017)].

dilute solutions, two polyelectrolyte chains of opposite charges spontaneously form complexes driven by counterion release. The complexation between two oppositely charged polyelectrolyte chains leads to a single dipolar chain. Such complexation mechanism occurs at higher polymer concentrations as well, and branched structures shown in Fig. 9.12 are more prevalent. In order to evaluate the role of spontaneous formation of dipolar domains on coacervate phase behavior, let us consider the dipolar theory of coacervation [Muthukumar (2017)].

Returning to the general situation introduced earlier, let a solution contain n_1 polyanion chains of N_1 segments, n_2 polycation chains of N_2 segments, n_+ cations and n_- anions from added salt, and n_0 solvent molecules. The degree of ionization of the polyelectrolyte chains is α. The volume fractions of the polyanion, polycation, cations and anions from the salt, and solvent are $\phi_1 = n_1 N_1 \ell^3 / V, \phi_2 = n_2 N_2 \ell^3 / V, \phi_+ = n_+ \ell^3 / V$, $\phi_- = n_- \ell^3 / V$, and $\phi_0 = n_0 \ell^3 / V$, respectively, where ℓ is the segment length and V is the total volume of the solution. In addition, there are counterions from the polymer chains. Let us assume that all small ions are monovalent. The volume fraction of the counterions from the polyanions is $\phi_{c-} = \alpha \phi_1$ and that from polycations is $\phi_{c+} = \alpha \phi_2$. The volume fraction of the solvent is ϕ_0. Furthermore, the solution is assumed to be incompressible.

In order to elicit the role of intermolecular complexes in coacervate phase behavior, consider a solution with asymmetric composition such that $N_2 \geq N_1$ and $n_2 \geq n_1$. In addition, let all n_1 chains of the minority component undergo complete pairwise complexation with polycations as illustrated in Fig. 9.19. Therefore, the solution is made of n_1 complexed chains, each with N_1 dipoles and $N_2 - N_1$ positive charges and $(n_2 - n_1)$ chains of N_2 positive charges.

The Helmholtz free energy density of this system follows as a generalization of Eqs. (6.3.33)-(6.3.38) derived for polyelectrolyte solutions as

$$f = f_S + f_H + f_{fl,i} + f_{fl,p}. \tag{9.6.11}$$

Here f_S represents the contribution from mixing entropy of polymer chains, small ions, and solvent, given as [Muthukumar (2017)]

$$f_S = \frac{\phi_1}{N_1} \ln \left[\phi_1 \left(1 + \frac{N_2}{N_1} \right) \right] + \left(\frac{\phi_2}{N_2} - \frac{\phi_1}{N_1} \right) \ln \left[\phi_2 - \frac{N_2}{N_1} \phi_1 \right]$$
$$+ \phi_1 \ln \phi_1 + [(1 - \alpha)\phi_1 + \phi_2] \ln [(1 - \alpha)\phi_1 + \phi_2]$$
$$+ \phi_+ \ln \phi_+ + \phi_- \ln \phi_- + \phi_0 \ln \phi_0. \tag{9.6.12}$$

This equation is obtained by treating the counterions to be distinct from the salt ions and assuming ideal mixing of the various species.

The enthalpic contribution in Eq. (9.6.11) is taken from the mean field values for dipole–dipole interactions among ion pairs in the dipolar domains, charge–charge interactions among segments in the uniformly charged domains, dipole–charge interactions between segments in dipolar and polycation domains, van der Waals type polymer–polymer interactions, and polymer–solvent interactions. By taking all combinations of pairwise interactions, f_H is given as

$$f_H = \frac{1}{2}v_{dd}\phi_1^2 + \frac{1}{2}v_{cc}(\phi_2 - \phi_1)^2 + v_{dc}\phi_1(\phi_2 - \phi_1) + \frac{1}{2}v_{pp}(\phi_2 - \phi_1)^2$$
$$+ v_{ds}\phi_1\phi_0 + v_{cs}(\phi_2 - \phi_1)\phi_0 + v_{ss}\phi_0^2, \tag{9.6.13}$$

where $v_{dd}, v_{cc}, v_{dc}, v_{pp}, v_{ds}, v_{cs}$, and v_{ss} are, respectively, the contact interaction energies between two dipoles, two charges, one dipole and one charge, two polymer segments interacting via van der Waals interactions, one dipolar segment and one solvent molecule, one polycation segment and one solvent molecule, and two solvent molecules. v_{pp}, v_{ds}, v_{cs}, and v_{ss} are the two-body excluded volume parameters described in Section 2.4. Analogous to these pseudo-potentials identified in the high temperature expansion limit, v_{cc}, v_{dc}, and v_{dd} are given by (see Appendix 3)

$$v_{cc} = \frac{4\pi\alpha^2\ell_B}{\kappa^2\ell^3}, \tag{9.6.14}$$

$$v_{dc} = -\frac{\pi}{3}\frac{\alpha^2\ell_B^2p^2}{\ell^4}e^{-2\kappa\ell}(2 + \kappa\ell), \tag{9.6.15}$$

$$v_{dd} = -\frac{\pi}{9}\frac{\ell_B^2p^4}{\ell^6}e^{-2\kappa\ell}\left[4 + 8\kappa\ell + 4(\kappa\ell)^2 + (\kappa\ell)^3\right], \tag{9.6.16}$$

where p is the dipole length and κ is the inverse Debye length given by

$$\kappa^2 = \frac{4\pi\ell_B}{\ell^3}\left[2\phi_1 + \alpha(\phi_2 - \phi_1) + \phi_s\right]. \tag{9.6.17}$$

In this equation, ϕ_s is the total volume fraction of salt including both cations and anions.

The fluctuation term $f_{fl,i}$ in Eq. (9.6.11) arising from the electrostatic correlations of all mobile ions is given by the Debye–Hückel theory as

$$f_{fl,i} = -\frac{1}{4\pi}\left[\ln(1 + \kappa\ell) - \kappa\ell + \frac{1}{2}\kappa^2\ell^2\right]. \tag{9.6.18}$$

The contribution $f_{fl,p}$ from conformational fluctuations can be calculated by implementing the same procedure as in Section 6.3, but can be ignored in strongly complexing systems.

Typical coacervate phase diagrams calculated using the dipolar theory is given in Fig. 9.20 as a plot of total ion concentration ϕ_i versus total polymer concentration ϕ_p at different temperatures [Adhikari *et al.* (2018)]. The coexistence curves in this figure correspond to symmetric solutions with $N_1 = N_2 = 100$ and $n_1 = n_2$ for $\alpha = 1/3, p = \ell = 0.55$ nm, and $\epsilon = 80$. In defining the temperature, Eqs. (8.3.10) and (8.4.1) are used with the choice of $a_\chi = 1.7$ corresponding to the value 0.434

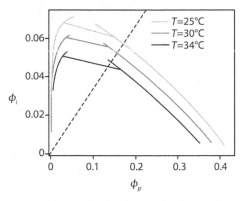

Figure 9.20 Coexistence curves calculated using the dipolar theory for a symmetric solution ($N_1 = N_2 = 100$ and $n_1 = n_2$) and $\alpha = 1/3$ at different temperatures. ϕ_i is the total ion concentration from counterions and salt, and ϕ_p is the total polymer concentration. The dashed line is $\phi_i = \alpha\phi_p$ [Adhikari *et al.* (2018)].

for the Flory–Huggins χ parameter. At each temperature below 47°C, a polymer-rich coacervate coexists with a dilute phase. The negative slope of the tie lines demonstrates that salt preferentially partitions into the dilute phase. The dashed line passing through the origin, with the degree of ionization α as the slope, represents the total concentration of all dissociated counterions without any contribution from added salt, which corresponds to the salt-free limit. Therefore, in experiments, only the region on the left-hand side of the dashed line is relevant and the region on the right-hand side is not accessible experimentally.

The predicted temperature range where coacervation occurs and the various features of coexistence curves in Fig. 9.20 are in qualitative agreement with experiments. The above model is able to capture the key aspects of coacervation in LCST systems as well [Adhikari *et al.* (2019)]. However, note that only ladder-like complexation is used in constructing the phase diagram in Fig. 9.20. In reality, all kinds of branched structures involving multiple dipolar domains for each chain can occur. A full theory incorporating such structures and associative physical gelation is yet to be developed to fully understand the coacervation phenomenon.

9.7 Membraneless Organelles

Transient liquid-like droplets made of RNA and proteins constitute a vast class of membraneless organelles, complementing the well-known lipid-bilayer-wrapped organelles such as the nucleus and mitochondria. In view of the recent recognition of occurrence of a plethora of membraneless organelles in nature [Banani *et al.* (2017)], active research is pursued to understand their origin, structure, and biological functions. As a result, a variety of conjectures, accompanied by experimental support, have emerged that invoke intermolecular complexation, liquid–liquid phase separation, coaceration, gelation, fibrillization, etc. However, an adequate understanding of this important class of organelles is only at its beginning state.

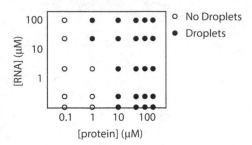

Figure 9.21 Phase diagram of RNA concentration versus protein concentration at 10°C and 150 mM NaCl. Filled symbols denote the formation of droplets and open symbols denote the absence of droplets. [Adapted from Molliex *et al.* (2015)].

One of the key properties of membraneless organelles is the significant role of the electrostatic interactions among the various components of the system. For example, consider the stress granules, which are enriched in RNA-binding proteins and mRNAs. Toward an understanding of molecular forces in stress granules, experiments have been carried out on solutions of hnRNPA1 (a mutant of heterogeneous nuclear ribonucleoprotein) and RNA. As described in Section 8.8.2 and Fig. 8.34, solutions of hnRNPA1 without RNA undergo LLPS. Further experiments have revealed that hnRNPA1 phase separation properties are strongly influenced by the presence of RNA, which is oppositely charged with respect to the hnRNPA1. For example, the phase diagram of hnRNPA1 as a function of protein concentration and RNA concentration is given in Fig. 9.21 based on experiments (performed in 50 mM HEPES, 150 mM NaCl, 5 mM DDT and 150 mg/mL Ficoll at 10°C). The open and filled symbols indicate the single-phase and two-phase regimes, respectively [Molliex *et al.* (2015)]. As seen in Fig. 9.21, RNA facilitates LLPS of hnRNPA1. This result demonstrates the important role of the electrostatic interactions between the protein and RNA resulting in intermolecular complexation and coacervate phase behavior.

The above conclusion regarding the role of electrostatically driven biomolecular condensates is not specific only to the stress granules. Experiments show that solutions of the negatively charged Nephrin intracellular domain (NICD) can form nuclear bodies as micron-size spherical droplets, which is facilitated by positively charged partners. The spontaneity of formation of nuclear bodies by NICD depends on the extent of positive charge on the partner. This effect is illustrated using supercharged GFP (scGFP) as the partner by varying its net charge [Pak *et al.* (2016)]. Images by fluorescence microscopy are given in Fig. 9.22 for solutions of 5 μM NICD and 5 μM scGFP (with charges +9, +15, +20, +25, and +36). As seen in the figure, higher the charge on scGFP higher is the propensity of the nuclear bodies. Thus, the spontaneity of formation of nuclear bodies by NICD depends on the extent of the positive charge on the partner (scGFP). It is clear from Fig. 9.22 that electrostatically driven complexation plays an important role in the birth of nuclear bodies.

Although only two examples are exhibited above, the role of electrostatic interactions in the formation of membraneless organelles is a universal feature. Furthermore, the compositions and sequences of the participating macromolecules play additional

10 Gels

A gel is a wonderful material of common occurrence in both biological contexts and soft materials. This ubiquitous material is composed of many polymer chains cross-linked together into a single gigantic macromolecule holding an enormous amount of solvent within. The cross-links bind the chains together as a single macroscopic structure resulting in retention of gel shape. Thus, gels exhibit solid-like behavior at large length scales and long time scales. The solvent prevents the polymer chains from collapsing on themselves, and the polymer network in turn prevents the solvent from flowing away. As a result, gels exhibit liquid-like behavior at small length scales and short time scales. The simultaneous liquid-like and solid-like behaviors endow gels with rich properties occurring at a hierarchy of length and time scales. This chapter develops the relationship between the microscopic dynamics at the monomeric level and the macroscopic elastic properties at the level of the whole gel sample, and describes various gel properties.

The polymer strands comprising the gel can either be electrically uncharged or charged with sequences of hydrophobic and ionic groups. For charged gels, the solvent is usually water, although it can be a mixture of polar solvents or ionic liquids. When water is the major component of the solvent, the gel is called hydrogel. The cross-links can be chemical cross-links, which are permanent in the gel. Alternatively, the cross-links can be physical cross-links made by hydrogen bonding, ion bridging, or stereocomplex formation. The gels made of physical cross-links are called "physical gels." In general, physical cross-links have a certain lifetime depending on their chemical nature. If the lifetime of the physical cross-links are sufficiently long in comparison with the characteristic time associated with the interrogation of the gel sample, then the physical gel can be treated as a chemically cross-linked gel. We treat only this class of gels in this chapter.

For gels made of charged molecules, the electrostatic repulsion between the charged monomers, and the translational entropy of solvent molecules, mobile counterions and other ions, lead to gel swelling. These forces are opposed by the elastic force arising from the cross-links. The surface of a gel acts essentially like a semipermeable membrane through which solvent molecules and ions exchange between the gel and its exterior solution. As a result of the two opposing effects of swelling and elasticity, the gel attains an equilibrium state and coexists with the external solution (Fig. 10.1). The symbiotic relation between the polymer network and the solvent inside the gel can be tuned with various experimental variables such as the solvent quality,

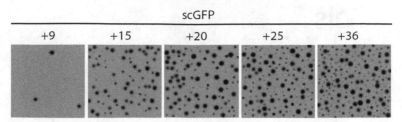

Figure 9.22 Fluorescence microscopy images of solutions containing 5 μM NICD and 5 μM supercharged GFP with charges +9, +15, +20, and +25, and +36. [Adapted from Pak *et al.* (2016)].

significant role [Nott *et al.* 2015]. As already mentioned in Section 8.8.2, intricate complexities such as physical gelation and fibrillization further decorate the structures and functions of membraneless organelles. The formation and properties of charged gels must also be addressed in addition to electrostatic complexation, which is the subject of the next chapter. The basic principle behind electrostatically driven self-assembly and phase separation described so far are only a few components to fully understand the membraneless biomolecular condensates. In order to fully appreciate the basic concepts and rich properties of membraneless organelles and analogous biological assemblies, it is necessary to self-consistently compound the various conceptual frameworks described so far and in the following chapter.

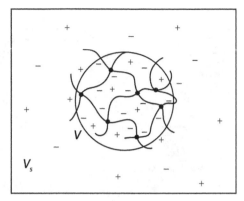

Figure 10.1 A gel of volume V is in an external solution of volume V_S. The counterions from the polymer, dissociated ions from the added salt, and solvent molecules move across the boundary of the gel and attain an equilibrium partitioning of the mobile species. Optimization between the swelling forces due to translational entropy of mobile species and intersegment electrostatic repulsion, and the contractile force due to elasticity of the network results in the equilibrated swelling state of the gel.

temperature, pH, charge density of the polyelectrolyte chains, and concentration of added low molar mass electrolyte. Even with a tiny change in these variables acting as external stimuli, gels can undergo dramatic discontinuous volume changes, easily by factors of 500 or more. The softness of polyelectrolyte gels and their sensitivity to large volume changes under external stimuli have led to numerous water-based materials for technological applications. Examples include super-water-absorbents, drug delivery systems, actuators, sensors, food preservatives, and materials for molecular separation using electrophoresis and chromatography.

Gels are also of common occurrence in the human body. The aforementioned volume transitions play a crucial role in the generation of mucus in various parts of the human body. Mucus is a self-assembled hierarchical aqueous polyelectrolyte gel, made of glycoproteins, multivalent ions, and other molecules. It serves as a barrier for pathogens, and aids in lubrication, hydration of epithelial cells, selective particle transport, and absorption of nutrients. Additionally, many diseases are connected to the behavior of gels. For example, production of unusually thicker and stickier mucus in the lungs and digestive system is associated with the hereditary disease cystic fibrosis. Another example is related to the human eye, where the dominant compartment is the vitreous attached to the retina. The vitreous is a hydrogel made primarily of hyaluronic acid and collagen. Volume changes in the vitreous gel can generate disadvantageous forces on the retina, leading to vitreal detachment and even retinal detachment. Prevention of blindness from these events, and more generally speaking, formulation of biomimetic therapeutics in health care, requires an understanding of swelling equilibrium, elasticity, and volume transitions of gels.

In this chapter, we describe the basic principles behind gel behavior. In practice, polyelectrolyte gels can be prepared by first forming an uncharged gel and then ionizing the gel. Alternatively, copolymer gels with unionizable and ionized monomers

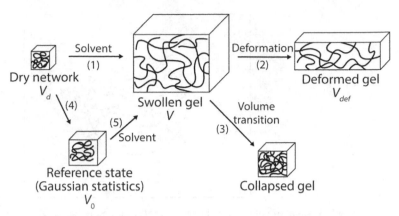

Figure 10.2 Gel behavior: (1) Swelling of a gel from its dry state to its final equilibrated swollen state. (2) Deformation of an equilibrated gel due to confinement or externally imposed tension. (3) The gel can undergo a dramatic volume transition due to delicate changes in stimuli such as pH, temperature, salt concentration, solvent quality, and electric field. In theoretical models, step (1) is taken as the sum of steps (4) and (5) by invoking a hypothetical reference state where the chains are assumed to obey Gaussian chain statistics.

can be synthesized in the presence of cross-linkers. In general, the synthesized polymer network is a complex architecture consisting of strands ending at branch points (cross-links), loops, dangling ends, entanglements, and heterogeneously distributed cross-links. In order to understand the macroscopic equilibrium properties of gels without getting into local chemical details and non-ergodic aspects of the system, we restrict ourselves to the simplest description of a gel.

The flow of content in this chapter is as follows (Fig. 10.2): (1) General description of model gels and key variables (Section 10.1), (2) Thermodynamics of gels and swelling equilibrium (Sections 10.2 & 10.3), (3) Gels under confinement (Section 10.4) and under tension (Section 10.5), (4) Gel elasticity (Section 10.6), (5) Dynamics of charged gels (Section 10.7), (6) Scaling laws for swollen gels (Section 10.7.4), (7) Kinetics of gel swelling (Section 10.8), (8) Volume transitions in gels (Section 10.9), and (9) Charge regularization in gels (Section 10.10). The preparation conditions for swollen gels and their subsequent treatment in terms of deformation and volume transition are shown in Fig. 10.2, where the key variables characterizing the gel volume are defined.

10.1 Model Gel and Definitions

Consider a uniformly charged polyelectrolyte gel in a polar solvent containing added salt. Let there be n strands, each containing N Kuhn segments, and let N_c be the number of cross-links. We take z_p as the number of ionized groups per Kuhn segment in the absence of any counterion adsorption. Let α be the degree of ionization after accounting for counterion adsorption. Therefore, there are $\alpha z_p n N$ monovalent counter ions. The counterion concentration is $\alpha z_p n N / V = \alpha z_p c$, where V is the volume of the gel and c is the monomer number concentration ($c = nN/V$). Let the number concentration of added monovalent salt be c_s.

When a gel is synthesized and purified, it has a certain initial volume. In some situations, this initial state is close to the dry state. Irrespective of the volume of the gel in the initial synthesis, let V_d denote the volume of the dry gel (Fig. 10.2). Since the network has n strands each with N segments,

$$V_d = nNv_1, \qquad (10.1.1)$$

where v_1 is the volume of a segment. As we have seen in previous chapters, there are several ways to choose the volume of a segment, and accordingly the value of N. For example, v_1 can be chosen as simply the volume carved out by the cube of the average Kuhn segment length ℓ^3, or alternatively as ℓb^2, where b is the thickness of the chain backbone. In the context of the Flory–Huggins theory used in the following text, v_1 is taken as the volume of the solvent molecule. The particular choice of the segment volume, accompanied by a proper adjustment of N, does not alter the physical conclusions but only the numerical prefactors in comparisons between theory and experiments.

Starting from the dry state of the gel, we shall discuss (1) swelling of the gel by a solvent to the equilibrated state of the gel with volume V, (2) deformation of a swollen gel under tension or confinement, and gel elasticity and dynamics, and (3) phase transitions of gels under external stimuli, as shown in Fig. 10.2. In order to develop a molecular theory of gel behavior, it is necessary to properly account for chain statistics. Since the chain statistics in the initial state of synthesized gel is unknown and since the ideal Gaussian chain statistics is the starting point in considering the roles of all effects on chain conformations, it becomes necessary to invoke a hypothetical reference state for a gel where the strands between the cross-links are assumed to obey Gaussian chain statistics (Chapter 2). The swelling process (1) is equivalent to the sum of steps (4) and (5) in Fig. 10.2. Let the volume of the gel in the reference state be V_0. The volume fraction of the polymer in the swollen state is ϕ and that in the hypothetical state is ϕ_0 given by

$$\phi = \frac{V_d}{V} \qquad \phi_0 = \frac{V_d}{V_0}, \qquad (10.1.2)$$

where ϕ_0 is an inevitable free parameter needed to connect experiments with theoretical models.

The number of chains and the number of cross-links in the gel are related to each other. This topological relation can be quite complicated depending on the frequency of defects such as dangling ends, ring-like closed circuits, and incomplete branching emanating from the cross-link points [Flory (1953a), Mark & Erman (1988)]. However, for a perfect network (without any defects), the number of f-functional cross-links N_c is related to the number of chains n by $N_c = 2n/f$. When the cross-links are tetrafunctional, a perfect network of n chains thus has $n/2$ cross-links,

$$N_c = \frac{n}{2}. \qquad (10.1.3)$$

10.2   Free Energy and Osmotic Pressure of Gels

The Helmholtz free energy of the gel ΔF, with respect to the Gaussian reference state, is generally given as the sum of a mean field contribution and a contribution arising from fluctuations,

$$\Delta F = \Delta F_{\text{mean field}} + \Delta F_{\text{fluctuations}}. \tag{10.2.1}$$

The second term on the right-hand side arises from fluctuations in conformations of the various strands, local monomer concentration, and distributions of small electrolyte ions in the gel. The first term on the right-hand side of Eq. (10.2.1) arises from the free energy of mixing between the polymer and solvent, electrostatic interactions between polymer segments, elasticity of the gel network, and the Donnan equilibrium for the salt ions:

$$\Delta F_{\text{mean field}} = \Delta F_{\text{mix}} + \Delta F_{\text{electrostatic}} + \Delta F_{\text{elastic}} + \Delta F_{\text{Donnan}}, \tag{10.2.2}$$

where the subscripts denote the various contributions mentioned above. We assume that these contributions are simply additive at the mean field level. The first term on the right-hand side is the sum of contributions from the entropy of mixing between polymer and solvent, and the enthalpy due to solvent quality. In accounting for the solvent quality, only short-ranged van der Waals-type interactions are included. The electrostatic contribution from the interactions among all charged segments mediated by mobile ions inside the gel is given by the second term on the right-hand side of Eq. (10.2.2). The various contributions in Eqs. (10.2.1) and (10.2.2) are given in the following subsections.

The net result is that ΔF is a function of polymer volume fraction ϕ, degree of ionization, salt concentration, the Flory–Huggins χ parameter, and cross-link density. In general, the osmotic pressure Π and the isothermal osmotic bulk modulus K are given by [de Gennes (1979), Landau & Lifshitz (1986), Onuki (2002)]

$$\Pi = -\left(\frac{\partial \Delta F}{\partial V}\right)_T = \phi^2 \frac{\partial}{\partial \phi}\left[\frac{\Delta F/V}{\phi}\right]_T; \quad K = -V\left(\frac{\partial \Pi}{\partial V}\right)_T = \phi\left(\frac{\partial \Pi}{\partial \phi}\right)_T. \tag{10.2.3}$$

The subscript T denotes the isothermal condition. The various contributions to the osmotic pressure of the gel arising from the free energy of mixing, electrostatic interactions between polymer segments, gel elasticity, Donnan equilibrium, and fluctuations are obtained from their corresponding expressions for the free energy and using Eq. (10.2.3). The osmotic pressure of the gel is the sum of these contributions,

$$\Pi = \Pi_{\text{mix}} + \Pi_{\text{electrostatic}} + \Pi_{\text{elastic}} + \Pi_{\text{Donnan}} + \Pi_{\text{fluctuations}}. \tag{10.2.4}$$

When the gel is in equilibrium, the osmotic pressure of the gel must be zero:

$$\Pi = 0. \qquad \text{(condition of equilibrium)} \tag{10.2.5}$$

We shall now derive expressions for the terms on the right-hand sides of Eqs. (10.2.1) and (10.2.4), based on simple models allowing analytical tractability. Suitable assumptions and approximations are invoked in order to track the conceptual basis for most of the major observations on gels. Incorporation of additional features into the models will be addressed in later Sections.

10.2.1   Free Energy of Mixing

The free energy of mixing for solutions of n chains, each with N segments, and n_1 solvent molecules in volume V is given by the Flory–Huggins theory (Chapter 6) as

$$\frac{\Delta F_{\text{mix}}}{k_B T} = \frac{V}{v_1} \left[\frac{\phi}{N} \ln \phi + (1 - \phi) \ln(1 - \phi) + \chi \phi (1 - \phi) \right], \tag{10.2.6}$$

where $1 - \phi = n_1 v_1 / V$, χ is the Flory–Huggins parameter (which captures the short-ranged interactions among the polymer segments and solvent molecules), and $k_B T$ is the Boltzmann constant times the absolute temperature. Here, the volume of the segment and that of a solvent molecule are taken as the same, as already noted. The first and second terms on the right-hand side of Eq. (10.2.6) are due to the translational entropy of polymer chains and solvent molecules, respectively. Since the gel can be treated as a single gigantic macromolecule with its center of mass not undergoing significant translational motion, its contribution to the entropy of mixing can be ignored in the above equation. Furthermore, since terms linear in ϕ inside the square brackets lead to constants upon multiplication by V, these terms can be ignored. Therefore, the free energy of mixing between the polymer network and solvent is obtained as

$$\frac{\Delta F_{\text{mix}}}{k_B T} = \frac{V}{v_1} \left[(1 - \phi) \ln(1 - \phi) - \chi \phi^2 \right]. \tag{10.2.7}$$

The osmotic pressure Π_{mix} corresponding to the above free energy of mixing follows from Eq. (10.2.3) as (see Eq. (8.2.3))

$$\frac{\Pi_{\text{mix}} v_1}{k_B T} = -\ln(1 - \phi) - \phi - \chi \phi^2. \tag{10.2.8}$$

10.2.2   Electrostatic Interaction Energy

The electrostatic interaction energy among all polymer segments constituting the gel is given by

$$\frac{F_{\text{electrostatic}}}{k_B T} = \frac{1}{2} \int_0^{nN} ds \int_0^{nN} ds' U \left[\mathbf{R}(s) - \mathbf{R}(s') \right], \tag{10.2.9}$$

where $\mathbf{R}(s)$ is the position vector of the sth segment and $U \left[\mathbf{R}(s) - \mathbf{R}(s') \right]$ is the electrostatic interaction energy between the sth and s'th segments. Assuming the Debye–Hückel potential for this interaction, we have

$$U \left[\mathbf{R}(s) - \mathbf{R}(s') \right] = \frac{\alpha^2 z_p^2 e^2}{4 \pi \epsilon_0 \epsilon k_B T} \frac{1}{|\mathbf{R}(s) - \mathbf{R}(s')|} \exp \left[-\kappa |\mathbf{R}(s) - \mathbf{R}(s')| \right], \tag{10.2.10}$$

where κ is the inverse Debye length given by

$$\kappa^2 = \frac{e^2}{\epsilon_0 \epsilon k_B T v_1} \left(\alpha z_p \phi + 2 c_s v_1 \right). \tag{10.2.11}$$

Performing the double integral in Eq. (10.2.9) in the limits of salt-free and high salt concentrations, an approximate extrapolation formula valid also for intermediate salt concentrations can be obtained, analogous to Eq. (6.3.38), as [Hua *et al.* (2012)]

$$\frac{F_{\text{electrostatic}}}{k_B T} = \frac{1}{2} \frac{V}{v_1} \frac{\alpha^2 z_p^2 e^2}{\epsilon_0 \epsilon \ell k_B T} \frac{(nN)^{2/3}}{\left[\kappa^2 \ell^2 (nN)^{2/3} + \frac{3^{4/3} \pi^{7/6}}{2^{5/3}} \phi^{2/3} \right]} \phi^2, \tag{10.2.12}$$

where ℓ is the Kuhn segmental length. In writing this expression, we have expressed all lengths in the system in units of the Kuhn length ℓ and we have chosen v_1 as ℓ^3. Different choices for v_1 can of course be made, and accordingly redefining the value of N. Since the polymer volume fraction is very small and the total number of segments in the gel is very large, $\phi^{2/3} << \kappa^2 \ell^2 (nN)^{2/3}$, and hence $F_{\text{electrostatic}}$ becomes

$$\frac{F_{\text{electrostatic}}}{k_B T} = \frac{1}{2} \frac{V}{v_1^2} \frac{\alpha^2 z_p^2 e^2}{\epsilon_0 \epsilon \kappa^2 k_B T} \phi^2 = \frac{1}{2} \frac{V}{v_1} \frac{\alpha^2 z_p^2 \phi^2}{\left(\alpha z_p \phi + 2c_s v_1 \right)}. \tag{10.2.13}$$

For the limits of no salt ($c_s = 0$) and high salt ($2c_s >> \alpha z_p c$), this reduces to

$$\frac{F_{\text{electrostatic}}}{k_B T} = \frac{V}{v_1} \begin{cases} \frac{1}{2} \alpha z_p \phi & (c_s = 0) \\ \frac{\alpha^2 z_p^2}{4 c_s v_1} \phi^2. & \text{(high salt)} \end{cases} \tag{10.2.14}$$

In view of the quadratic dependence of $F_{\text{electrostatic}}$ on ϕ^2 in the high salt limit, the electrostatic contribution can be combined with the χ parameter in Eq. (10.2.7) to get an effective χ parameter as (Eq. (8.4.19))

$$\chi_{\text{eff}} = \chi - \frac{\alpha^2 z_p^2}{4 c_s v_1}. \tag{10.2.15}$$

Combining Eqs. (10.2.3) and (10.2.13), and noting that $c_s \sim V^{-1}$, the osmotic pressure $\Pi_{\text{electrostatic}}$ corresponding to the electrostatic interaction component at a given salt concentration c_s is

$$\frac{\Pi_{\text{electrostatic}} v_1}{k_B T} = 0. \tag{10.2.16}$$

10.2.3   Free Energy of Deformation

Consider an isotropic cube of undeformed gel with linear dimension L_0 (Fig. 10.3) representing the reference state. The gel consists of n long strands, each terminated by two cross-link points. These elastically effective strands are made of n_p strands each with p Kuhn segments, such that $n = \sum_p n_p$. We allow the gel to undergo a deformation so that the linear dimensions of the gel along the three Cartesian directions are L_1, L_2, and L_3. The stretching (or contracting) ratios along the three directions are defined as

$$\lambda_1 = \frac{L_1}{L_0}, \lambda_2 = \frac{L_2}{L_0}, \lambda_3 = \frac{L_3}{L_0}. \tag{10.2.17}$$

In addition, we make the following important assumptions. In both the deformed and the undeformed state, each cross-link point is regarded as fixed at its mean

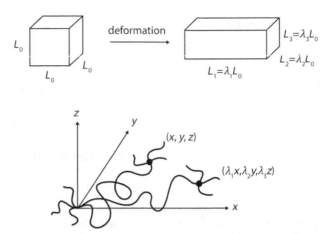

Figure 10.3 Affine deformation of a network. The components of the end-to-end vector of a chain change with the same stretching ratios as the macroscopic network.

position. The effect of the deformation is to change the dimensions of each strand in the same ratio as the corresponding dimensions of the bulk gel. This assumption, called the "affine deformation," is sketched in Fig. 10.3. Furthermore, in this simple model, the strands are assumed to obey the Gaussian statistics. The elastic contribution of the free energy change accompanying the deformation is only entropic in origin and the enthalpic change is assumed to be negligible.

In calculating $\Delta F_{\text{elastic}}$ of the system, let us first consider the change in entropy of a single strand of p segments undergoing an affine deformation, as cartooned in Fig. 10.3, where one end is at the origin of the coordinate system, and the other end undergoes a deformation from the initial position (x, y, z) to its deformed final position $(\lambda_1 x, \lambda_2 y, \lambda_3 z)$. For this one strand, which is assumed to obey Gaussian chain statistics, the entropy S_i in the undeformed state is given by (Eq. (2.1.11))

$$S_i = \text{constant} - \frac{3k_B}{2p\ell^2}(x^2 + y^2 + z^2), \tag{10.2.18}$$

where $(x^2 + y^2 + z^2)$ is the square of the end-to-end distance, ℓ is the Kuhn length, and k_B is the Boltzmann constant. The entropy in the final state (with the assumption that the Gaussian chain statistics are still valid for the deformed strand) is

$$S_f = \text{constant} - \frac{3k_B}{2p\ell^2}(\lambda_1^2 x^2 + \lambda_2^2 y^2 + \lambda_3^2 z^2). \tag{10.2.19}$$

Therefore, the change in entropy ΔS_1 due to the deformation of this one strand is

$$\Delta S_1 = S_f - S_i = -\frac{3k_B}{2p\ell^2}[(\lambda_1^2 - 1)x^2 + (\lambda_2^2 - 1)y^2 + (\lambda_3^2 - 1)z^2]. \tag{10.2.20}$$

The change in entropy ΔS_p for n_p strands, each with p segments is

$$\Delta S_p = -\frac{3k_B}{2p\ell^2}\left[(\lambda_1^2 - 1)\sum_{i=1}^{n_p} x_i^2 + (\lambda_2^2 - 1)\sum_{i=1}^{n_p} y_i^2 + (\lambda_3^2 - 1)\sum_{i=1}^{n_p} z_i^2\right]. \tag{10.2.21}$$

For Gaussian chains,

$$< x_i^2 >=< y_i^2 >=< z_i^2 >= \frac{1}{3} < R_i^2 >= \frac{1}{3} p\ell^2, \qquad (10.2.22)$$

where $< R_i^2 >$ is the mean square end-to-end distance of the ith strand. The change in the Helmholtz free energy due to deformation follows from Eq. (10.2.21), after averaging over all chain conformations, as (because there is no energy change)

$$\frac{\Delta F_{\text{elastic}}}{k_B T} = \frac{n}{2} \left[\lambda_1^2 + \lambda_2^2 + \lambda_3^2 - 3 - \ln(\lambda_1 \lambda_2 \lambda_3) \right]. \qquad (10.2.23)$$

The additional logarithmic term in this equation was derived by Flory with a more careful accounting of chain statistics in the deformed state in comparison with the undeformed state, and the entropy associated with the formation of cross-links in the network [Flory (1953a)].

One important feature of Eq. (10.2.23) is that the number of segments per strand between the cross-link points does not appear. Although the last term is zero for incompressible systems, we shall see its utility in compressible systems such as gels.

10.2.3.1   Elastic Contribution in Isotropic Swelling

If the swelling of the gel is isotropic, $\lambda_1 = \lambda_2 = \lambda_3 = \lambda$, we get

$$\frac{\Delta F_{\text{elastic}}}{k_B T} = \frac{3}{2} n \left(\lambda^2 - 1 - \ln \lambda \right). \qquad (10.2.24)$$

Noting that $\lambda^3 = V/V_0, \phi = V_d/V, \phi_0 = V_d/V_0$, and $V_d = nN v_1$, we obtain

$$\frac{\Delta F_{\text{elastic}}}{k_B T} = \frac{3}{2} n \left[\left(\frac{\phi_0}{\phi} \right)^{2/3} - 1 - \frac{1}{3} \ln \left(\frac{\phi_0}{\phi} \right) \right]. \qquad (10.2.25)$$

Substituting this equation into Eq. (10.2.3), the osmotic pressure from elasticity associated with isotropic gel swelling is given as

$$\frac{\Pi_{\text{elastic}} v_1}{k_B T} = -\frac{1}{N} \left(\phi_0^{2/3} \phi^{1/3} - \frac{\phi}{2} \right). \qquad (10.2.26)$$

The elastic contribution to the osmotic pressure of the gel is thus negative and works against the swelling.

Using Eqs. (10.1.1)–(10.1.3) and $\lambda^3 = V/V_0$, Π_{elastic} can be expressed in different but equivalent forms as derived independently by Flory (1953a), Dusek and Patterson (1968), and Tanaka (1978). For example, using $N_c = n/2$ and $V_d = nNv_1$, Eq. (10.2.26) gives Π_{elastic} in terms of the reduced polymer concentration ϕ/ϕ_0 as

$$\frac{\Pi_{\text{elastic}} v_1}{k_B T} = -\frac{2 N_c v_1}{V_0} \left[\left(\frac{\phi}{\phi_0} \right)^{1/3} - \frac{1}{2} \left(\frac{\phi}{\phi_0} \right) \right]. \qquad (10.2.27)$$

The factor $2N_c v_1/V_0$ is a measure of the cross-link density of the gel in the reference state, and we define it as

$$\frac{2 N_c v_1}{V_0} \equiv S \phi_0^3, \qquad (10.2.28)$$

where S is a parameter representing the cross-link density. In terms of S, Eq. (10.2.27) is written as

$$\frac{\Pi_{\text{elastic}} v_1}{k_B T} = -S\phi_0^3 \left[\left(\frac{\phi}{\phi_0} \right)^{1/3} - \frac{1}{2} \left(\frac{\phi}{\phi_0} \right) \right]. \tag{10.2.29}$$

Comparing Eq. (10.2.26) and Eq. (10.2.29), the cross-link density parameter S and the average number of segments in a chain are related by

$$S\phi_0^2 = \frac{1}{N}. \tag{10.2.30}$$

We will use this equivalence in the discussion of volume transitions in gels.

The above formulas are based on the effective chains being sufficiently long and the net stretching forces sufficiently low to warrant the applicability of Gaussian statistics given by Eq. (10.2.18). However, if the strands are short enough such that the root mean square end-to-end distance of an effective chain is comparable to its contour length, the finite extensibility of the chain must be included, as addressed in Chapter 2.

10.2.4 Osmotic Pressure from Mobile Ions and Donnan Equilibrium

When a charged gel is exposed to an electrolyte solution, the electrolyte ions exchange between the interior and exterior of the gel until an equilibrium distribution of ions is attained. At equilibrium (Donnan equilibrium), the interior and exterior are each charge neutral and the chemical potential of the electrolyte is uniform in the system. We follow the same derivation of the Donnan equilibrium as in Chapter 5 and Appendix 4 for the present situation. Here, a gel of volume V is immersed in an electrolyte solution of volume V_s (Fig. 10.1). The total volume of the system is $V_t = V + V_s$. Let α be the degree of ionization of the uniformly charged gel with n strands, each containing N segments. Each segment carries z_p ionizable groups and its segmental volume is assumed to be v_1. The counterion is taken as monovalent. The volume fraction of the polymer in the gel is $\phi = nNv_1/V$ and the charge concentration of the polymer in the gel is $\alpha z_p nN/V = \alpha z_p c$, where $c = nN/V = \phi/v_1$. Let the external solution have a monovalent strong electrolyte which is fully dissociated. As already mentioned, the boundary of the gel acts as a semipermeable membrane allowing exchange of the counterions from the polymer and the dissociated ions of the strong electrolyte between the gel and the external solution.

As a specific example, let the polymer be negatively charged and the chemical identity of the counterion from the polymer be the same as the cation from the strong electrolyte. The electrolyte ions and the counterions from the polymer exchange through the semipermeable boundary of the gel until osmotic equilibrium is established. At equilibrium, let the concentrations of mobile cations and anions in the gel be c_{+g} and c_{-g}. The corresponding concentrations in the external solution are c_+ and c_-, respectively. The electroneutrality condition for the gel is

$$\alpha z_p c + c_{-g} = c_{+g}, \tag{10.2.31}$$

and for the external solution

$$c_- = c_+ = c_s. \tag{10.2.32}$$

Thus, at equilibrium, the electrolyte concentration in the external solution is $c_s = c_+ = c_-$ and the gel has acquired a concentration of c_{-g} from the electrolyte.

The osmotic pressure of the gel due to mobile ions inside the gel relative to those outside follows from the ideal law of osmotic pressure as

$$\frac{\Pi_{\text{ion}}}{k_B T} = \left(c_{+g} + c_{-g}\right) - (c_+ + c_-). \tag{10.2.33}$$

Expressions for c_{+g} and c_{-g} are obtained by stipulating the Donnan criterion of equating the activities (exponential of chemical potentials) of the electrolyte in the gel and the external solution [Robinson & Stokes (1959)],

$$c_{+g} c_{-g} = c_+ c_-, \tag{10.2.34}$$

where the activities are replaced by the concentrations under the presumed ideal conditions. Since $c_+ = c_- = c_s$, the above equation gives

$$c_{+g} = \frac{c_s^2}{c_{-g}}. \tag{10.2.35}$$

Substituting this into Eq. (10.2.31), we get the quadratic equation

$$c_{-g}^2 + \alpha z_p c c_{-g} - c_s^2 = 0. \tag{10.2.36}$$

The physically allowed solution of this equation is

$$c_{-g} = \frac{1}{2}\left[-\alpha z_p c + \sqrt{\alpha^2 z_p^2 c^2 + 4c_s^2}\right]. \tag{10.2.37}$$

Similarly c_{+g} can be obtained by substituting for c_{-g} from Eq. (10.2.34) into Eq. (10.2.31) as

$$c_{+g} = \frac{1}{2}\left[\alpha z_p c + \sqrt{\alpha^2 z_p^2 c^2 + 4c_s^2}\right]. \tag{10.2.38}$$

Insertion of Eq. (10.2.37) and Eq. (10.2.38) into Eq. (10.2.33) yields

$$\frac{\Pi_{\text{ion}}}{k_B T} = \sqrt{\alpha^2 z_p^2 c^2 + 4c_s^2} - 2c_s, \tag{10.2.39}$$

since $c_+ = c_- = c_s$. It should be noted that the total concentration of mobile ions inside the gel at equilibrium exceeds that in the external solution in order to balance the charges on the polymer chains of the gel.

If the external solution is salt-free ($c_s = 0$), the contribution of ions to the osmotic pressure is solely from the counterions of the polymer, since all of them will be inside the gel to maintain electroneutrality. If the concentration of the strong electrolyte in the external solution is high, the ionic contribution to the osmotic pressure of the gel becomes negligible, since the difference in the salt concentration between the gel and the surrounding solution becomes very small. This can be seen by expanding the right-hand side of Eq. (10.2.37) in $\alpha z_p c / 4c_s$. The limits are

$$\frac{\Pi_{\text{ion}}}{k_B T} = \begin{cases} \alpha z_p c, & c_s = 0 \\ \frac{\alpha^2 z_p^2 c^2}{4c_s}[1 - \frac{1}{16}\frac{\alpha^2 z_p^2 c^2}{c_s^2} + \cdots], & c_s \gg \alpha c. \end{cases} \qquad (10.2.40)$$

By writing the polymer concentration c in terms of volume fraction $\phi = cv_1$, Eq. (10.2.39) becomes

$$\frac{\Pi_{\text{ion}} v_1}{k_B T} = \sqrt{\alpha^2 z_p^2 \phi^2 + 4v_1^2 c_s^2} - 2v_1 c_s. \qquad (10.2.41)$$

10.2.5 Fluctuations

An analogy between a swollen gel and a semidilute solution near the overlap concentration can be drawn such that the free energy of the system per correlation volume ξ^3 is the thermal energy $k_B T$, where ξ is the correlation length for monomer concentration fluctuations arising from chain connectivity. As described in Section 6.2, the free energy contribution from conformational fluctuations is given as

$$\frac{\Delta F_{\text{fluc}}}{k_B T} \sim \frac{V}{\xi^3}, \qquad (10.2.42)$$

where the known prefactor depends on the salt concentration and ξ is self-consistently related to the Debye screening length κ^{-1}. As described in Chapter 6, $\xi \sim \phi^{-3/4}$ in the high salt limit and $\xi \sim \phi^{-1/2}$ in the salt-free limit. Furthermore, the free energy contribution from the electrostatic correlations of mobile ions is given by the Debye–Hückel theory (Chapter 3) as

$$\frac{\Delta F_{\text{DH}}}{k_B T} = -\left(\frac{V}{v_1}\right)\frac{1}{4\pi}\left[\ln(1 + \kappa\ell) - \kappa\ell + \frac{1}{2}\kappa^2\ell^2\right]. \qquad (10.2.43)$$

In addition, fluctuations in local polymer concentrations also contribute to the free energy of the gel. This contribution is particularly significant near critical points.

In the derivation of swelling equilibrium discussed in the next Sub-section, we do not address the role of fluctuations, and restrict ourselves only to the mean field theory.

10.3 Equilibrium of Isotropically Swollen Gels

The net osmotic pressure of the gel arising from free energy of mixing, electrostatic interactions, elasticity, and Donnan equilibrium is obtained from Eqs. (10.2.8), (10.2.16), (10.2.26), and (10.2.41) as

$$\frac{\Pi v_1}{k_B T} = -\ln(1 - \phi) - \phi - \chi\phi^2 + \sqrt{\alpha^2 z_p^2 \phi^2 + 4c_s^2 v_1^2} - 2c_s v_1 - \frac{1}{N}\left(\phi_0^{2/3}\phi^{1/3} - \frac{\phi}{2}\right). \qquad (10.3.1)$$

At the swelling equilibrium, $\Pi = 0$, so that the condition of swelling equilibrium of charged gels follows from Eq. (10.3.1) as

$$-\ln(1-\phi) - \phi - \chi\phi^2 = -\sqrt{\alpha^2 z_p^2 \phi^2 + 4c_s^2 v_1^2} + 2c_s v_1 + \frac{1}{N}\left(\phi_0^{2/3}\phi^{1/3} - \frac{\phi}{2}\right). \quad (10.3.2)$$

The polymer concentration of the gel in equilibrium is obtained by solving Eq. (10.3.2) in terms of α, N, χ, and c_s. The above equation is the generalization of the Flory–Rehner theory [Flory & Rehner (1943)] of swelling equilibrium of uncharged gels to polyelectrolyte gels. We now proceed to derive the limiting behaviors of swelling equilibrium in the salt-free and high salt limits.

10.3.1   Salt-Free Gels

When $c_s = 0$, the swelling equilibrium for isotropically swollen gels follows from Eq. (10.3.2) as

$$-\ln(1-\phi) - \phi - \chi\phi^2 + \alpha z_p \phi = \frac{1}{N}\left(\phi_0^{2/3}\phi^{1/3} - \frac{\phi}{2}\right). \quad (10.3.3)$$

Expanding the logarithmic term for $\phi \ll 1$, and for $\alpha z_p \gg 1/2N$, we get

$$\phi^{2/3} = \frac{\phi_0^{2/3}}{\alpha z_p N}. \quad (10.3.4)$$

The ratio of the gel volume to the volume of the dry network, called the swelling ratio, follows as

$$\frac{V}{V_d} = \frac{1}{\phi} = \frac{(\alpha z_p N)^{3/2}}{\phi_0}. \quad (10.3.5)$$

Here ϕ_0 is the parameter defined in Eq. (10.1.2) due to the necessity of invoking a reference state for the gel with Gaussian chain statistics. The swelling ratio of the volume of the gel is proportional to the $3/2$ power of the effective number of counterions per chain in the gel. In the salt-free strong swelling regime, the swelling ratio is independent of the χ parameter.

The dependence of the swelling ratio on the degree of ionization as given by Eq. (10.3.2) for $c_s = 0$ is given in Fig. 10.4. As the average number of segments per chain increases, the swelling ratio increases. The curvature of the traces in Fig. 10.4 is due to the $3/2$ power law given by Eq. (10.3.5). As seen from the figure, the swelling ratio can be substantial even for very small degrees of ionization. For such strong swelling, the assumed Gaussian statistics cannot be adequate and the inverse Langevin function (Section 2.1.7) must be used to correctly predict the swelling equilibrium.

10.3.2   Gels with High Salt

For $c_s v_1 \gg \alpha z_p \phi$, Eq. (10.3.2) gives

$$-\ln(1-\phi) - \phi - \chi_{\text{eff}}\phi^2 + O(\phi^3) = \frac{1}{N}\left(\phi_0^{2/3}\phi^{1/3} - \frac{\phi}{2}\right) \quad (10.3.6)$$

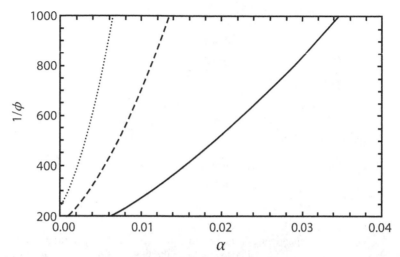

Figure 10.4 Dependence of the swelling ratio on degree of ionization and chain length for $c_S = 0$, $\chi = 0$, $\phi_0 = 0.008$ and $N = 100$ (solid), 250 (dashed), and 500 (dotted).

where

$$\chi_{\text{eff}} = \chi - \frac{1}{4}\frac{\alpha^2 z_p^2}{c_s v_1}. \tag{10.3.7}$$

Note that χ_{eff} is of the same form given in Eq. (10.2.15). The modification of χ by inter-segment electrostatic interactions does not contribute to the osmotic pressure as given in Eq. (10.2.16). However, χ is modified into χ_{eff} due to the additional contribution from mobile ions in maintaining the Donnan equilibrium. Thus at higher salt concentrations, the contribution from the mobile ions to the properties of the gel is negligible due to electrostatic screening and alleviation of Donnan pressure. As a result, the effective χ is essentially the same as the χ value for the polymer–solvent combination which usually leads to phase separation. When c_s is decreased, χ_{eff} becomes more negative, making the background fluid of the gel a better solvent. If the bare χ is sufficiently positive, so that the uncharged gel is in a poor solvent and thus can collapse, charging the gel will stabilize the swollen gel with fewer strong electrolyte ions. However, upon addition of more salt, the χ_{eff} approaches the bare χ leading to the collapse of the gel.

For small values of polymer concentration, $\phi \ll 1$, which is typical for swollen gels, the logarithmic term of Eq. (10.3.6) can be expanded to give the swelling equilibrium as

$$\phi^{5/3} \simeq \frac{\phi_0^{2/3}}{N\left(\frac{1}{2} - \chi_{\text{eff}}\right)}. \tag{10.3.8}$$

Therefore, the swelling ratio for the gel in the limit of high salt is

$$\frac{1}{\phi} \simeq \left(\frac{N(\frac{1}{2} - \chi_{\text{eff}})}{\phi_0^{2/3}}\right)^{3/5}. \tag{10.3.9}$$

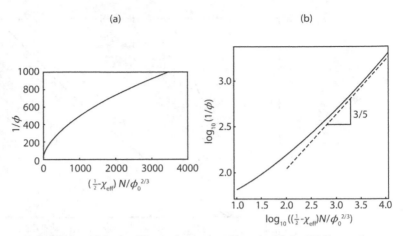

(a)   (b)

Figure 10.5 Dependence of the swelling ratio on strand length: (a) double linear plot
($\chi_{eff} = 0$, $\phi_0 = 0.008$), (b) double logarithmic plot. (Eq. (10.3.6) with $\chi_{eff} = 0$, $\phi_0 = 0.008$,
solid curve; Eq. (10.3.9) with $\phi_0 = 0.008$, dashed curve; slope of the dashed curve is 3/5.)

Thus the swelling ratio of the gel in the high salt limit is proportional to the 3/5 power
of the average number N of segments in the elastically active strands between cross-
links. As N increases (that is, as the cross-link density decreases), the swelling ratio
increases.

The dependence of the swelling ratio on $(\frac{1}{2} - \chi_{eff})N/\phi_0^{2/3}$ is plotted in Fig. 10.5a
for $\phi_0 = 0.008$ and $\chi_{eff} = 0$. As the chain length increases, the swelling ratio is higher.
For large values of $(\frac{1}{2} - \chi_{eff})N/\phi_0^{2/3}$, the behavior is in accordance with the 3/5 power
law. This is illustrated with the double logarithmic plot given in Fig. 10.5b, where
the solid curve is from Eq. (10.3.6), and the dashed line is from Eq. (10.3.9). As the
salt concentration is reduced, χ_{eff} becomes more negative, which results in enhanced
swelling of the gel.

The above results are the generalization of the Flory–Rehner theory [Flory &
Rehner (1943)] of uncharged gels, where χ_{eff} is simply χ.

10.3.3   Crossover Behavior with Salt

The equilibrium swelling ratio $1/\phi$ is given by Eq. (10.3.2) in terms of χ, α, and c_s.
As noted already, ϕ_0 is a parameter connecting the dry state and the reference state
of the gel. A typical result is illustrated in Fig. 10.6a, where the swelling ratio is
plotted against the degree of ionization for different salt concentrations. As the salt
concentration is reduced, the swelling ratio increases drastically for a fixed degree of
ionization. The swelling ratio is higher at higher degrees of ionization. Although the
curves in Figs. 10.4, 10.5, and 10.6 are calculated with the assumption of fixed degrees
of ionization at all swelling conditions, these figures clearly show that even a slight
change in α can lead to significant changes in the equilibrium swelling ratio. Early
experimental data [Kuhn *et al.* (1950), Katchalsky *et al.* (1951)] on the dependence of
swelling ratio on the degree of ionization are presented in Fig. 10.6b. The experimental

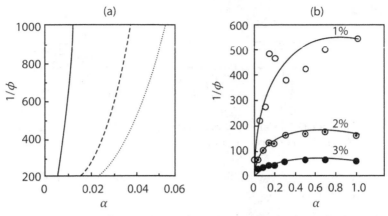

Figure 10.6 (a) Dependence of the swelling ratio on degree of ionization and salt concentration for $\chi = 0, \phi_0 = 0.008$ and $N = 100$. $c_s v_1 = 0.001$ (solid), 0.01 (dashed), and 0.02 (dotted). (b) Experimental data on swelling ratio $1/\phi$ of poly(methacrylic acid) gels versus degree of ionization α. The cross-link concentrations are 1%, 2%, and 3%, with the corresponding average number of segments 750, 370, and 190, respectively, from the top curve to the bottom curve [From Kuhn *et al.* (1950)].

system is aqueous poly(methacrylic acid) gel with three cross-link concentrations, 1%, 2%, and 3%, with the corresponding values of strand length $N = 750$, 370, and 190, respectively. As seen in the figure, the swelling ratio for the 1% gel is about 550. The swelling ratio increases with degree of ionization and decreases with the degree of cross-linking.

The generic features of gel swelling predicted by Eq. (10.3.2) are in qualitative agreement with experimental results for lower degrees of ionization. In terms of quantitative details, there are deviations. The notable difference is for very high swelling, where the swelling ratio approaches a plateau corresponding to the maximum finite stretching of the chains, as seen in Fig. 10.6b. The discrepancy between this observed feature and the predictions from Eq. (10.3.2) is due to the use of Gaussian chain statistics whereby the chains can extend unrealistically to infinite lengths. This can be corrected by using finitely extensible chain statistics (Section 2.1.7). The other aspect is the self-consistent change of the effective degree of ionization as the gel swells, as will be discussed in Section 10.10.

10.4 Gel Swelling under Confinement

When a gel is confined by rigid boundaries, the swelling can occur only along the directions perpendicular to the confining direction. Two simple examples are the one-dimensional swelling of a cylinder-like gel, as sketched in Fig. 10.7a, and the two-dimensional swelling of a disk-like gel, as sketched in Fig. 10.7b. By implementing the geometrical constraints of these situations in Eq. (10.2.17), the free energy change associated with gel deformation is expressed from Eq. (10.2.23). From this derived

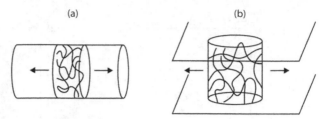

Figure 10.7 Gel under confinement. (a) One-dimensional swelling and (b) two-dimensional swelling in an external electrolyte solution.

result, the swelling equilibrium and the osmotic modulus can be obtained as derived in the following text.

10.4.1   One-Dimensional Swelling

For the geometry of Fig. 10.7a, let λ be the stretching ratio in the allowed direction and $\lambda_\perp$ be the constant stretching (or contracting) ratio in the orthogonal direction with respect to the reference state. The volume fraction of the polymer in the gel follows from Eq. (10.1.2) as

$$\phi = \phi_0 \frac{V_0}{V} = \frac{\phi_0}{\lambda \lambda_\perp^2}. \tag{10.4.1}$$

Therefore the stretching ratio can be written in terms of the constant $\lambda_\perp$ and the volume fraction of the polymer as

$$\lambda = \frac{1}{\lambda_\perp^2} \frac{\phi_0}{\phi}. \tag{10.4.2}$$

Substituting $\lambda_1 = \lambda$ and $\lambda_2 = \lambda_3 = \lambda_\perp$ in Eq. (10.2.23), we get

$$\frac{\Delta F_{\text{elastic}}}{k_B T} = \frac{n}{2} \left[\frac{1}{\lambda_\perp^4} \left(\frac{\phi_0}{\phi} \right)^2 + 2\lambda_\perp^2 - 3 + \ln\left(\frac{\phi}{\phi_0} \right) \right]. \tag{10.4.3}$$

Substituting this result in Eq. (10.2.3), the elastic contribution to the osmotic pressure of the gel follows as

$$\frac{\Pi_{\text{elastic}} v_1}{k_B T} = -\frac{1}{N} \left[\frac{1}{\lambda_\perp^4} \left(\frac{\phi_0}{\phi} \right)^2 - \frac{1}{2} \right] \phi. \tag{10.4.4}$$

Combining Eqs. (10.2.8), (10.2.16), (10.2.41), and (10.4.4), the net osmotic pressure of the anisotropically swollen gel is given by

$$\frac{\Pi v_1}{k_B T} = -\ln(1 - \phi) - \phi - \chi\phi^2 + \sqrt{\alpha^2 z_p^2 \phi^2 + 4 c_s^2 v_1^2} - 2 c_s v_1 - \frac{1}{N} \left[\frac{1}{\lambda_\perp^4} \left(\frac{\phi_0}{\phi} \right)^2 - \frac{1}{2} \right] \phi. \tag{10.4.5}$$

In view of Eq. (10.4.2), this result can be alternatively written as

$$\frac{\Pi v_1}{k_B T} = -\ln(1 - \phi) - \phi - \chi\phi^2 + \sqrt{\alpha^2 z_p^2 \phi^2 + 4 c_s^2 v_1^2} - 2 c_s v_1 - \frac{1}{N} \left(\lambda^2 - \frac{1}{2} \right) \phi. \tag{10.4.6}$$

Note that λ is not a constant, but a function of ϕ. For gels at equilibrium, $\Pi = 0$. In the salt-free situation, the above equation yields for small ϕ

$$\left(\alpha z_p + \frac{1}{2N}\right) + \cdots = \frac{\lambda^2}{N}, \tag{10.4.7}$$

which leads to the limiting behavior for small ϕ and large N,

$$\alpha z_p N \simeq \lambda^2. \tag{10.4.8}$$

For high salt concentrations, as we have seen in Section 10.3.2, the effects of the partitioning of mobile ions between the gel and the external solution and the electrostatic interaction among polymer segments can be expressed in terms of a modified Flory–Huggins parameter, $\chi_{\rm eff}$. Thus, Eq. (10.4.6) yields

$$\left(\frac{1}{2} - \chi_{\rm eff}\right) N = \frac{\left(\lambda^2 - \frac{1}{2}\right)}{\phi}. \tag{10.4.9}$$

The crossover behavior between the limits of Eq. (10.4.8) for $c_s = 0$ and Eq. (10.4.9) for high salt concentrations is given by Eq. (10.4.6) with $\Pi = 0$.

10.4.2   Two-Dimensional Swelling

For the geometry of Fig. 10.7b, let λ be the stretching ratio in the lateral directions (x and y) normal to the confining direction (z), along which the stretching (or contracting) ratio is taken as a fixed value $\lambda_\perp$. The volume fraction of the polymer in the gel follows from Eq. (10.1.2) as

$$\phi = \phi_0 \frac{V_0}{V} = \frac{\phi_0}{\lambda_\perp \lambda^2}. \tag{10.4.10}$$

Therefore the stretching ratio can be written in terms of the constant $\lambda_\perp$ and the volume fraction of the polymer as

$$\lambda^2 = \frac{1}{\lambda_\perp} \frac{\phi_0}{\phi}. \tag{10.4.11}$$

Substituting $\lambda_1 = \lambda_2 = \lambda$ and $\lambda_3 = \lambda_\perp$ in Eq. (10.2.23), we get

$$\frac{\Delta F_{elastic}}{k_B T} = \frac{n}{2} \left[\lambda_\perp^2 + \frac{2}{\lambda_\perp}\left(\frac{\phi_0}{\phi}\right) - 3 + \ln\left(\frac{\phi}{\phi_0}\right) \right]. \tag{10.4.12}$$

Combining Eqs. (10.2.8), (10.2.16), (10.2.41), and (10.4.12), the osmotic pressure of the gel is obtained as

$$\frac{\Pi v_1}{k_B T} = -\ln(1 - \phi) - \phi - \chi\phi^2 + \sqrt{\alpha^2 z_p^2 \phi^2 + 4c_s^2 v_1^2} - 2c_s v_1 - \frac{1}{N}\left[\frac{1}{\lambda_\perp}\left(\frac{\phi_0}{\phi}\right) - \frac{1}{2}\right]\phi. \tag{10.4.13}$$

In view of Eq. (10.4.11), this result can be alternatively written as

$$\frac{\Pi v_1}{k_B T} = -\ln(1 - \phi) - \phi - \chi\phi^2 + \sqrt{\alpha^2 z_p^2 \phi^2 + 4c_s^2 v_1^2} - 2c_s v_1 - \frac{1}{N}\left(\lambda^2 - \frac{1}{2}\right)\phi. \tag{10.4.14}$$

This result is the same as for the one-dimensional swelling (Eq. (10.4.6)), and the limiting results for two-dimensionally equilibrated swollen gels are given by Eqs. (10.4.8) and (10.4.9) for salt-free and high salt conditions.

10.5   Gel Swelling under Tension

One of the very useful properties of gels is their capacity to absorb lots of water upon stretching, as experienced with super-absorbent sanitary products used daily in the world. In this Section, we describe the quantitative aspects of this phenomenon, by applying tension to a gel while it is submerged in a background bath.

Let us consider the situation of a gel in equilibrium with an external electrolyte solution stretched uniaxially under a tensile force. We choose the initial volume of the dry polymer as $V_d = L_d^3$ and let the tension impose the anisotropic shape change of the stretched gel, as sketched in Fig. 10.2, with L_1 the dimension along the stretching direction and $L_2 = L_3 = L_\perp$ in the orthogonal directions. We derive below the extent of water absorbance upon gel stretching and the stress–strain relations.

10.5.1   Solvent Uptake with Gel Stretching

Defining the stretching ratios,

$$\lambda = \frac{L_1}{L_0}, \qquad \lambda_\perp = \frac{L_\perp}{L_0}, \tag{10.5.1}$$

the swelling ratio becomes

$$\frac{1}{\phi} = \frac{L_1 L_\perp^2}{L_d^3} = \frac{\lambda \lambda_\perp^2}{\phi_0}, \tag{10.5.2}$$

where ϕ_0 is defined in Eq. (10.1.2). Therefore, $\lambda_\perp$ can be written in terms of the stretching ratio and the volume fraction of the polymer in the stretched gel as

$$\lambda_\perp = \sqrt{\frac{\phi_0}{\lambda \phi}}. \tag{10.5.3}$$

Substituting λ for λ_1 and $\lambda_\perp$ for λ_2 and λ_3 in the right-hand side of Eq. (10.2.23) for the free energy of the gel due to deformation, we obtain

$$\frac{\Delta F_{elastic}}{k_B T} = \frac{n}{2} \left[\lambda^2 + \frac{2\phi_0}{\lambda \phi} - 3 + \ln(\frac{\phi}{\phi_0}) \right]. \tag{10.5.4}$$

Although the formalism is analogous to one-dimensional swelling, the stretching ratio λ here is an externally imposed constraint, and the swelling ratio of the gel under stretching is determined by equilibrating the gel under stretching. Substituting Eq. (10.5.4) in Eq. (10.2.3), the osmotic pressure due to elasticity $\Pi_{elastic}$ is given by

$$\frac{\Pi_{elastic} v_1}{k_B T} = -\frac{1}{N} \left(\frac{\phi_0}{\lambda} - \frac{\phi}{2} \right). \tag{10.5.5}$$

The net osmotic pressure is the sum of Π_{mix}, $\Pi_{electrostatic}$, $\Pi_{elastic}$, and Π_{ion}. The contributions from the entropy of mixing of mobile ions and solvent, electrostatic interactions among polymer segments, and the distribution of ions in accordance with the Donnan equilibrium are the same as in Eqs. (10.2.8), (10.2.16), and (10.2.41), respectively. $\Pi_{mix} + \Pi_{electrostatic} + \Pi_{ion}$ is positive tending to swell the gel. This is countered by $\Pi_{elastic}$ which is negative. At the swelling equilibrium, the net osmotic pressure is zero, and the equation of state in equilibrium is given by this condition as

$$ -\ln(1-\phi) - \phi - \chi\phi^2 = -\sqrt{\alpha^2 z_p^2 \phi^2 + 4c_s^2 v_1^2} + 2c_s v_1 + \frac{1}{N}\left(\frac{\phi_0}{\lambda} - \frac{\phi}{2} \right). \quad (10.5.6) $$

In the salt-free situation, Eq. (10.5.6) yields for low polymer concentrations

$$ (\alpha z_p + \frac{1}{2N})\phi + \cdots = \frac{\phi_0}{\lambda N}. \quad (10.5.7) $$

If $\alpha z_p N \gg 1/2$, the swelling ratio is given by

$$ \frac{1}{\phi} = \frac{\alpha z_p N}{\phi_0}\lambda. \qquad (c_s = 0) \quad (10.5.8) $$

Thus the swelling ratio is proportional to the stretching ratio in the salt-free case. The gels absorb more solvent upon stretching.

In the limit of high salt concentrations and small volume fraction of polymer in the gel, we get from Eq. (10.5.6)

$$ \left(\frac{1}{2} - \chi_{eff} \right)\phi^2 \simeq \frac{1}{N}\left(\frac{\phi_0}{\lambda} - \frac{\phi}{2} \right), \quad (10.5.9) $$

where χ_{eff} is given by Eq. (10.3.7). Furthermore, the second term on the right-hand side is negligible for $\phi \to 0$. With these approximations, the swelling ratio in the presence of higher levels of salt concentration becomes

$$ \frac{1}{\phi} = \left(\frac{\left(\frac{1}{2} - \chi_{eff}\right)N}{\phi_0} \right)^{1/2} \sqrt{\lambda}. \qquad \text{(high salt)} \quad (10.5.10) $$

In this limit, the swelling ratio is proportional to the square root of the stretching ratio, in contrast to the salt-free case. In general, when the gel is stretched ($\lambda > 1$), the volume fraction of the polymer decreases. In other words, the gel absorbs solvent during stretching.

10.5.2 Length–Force Relation

The retractive force τ required to maintain the gel with the dimension L_1 along the x-direction is given by the thermodynamic relation

$$ \tau = \left(\frac{\partial \Delta F}{\partial L_1} \right)_{T,n_1,n}, \quad (10.5.11) $$

where ΔF is the free energy of formation of the gel in the final deformed state, and n_1 and n are the number of solvent molecules and the number of effective chains, respectively. ΔF is given by Eqs. (10.2.1) and (10.2.2). Since the partial derivative in Eq. (10.5.11) is taken by keeping n and n_1 fixed, only the $\Delta F_{\text{elastic}}$ part contributes to the retractive force in equilibrium,

$$\tau = \left(\frac{\partial \Delta F_{\text{elastic}}}{\partial L_1} \right)_{T,n_1,n} = \frac{k_B T}{L_0} \left(\frac{\partial \left(\Delta F_{\text{elastic}}/k_B T \right)}{\partial \lambda} \right)_{T,n_1,n}, \qquad (10.5.12)$$

where $L_1 = \lambda L_0$ is used. Substitution of Eq. (10.5.4) in the above equation gives

$$\tau = \frac{k_B T n}{L_0} \left(\lambda - \frac{\phi_0}{\lambda^2 \phi} \right). \qquad (10.5.13)$$

Writing the tensile force in dimensionless units, we get

$$\tilde{\tau} \equiv \frac{\tau L_0}{n k_B T} = \left(\lambda - \frac{\phi_0}{\lambda^2 \phi} \right). \qquad (10.5.14)$$

For a given λ, Eq. (10.5.6) is solved to obtain the corresponding equilibrium value of ϕ, which is then substituted in the above equation to get $\tilde{\tau}$ in terms of λ. The above equation is the generalization of the equation of state of the gel in swelling equilibrium to the presence of tension. If $\tilde{\tau} = 0$, the above equation reduces to the condition of isotropic swelling,

$$\frac{\phi_0}{\phi} = \lambda^3 = \frac{V}{V_0}, \qquad (10.5.15)$$

in accordance with the definitions given in Eq. (10.1.2).

In the salt-free limit for swollen gels ($\phi \to 0$), substitution of Eq. (10.5.8) in Eq. (10.5.14) gives

$$\tilde{\tau} = \lambda - \frac{\alpha z_p N}{\lambda}. \qquad (c_s = 0) \qquad (10.5.16)$$

In the high salt limit, substitution of $\phi_0/\lambda^2 \phi$ from Eq. (10.5.14) into Eq. (10.5.9) yields

$$\left(\frac{1}{2} - \chi_{\text{eff}} \right) N = \frac{\lambda^2 (\lambda - \tilde{\tau})}{\phi_0} \left[\lambda (\lambda - \tilde{\tau}) - \frac{1}{2} \right]. \qquad (10.5.17)$$

A plot of the stretching ratio λ given by Eq. (10.5.17) is presented in Fig. 10.8 as a function of $(1/2 - \chi_{\text{eff}})\phi_0 N$ for three different values of the dimensionless tension $\tilde{\tau}$. In the presence of the tensile force, the extension is clearly higher for a fixed value of $(1/2 - \chi_{eff})\phi_0 N$. From such plots, the effect of changing the degree of ionization, salt concentration, solvent quality, and the cross-link density of the gel on the force–extension relation can be readily discerned. For low polymer concentrations, the factor $1/2$ on the right-hand side of Eq. (10.5.17) can be ignored, resulting in

$$\tilde{\tau} = \lambda - \frac{\sqrt{\left(\frac{1}{2} - \chi_{\text{eff}} \right) \phi_0 N}}{\lambda^{3/2}}. \qquad \text{(high salt)} \qquad (10.5.18)$$

Thus, $(\lambda - \tilde{\tau})$ for the gel is proportional to N/λ and $\sqrt{N}/\lambda^{3/2}$, respectively, in the low and high salt concentration limits in the external solution surrounding the gel. Eqs. (10.5.16) and (10.5.18) are the stress–strain relations for salt-free gels and gels with higher concentrations of added salt.

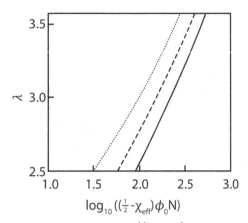

Figure 10.8 Dependence of the stretching ratio on $\left(\frac{1}{2} - \chi_{\text{eff}}\right)\phi_0 N$ and dimensionless tension $\tilde{\tau}$ in the high salt limit. $\tilde{\tau} = 0$ (solid), 0.5 (dashed), and 1.0 (dotted).

10.6 Elasticity of Swollen Gels

Consider the situation where a swollen gel is pulled out of the liquid and then subjected to deformation by keeping the polymer volume fraction fixed. The various elastic properties of equilibrated gels depend on polyelectrolyte concentration, degree of ionization, pH, and salt concentration. It is desirable to formulate design rules to achieve desired elastic properties of synthetic hydrogels as well as to understand the basic features of natural gels. Toward this goal, we address in this section the connection between the macroscopic elastic properties and the molecular aspects of gels. By considering simple elongation and simple shear, we shall present stress–strain relations and molecular expressions for Young's modulus, shear modulus, bulk modulus, and the Poisson ratio of equilibrated swollen gels.

10.6.1 Simple Elongation

Consider a simple uniaxial stretching of a swollen gel, as cartooned in Fig. 10.9a, where the stretching is along the longitudinal direction 1, and the two transverse directions, 2 and 3, are equivalent. Let $\lambda_s = L_1/L$ be the stretching ratio with respect to the unstretched swollen gel, and $\lambda_1 = L_1/L_0$ be the stretching ratio with respect to the reference state (due to both swelling and stretching). Using $L/L_0 = (\phi_0/\phi)^{1/3}$ (Eq. (10.1.2)), we get

$$\lambda_1 = \frac{L_1}{L_0} = \frac{L_1}{L}\frac{L}{L_0} = \lambda_s \left(\frac{\phi_0}{\phi}\right)^{1/3}. \tag{10.6.1}$$

Now ϕ_0/ϕ is a constant and does not depend on deformation. Assuming that there is no volume change during the stretching of the swollen gel, we have

$$\lambda_1\lambda_2\lambda_3 = \frac{V}{V_0} = \frac{\phi_0}{\phi}. \tag{10.6.2}$$

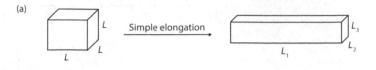

Figure 10.9 (a) Simple elongation, $L_2 = L_3$. (b) Shear deformation.

Therefore,

$$\lambda_2 = \lambda_3 = \sqrt{\frac{\phi_0}{\lambda_1 \phi}} = \frac{1}{\sqrt{\lambda_s}}\left(\frac{\phi_0}{\phi}\right)^{1/3}. \tag{10.6.3}$$

Hence, from Eq. (10.2.23), we get the free energy change due to elastic deformation as

$$\Delta F_{\text{elastic}} = \frac{nk_BT}{2}\left[\lambda_s^2\left(\frac{\phi_0}{\phi}\right)^{2/3} + \frac{2}{\lambda_s}\left(\frac{\phi_0}{\phi}\right)^{2/3} - 3 - \ln\left(\frac{\phi_0}{\phi}\right)\right]. \tag{10.6.4}$$

The tensile stress Σ_t is given by

$$\Sigma_t = \frac{1}{L^2}\frac{\partial}{\partial L_1}\Delta F_{\text{elastic}} = \frac{1}{L^3}\frac{\partial}{\partial \lambda_s}\Delta F_{\text{elastic}}. \tag{10.6.5}$$

Combining Eqs. (10.6.4) and (10.6.5), we get

$$\Sigma_t = \frac{nk_BT}{V}\left(\frac{\phi_0}{\phi}\right)^{2/3}\left(\lambda_s - \frac{1}{\lambda_s^2}\right). \tag{10.6.6}$$

Since $\phi = V_d/V$ and $V_d = nNv_1$,

$$\Sigma_t = \frac{k_BT}{Nv_1}\phi_0^{2/3}\phi^{1/3}\left(\lambda_s - \frac{1}{\lambda_s^2}\right). \tag{10.6.7}$$

The Young's modulus E follows from this equation as

$$E = \left(\frac{\partial \Sigma_t}{\partial \lambda_s}\right)_{\lambda_s \to 1} = \frac{3k_BT}{Nv_1}\phi_0^{2/3}\phi^{1/3}. \tag{10.6.8}$$

10.6.2   Simple Shear

In simple shear, one plane slides parallel to a given plane by an amount proportional to the separation distance (Fig. 10.9b). The amount of shear is $\gamma = \tan\theta$, where θ is defined in Fig. 10.9b. For an incompressible system, γ is related to the deformation λ_s as [Treloar (1958)]

$$\gamma = \lambda_s - \frac{1}{\lambda_s}, \tag{10.6.9}$$

where the deformation along the principal axes (not simply related to the direction of sliding) are

$$L_1 = \lambda_s L, \qquad L_2 = \frac{1}{\lambda_s} L, \qquad L_3 = L. \qquad (10.6.10)$$

Therefore,

$$\lambda_1 = \frac{L_1}{L_0} = \frac{L_1}{L} \frac{L}{L_0} = \lambda_s \left(\frac{\phi_0}{\phi}\right)^{1/3}; \quad \lambda_2 = \frac{1}{\lambda_s} \left(\frac{\phi_0}{\phi}\right)^{1/3}; \quad \lambda_3 = \left(\frac{\phi_0}{\phi}\right)^{1/3}. \qquad (10.6.11)$$

Note that $\lambda_1 \lambda_2 \lambda_3 = \phi_0/\phi = V/V_0$, as expected.

The free energy change due to simple shear follows from Eqs. (10.2.23) and (10.6.11) as

$$\Delta F_{\text{elastic}} = \frac{nk_B T}{2} \left[\lambda_s^2 \left(\frac{\phi_0}{\phi}\right)^{2/3} + \frac{1}{\lambda_s^2} \left(\frac{\phi_0}{\phi}\right)^{2/3} + \left(\frac{\phi_0}{\phi}\right)^{2/3} - 3 - \ln\left(\frac{\phi_0}{\phi}\right) \right]. \qquad (10.6.12)$$

Using Eq. (10.6.9), Eq. (10.6.12) is rewritten in terms of the strain γ as

$$\Delta F_{\text{elastic}} = \frac{nk_B T}{2} \left(\frac{\phi_0}{\phi}\right)^{2/3} \gamma^2 + \text{terms in } (\phi_0/\phi), \qquad (10.6.13)$$

where γ is defined in Eq. (10.6.9).

The shear stress Σ_s, defined as $\Sigma_s = \partial(\Delta F_{elastic}/V)/\partial\gamma$, follows from Eq. (10.6.13) as

$$\Sigma_s = \frac{nk_B T}{V} \left(\frac{\phi_0}{\phi}\right)^{2/3} \gamma \qquad (10.6.14)$$

so that the shear modulus G_s is

$$G_s = \frac{nk_B T}{V} \left(\frac{\phi_0}{\phi}\right)^{2/3}. \qquad (10.6.15)$$

In view of $V = V_d/\phi = nN v_1/\phi$, we get

$$G_s = \frac{k_B T}{N v_1} \phi_0^{2/3} \phi^{1/3}. \qquad (10.6.16)$$

Comparing Eqs. (10.6.7) and (10.6.16), the tensile stress can be written in terms of shear modulus as

$$\Sigma_t = G_s \left(\lambda_s - \frac{1}{\lambda_s^2}\right). \qquad (10.6.17)$$

The expression for the shear modulus given by Eq. (10.6.16), derived for incompressible gels, is a general expression that does not require equilibrium. The validity of Eq. (10.6.16) can be examined experimentally by measuring the shear modulus of nearly ideal gels with fixed architecture [Horkay et al. (2017)]. On the other hand, if different gels of unknown structure were prepared, then the prefactor representing the structure (proportional to $1/N$) is not known. However, the factor containing N in Eq. (10.6.16) can be eliminated using the condition for swelling equilibrium given by Eq. (10.3.2) [Morozova & Muthukumar (2017), Jia & Muthukumar (2020)].

Using the expression for $\phi_0^{2/3}\phi^{1/3}/N$ from Eq. (10.3.2) into Eq. (10.6.16), the shear modulus is given as

$$\frac{G_s v_1}{k_B T} = -\ln(1-\phi) - \phi - \chi\phi^2 + \sqrt{\alpha^2 z_p^2 \phi^2 + 4c_s^2 v_1^2} - 2c_s v_1 + \frac{\phi}{2N}. \qquad (10.6.18)$$

For $\phi \ll 1$ and for N so large that $\phi/(2N)$ is negligible, the shear moduli in the salt-free and high salt limits follow as

$$\frac{G_s v_1}{k_B T} = \begin{cases} \left(\alpha z_p\right)\phi & (c_s = 0) \\ \left(\frac{1}{2} - \chi + \frac{\alpha^2 z_p^2}{4c_s v_1}\right)\phi^2. & \text{(high salt)} \end{cases} \qquad (10.6.19)$$

Note that ϕ is determined at the swelling equilibrium, since the derived shear modulus is applicable for equilibrated gels.

For polyelectrolyte gels without added salt and $N \gg 1$, the shear modulus is determined by the osmotic pressure from counterions, $\alpha z_p \phi$, as given by Eq. (10.6.19). Now, G_s is directly proportional to the polymer concentration. At high enough concentrations of added salt, the shear modulus depends quadratically on the polymer concentration. The predicted quadratic dependence, $G_s \sim \phi^2$, is observed in experiments on weakly cross-linked hyaluronic acid hydrogels with concentrations of added NaCl ranging from 10^{-3}M to 1M [Morozova & Muthukumar (2017)]. This quadratic scaling behavior is also observed for synthetic gels as shown in Fig. 10.10a [Jia & Muthukumar (2020)], where the data are for poly(acrylamide-co-acrylate) hydrogels with 10% charge density containing 0.005, 0.01, 0.1, and 1.0 M NaCl. In these experiments, the different polymer concentrations of the swollen gels are obtained by choosing different values of cross-link density and allowing the gels to attain their ultimate equilibrium, while keeping the salt concentration fixed. The quadratic dependence $G_s \sim \phi^2$ derived in the high salt limit is observed even at the salt concentration of 0.005 M NaCl, indicating that even for such low salt concentrations, theoretical expressions derived for the high salt limit may be applicable.

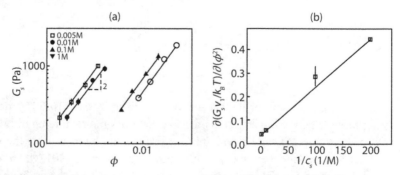

Figure 10.10 (a) Quadratic dependence of shear modulus on polymer concentration for poly(acrylamide-co-acrylate) hydrogels equilibrated at different NaCl concentrations. (b) Determination of degree of ionization and χ parameter from shear modulus [Adapted from Jia & Muthukumar (2020)].

The slope of the plot of $G_s v_1/k_B T$ versus ϕ^2 is given by Eq. (10.6.19) as

$$\frac{\partial(G_s v_1/k_B T)}{\partial \phi^2} = \frac{1}{2} - \chi + \frac{\alpha^2}{4c_s v_1},$$

(10.6.20)

where χ is the Flory–Huggins parameter accounting for only short-ranged van der Waals interactions and the last term arising from the intersegment electrostatic interactions and the Donnan equilibrium. A plot of the above slope against $1/c_s$ allows an experimental method to determine the value of χ and the effective degree of ionization in charged gels. The slopes from Fig. 10.10a are plotted in Fig. 10.10b against the reciprocal of salt concentration for poly(acrylamide-co-acrylate) gels. The χ parameter and the effective degree of ionization α are obtained from the intercept and the slope as $\chi = 0.46 \pm 0.01$ and $\alpha = 0.012 \pm 0.003$. The degree of ionization is essentially independent of salt concentration for the weakly charged hydrogels used in these experiments.

In general, Eq. (10.6.18) can be used to relate a macroscopic elastic property such as shear modulus to molecular characteristics such as degree of ionization and strand length. Eq. (10.6.18) provides design rules for tuning the shear modulus of hydrogels.

10.6.3 Osmotic Bulk Modulus

The bulk modulus K of isotropically swollen gels follows from Eqs. (10.2.3) and (10.3.1) as

$$\frac{K v_1}{k_B T} = \frac{\phi}{1-\phi} - \phi - 2\chi\phi^2 + \sqrt{\alpha^2 z_p^2 \phi^2 + 4c_s^2 v_1^2} - 2c_s v_1 - \frac{1}{3N}\left(\phi_0^{2/3}\phi^{1/3} - \frac{3}{2}\phi\right).$$

(10.6.21)

This expression reduces to simple laws for the concentration dependence of the osmotic bulk modulus of swollen gels for salt-free and high salt limits as given below.
(i) Salt-Free

For $c_s = 0$ and $\phi \ll 1$, the above equation simplifies to

$$\frac{K v_1}{k_B T} \simeq \left(\alpha z_p + \frac{1}{2N}\right)\phi - \frac{1}{3N}\phi_0^{2/3}\phi^{1/3},$$

(10.6.22)

where terms of order ϕ^2 are ignored. Substituting the value of $\phi_0^{2/3}\phi^{1/3}/N$ from the swelling equilibrium given in Eq. (10.3.3), we get

$$\frac{K v_1}{k_B T} = \frac{2}{3}\left(\alpha z_p + \frac{1}{2N}\right)\phi.$$

(10.6.23)

For $\alpha z_p \gg 1/2N$, this result becomes

$$\frac{K v_1}{k_B T} \simeq \frac{2}{3}\alpha z_p \phi.$$

(10.6.24)

Therefore, according to the mean field theory used in deriving the above results, the osmotic bulk modulus in the salt-free situation is proportional to $\alpha\phi$, just as the shear modulus. In addition, the prediction from such a simple theory is that the osmotic bulk modulus is 2/3 of the shear modulus in this limit.

(ii) High salt

For $2c_s v_1 \gg \alpha z_p \phi$ and $\phi \ll 1$, Eq. (10.6.21) becomes

$$\frac{Kv_1}{k_B T} = \left(1 - 2\chi + \frac{\alpha^2 z_p^2}{2c_s v_1}\right)\phi^2 - \frac{1}{3N}\phi_0^{2/3}\phi^{1/3}. \qquad (10.6.25)$$

Combining Eqs. (10.3.8) and (10.6.25), we get

$$\frac{Kv_1}{k_B T} = \frac{5}{3}\left(\frac{1}{2} - \chi + \frac{1}{4}\frac{\alpha^2 z_p^2}{c_s v_1}\right)\phi^2. \qquad (10.6.26)$$

Using Eq. (10.3.7), the osmotic bulk modulus can be alternatively written as

$$\frac{Kv_1}{k_B T} = \frac{5}{3}\left(\frac{1}{2} - \chi_{\text{eff}}\right)\phi^2. \qquad (10.6.27)$$

In view of the derived expression for the shear modulus G_s, Eq. (10.6.19), the relation between the bulk modulus and the shear modulus in the high salt limit follows from Eq. (10.6.27) as

$$K = \frac{5}{3}G_s. \qquad (10.6.28)$$

Note that both K and G_s are proportional to ϕ^2 at high salt concentrations and in equilibrium, according to the mean-field theory presented above.

10.6.4   Poisson Ratio

The response of the aspect ratio of an elastic body to an externally imposed deformation is given by the Poisson ratio Σ_P, which is the ratio of the transverse compression to the longitudinal extension. The relations between Young's modulus E, shear modulus G_s, and bulk modulus K of an elastic body can be written in terms of the Poisson ratio as [Landau & Lifshitz (1986)]

$$G_s = \frac{E}{2(1 + \Sigma_P)}, \quad K = \frac{E}{3(1 - 2\Sigma_P)}, \quad \frac{K}{G_s} = \frac{2(1 + \Sigma_P)}{3(1 - 2\Sigma_P)}, \qquad (10.6.29)$$

and the Poisson ratio is written in terms of G_s and K as

$$\Sigma_P = \frac{1}{2}\frac{(3K - 2G_s)}{(3K + G_s)}. \qquad (10.6.30)$$

Combining Eqs. (10.6.28) and (10.6.30), the Poisson ratio for swollen gels at high salt concentrations is given as

$$\Sigma_P = \frac{1}{4}. \qquad (10.6.31)$$

When the salt concentration in the swollen gel is reduced, the Poisson ratio approaches zero, as seen from Eqs. (10.6.19) and (10.6.24). These results are to be contrasted with the value of $\Sigma_P = 1/2$ for an incompressible network where $K \to \infty$ or $K \gg G_s$.

The results derived in previous sections on the various elastic properties of equilibrated swollen gels are general. However, these results are based only on the mean field theory for fully equilibrated swollen charged gels as derived in Section 10.3. In spite of the success of the above results in capturing a large body of experimental results, discrepancies between theoretical predictions and experimental findings can emerge due to several factors that are left out in the mean-field theory. A few of these factors include structural inhomogeneity in the gel, change in degree of ionization during elastic deformation, entropy change due to reorientation of water around polymer segments, and concentration fluctuations. In view of these considerations, deviations from the predictions of the mean-field theory are in principle expected. Nevertheless, the formulas derived above are good guidelines, and provide a solid conceptual framework to understand the swelling equilibrium and elasticity of charged gels.

10.7 Dynamics of Charged Gels

At ambient temperatures, the various polymer strands constituting the hydrogel undergo significant conformational fluctuations mimicking semidilute conditions, because the gel is predominantly water. These fluctuations are controlled by chain connectivity, segmental friction against the solvent, relatively immobile cross-link constraints, and any externally imposed forces. These molecular aspects of the gel are manifest in the macroscopic elastic properties described in the preceding section. In this section, we describe the dynamics and correlations of collective modes of polymer segments in a gel as interrogated by scattering techniques, and make connections to the elastic properties of gels.

10.7.1 Elasticity Equations

Consider an equilibrated aqueous gel at ambient temperatures, where polymer strands undergo conformational fluctuations as sketched in the cartoon of Fig. 10.11. As a typical representation, the position $\mathbf{r}'$ of a segment fluctuates around its equilibrium position $\mathbf{r}$. Let us define the displacement vector of this segment as

$$\mathbf{u} = \mathbf{r}' - \mathbf{r}. \tag{10.7.1}$$

The divergence of $\mathbf{u}$ can be directly related to the fluctuation δc in the local polymer concentration as [Landau & Lifshitz (1986)]

$$\nabla \cdot \mathbf{u} = -\frac{\delta c}{c}, \tag{10.7.2}$$

where c is the average polymer concentration.

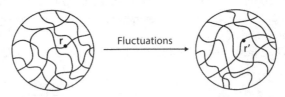

Figure 10.11 Cartoon of a gel in equilibrium with concentration fluctuations.

The equation of motion for the displacement vector is [Landau & Lifshitz (1986)]

$$\rho \frac{\partial^2 \mathbf{u}}{\partial t^2} = -f \frac{\partial \mathbf{u}}{\partial t} + \nabla \cdot \Sigma. \tag{10.7.3}$$

The term on the left-hand side is from inertia with ρ being the mass density. The first term on the right-hand side is due to friction, with f being the friction coefficient of the polymer network against the background fluid per unit volume. The last term is the force from stress, with Σ being the stress tensor. Using the theory of elasticity for the stress, the equation of motion for $\mathbf{u(t)}$ is given by [Landau & Lifshitz (1986)]

$$\rho \frac{\partial^2 \mathbf{u}}{\partial t^2} = -f \frac{\partial \mathbf{u}}{\partial t} + G_s \nabla^2 \mathbf{u} + (K + \frac{1}{3} G_s) \nabla (\nabla \cdot \mathbf{u}), \tag{10.7.4}$$

where G_s is the shear modulus and K is the bulk modulus.

Choosing the Cartesian coordinate system $(x, y, \text{and } z)$ for positioning the displacement vector, let the elastic wave $\mathbf{u}$ propagate along the x-direction. The direction x is the longitudinal direction; y, and z directions are the transverse directions. The x-component of the displacement vector, u_ℓ, denotes the longitudinal mode and the y- and z-components of the displacement vector, $u_\perp$, denote the transverse modes. u_ℓ and $u_\perp$ are independent of y and z and depend only on x.

Resolving the elasticity equation, Eq. (10.7.4), into the longitudinal and transverse modes enables separate measurements of the shear modulus and bulk modulus of the gel using scattering techniques [Tanaka *et al.* (1973)], as shown in the following text.

10.7.1.1 Longitudinal Mode

For the longitudinal mode, where $\nabla^2 \mathbf{u} = \partial^2 u_\ell / \partial x^2$ and $\nabla (\nabla \cdot \mathbf{u}) = \partial^2 u_\ell / \partial x^2$, Eq. (10.7.4) becomes [Tanaka *et al.* (1973)]

$$\rho \frac{\partial^2 u_\ell}{\partial t^2} + f \frac{\partial u_\ell}{\partial t} = \left(K + \frac{4}{3} G_s \right) \frac{\partial^2 u_\ell}{\partial x^2}. \tag{10.7.5}$$

If frictional force is negligible, the gel supports non-attenuated sound waves, with the longitudinal sound velocity c_ℓ given by

$$c_\ell = \sqrt{\frac{K + \frac{4}{3} G_s}{\rho}}. \tag{10.7.6}$$

Sound travels slowly in gels. A typical value of speed of sound in gel-like media is about 10 m/s, in comparison with about 1500 m/s in water and about 343 m/s in air. Although the gel is mainly water, the speed of sound is significantly attenuated due to the presence of polymer strands.

When inertia is negligible, Eq. (10.7.5) appears as a "diffusion equation,"

$$\frac{\partial u_\ell}{\partial t} = \left(\frac{K + \frac{4}{3} G_s}{f} \right) \frac{\partial^2 u_\ell}{\partial x^2}. \tag{10.7.7}$$

In view of the similarity to the diffusion equation, we may call the front factor on the right-hand side as the "gel diffusion coefficient," defined as

$$D_g = \left(\frac{K + \frac{4}{3}G_s}{f} \right) \equiv \frac{M_\ell}{f}, \qquad (10.7.8)$$

where M_ℓ is known as the longitudinal modulus defined by

$$M_\ell = K + \frac{4}{3}G_s. \qquad (10.7.9)$$

We note that no element inside the gel actually undergoes diffusion, in contrast with diffusion of molecules in solutions. The gel diffusion coefficient is only an analogy and it denotes a combination of bulk modulus, shear modulus, and gel friction coefficient. Nevertheless, in view of the connection between the divergence of the displacement vector and local polymer concentration given by Eq. (10.7.2), D_g is the cooperative diffusion coefficient related to the correlation length of monomer concentrations in the gel, analogous to that in semidilute solutions:

$$D_g = \frac{k_B T}{6\pi\eta_0 \xi}, \qquad (10.7.10)$$

where ξ is the correlation length for the distribution of polymer monomers. It is common in the literature to geometrically identify the correlation length ξ as the mesh size [de Gennes (1979)]. For swollen gels, which may be imagined to be a three-dimensional structure made of meshes of volume comparable to ξ^3, we call ξ as the mesh size of the gel, interchangeably with the correlation length for monomer concentration.

As we shall see in Section 10.7.2, the longitudinal modulus and the gel diffusion coefficient can be measured using polarized light scattering.

10.7.1.2 Transverse Mode

For the transverse modes, where $\nabla^2\mathbf{u} = \partial^2 u_\perp / \partial x^2$ and $\nabla(\nabla \cdot \mathbf{u}) = 0$, the equation of motion for the displacement follows from Eq. (10.7.4) as

$$\rho \frac{\partial^2 u_\perp}{\partial t^2} + f \frac{\partial u_\perp}{\partial t} = G_s \frac{\partial^2 u_\perp}{\partial x^2}, \qquad (10.7.11)$$

with the transverse sound velocity $c_\perp$ and the transverse gel diffusion coefficient $D_\perp$ given as

$$c_\perp = \sqrt{\frac{G_s}{\rho}}, \qquad (10.7.12)$$

$$D_\perp = \frac{G_s}{f}. \qquad (10.7.13)$$

The transverse gel diffusion coefficient $D_\perp$ can be determined using depolarized light scattering (Section 10.7.2).

10.7.1.3   Gel Friction Coefficient

In order to determine the shear modulus and bulk modulus from the measurements of D_g and $D_\perp$, it is necessary to know the value of the gel friction coefficient, as evident from Eqs. (10.7.8) and (10.7.13). The gel friction coefficient f can be directly measured by monitoring the uniform velocity of water flow through the hydrogel [Tokita & Tanaka (1991)]. On the other hand, it is difficult to calculate f. From a theoretical point of view, the friction coefficient appearing in the continuum elasticity equation, Eq. (10.7.5), is the friction coefficient of the polymer network against the flow of water per unit volume. Toward an estimate of f, we imagine that the gel network is a three-dimensional assembly of meshes, each with an average linear size comparable to the correlation length ξ of monomer concentration fluctuations in the gel. This implies that the average end-to-end distance of a strand, connected to two cross-links at its ends, is comparable to ξ. Assuming that the Stokes law (Chapter 7) of friction is applicable to each strand, its friction coefficient is $6\pi\eta_0\xi$, where η_0 is the viscosity of the solution in which the gel matrix is embedded. Therefore, the gel friction coefficient per unit volume is given as

$$f = \frac{6\pi\eta_0\xi}{\left(\frac{4}{3}\pi\xi^3\right)} = \frac{9}{2}\frac{\eta_0}{\xi^2}, \tag{10.7.14}$$

where orientational and related geometric factors affecting the hydrodynamics around the strands are ignored.

Using direct measurement of f or its theoretical estimate from the experimental knowledge of the mesh size of the gel, the shear modulus and bulk modulus can be determined from D_g and $D_\perp$. These relations form the link between the elastic properties and microscopic structures of gels.

10.7.2   Determination of Gel Moduli Using Scattering Experiments

Scattering techniques with either light or neutrons can be used to determine the structure and elastic constants of charged gels. Both static and dynamic scattering methods are used in obtaining the elastic properties of gels in terms of their microscopic structures.

10.7.2.1   Static Scattering

A typical example is given in Fig. 10.12a from small angle neutron scattering (SANS) studies on poly(acrylic acid) gels in water containing NaBr salt [Schosseler *et al.* (1991)]. In this figure, the ratio of the excess scattered intensity $\Delta I(k)$ to the polymer concentration c is plotted against the scattering wave vector $k = \frac{4\pi}{\lambda}\sin\frac{\theta}{2}$ (λ is the wavelength of the incident beam and θ is the scattering angle). The SANS data for different salt concentrations are included in the figure where polymer concentration (which is proportional to the polymer volume fraction ϕ) $c = 0.707$ M and the degree of ionization $\alpha = 0.101$. At low salt concentrations, a peak is observed reminiscent of the polyelectrolyte peak described in Chapter 6. The position of the scattering peak

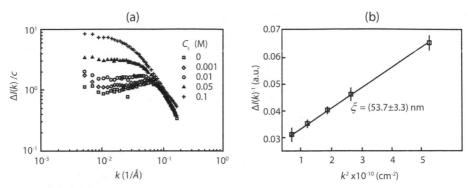

Figure 10.12 (a) Plot of SANS excess scattering intensity versus scattering wave vector at different salt concentrations for poly(acrylic acid) gels with polymer concentration $c = 0.707$ M and the degree of ionization $\alpha = 0.101$ [From Schossler *et al.* (1991)]. (b) Ornstein–Zernike plot of excess scattering intensity in static light scattering versus the square of the scattering wave vector for poly(acrylamide-co-acrylate) gels with 0.2% cross-link density and 10% charge density in 0.01 M NaCl solution. The mesh size is 53.7±3.3 nm and it obeys the scaling law, $\xi \sim c^{-2/3}$ [Jia & Muthukumar (2020)].

for polyelectrolyte gels depends on polymer concentration, salt concentration, and temperature in a manner analogous to the behavior in semidilute polyelectrolyte solutions (Chapter 6). As seen in Fig. 10.12a, the polyelectrolyte peak disappears as the added salt concentration is increased. Under such high salt conditions, and for small scattering wave vector k such that $k\xi \ll 1$, the excess scattered intensity is given by the Ornstein–Zernike equation (Chapter 6),

$$\Delta I(k) = \frac{\Delta I(k = 0).}{1 + k^2 \xi^2} \qquad (10.7.15)$$

As described in Chapter 6, the correlation length of concentration fluctuations is obtained as the square root of the slope/intercept ratio from a linear plot of $1/\Delta I(k)$ versus k^2. A typical example of measurement of the correlation length (mesh size) ξ is given in Fig. 10.12b. The data in this figure are from static light scattering on poly(acrylamide-co-acrylate) gels with 0.2% cross-link density and 10% charge density in 0.01 M NaCl solution at 25°C [Jia & Muthukumar (2020)]. For this particular system, the correlation length ξ is 53.7 ± 3.3 nm, and it scales with polymer concentration c according to $\xi \sim c^{-2/3}$ [Jia & Muthukumar (2020)].

Since the elastic properties given by the elasticity equations (Section 10.7.1) are at larger length scales in comparison with the scales associated with a single mesh, we focus on the scattering intensity at very small values of the scattering wave vector. According to the theory of light scattering from gels by Tanaka *et al.* (1973), the excess intensity ΔI scattered from concentration fluctuations in the gel is given by

$$\Delta I(k \to 0) = A_0 k_B T \left(\frac{\partial n_r}{\partial c}\right)^2 \frac{c^2}{M_\ell}, \qquad (10.7.16)$$

where A_0 is a constant specific to the experimental apparatus and the polarization of the incident light, and n_r is the refractive index of the scattering medium. Using the

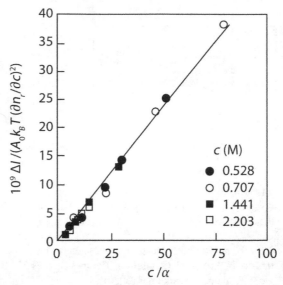

Figure 10.13 Plot of $\Delta I/(A_0 k_B T(\partial n_r/\partial c)^2$ versus c/α in agreement with Eq. (10.7.18) for salt-free aqueous poly(acrylic acid) gels [From Schosseler *et al.* (1991)].

expression for the longitudinal modulus given by Eq. (10.7.9), the excess scattered intensity with polarized light at $k \to 0$ is given as

$$\frac{\Delta I(k \to 0)}{A_0 k_B T \left(\frac{\partial n_r}{\partial c}\right)^2} = \frac{c^2}{\left(K + \frac{4}{3}G_s\right)}, \qquad (10.7.17)$$

where the shear modulus G_s and the osmotic modulus K are given in Eqs. (10.6.18) and (10.6.21).

10.7.2.1.1 Salt-Free Gels

Substituting Eqs. (10.6.19) and (10.6.23) in Eq. (10.7.17), we get

$$\frac{\Delta I(k \to 0)}{A_0 k_B T \left(\frac{\partial n_r}{\partial c}\right)^2} \sim \frac{c}{\alpha}, \qquad (10.7.18)$$

where we have assumed $\alpha z_p \gg 1/2N$. Therefore a plot of $\Delta I/(A_0 k_B T(\partial n_r/\partial c)^2$ versus c/α should be a straight line. This prediction is validated in experiments on poly(acrylic acid) gels as demonstrated in Fig. 10.13, where the data are from four different polymer concentrations and different levels of ionization [Schosseler *et al.* (1991)]. Thus, the scattered intensity decreases with an increase in release of counterions from the polyelectrolyte in salt-free conditions.

10.7.2.1.2 Gels with Added Salt

The excess scattering intensity for gels with added salt is given generally by substituting Eqs. (10.6.18) and (10.6.21) in Eq. (10.7.17). If the added salt concentration is sufficiently high, $K + \frac{4}{3}G_s$ simplifies from Eqs. (10.6.26), (10.6.28), and (10.6.19) as

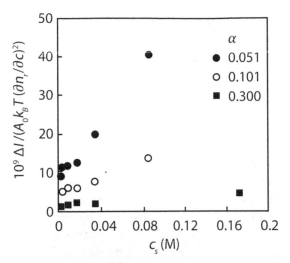

Figure 10.14 Increase of excess scattered intensity with concentration of added NaBr salt in aqueous poly(acrylic acid) gels at polymer concentration $c = 0.707$ M. The excess scattered intensity decreases with degree of ionization. [From Schosseler *et al.* (1991)].

$$K + \frac{4}{3}G_s = \frac{3k_BT}{v_1}\left(\frac{1}{2} - \chi + \frac{\alpha^2}{4v_1c_s}\right)\phi^2. \tag{10.7.19}$$

Substituting this result into Eq. (10.7.17), we get the excess scattering intensity $\Delta I(k \to 0)$ as

$$\frac{\Delta I}{A_0k_BT\left(\frac{\partial n_r}{\partial c}\right)^2} = \frac{1}{3k_BTv_1\left(\frac{1}{2} - \chi + \frac{\alpha^2}{4v_1c_s}\right)}. \tag{10.7.20}$$

This asymptotic result is valid only at very high salt concentrations. In general, it is seen from Eq. (10.7.20) that the denominator on the right-hand side is a decreasing function of c_s and hence the excess scattering intensity is predicted to increase with salt concentration. Similarly, ΔI is predicted by Eq. (10.7.20) to decrease with an increase in the degree of ionization of the gel. These predictions are in qualitative agreement with experiments as seen in Fig. 10.14, where $\Delta I/(A_0k_BT(\partial n_r/\partial c)^2$ is plotted against c_s for poly(acrylic acid) gels with $\alpha = 0.051$, 0.101, and 0.3, and polymer concentration $c = 0.707$ M [Schosseler *et al.* (1991)].

10.7.2.2 Dynamic Light Scattering

In dynamic light scattering, the experimentally measured time correlation function of scattered intensity at a scattering wavevector $\mathbf{k}$ is related to the theoretically derived electric field correlation function at $\mathbf{k}$. In view of the distinction between the longitudinal and transverse modes inside the gel, let $\mathbf{k}$ be directed along the longitudinal direction, i.e., $\mathbf{k} = (k,0,0)$. According to the theory of light scattering from gels [Tanaka *et al.* (1973)], the electric field correlation function $\langle \mathbf{E}(\mathbf{k},t) \cdot \mathbf{E}(\mathbf{k},0)\rangle$ is calculated directly from the elasticity equation Eq. (10.7.4) for the displacement vector. In experiments with polarized light, the time correlation function of the electric field scattered by longitudinal fluctuations is measured; with depolarized light, transverse fluctuations are measured [Tanaka *et al.* (1973)]. Using Eqs. (10.7.5) and (10.7.11) and ignoring the inertia term, the electric field correlation functions are

$$\langle \mathbf{E}_{\mathrm{pol}}(\mathbf{k},t) \cdot \mathbf{E}_{\mathrm{pol}}(\mathbf{k},0) \rangle = \langle E_{\mathrm{pol}}^2 \rangle \exp\left(-\frac{M_\ell}{f}k^2 t\right)$$ (10.7.21)

for polarized scattering, and

$$\langle \mathbf{E}_{\mathrm{dep}}(\mathbf{k},t) \cdot \mathbf{E}_{\mathrm{dep}}(\mathbf{k},0) \rangle = \langle E_{\mathrm{dep}}^2 \rangle \exp\left(-\frac{G_s}{f}k^2 t\right)$$ (10.7.22)

for depolarized scattering. M_ℓ is the longitudinal modulus given by Eq. (10.7.9) and G_s is the shear modulus.

The electric field correlation function can also be written in terms of the correlation function of fluctuations in the local monomer concentration. Since the divergence of the displacement vector inside a gel is related to the local monomer concentration δc as $\nabla \cdot \mathbf{u} = -\delta c(\mathbf{r},t)/c$, Eq. (10.7.2), the time evolution of $\delta c(\mathbf{r},t)$ can be obtained from Eq. (10.7.4). Taking the Fourier transform of the resulting equation for the longitudinal mode, we can get

$$\frac{\partial \delta \hat{c}(\mathbf{k},t)}{\partial t} = -\left(\frac{K + \frac{4}{3}G_s}{f}\right)k^2 \delta \hat{c}(\mathbf{k},t),$$ (10.7.23)

where the inertia term is ignored. The time correlation function for $\delta \hat{c}(\mathbf{k},t)$ follows from Eq. (10.7.23) as

$$\langle \delta \hat{c}(\mathbf{k},t) \delta \hat{c}(\mathbf{k},0) \rangle = \langle (\delta \hat{c}(\mathbf{k},t))^2 \rangle \exp(-D_g k^2 t),$$ (10.7.24)

where the gel diffusion coefficient is $D_g = (K + \frac{4}{3}G_s)/f$.

The experimentally measured correlation function of the scattered intensity at scattering wavevector k, $\langle I(k,t_0)I(k,t_0 + \tau) \rangle$ (where t_0 is the initial time and τ is the delay time) is related to the normalized electric field correlation function $g^{(1)}(k,\tau) = \langle E(k,t_0)E^*(k,t_0 + \tau) \rangle / \langle E^2(k,t_0) \rangle$ through the Seigert relation [Chu (1991), Brown (1993), Han & Akcasu (2011)]

$$\frac{\langle I(k,t_0)I(k,t_0 + \tau) \rangle}{\langle I(k,t_0) \rangle^2} = 1 + \beta |g^{(1)}(k,\tau)|^2,$$ (10.7.25)

where the angular brackets denote time averages and β is the coherence factor of the detection optics. Using Eqs. (10.7.8) and (10.7.21) and the definition of $g^{(1)}$, the normalized electric field correlation function for the longitudinal mode is given by

$$|g^{(1)}(k,\tau)|^2 = e^{-2D_g k^2 \tau},$$ (10.7.26)

where the gel diffusion coefficient D_g is given by Eq. (10.7.8). Similarly, the normalized electric field correlation function in the depolarized light scattering experiments is

$$|g^{(1)}(k,\tau)|^2 = e^{-2D_\perp k^2 \tau},$$ (10.7.27)

where $D_\perp = G_s/f$ is the transverse gel diffusion coefficient. Therefore, by measuring the intensity correlation functions, the gel diffusion coefficient D_g and the transverse gel diffusion coefficient $D_\perp$ can be determined by using the above relations.

In practice, determination of the gel diffusion coefficients can be difficult due to the presence of inherent structural inhomogeneities. These inhomogeneities give rise to static electric fields which interfere directly with the scattered electric field from the gel modes. In this case, it can be shown that the intensity correlation functions are [Joosten *et al.* (1991), Norisuye *et al.* (2004)]

$$g^{(2)}(k,\tau) = 1 + X^2 \left(g^{(1)}(k,\tau)\right)^2 + 2X(1-X)g^{(1)}(k,\tau), \tag{10.7.28}$$

where $g^{(2)}(k,\tau)$ is the normalized intensity correlation function defined as the left-hand side of Eq. (10.7.25). The factor $X = \frac{\langle I \rangle_F}{\langle I \rangle_S + \langle I \rangle_F}$ is the ratio of the intensity due to the fluctuating component $\langle I \rangle_F$ to the total intensity $\langle I \rangle_S + \langle I \rangle_F$, which is the sum of the intensity due to the fluctuating component and the static intensity arising from inhomogeneities. It has been shown that for these systems [Norisuye *et al.* (2004)]

$$g^{(2)}(k,\tau) = 1 + \beta e^{-2D_A k^2 \tau}, \tag{10.7.29}$$

where β is the coherence factor and D_A is the apparent elastic diffusion coefficient related to the true elastic diffusion coefficient D_g by the value X as

$$D_A = \frac{D_g}{2 - X}. \tag{10.7.30}$$

Substituting Eqs. (10.6.18), (10.6.21), and (10.7.14) in Eq. (10.7.8), the gel diffusion coefficient D_g can be expressed as a general result in terms of polymer volume fraction, degree of ionization, solvent quality, and added salt concentration. In the high salt limit, this result simplifies to

$$D_g = \frac{2}{3} \frac{k_B T}{v_1 \eta_0} \xi^2 \left(\frac{1}{2} - \chi + \frac{\alpha^2 z_p^2}{4c_s v_1}\right) \phi^2. \tag{10.7.31}$$

As shown below in Section 10.7.4, the correlation length ξ is inversely related to $\left(\frac{1}{2} - \chi + \frac{\alpha^2 z_p^2}{4c_s v_1}\right) \phi^2$. However, $\xi^2 \left(\frac{1}{2} - \chi + \frac{\alpha^2 z_p^2}{4c_s v_1}\right) \phi^2$ is an increasing function of $\left(\frac{1}{2} - \chi + \frac{\alpha^2 z_p^2}{4c_s v_1}\right) \phi^2$. Therefore, according to Eq. (10.7.31), D_g decreases with c_s, and increases with α and ϕ. These predicted trends are experimentally observed, as illustrated in Fig. 10.15. The dependence of D_g on polymer volume fraction is given in Fig. 10.15a, for poly(acrylamide-co-acrylate) hydrogels with 10% charge density in 0.01 M NaCl solution, where the polymer volume fraction is tuned by cross-link density. For this system, D_g has been found to be proportional to the 2/3 power law of the polymer volume fraction, $D_g \sim \phi^{2/3}$ [Jia & Muthukumar (2020)]. The dependence of D_g on the salt concentration is given in Fig. 10.15b for poly(acrylamide-co-acrylate) hydrogels with 0.2% cross-link density and 10% charge density [Jia & Muthukumar (2020)]. As evident from the figure, an increase in salt concentration slows the elastic diffusion of hydrogels. The dependence of D_g on the degree of ionization is shown in Fig. 10.15c, where the data are for poly(acrylic acid) gels with polymer concentration $c = 0.707$ M and the degree of ionization $\alpha = 0.051, 0.101$, and 0.3 [Schosseler *et al.* (1991)]. As the degree of ionization increases, the gel diffusion coefficient increases. The experimental results portrayed in Fig. 10.15 are in qualitative agreement with the

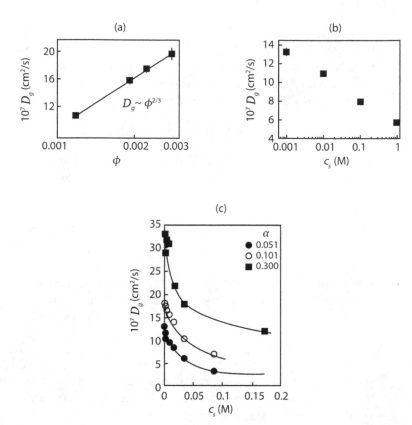

Figure 10.15 (a) Dependence of gel diffusion coefficient D_g on polymer volume fraction for poly(acrylamide-co-acrylate) gels with 10% charge density in 0.01 M NaCl solution, where the polymer volume fraction is tuned by cross-link density [Jia & Muthukumar (2020)]. (b) Dependence of D_g on salt concentration for poly(acrylamide-co-acrylate) gels with 0.2% cross-link density and 10% charge density [Jia & Muthukumar (2020)]. (c) Dependence of D_g on degree of ionization at different salt concentrations for poly(acrylic acid) gels with 0.707 M polymer concentration [From Schosseler *et al.* (1991)]. All of these figures are in qualitative accordance with Eq. (10.7.31).

predictions of Eq. (10.7.31). Hence, Eq. (10.7.31), which is a microscopic representation of the macroscopic elastic properties of hydrogels, offers design rules for tuning the elastic moduli of equilibrated swollen gels in view of Eq. (10.7.8).

10.7.3   Coupling between Gel Dynamics and Counterion Dynamics

Analogous to the dynamics of polyelectrolyte solutions in the context of the "ordinary-extraordinary" transition, where the counterion cloud is generally coupled to the segmental dynamics as described in Chapter 7, the segmental dynamics of charged gels is coupled to the dynamics of the counterion cloud surrounding the segments. Generalizing the equations in Section 7.6.3.1 and Appendix 6.5.3 for polyelectrolyte solutions to gels, where the cooperative diffusion coefficient in solutions is replaced by the gel diffusion coefficient, we get the following coupled equations for salt-free gels [Sasaki & Schipper (2001), Jia & Muthukumar (2021b)],

$$\frac{\partial \delta \hat{c}_1}{\partial t} = -\left(\frac{M_\ell}{f}\right) k^2 \delta \hat{c}_1 - \left(\frac{c_1^0}{f}\right) \frac{\alpha z_p e^2 c_1^0}{\epsilon_0 \epsilon}(\alpha z_p \delta \hat{c}_1 + z_c \delta \hat{c}_2) \tag{10.7.32}$$

$$\frac{\partial \delta \hat{c}_2}{\partial t} = -D_2 k^2 \delta \hat{c}_2 - D_2 \frac{z_c e^2 c_2^0}{\epsilon_0 \epsilon k_B T}(\alpha z_p \delta \hat{c}_1 + z_c \delta \hat{c}_2). \tag{10.7.33}$$

Here $\delta \hat{c}_1$ is the Fourier transform of the fluctuation in the local polymer concentration from its average value c_1^0 and $\delta \hat{c}_2$ is the Fourier transform of the fluctuation in the local counterion concentration from its average value c_2^0. D_2 is the cooperative diffusion coefficient of the counterion without any coupling to the polymer matrix. z_p and z_c are the valencies of the segment and counterion, respectively, α is the degree of ionization, e is the electronic charge, ϵ_0 is the permittivity of vacuum, and ϵ is the dielectric constant of the gel medium. In the presence of added salt, additional equations similar to Eq. (10.7.33) appear for each electrolyte ionic species in the gel as seen in Section 7.7.4.2.

The first term on the right-hand side of Eq. (10.7.32) is due to the diffusive flux and the second term is due to the electrostatic coupling between the charged segments and their counterion clouds. Following the same procedure as described in Chapter 7 for polyelectrolyte solutions, and assuming that the counterion clouds relax much faster than the gel, the rate of change of fluctuation in local polymer concentration is given by

$$\frac{\partial \delta \hat{c}_1(k,t)}{\partial t} = -D_{g,\text{coupled}} k^2 \delta \hat{c}_1(k,t), \tag{10.7.34}$$

where

$$D_{g,\text{coupled}} = \frac{M_\ell}{f} + \frac{\alpha^2 z_p^2 e^2 (c_1^0)^2}{f \epsilon_0 \epsilon \kappa^2}, \tag{10.7.35}$$

where κ is the inverse Debye length. For monovalent salt ions and $z_p = 1$, κ^2 is given by $\kappa^2 = 4\pi \ell_B (\alpha c_1^0 + 2c_s)$, with ℓ_B being the Bjerrum length.

Therefore, the coupling between the ion cloud and gel leads to an additional contribution to the enhancement of D_g with decreased salt concentration and increased degree of ionization as seen in Fig. 10.15. In any quantitative comparison with experiments, Eq. (10.7.35) needs to be employed.

10.7.4   Scaling Laws

So far in this chapter, we have derived only the mean-field theory of charged gels without accounting for concentration and conformational fluctuations in the system. The advantage of the above derivations is the ability to get closed form analytical formulas enabling direct comparison with the numerical values of experimental data. Although fluctuations are ignored above, the derived results are in qualitative agreement with experimental findings. Nevertheless, it is useful to extract scaling laws, such as power-law dependence of the elastic moduli on polymer concentration, from the above equations. Before doing this, let us revisit the role of fluctuations as introduced in Eqs. (10.2.1) and (10.2.42).

The free energy of a gel can be written in general as the sum of a mean-field component and a component arising from concentration fluctuations in a manner analogous to the treatment of polyelectrolyte solutions,

$$\Delta F = \Delta F_{\text{mean-field}} + \Delta F_{\text{fluctuations}}. \tag{10.7.36}$$

The contribution to the free energy from concentration fluctuations can be written using scaling arguments as

$$\Delta F_{\text{fluctuations}} \simeq k_B T \frac{V}{\xi^3}, \tag{10.7.37}$$

where ξ is the correlation length for monomer concentration fluctuations. This is basically equivalent to the argument in Chapter 6 that the energy of the system per correlation volume is the thermal energy $k_B T$. If the fluctuations dominate the free energy of the gel over the mean field contribution, the scaling laws for the osmotic pressure and osmotic modulus are

$$\Pi \sim \frac{T}{\xi^3}, \tag{10.7.38}$$

$$K = c \frac{\partial \Pi}{\partial c} \sim \frac{T}{\xi^3}. \tag{10.7.39}$$

As we have already noted in Eq. (10.7.14), the scaling law for the gel friction coefficient is

$$f \sim \frac{\eta_0}{\xi^2}. \tag{10.7.40}$$

Assuming that the scaling behavior of the shear modulus G_s is proportional to that of the bulk modulus K, we get the scaling law for the gel diffusion coefficient as

$$D_g = \frac{M_\ell}{f} \sim \frac{T}{\eta_0 \xi}. \tag{10.7.41}$$

This result is internally self-consistent with the generalized Stokes–Einstein law for the cooperative diffusion coefficient given by Eq. (10.7.10).

A common assumption in the literature is that the scaling behavior in swollen gels containing salt is the same as the scaling behavior in semidilute solutions under similar salt conditions [de Gennes (1979), Shibayama & Tanaka (1993), Candau *et al.* (1982), Rubinstein & Colby (2003)]. As seen in Chapter 6, for sufficiently high salt concentrations, the polyelectrolyte solutions behave as good solutions and hence the correlation length obeys the scaling law $\xi \sim c^{-3/4}$ by using the size exponent $\nu = 3/5$ for good solutions in the scaling law, $\xi \sim c^{-\nu/(3\nu-1)}$. This would imply the following scaling laws for swollen gels with high salt,

$$\xi \sim c^{-3/4}, \quad K \sim c^{9/4}, \quad f \sim c^{3/2}, \quad \text{and} \quad D_g \sim c^{3/4}. \tag{10.7.42}$$

However, systematic experiments on weakly cross-linked hyaluronic acid gels and aqueous poly(acrylamide-co-acrylate) gels containing salt show the following scaling laws [Morozova & Muthukumar (2017), Jia & Muthukumar (2020)]

$$\xi \sim c^{-2/3}, \quad K \sim c^2, \quad f \sim c^{4/3}, \quad \text{and} \quad D_g \sim c^{2/3}, \tag{10.7.43}$$

where $\xi, G_s, f,$ and D_g are determined independently using static light scattering, rheology, water permeation, and dynamic light scattering, respectively, for equilibrated swollen gels [Morozova & Muthukumar (2017), Jia & Muthukumar (2020)]. These laws are precisely the results already mentioned, predicted by the mean-field theory described in this chapter, and are internally self-consistent by satisfying Eqs. (10.7.39)–(10.7.41). Note that this self-consistency among the scaling laws is observed, although fluctuations are completely ignored in the theory.

Since $\xi \sim c^{-\nu/(3\nu-1)}$ under semidilute conditions, and $\xi \sim c^{-2/3}$ in Eq. (10.7.43), the effective size exponent for the weakly cross-linked gels investigated in the above experiments is 2/3 and not the value of 3/5 nominally used in good solutions. With the value of 2/3 for ν, the scaling law for the dependence of the correlation length on $\chi, \alpha, c_s,$ and c is given by (analogous to the derivation in Chapter 6)

$$\xi \sim \left(\frac{1}{2} - \chi + \frac{\alpha^2 z_p^2}{4 c_s v_1} \right)^{-1/3} c^{-2/3}. \tag{10.7.44}$$

Substituting this result in Eq. (10.7.31), the dependence of the elastic diffusion coefficient of gels on $\chi, c_s, \alpha,$ and c is given by

$$D_g \sim \frac{T}{\eta_0} \left(\frac{1}{2} - \chi + \frac{\alpha^2 z_p^2}{4 c_s v_1} \right)^{1/3} c^{2/3}. \tag{10.7.45}$$

The predictions of the above equation are borne out to be valid as seen from the experimental results in Fig. 10.15.

The demonstrated consistency of predictions of the mean-field theory with experiments provides confidence in using the derived equations as design principles to tune the desired elastic properties of charged hydrogels. Deviations are expected if the contribution from fluctuations to the free energy of the gel dominates over the mean-field component addressed in Section 10.1, and if structural inhomogeneities inside the gel and charge regularization are significant.

10.8 Kinetics of Gel Swelling

When a freshly synthesized gel is allowed to swell in a solution with a controllable amount of salt, the swelling is usually a slow process and the equilibration might take days, even weeks. The characteristic time for swelling kinetics depends on the cross-link density, degree of ionization, polymer concentration, and salt concentration. In fact, the characteristic time for gel swelling is dictated by the gel diffusion coefficient as shown in the following text.

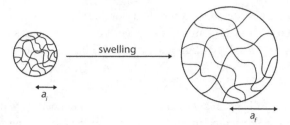

Figure 10.16 Cartoon of a spherical gel swelling from an initial radius a_i to the final radius a_f.

Consider a uniformly cross-linked gel of spherical shape with initial radius a_i, which is allowed to swell to a spherical gel of final radius a_f, as depicted in the cartoon of Fig. 10.16. At time t during the swelling process, let the radius of the gel be $a(t)$. In order to get an expression for $a(t)$ in terms of the molecular parameters of the gel, we need to solve the elasticity equation for the displacement vector, Eq. (10.7.4),

$$f\frac{\partial \mathbf{u}}{\partial t} = G_s \nabla^2 \mathbf{u} + \left(K + \frac{1}{3}G_s\right) \nabla (\nabla \cdot \mathbf{u}),\tag{10.8.1}$$

with the boundary condition that the normal stress at the gel surface is zero. In Eq. (10.8.1), the inertia term of Eq. (10.7.4) is ignored. The result is [Li & Tanaka (1990)]

$$\frac{a_f - a(t)}{a_f - a_i} = \frac{6}{\pi^2} \sum_{p=1}^{\infty} \frac{1}{p^2} \exp\left(-\frac{p^2 t}{\tau}\right),\tag{10.8.2}$$

where the characteristic time for the swelling kinetics is

$$\tau = \frac{a_f^2}{D_g}.\tag{10.8.3}$$

It is evident from Eq. (10.8.2) that, for long enough times (such that the sum is dominated by the lowest mode $p = 1$), the swelling kinetics is a simple exponential growth in time with the characteristic time $\tau \sim 1/D_g$.

The above general premise of gel swelling kinetics has been validated for uncharged gels [Li & Tanaka (1990)]. For charged gels, all results derived in the preceding sections for the gel diffusion coefficient can be used to describe the swelling kinetics.

10.9   Phase Transitions in Gels

Polyelectrolyte gels can undergo discontinuous volume transitions as the degree of ionization of the chains or the solvent quality changes. The approximate theory of gels described in the preceding sections is capable of providing insight into this phenomenon, which we shall discuss now. The osmotic pressure of the gel is given by Eq. (10.3.1) for a given polymer concentration and salt concentration in terms of χ, the average number of segments N in each of the elastically active chains, and the parameter ϕ_0. Expanding the logarithmic term in ϕ in the limit of $\phi \to 0$, and keeping the leading terms, we get

$$\frac{\Pi v_1}{k_B T} = \left(\frac{1}{2} - \chi\right) \phi^2 + \frac{\phi^3}{3} + \sqrt{\alpha^2 z_p^2 \phi^2 + 4 c_s^2 v_1^2} - 2 c_s v_1 - \frac{1}{N} \left(\phi_0^{2/3} \phi^{1/3} - \frac{\phi}{2}\right). \quad (10.9.1)$$

The salt-free and high salt limits of this equation are

$$\frac{\Pi v_1}{k_B T} = \begin{cases} \left(\frac{1}{2} - \chi\right) \phi^2 + \frac{\phi^3}{3} + \alpha z_p \phi - \frac{1}{N} \left(\phi_0^{2/3} \phi^{1/3} - \frac{\phi}{2}\right) & (c_s = 0) \\ \left(\frac{1}{2} - \chi_{\text{eff}}\right) \phi^2 + \frac{\phi^3}{3} - \frac{1}{N} \left(\phi_0^{2/3} \phi^{1/3} - \frac{\phi}{2}\right), & \text{(high salt)} \end{cases} \quad (10.9.2)$$

where χ_{eff} is defined in Eq. (10.3.7).

Defining ϕ/ϕ_0 as the reduced polymer concentration ψ,

$$\psi = \frac{\phi}{\phi_0}, \quad (10.9.3)$$

and using the cross-link density parameter $S = 1/N\phi_0^2$ defined in Eq. (10.2.30), Eq. (10.9.2) becomes

$$\frac{\Pi v_1}{k_B T \phi_0^3} = \begin{cases} \frac{1}{\phi_0} \left(\frac{1}{2} - \chi\right) \psi^2 + \frac{\psi^3}{3} + \left(\frac{\alpha z_p}{\phi_0^2} + \frac{S}{2}\right) \psi - S \psi^{1/3} & (c_s = 0) \\ \frac{1}{\phi_0} \left(\frac{1}{2} - \chi_{\text{eff}}\right) \psi^2 + \frac{\psi^3}{3} - S \left(\psi^{1/3} - \frac{\psi}{2}\right). & \text{(high salt)} \end{cases} \quad (10.9.4)$$

The dependence of the osmotic pressure of a gel on the dimensionless polymer concentration ψ is dictated by the Flory–Huggins χ parameter, degree of ionization, salt concentration, and the cross-link density, as given by the above equation. As we expect the onset of thermodynamic instability and consequent phase transitions in gels at conditions of weak electrostatic repulsion among polymer segments, we now consider the high salt limit of Eq. (10.9.4). The values of the parameters $\chi_{\text{eff}}, \phi_0$, and S are chosen as follows. Defining the temperature dependence of the Flory–Huggins parameter in terms of the ideal temperature Θ (for UCST systems) as

$$\chi = \frac{\Theta}{2T}, \quad (10.9.5)$$

$(1/2 - \chi_{\text{eff}})$ can be written as

$$\frac{1}{2} - \chi_{\text{eff}} = \frac{1}{2} \left(1 - 2\chi_{\text{ion}} - \frac{\Theta}{T}\right), \quad (10.9.6)$$

where χ_{ion} follows from Eq. (10.3.7) as

$$\chi_{\text{ion}} = -\frac{\alpha^2 z_p^2}{4 c_s v_1}. \quad (10.9.7)$$

The parameter ϕ_0 is the volume fraction of the polymer in the reference state with chains obeying Gaussian statistics. Instead of assuming that the volume of a segment is simply ℓ^3, as tacitly assumed in the preceding sections, the segmental volume can be taken as ℓb^2, where b is a measure of the thickness of the chain backbone. In the reference state, the volume occupied by each chain scales as R_g^3 which is about $N^{3/2} \ell^3$ (assuming Gaussian statistics). Therefore, we expect

$$\phi_0 \sim \frac{N\ell b^2}{N^{3/2}\ell^3} \sim \frac{1}{\sqrt{N}}\left(\frac{b}{\ell}\right)^2. \tag{10.9.8}$$

As a result, we expect ϕ_0 to be considerably less than unity. Similarly, the cross-link density defined in Eq. (10.2.28) follows from Eqs. (10.2.30) and (10.9.8) as

$$\frac{2N_c v_1}{V_0} \sim \frac{\ell b^2}{N^{3/2}\ell^3} \sim \frac{1}{N^{3/2}}\left(\frac{b}{\ell}\right)^2, \tag{10.9.9}$$

indicating that one cross-link is expected on average within the volume of one chain. Combining Eq. (10.2.28) and the above two equations, we get

$$S = \frac{2N_c v_1}{V_0 \phi_0^3} \sim \left(\frac{\ell}{b}\right)^4. \tag{10.9.10}$$

Since ℓ is twice the persistence length of the chain, ℓ/b is in the range of 5–10 for flexible chains. As a result, the typical values of the cross-link parameter S is in the hundreds to thousands [Tanaka (1978)].

A plot of $\Pi v_1/k_B T$ given by Eq. (10.9.4), at high salt concentrations, against the reduced polymer concentration ψ is given in Fig. 10.17 for different temperatures. The values of the parameters used in this figure are $\Theta = 410$ K, $\chi_{ion} = -0.01$, $\phi_0 = 0.01$, and $S = 600$. The curves correspond to $T = 325.0$ K, 306.5 K, 302.0 K, 299.5 K, and 296.0 K from the top to bottom in the figure. The thermodynamically allowed states at each temperature can be inferred from these curves. In thermodynamic equilibrium, the osmotic pressure must be either zero or positive. In addition, the osmotic modulus must be positive, because the osmotic pressure must increase as polymer concentration is increased (see Eq. (10.2.3)). This means that regions with negative slopes in plots of Π versus ϕ are not physically allowed.

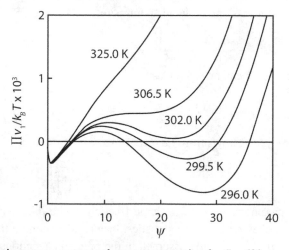

Figure 10.17 Osmotic pressure versus polymer concentration for $S = 600$, $\phi_0 = 0.01$, $\chi_{ion} = -0.01$, and $\Theta = 410$ K. $T = 325.0$ K, 306.5 K, 302.0 K, 299.5 K, and 296.0 K from the top to bottom. The critical temperature is $T_c = 306.5$ K and volume transition temperature is $T_t = 299.5$ K.

At $T = 325.0$ K, all states with $\Pi \geq 0$ are allowed. However, at this temperature, the equilibrium state corresponds to $\Pi = 0$ and a gel with the particular polymer concentration corresponding to the intersection of the curve with the abscissa coexists with the pure fluid phase.

As the temperature is reduced below a critical value ($T_c = 306.5$ K), instability sets in, exhibiting the van der Waals loop as seen in Chapter 8 in the context of phase separation in solutions. When a van der Waals loop occurs, two phases can in principle coexist. The compositions of these two coexisting phases in a two-component system are obtained by satisfying the condition that the chemical potentials of each component are the same in the two phases. Equivalently, a tie line is drawn within the van der Waals loop such that the Maxwell criterion of equal areas for the upper and lower lobes about the tie line within the loop is satisfied. At $T = 302.0$ K, an inspection of the curve shows that two coexisting gel phases corresponding to the tie line are predicted and that the osmotic pressure of these two gels is positive and must be the same. The region of polymer concentration bounded by the maximum and the minimum of the van der Waals loop is unstable, as is the region of spinodal decomposition. Outside this unstable region, but within the compositions of the coexisting phases, the system is metastable. Furthermore, for this temperature, there is a gel phase at a lower concentration than those of the coexisting phases, with zero osmotic pressure. As the temperature is reduced further to $T_t = 299.5$ K, the tie line within the van der Waals loop coincides with the abscissa ($\Pi = 0$). In this situation, two gel phases coexist, each with zero osmotic pressure, occurring at the volume transition temperature T_t. If the temperature is further decreased to $T = 296.0$ K, the tie line of the van der Waals loop would correspond to negative osmotic pressure, which is thermodynamically forbidden. However, there exists a concentration for the gel phase in the stable branch of the van der Waals loop at higher concentrations at which the osmotic pressure is zero. In this situation, a gel with higher polymer concentration coexists with the pure fluid phase. Therefore, $T_t = 299.5$ K denotes the first-order volume transition between the swollen gel at higher temperatures to a contracted gel at lower temperatures.

Based on the above results, the phase diagram of polyelectrolyte gels can be constructed as depicted in Fig. 10.18. At higher temperatures above the critical temperature T_c, the gel is swollen and coexists with the external solution. As the temperature is reduced, this gel becomes slightly denser and at the volume transition temperature T_t, it undergoes a discontinuous volume transition to a gel with a very high polymer concentration. For temperatures below T_t, the shrunken gel is in equilibrium with the external solution. In addition to the volume transition depicted by the thick curves in Fig. 10.18, the theory predicts a critical phenomenon. Between the critical temperature T_c and the volume transition temperature T_t, there are two coexisting gel phases. The coexistence curve is denoted by thin curves in Fig. 10.18. Since the gel is a single connected entity, creation of two phases inside this entity, one with lower polymer concentration and the other with higher polymer concentration, is prohibited. As a result, such coexisting gel phases are never observed in experiments, although the gels develop significant structural inhomogeniety under certain conditions, as seen in scattering experiments.

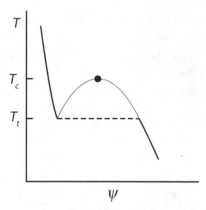

Figure 10.18 Sketch of the phase diagram for polyelectrolyte gels. Thick curves denote the discontinuous volume transition and the thin curve denotes the coexistence curve.

The critical point for the occurrence of volume transitions is given by the thermodynamic condition

$$\frac{\partial^2 \Delta F}{\partial \phi^2} = 0 = \frac{\partial^3 \Delta F}{\partial \phi^3}. \tag{10.9.11}$$

At the critical point, the values of the polymer concentration, temperature, and the cross-link density are ψ_c, T_c, and S_c, respectively. We shall calculate these values separately for salt-free gels and at high salt concentrations, starting from the Helmholtz free energy of the gel.

10.9.1   Critical Point at High Salt Concentrations

In order to find the second and third derivatives of the free energy of the gel required in Eq. (10.9.11), we start from the osmotic pressure given by Eq. (10.9.2). In the high salt limit and for $\phi \ll 1$, we get from Eq. (10.2.3)

$$\frac{\partial}{\partial \phi}\left(\frac{\Delta F}{k_B T n}\right) = N\left(\frac{1}{2} - \chi_{\text{eff}}\right) + \frac{N\phi}{3} - \frac{\phi_0^{2/3}}{\phi^{5/3}} + \frac{1}{2\phi}, \tag{10.9.12}$$

where $nN = V\phi/v_1$ is used. In view of $S\phi_0^2 N = 1$ (Eq. (10.2.30)) and $\psi = \phi/\phi_0$ (Eq. (10.9.3)), we get

$$\frac{\partial}{\partial \psi}\left(\frac{\Delta F}{k_B T n}\right) = \frac{1}{2}\frac{(1 - 2\chi_{\text{eff}})}{S\phi_0} + \frac{1}{3S}\psi - \frac{1}{\psi^{5/3}} + \frac{1}{2\psi}. \tag{10.9.13}$$

This result is proportional to the osmotic pressure. Taking the second and third derivatives of $\Delta F/nk_B T$ with respect to ψ and equating them to zero, we obtain

$$\frac{1}{3S} + \frac{5}{3}\frac{1}{\psi^{8/3}} - \frac{1}{2}\frac{1}{\psi^2} = 0 \tag{10.9.14}$$

and

$$\frac{1}{\psi^3} - \frac{40}{9}\frac{1}{\psi^{11/3}} = 0. \tag{10.9.15}$$

Solving the above two equations for the critical values of $\psi = \psi_c$ and $S = S_c$, we obtain

$$\psi_c = \left(\frac{40}{9}\right)^{3/2} \simeq 9.37 \tag{10.9.16}$$

and

$$S_c = \frac{3}{5}\left(\frac{40}{9}\right)^4 \simeq 234. \tag{10.9.17}$$

The critical temperature is obtained from the equilibrium condition $\Pi = 0$. Substituting the critical values of S_c and ψ_c in Eq. (10.9.13) and equating the result to zero, we get

$$1 - 2\chi_{\text{ion}}, -\frac{\Theta}{T_c} = -\frac{32}{15}\psi_c\phi_0, \tag{10.9.18}$$

where Eq. (10.9.6) is used. Rearranging this result and using Eq. (10.9.7), the critical temperature follows as

$$T_c = \frac{\Theta}{\left(1 + \frac{\alpha^2}{c_s v_1} + \frac{32}{15}\psi_c\phi_0\right)}. \tag{10.9.19}$$

Thus, either a decrease in the salt concentration or an increase in the degree of ionization leads to a decrease in the critical temperature. Also, as the solvent quality decreases, as represented by a higher value of Θ, the volume transition is more feasible. The above predictions are in good qualitative agreement with experiments. An example is given in Fig. 10.19 showing that a volume change of poly(acrylic acid) gels can be induced by decreasing the solvent quality with a mixture of water and acetone. The composition of the solvent mixture required for the occurrence of the discontinuous volume transition of the gel depends crucially on the electrolyte concentration, in accordance with Eq. (10.9.19). Thus, there is an interplay between the chemical mismatch parameter for the polymer–solvent interaction (Flory–Huggins χ parameter) and the electrostatic interactions mediated by gel elasticity.

10.9.2 Critical Point for Salt-Free Gels

The critical point for a gel in a salt-free solution is derived by following the same steps as above for the high salt condition. Using Eq. (10.2.3), the free energy of the gel is obtained by integrating Eq. (10.9.2), for salt-free condition at low polymer concentrations, as

$$\frac{\Delta F}{k_B T n} = \left(\frac{1}{2} - \chi\right)N\phi + \frac{1}{6}N\phi^2 + \frac{3}{2}\left(\frac{\phi_0}{\phi}\right)^{2/3} + \left(\frac{1}{2} + \alpha z_p N\right)\ln\phi. \tag{10.9.20}$$

The first derivative of this expression with respect to ψ, which is proportional to the osmotic pressure, is

$$\frac{\partial}{\partial\psi}\left(\frac{\Delta F}{k_B T n}\right) = \left(\frac{1}{2} - \chi\right)N\phi_0 + \frac{1}{3}N\phi_0^2\psi - \frac{1}{\psi^{5/3}} + \frac{\left(\frac{1}{2} + \alpha z_p N\right)}{\psi}. \tag{10.9.21}$$

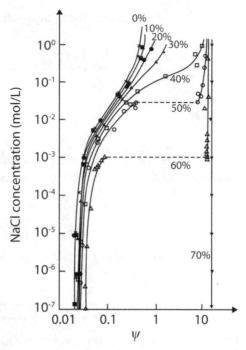

Figure 10.19 Reduced polymer concentration $\psi = \phi/\phi_0$ versus NaCl concentration for various acetone concentrations (denoted by volume percent) in the acetone–water mixture. [From Ohmine & Tanaka (1982)].

At the swelling equilibrium, Eq. (10.3.3) gives

$$1 - 2\chi = 1 + \frac{2}{\phi_0\psi} + \frac{2\ln(1 - \phi_0\psi)}{\phi_0^2\psi^2} - \frac{2S\phi_0}{\psi^2}\left[\left(\frac{1}{2} + \alpha z_p N\right)\psi - \psi^{1/3}\right], \quad (10.9.22)$$

where $\psi = \phi/\phi_0$ and $S = 1/N\phi_0^2$ is the cross-link density parameter defined in Eq. (10.2.30). A plot of the reduced temperature $1 - 2\chi$ versus $\psi = \phi/\phi_0$, given by the above equation, is presented in Fig. 10.20 where $\phi_0 = 0.05$ and $S = 10$. Different curves correspond to different values of $\alpha z_p N$. As seen from the figure, discontinuous volume transitions can easily be induced upon ionization of the gel. Only a few charges are required for the onset of a volume transition.

The discontinuous volume transition predicted by Eq. (10.9.22), Fig. 10.20, has been seen in experiments on polyelectrolyte hydrogels [Hirotsu *et al.* (1987)]. Fig. 10.21 shows the volume transitions of a series of poly(N-isopropylacrylamide-co-acrylate) hydrogels with a varying molar concentration of the acrylate content, ranging from 7 to 70 mM whereas the total molar concentration of the N-propylacrylamide (NIPA) content is kept at 700 mM. As seen in Fig. 10.21, the temperature of the volume transition of the gel increases with an increase in the fraction of ionized groups in the gel. Note that NIPA in water is an example of lower critical solution temperature (LCST) behavior. Therefore, the volume transition of the poly(NIPA-co-acrylate)

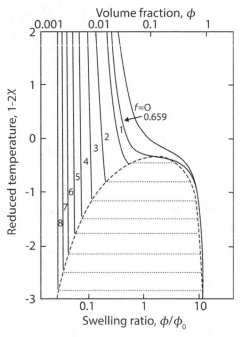

Figure 10.20 Only a low level of ionization is required for the occurrence of a discontinuous volume transition. Plot of the reduced temperature versus swelling ratio for $f \equiv \alpha z_p N$ with ϕ_0 = 0.05 and the cross-link density parameter $S = 10$. [From Tanaka *et al.* (1980)].

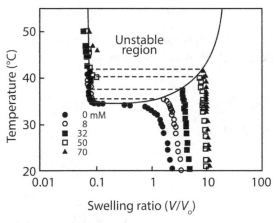

Figure 10.21 Equilibrium volume of ionized N-isopropylacrylamide (NIPA) gels in water exhibits discontinuous volume transition in response to temperature (lower solution critical temperature, LCST, behavior). The values shown indicate the amount of ionizable group (sodium acrylate) incorporated in 700 mM NIPA. [Adapted from Hirotsu *et al.* (1987)].

hydrogels occurs as the temperature is increased. Furthermore, the jump in the discontinuity of the gel volume increases with the charge fraction of the gel, in accordance with the predictions from Eq. (10.9.22) and Fig. 10.20.

The second and third derivatives of ΔF with respect to ψ are zero at the critical point, and Eq. (10.9.20) gives

$$\frac{1}{3}N\phi_0^2 + \frac{5}{3}\frac{1}{\psi^{8/3}} - \frac{\left(\frac{1}{2} + \alpha z_p N\right)}{\psi^2} = 0, \tag{10.9.23}$$

$$\frac{\left(\frac{1}{2} + \alpha z_p N\right)}{\psi^3} - \frac{20}{9}\frac{1}{\psi^{11/3}} = 0. \tag{10.9.24}$$

Solving the above two equations, the critical condition is obtained as

$$\frac{\left(\alpha z_p N + \frac{1}{2}\right)}{N^{1/4}\phi_0^{1/2}} > \frac{4}{3}\left(\frac{5}{3}\right)^{3/4} \simeq 1.956. \tag{10.9.25}$$

By fixing N, the degree of ionization α can be varied experimentally. A first order volume transition can occur only if α is greater than a critical value α_c given by the above equation,

$$\alpha > \alpha_c \simeq \frac{1.956\phi_0^{1/2}}{z_p N^{3/4}} - \frac{1}{2z_p N}. \tag{10.9.26}$$

Since N is usually large in gels, even a very small extent of ionization can lead to discontinuous volume transitions, as seen in Fig. 10.20.

By fixing the degree of ionization, the result of Eq. (10.9.26) can be written equivalently in terms of the critical value of the cross-link parameter S_c as

$$\frac{\left(S_c + \frac{2\alpha z_p}{\phi_0^2}\right)^4}{S_c^3} = \frac{3}{5}\left(\frac{40}{9}\right)^4 \simeq 234, \tag{10.9.27}$$

where Eq. (10.2.30) is used. At the critical point, the polymer concentration follows from Eqs. (10.9.23) and (10.9.24) as

$$\psi_c = \left(\frac{5}{3N\phi_0^2}\right)^{3/8}. \tag{10.9.28}$$

The critical value of χ is obtained from $\Pi = 0$ given by Eq. (10.9.22) as

$$\left(\frac{1}{2} - \chi_c\right)N\phi_0 = -\frac{16}{9}\psi_c^{-5/3}, \tag{10.9.29}$$

so that

$$\chi_c = \frac{1}{2} + \frac{16}{9}\left(\frac{3}{5}\right)^{5/8}N^{-3/8}\phi_0^{1/4}, \tag{10.9.30}$$

where Eq. (10.9.28) is used. If χ obeys the temperature dependence given in Eq. (10.9.5), the critical temperature is given by

$$T_c = \frac{\Theta}{1 + \frac{32}{9}\left(\frac{3}{5}\right)^{5/8}N^{-3/8}\phi_0^{1/4}}. \tag{10.9.31}$$

The above predictions are in good qualitative agreement with experiments [Candau *et al.* (1982), Shibayama & Tanaka (1993)].

A comparison of the above results with those in the preceding section shows that there are marked qualitative differences between the salt-free limit and the high salt concentration limit. Whereas ψ_c and T_c depend on the chain length in the salt-free limit, these critical values are independent of N in the limit of high salt concentration.

10.10   Charge Regularization

All of the preceding results are based on a model where analytical calculations can be performed providing insights into the swelling equilibrium and volume transitions of polyelectrolyte gels immersed in an external electrolyte solution. Although most of the qualitative trends observed in experiments involving salt-free and monovalent electrolyte solutions are captured by the above theory, quantitative differences between the predictions of the model and experimental data can be substantial. Furthermore, there are some fundamentally significant issues that are not accounted for in the model described above. The most important issue is the omnipresent correlation between the effective degree of ionization and the volume of the gel. As the gel swells, the extent of counterion adsorption on the chains decreases, since there is a gain in translational entropy of the ionized counterions in the more swollen volume of the gel. The dissociated ions from the polyelectrolyte chains and the strong electrolyte in the external solution are not point charges, but have finite volumes. Furthermore, the polarizability of the solvent around the ions and in the neighborhood of the ion pairs formed by adsorbing counterions can be quite different from that of the pure solvent as we have seen in Chapter 3. The correlations of the charge distributions and polymer density fluctuations also contribute to the free energy of the gel. All of these issues are discussed in the preceding chapters. In this section, we combine these aspects with the elasticity theory presented in this chapter.

As in the preceding sections, we consider a gel immersed in an external solution containing monovalent strong electrolytes. Let the polyelectrolyte chains in the gel be negatively charged and the counterion of the polymer be the same as the cation of the strong electrolyte. We define the free energy density of the system as the Helmholtz free energy of the system, consisting of the gel and the solution, per unit volume in units of $k_B T$:

$$f \equiv \frac{\Delta F v_1}{k_B T V_t},   (10.10.1)$$

where V_t is the total volume of the system, $V_t = V + V_s$, as described in Section 10.2.4. Denoting x_{gel} as the volume fraction of the system constituting the gel phase, the free energy density is given in terms of its contributions from the gel and the solution, f_{gel} and $f_{solution}$, respectively, as

$$f = f_{gel} x_{gel} + f_{solution}(1 - x_{gel}).   (10.10.2)$$

The theory of polyelectrolyte gels considered here is the generalization of the theory of polyelectrolyte solutions discussed in the previous chapters by including the $\Delta F_{elastic}$ derived in Section 10.2.3. As a summary, the free energy density of the gel is composed of (i) translational entropy of the dissociated counterions and the solvent molecules (f_s); (ii) the entropy of adsorbed counterions along the polymer backbone (f_{sa}); (iii) fluctuations of the dissociated counterions ($f_{fl,i}$); (iv) the van der Waals interactions between the polymer and the solvent (f_χ); (v) the attractive energy corresponding to the formation of ion pairs by the adsorbed counterions (f_{ad}); (vi) electrostatic interactions among polymer segments ($f_{electrostatic}$); and (vii) elastic energy of the gel network ($f_{elastic}$).

The free energy density due to the entropy of mixing of the dissociated counterions, the cations and anions from the dissociation of the salt, and the solvent molecules is given by

$$f_s = \left(\phi_{+g} + \alpha\phi\right)\ln\left(\phi_{+g} + \alpha\phi\right) + \phi_{-g}\ln\phi_{-g}$$
$$+ \left[\left(1 - \phi - \alpha\phi - \phi_{+g} - \phi_{-g}\right) \times \ln\left(1 - \phi - \alpha\phi - \phi_{+g} - \phi_{-g}\right)\right], \quad (10.10.3)$$

where $\phi_{\{+/-\}g}$ represents the volume fraction of the positive/negative ion species from the dissociated salt in the gel phase, ϕ is the volume fraction of the polymer gel, and α is the degree of ionization of the gel. In writing the above expression we have used the incompressibility constraint whereby

$$\phi + \alpha\phi + \phi_{+g} + \phi_{-g} + \phi_s = 1, \quad (10.10.4)$$

where ϕ_s is the volume fraction of the solvent. Also, the effective volume occupied by the monomer-counterion ion pair is assumed to be the same as that of the monomer alone. However, once the counterion dissociates, its volume is taken to be comparable to that of the solvent, within the usual premise of the Flory–Huggins theory. The counterions adsorbed to the polymer backbone also have a translational entropy along the backbone. The free energy density due to this effect is (see Eq. (5.7.1))

$$f_{sa} = [\alpha\ln\alpha + (1 - \alpha)\ln(1 - \alpha)]\,\phi. \quad (10.10.5)$$

The contribution from the density fluctuations of all mobile ions is given by (Eq. (10.2.43),

$$f_{fl,i} = -\frac{1}{4\pi}\left[\ln(1 + \tilde\kappa) - \tilde\kappa + \frac{1}{2}\tilde\kappa^2\right], \quad (10.10.6)$$

where the effective inverse Debye length ($\tilde\kappa$) is given in dimensionless units as

$$\tilde\kappa^2 = 4\pi\tilde\ell_B\left(\alpha\phi + \phi_{+g} + \phi_{-g}\right), \quad (10.10.7)$$

with $\tilde\ell_B = \tilde\ell_B/\ell$ and $\tilde\kappa = \kappa\ell$.

The van der Waals interactions between the polymer segments and the solvent are modeled via the Flory–Huggins term,

$$f_\chi = \chi\phi\phi_s. \quad (10.10.8)$$

The free energy density associated with all ion pairs on the polymer chains corresponding to all adsorbed counterions is (see Eq. (5.7.6))

$$f_{ad} = -(1 - \alpha)\phi\tilde\ell_B\delta, \quad (10.10.9)$$

as discussed in Chapter 5. The parameter δ is unique to a particular pair of the counterion and the ionizable group of the polymer and solvent. It is related to the intrinsic ionization constant of an isolated charged group in the solvent. The free energy density contribution from the intersegment electrostatic interactions, from Eq. (10.2.12) is

$$f_{electrostatic} = 2\pi \alpha^2 \tilde{l}_B \phi^2 \frac{(nN)^{2/3}}{\frac{3^{4/3}\pi^{7/6}}{2^{5/3}}\phi^{2/3} + \tilde{\kappa}^2 (nN)^{2/3}}. \tag{10.10.10}$$

Due to the fact that $\tilde{\kappa}$ is never zero, the second term in the denominator dominates so that the right-hand side is essentially independent of N for large values of N. As a result, the calculated phase diagrams are insensitive to N appearing in this equation.

As derived in Section 10.2.3.1, the elasticity contribution to the free energy density is

$$f_{elastic} = \frac{3}{2}S\phi_0^3 \left[(\frac{\phi}{\phi_0})^{1/3} - \frac{\phi}{\phi_0} + \frac{1}{3}\frac{\phi}{\phi_0}\ln\frac{\phi}{\phi_0} \right]. \tag{10.10.11}$$

The free energy density of the gel is obtained by adding the above seven contributions given by Eqs. (10.10.2)–(10.10.11):

$$f_{gel} = f_s + f_{sa} + f_{fl,i} + f_\chi + f_{ad} + f_{electrostatic} + f_{elastic}. \tag{10.10.12}$$

The free energy density of the solution phase consists of only two contributions, the translational entropy of the mobile ions (f_s) and the contributions from the fluctuations in the density of these ions ($f_{fl,i}$). The translational entropy in this case is given by

$$f_{s,solution} = \phi_{+s}\ln\phi_{+s} + \phi_{-s}\ln\phi_{-s}$$
$$+ (1 - \phi_{+s} - \phi_{-s})\ln(1 - \phi_{+s} - \phi_{-s}), \tag{10.10.13}$$

where ϕ_{+s} represents the combined volume fraction of all positive ion species, both from dissociated salt and the dissociated counterions, in the solution phase, while ϕ_{-s} represents the volume fraction of negative salt ions in solution phase. The fluctuation free energy is given by Eq. (10.10.6),

$$f_{fl,i,solution} = -\frac{1}{4\pi}\left[\ln(1 + \tilde{\kappa}) - \tilde{\kappa} + \frac{1}{2}\tilde{\kappa}^2\right], \tag{10.10.14}$$

except that the effective inverse Debye length in this case is given by

$$\tilde{\kappa}^2 = 4\pi\tilde{l}_B(\phi_{+s} + \phi_{-s}). \tag{10.10.15}$$

The free energy of the solution phase is then given by

$$f_{solution} = f_{s,solution} + f_{fl,i,solution}. \tag{10.10.16}$$

Substituting Eqs. (10.10.12) and (10.10.16) into Eq. (10.10.2) and minimizing the free energy density of the system, the phase diagram is constructed.

As discussed in Chapter 8, we combine all the temperature effects by defining a reduced temperature t as

$$t = \frac{\ell}{4\pi\ell_B}. \tag{10.10.17}$$

As an example of the temperature dependence of χ, let us choose the form of Eq. (10.9.5). As discussed in Chapter 8, the temperature dependence of χ is written in terms of the reduced temperature t by defining the parameter a_χ:

$$a_\chi = 20\pi t \chi = \frac{10\pi \epsilon \epsilon_0 k_B \Theta \ell}{e^2}. \tag{10.10.18}$$

The free energy density of the system is minimized, for a given set of values of the effective interaction parameter a_χ and the dielectric mismatch parameter δ, in the four-dimensional space of temperature, polymer volume fraction, degree of ionization, and salt concentration. Minimization of the free energy density is equivalent to maintaining the Donnan equilibrium between the gel phase and the external solution. This multi-dimensional minimization is performed numerically. Typical results from such numerical calculations [Hua et al. (2012)] are given in the following text. For a gel immersed in a pure solvent without any added electrolyte, the computed volume transition is presented in Fig. 10.22 as a plot of temperature versus polymer volume fraction. The self-consistent variation of the degree of ionization accompanying the volume transition is also included in the figure. The values of the parameters used in constructing this plot are $a_\chi = 20.0, \delta = 7.6, \phi_0 = 5 \times 10^{-5}, S = 400$, and N in Eq. (10.10.10) is 1000. For each temperature T, the reduced temperature t is calculated using Eq. (10.10.17) with $\epsilon = 80$, and $\ell = 0.9$ nm. The choice of a_χ corresponds to specifying the ideal temperature Θ as defined by Eq. (10.10.18). At higher temperatures, the gel is swollen with the polymer volume fraction close to zero. As the temperature is decreased, the gel undergoes a discontinuous volume transition at a transition temperature of about 300 K. At this temperature, the gel shrinks abruptly with the polymer volume fraction approaching a value close to unity, corresponding to the completely collapsed gel. The volume transition is coupled with the charge regularization on the polyelectrolyte chains. When the gel is swollen, the counterions

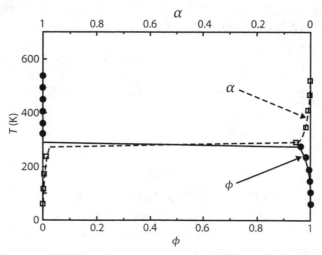

Figure 10.22 Volume transition of a charged gel in a salt-free solution. The degree of ionization undergoes a discontinuous change concomitant with the discontinuous volume transition. [Adapted from Hua et al. (2012)].

are favored to be in the fluid of the expanded gel due to their translational entropy, hence the degree of ionization approaches unity. As the temperature is reduced, the degree of ionization also undergoes a discontinuous change simultaneously with the volume transition. As the gel collapses, the counterions condense on the polymer, and the degree of ionization approaches zero. Analysis of the various contributions to the free energy at different stages of the computed volume transition shows that the cooperative adsorption of counterions provides the driving force for the abrupt volume transition of the charged gel.

As seen in Section 10.9.1, volume transitions of charged gels from a swollen state to a collapsed state are experimentally observed upon an increase in the concentration of an added salt. The required amount of salt for executing this volume transition depends on the solvent quality. An example of the coupled effect of solvent quality and salt concentration on the volume transition of a gel given by the above model is provided in Fig. 10.23. Here the salt is monovalent and the values of the parameters are $T = 300$ K, $\delta = 7.6$, $\phi_0 = 5 \times 10^{-4}$, $S = 400$, $\epsilon = 80$, $\ell = 0.9$ nm, and N in Eq. (10.10.10) is 1000. In Fig. 10.23, the molar concentration of monovalent salt is plotted against the volume fraction of the polymer for different values of solvent quality parameter a_χ. For a given solvent quality, volume transition from a swollen state to a collapsed state occurs at a particular salt concentration. The degree of ionization also undergoes a simultaneous discontinuous change, analogous to Fig. 10.22. Specifically, for $a_\chi = 5.0$, the volume transition occurs at about $c_s = 0.7$ M. As the Flory–Huggins χ parameter increases (with the corresponding increase in a_χ), the discontinuity associated with the gel collapse increases. Furthermore, the salt concentration required to induce the volume transition is lower for poor solvents (which corresponds to higher values of the solvent quality parameter a_χ).

It should be recalled from Section 5.2 that the dielectric mismatch parameter δ represents the intrinsic ionization equilibrium of the ion pair formed by the adsorbing counterion and the ionized monomer on the chain backbone. As discussed already, the specificity of the counterions can be phenomenologically incorporated in this

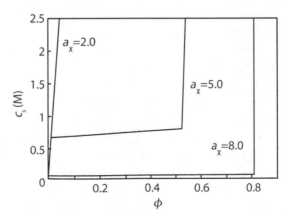

Figure 10.23 Effect of solvent quality and salt concentration on the volume transition of a gel. [From Hua *et al.* (2012)].

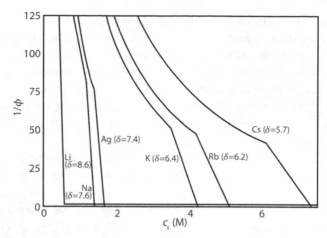

Figure 10.24 Effect of counterion identity on the dependence of the swelling ratio on salt concentration. [From Hua *et al.* (2012)].

parameter. The size of the ion and the effective ion-pair distance d in Eq. (5.2.13) are encapsulated in δ, with a larger value of δ implying smaller values of the ionic radius and d. The effect of sizes of monovalent counterions on the volume transition is given in Fig. 10.24. The values of the parameters are $a_\chi = 5.0$, T = 300 K, $\phi_0 = 5 \times 10^{-4}$, $\epsilon = 80$, $\ell = 1.0$ nm, S = 400, and N in Eq. (10.10.10) is 1000. In order to capture the qualitative trend of the effect of ion size on the volume transition, we set the value of the dielectric mismatch parameter for the sodium ion arbitrarily as $\delta_{Na^+} = 7.6$ as a reference point. With reference to this value, the values of other monovalent cations are obtained from the known ionic radii. Using these values of δ the dependence of the swelling ratio on the salt concentration is presented in Fig.10.24 for the various monovalent cations. Larger values of δ, or equivalently smaller ionic radii and d, cause the volume transition at lower salt concentrations.

Although the dielectric mismatch parameter δ is related in the above discussion to the ionic radii and d of counterions, an accurate description of the polarizability of the medium near the adsorption sites for the counterions is necessary in matching theoretical predictions with experimental data. The above discussed charge regularization is further complicated if the counterion is multivalent. In general, multivalent counterions induce the volume transition more readily than monovalent counterions, primarily due to bridging between several polymer segments. In addition, the ion specificity of the counterion can significantly affect the electrolyte concentration required to execute the discontinuous volume transitions, although the valency is the same. Representative results [Horkay *et al.* (2001)] are given in Fig. 10.25. First, equilibrated sodium polyacrylate hydrogels were prepared in 40 mM sodium chloride solutions, with a swelling ratio of about 115. Different salts are then added to the solution surrounding the gel. The dependence of the swelling ratio on the concentration of added salt, c_{salt}, is given in Fig. 10.25. The swelling behavior of the gels exhibits substantial differences for 1:1 salt (NaCl), 2:1 salts (CaCl$_2$, SrCl$_2$, NiCl$_2$, CoCl$_2$), and 3:1 salts (CeCl$_3$, LaCl$_3$). While the variation in the degree of swelling is weak for the monovalent salt, the

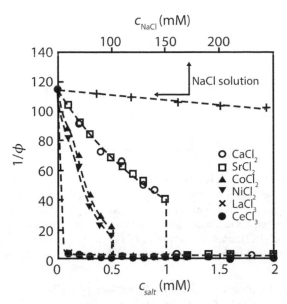

Figure 10.25 Dependence of the swelling ratio $1/\phi$ of sodium polyacrylate hydrogels on the salt concentration in the surrounding solution. The volume transition depends on the specificity of the counterion. The concentration of NaCl solution is given in a different scale (top) from the rest (bottom) [Horkay *et al.* (2001)].

multivalent counterions promote discontinuous volume transitions. The threshold salt concentration required to induce the discontinuous volume transition decreases as the valency of the counterion increases. Fig. 10.25 also shows the effect arising from the specificity of the counterions with the same valency, as evident from a comparison between Ca^{2+}, Sr^{2+}, Co^{2+}, and Ni^{2+}. This effect is related to the Hofmeister series on the specific features of electrolyte ions [Marcus (2015)]. The various features of gels arising from multivalent counterions with their specific effects are not yet fully explored theoretically.

10.10.1   Overcharging and Reentrant Volume Transition

A natural consequence of charge regularization accompanying conformational changes of a gel is overcharging of the gel. When a polyelectrolyte gel is immersed in a salt solution with multivalent counterions, the net charge of the gel can be reversed due to adsorption of the counterions. This is analogous to the cascade of accumulation of counterions during the coil–globule transition discussed in Chapter 5. The charge reversal, referred to as overcharging, results in a reentrant expanded state of the gel. This phenomenon is illustrated in Fig. 10.26, where the temperature dependence of polymer volume fraction of the gel is sketched. To begin with, let us consider a high temperature (for UCST systems) where the gel is in its fully swollen state due to a balance among intersegment electrostatic repulsion, entropy of all mobile species, and elasticity of the polymer network. As the temperature T or the dielectric constant ϵ, or equivalently their product $T\epsilon$, is reduced, the multivalent counterions adsorb more

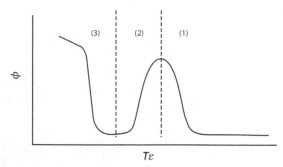

Figure 10.26 Charge reversal and reentrant volume transition. Regime (1): Polymer volume fraction ϕ increases with lowering of $T\epsilon$ as the gel shrinks due to adsorption of multivalent counterions, until electroneutrality is reached. Regime (2): Further adsorption of multivalent counterions results in charge reversal accompanied by reduction in ϕ due to gel swelling. Regime (3): Upon further lowering of $T\epsilon$, coions adsorb on the multipoles of segment-counterion pairs and neutralize the charge of the gel, resulting in gel collapse.

and more progressively until the net charge of the gel is zero. This process is marked as regime (1) in Fig. 10.26. Now the gel is compact, since the intersegment electrostatic repulsion is mostly mitigated. At this point of net zero charge, there are still many segments without adsorbed ions due to the fact that the adsorbing counterions are multivalent, whereas the polymer segments are taken to be monovalent. If $T\epsilon$ is reduced even further, more multivalent counterions adsorb making the net charge of the gel opposite to that of the initial gel. This is the overcharging of the gel.

Once the net charge is reversed, intersegment electrostatic repulsion will emerge again and, as a consequence, the gel volume will increase into its swollen state. This process is denoted as regime (2) in Fig. 10.26. Upon further reduction in $T\epsilon$, even the coions bind to the multipoles formed by polymer segments and adsorbed multivalent counterions. Hence, the charge would eventually be neutralized (regime (3)). The final state of regime (3) is a collapsed gel. The degree of ionization of the gel alternates its sign in a self-consistent manner with the overcharging and the reentrant phenomenon portrayed in Fig. 10.26 [Hua *et al.* (2012)]. The ranges of the three regimes and the nature of the collapsed and swollen gels depend on the identity and amount of the added salt and the chelating capacity of the multivalent ions [Besteman *et al.* (2007), Grosberg (2002), Hua *et al.* (2012), Nguyen *et al.* (2000)].

10.11   Summary

In this chapter, we have presented the basic conceptual framework for the three processes outlined in Fig. 10.2. These are (1) attainment of gel equilibrium, (2) deformation of gels due to confinement and external forces, and gel dynamics, and (3) volume phase transitions. The properties associated with these processes of charged gels depend on intersegment electrostatic and hydrophobic interactions, translational entropy of all mobile species in the gel, elasticity of the polymer network, and the Donnan equilibrium. Only elementary treatments of these contributing factors are presented at the level of mean field theories, but are found to describe the gel phenomenology very well. Closed-form formulas are derived for charged hydrogels to describe their equilibrium state, elastic moduli (Young's modulus, shear modulus, and bulk modulus), stress–strain equations, relation between microscopic properties such as gel diffusion and macroscopic properties such as elasticity, and criteria for dramatic volume transitions. The key experimental parameters appearing in these formulas are polymer concentration, cross-link concentration, degree of ionization, temperature and solvent quality, and salt concentration. The issue of charge regularization accompanying the volume changes of the gel is also addressed, although the predictions can only be numerically made. Advanced approaches [Muthukumar (1989), Rabin & Panyukov (1997), Shibayama *et al.* (1997)] to treat strong composition fluctuations and different time scales for the relaxation of incipient heterogeneities inside the gel are outside the scope of the present chapter. However, the formulas presented here are useful to adequately capture the essentials of the physics of charged gels and to design new hydrogels with specified elastic properties and volume transition capabilities.

11 Epilogue

The preceding chapters present only the most basic conceptual building units to understand the various complex behaviors of charged macromolecules. One example of how the numerous concepts are in collective display is a mammalian cell cartooned in Fig. 11.1. Inside a cell, which is constituted by many functional compartments, each with charged macromolecules and electrolyte ions under crowded conditions, the various units communicate with each other and work together to exhibit the process of living. The fundamental forces operative in this highly crowded and cooperative ensemble arise from charges embedded on the various molecules, van der Waals forces, organization of water molecules, and the topological correlations due to macromolecular connectivity. It is not an exaggeration if we call the cell a **Coulomb soup**. The functions of the constituents of this Coulomb soup are manifestations of the various concepts built through the previous chapters. These concepts revolve around how electrolyte ions work together, how they are coupled to the formation of structure and movement of charged macromolecules, how macromolecules organize into their assemblies, and how charged gel-like structures behave elastically, etc. The present challenge is to address real highly cooperative systems such as the one displayed in

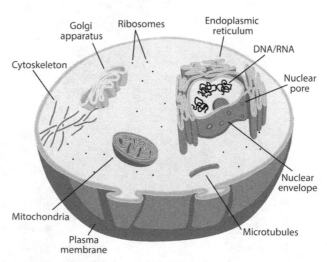

Figure 11.1 Sketch of a mammalian cell. Its function is a confluence of the various concepts presented in the preceding chapters operating under nonequilibrium conditions.

Fig. 11.1 by building on the concepts developed in this book. Parallel to the challenges in understanding living matter, there are also challenges to advance technologies such as macromolecule-based batteries. It is hoped that the various concepts and phenomenological details on the individual units (solutions, isolated assemblies, gels, etc.) presented so far will be put together to advance our understanding of living matter and to make new materials for societal benefit.

Appendix 1

A1.1 Line Charge and Manning Condensation

For the line charge given in Fig. 4.12, the electric field at r in the absence of electrolyte follows from the Gauss law (Eq. (4.3.15)), as

$$E(r) = \frac{e z_p}{2\pi \epsilon_0 \epsilon \ell r}, \tag{A1.1}$$

and (because $E = -d\psi/dr$)

$$\psi(r) = A - \frac{e z_p}{2\pi \epsilon_0 \epsilon \ell} \ln r, \tag{A1.2}$$

where A is an integration constant. The potential energy W for bringing a counterion of valency z_c to the location at r is

$$W(r) = (-e z_c)\psi(r). \tag{A1.3}$$

Assuming the Boltzmann distribution, the number of counterions at r is proportional to

$$\exp\left(-\frac{W(r)}{k_B T}\right) = \exp\left(\frac{e z_c A}{k_B T}\right) \exp\left(-\frac{e^2 z_p z_c}{2\pi \epsilon_0 \epsilon \ell k_B T} \ln r\right). \tag{A1.4}$$

By combining the exponential and the logarithm in the last factor, we get

$$\exp\left(-\frac{W(r)}{k_B T}\right) = \frac{1}{r^{2\Gamma z_p z_c}} \exp\left(\frac{e z_c A}{k_B T}\right), \tag{A1.5}$$

where Γ is the charge density parameter defined in Eq. (4.3.19).

The number of counterions inside a cylinder of radius r_0 (of length ℓ) is

$$n_{r_0} \sim \int_0^{r_0} (2\pi r) e^{-W(r)/k_B T} \, dr. \tag{A1.6}$$

Combining Eqs. (A1.5) and (A1.6) yields

$$n_{r_0} \sim \int_0^{r_0} r \frac{1}{r^{2\Gamma z_p z_c}} \, dr \sim \frac{1}{r^{2(\Gamma z_p z_c - 1)}} \Big|_0^{r_0}. \tag{A1.7}$$

Appendix 2

A2.1   Chain Swelling Due to Excluded Volume and Electrostatic Interactions

The derivation of Eqs. (5.6.9), (5.6.17), and (5.6.33) is given in this Appendix. The probability $P(\mathbf{R}, N)$ of a uniformly charged chain of N segments and end-to-end distance $\mathbf{R}$ follows from Eq. (5.3.4) as

$$P(\mathbf{R}, N) = \int_0^{\mathbf{R}} \mathcal{D}[\mathbf{R}(s)] \exp\left[-\left(\frac{3}{2\ell^2}\right) \int_0^N ds \left(\frac{\partial \mathbf{R}(s)}{\partial s}\right)^2 \right.$$
$$\left. -\frac{1}{2} \int_0^N ds \int_0^N ds' u[\mathbf{R}(s) - \mathbf{R}(s')] \right], \tag{A2.1}$$

where

$$u[\mathbf{R}(s) - \mathbf{R}(s')] = u_{\text{exc}}[\mathbf{R}(s) - \mathbf{R}(s')] + u_{\text{elec}}[\mathbf{R}(s) - \mathbf{R}(s')], \tag{A2.2}$$

with

$$u_{\text{exc}}(\mathbf{r}) = v\ell^3 \delta(\mathbf{r}), \tag{A2.3}$$

$$u_{\text{elec}}(\mathbf{r}) = \alpha^2 z_p^2 \ell_B \frac{e^{-\kappa r}}{r}. \tag{A2.4}$$

Introducing the Fourier transform

$$u(\mathbf{r}) = \int \frac{d^3 k}{(2\pi)^3} e^{i\mathbf{k} \cdot \mathbf{r}} u_{\mathbf{k}}, \tag{A2.5}$$

we get

$$u_{\text{exc}}(\mathbf{r}) = v\ell^3 \delta(\mathbf{r}) = v\ell^3 \int \frac{d^3 k}{(2\pi)^3} e^{i\mathbf{k}}, \tag{A2.6}$$

$$u_{\text{elec}}(\mathbf{r}) = \alpha^2 z_p^2 \ell_B \int \frac{d^3 k}{(2\pi)^3} \frac{4\pi}{k^2 + \kappa^2} e^{i\mathbf{k}}. \tag{A2.7}$$

The two limits of high salt condition, $\kappa \to \infty$, and zero-salt limit ($\kappa \to 0$) follow from Eq. (A2.7) as

$$u_{\text{elec}}(\mathbf{r}) = \begin{cases} \frac{4\pi\alpha^2 z_p^2 \ell_B}{\kappa^2} \delta(\mathbf{r}), & \kappa \to \infty \\ \alpha^2 z_p^2 \ell_B \frac{1}{r}, & \kappa = 0, \end{cases} \tag{A2.8}$$

where the definition of $\delta(\mathbf{r})$ is used from Eq. (A2.5).

Combining Eqs. (A2.1)–(A2.7), we get

$$\begin{aligned} P(\mathbf{R}, N) &= \int_0^{\mathbf{R}} \mathcal{D}[\mathbf{R}(s)] \exp\left\{ -\frac{3}{2\ell^2} \int_0^N ds \left(\frac{\partial \mathbf{R}(s)}{\partial s} \right)^2 \right. \\ &\left. -\frac{1}{2} \int_0^N ds \int_0^N ds' \int \frac{d^3 k}{(2\pi)^3} u_{\mathbf{k}} e^{i\mathbf{k} \cdot [\mathbf{R}(s) - \mathbf{R}(s')]} \right\}, \end{aligned} \tag{A2.9}$$

where $u_{\mathbf{k}}$ is

$$u_{\mathbf{k}} = v\ell^3 + \frac{4\pi\alpha^2 z_p^2 \ell_B}{k^2 + \kappa^2}. \tag{A2.10}$$

Let us introduce the Fourier variable q conjugate to the arc length variable s defined through the pair

$$\mathbf{R}(s) = \int_{-\infty}^{\infty} \frac{dq}{2\pi} e^{iqs} \hat{\mathbf{R}}_q, \tag{A2.11}$$

$$\hat{\mathbf{R}}_q = \int_0^N ds \, e^{-iqs} \mathbf{R}(s). \tag{A2.12}$$

Substitution of Eq. (A2.11) in Eq. (A2.1) for the chain connectivity part (Gaussian chain) gives

$$\left(\frac{3}{2\ell^2} \right) \int_0^N ds \left(\frac{\partial \mathbf{R}(s)}{\partial s} \right)^2 = \int_{-\infty}^{\infty} \frac{dq}{2\pi} \frac{\hat{\mathbf{R}}_q^2}{g_0(q)}, \tag{A2.13}$$

where

$$g_0(q) = \frac{2\ell^2}{3q^2}. \tag{A2.14}$$

Discretizing the continuous variable q as $2\pi p/N$ (where p is the Rouse mode, see Chapter 7), Eq. (A2.13) becomes

$$\left(\frac{3}{2\ell^2} \right) \int_0^N ds \left(\frac{\partial \mathbf{R}(s)}{\partial s} \right)^2 = \frac{3}{2\ell^2} \frac{1}{N} \sum_{p=-\infty}^{\infty} q^2 \hat{\mathbf{R}}_q^2, \tag{A2.15}$$

so that

$$\langle \hat{\mathbf{R}}_q^2 \rangle = \frac{N\ell^2}{q^2} = \frac{3}{2} N g_0(q). \tag{A2.16}$$

Furthermore, as we know for Gaussian chains, we get the mean square end-to-end distance of the chain directly from the first term in the argument of the exponential in Eq. (A2.1) as

$$\langle R^2 \rangle_0 = N\ell^2, \tag{A2.17}$$

where the subscript indicates Gaussian chain.

Using the collective coordinates $\{\hat{\mathbf{R}}_q\}$, Eq. (A2.1) becomes

$$P(\mathbf{R}, N) = \int \prod_q d\hat{\mathbf{R}}_q G(\{\hat{\mathbf{R}}_q\}, N), \tag{A2.18}$$

where

$$G(\{\hat{\mathbf{R}}_q\}, N) = \exp\left\{ -\int_{-\infty}^{\infty} \frac{dq}{2\pi} \frac{\hat{\mathbf{R}}_q^2}{g_0(q)} - \frac{1}{2} \int_0^N ds \int_0^N ds' \int \frac{d^3 k}{(2\pi)^3} u_{\mathbf{k}} \right.$$

$$\left. \times \exp\left[i\mathbf{k} \cdot \int_{-\infty}^{\infty} \frac{dq'}{2\pi} \hat{\mathbf{R}}_{q'} \left(e^{iq's} - e^{iq's'} \right) \right] \right\}, \tag{A2.19}$$

The key idea behind evaluating the above integral is to guess an effective term that looks like the leading Gaussian term and treat the deviation from this effective term as small, and then evaluate the effective term by maximizing the entropy $S = k_B \ln P(\mathbf{R}, N)$ of the chain. Let us choose this effective term as (analogous to the Gaussian term given in Eq. (A2.13))

$$\int_{-\infty}^{\infty} \frac{dq}{2\pi} \frac{\hat{\mathbf{R}}_q^2}{g(q)}, \tag{A2.20}$$

where $g(q)$ is to be determined self-consistently. Adding and subtracting this term in the argument of the first exponential in Eq. (A2.19), we obtain

$$G(\{\hat{\mathbf{R}}_q\}, N) = \exp\left\{ -\int_{-\infty}^{\infty} \frac{dq}{2\pi} \frac{\hat{\mathbf{R}}_q^2}{g(q)} - X(g, \{\hat{\mathbf{R}}_q\}) \right\}, \tag{A2.21}$$

where

$$X(g, \{\hat{\mathbf{R}}_q\}) = \left\{ \int_{-\infty}^{\infty} \frac{dq}{2\pi} \hat{\mathbf{R}}_q^2 \left(\frac{1}{g_0(q)} - \frac{1}{g(q)} \right) + \frac{1}{2} \int_0^N ds \int_0^N ds' \int \frac{d^3 k}{(2\pi)^3} u_{\mathbf{k}} \right.$$

$$\left. \times \exp\left[i\mathbf{k} \cdot \int_{-\infty}^{\infty} \frac{dq'}{2\pi} \hat{\mathbf{R}}_{q'} \left(e^{iq's} - e^{iq's'} \right) \right] \right\}, \tag{A2.22}$$

Multiplying and dividing Eq. (A2.18) by $\int d\hat{\mathbf{R}}_q \exp[-\int (dq/2\pi)\hat{\mathbf{R}}_q^2/g(q)]$, we get

$$P(\mathbf{R}, N) = \left[\int \prod_q d\hat{\mathbf{R}}_q \exp\left[-\int_{-\infty}^{\infty} \frac{dq}{2\pi} \frac{\hat{\mathbf{R}}_q^2}{g(q)} \right] \right] \langle e^{-X} \rangle_g, \tag{A2.23}$$

where the angular brackets with subscript g denotes averaging over $\exp[-\int (dq/2\pi)\hat{\mathbf{R}}_q^2/g(q)]$. Since

$$e^x = e^{c+x-c} \geq e^c [1 + x - c] \tag{A2.24}$$

for arbitrary choice of c, and choosing $c = \langle x \rangle$, we get

$$\langle e^{-x} \rangle \geq e^{\langle -x \rangle}. \tag{A2.25}$$

Therefore, the chain entropy $S = k_B \ln P$ is maximized at

$$\frac{S}{k_B} \geq \ln \left\{ \int \prod_q d\hat{\mathbf{R}}_q \exp \left[-\int_{-\infty}^{\infty} \frac{dq}{2\pi} \frac{\hat{\mathbf{R}}_q^2}{g(q)} \right] \right\} - \langle X \rangle_g. \tag{A2.26}$$

We determine the desired value of $g(q)$ to be used in Eq. (A2.20) by evaluating it at the extremum of the entropy S given by

$$\frac{\delta(S/k_B T)}{\delta g(q)} \bigg|_{g(q)=g^*(q)} = 0. \tag{A2.27}$$

Performing the average in Eq. (A2.26) using Eq. (A2.20) and taking the functional derivative, Eqs. (A2.26) and (A2.27) yield

$$\frac{3}{2} \frac{N}{g^*} - \frac{3}{2} \frac{N}{g_0} + \frac{1}{2} \int_0^N ds \int_0^N ds' \int \frac{d^3k}{(2\pi)^3} u_{\mathbf{k}} k^2 \sin^2 \left(\frac{q \mid s - s' \mid}{2} \right)$$
$$\times \exp \left[-k^2 \int_{-\infty}^{\infty} \frac{dq'}{2\pi} g^*(q') \sin^2 \left(\frac{q' \mid s - s' \mid}{2} \right) \right] = 0. \tag{A2.28}$$

The g^* given by this equation can be determined only numerically. However, focusing on large length scale properties such as the mean square end-to-end distance, an analytically tractable solution can be found. Analogous to $g_0(q)$ of Eq. (A2.14), writing g^* as

$$g^* = \frac{2\ell \ell_{\text{eff}}(q)}{3q^2}, \tag{A2.29}$$

the effective distribution function becomes

$$P(\mathbf{R}, N) = \left[\int \prod_q d\hat{\mathbf{R}}_q \exp \left[-\frac{3}{2} \int_{-\infty}^{\infty} \frac{dq}{2\pi} \frac{q^2 \hat{\mathbf{R}}_q^2}{\ell \ell_{\text{eff}}(q)} \right] \right], \tag{A2.30}$$

so that

$$\langle \hat{\mathbf{R}}_q^2 \rangle = \frac{N \ell \ell_{\text{eff}}(q)}{q^2}, \tag{A2.31}$$

analogous to Eq. (A2.16). Since we are interested in small q behavior, let us assume that only the lowest mode ($q \to 2\pi/N$) contributes to $\ell_{\text{eff}}(q)$. Therefore, we can take $\ell_{\text{eff}}(q)$ as q-independent, but N-dependent, and write it simply as ℓ_{eff}. Using this approximation, which is equivalent to the assumption of uniform swelling of the chain, and taking $q \to 0$ limit, Eq. (A2.28) becomes

$$\frac{3}{2} \left(\frac{1}{\ell_{\text{eff}}} - \frac{1}{\ell} \right) + \frac{\ell}{12N} \int_0^N ds \int_0^N ds' \int \frac{d^3k}{(2\pi)^3} \left(v\ell^3 + \frac{4\pi\alpha^2 z_p^2 \ell_B}{k^2 + \kappa^2} \right) k^2 \mid s - s' \mid^2$$
$$\times \exp \left[-\frac{k^2 \ell \ell_{\text{eff}} \mid s - s' \mid}{6} \right] = 0. \tag{A2.32}$$

Furthermore, comparing Eqs. (A2.15) and (A2.31), Eq. (A2.17) is now generalized to give the mean square end-to-end distance of the chain with excluded volume and electrostatic interactions as

$$\langle R^2 \rangle = N \ell \ell_{\text{eff}}, \tag{A2.33}$$

where ℓ_{eff} is obtained by performing the integrals in Eq. (A2.32) as

$$\left(\frac{\ell_{\text{eff}}}{\ell} \right)^{5/2} - \left(\frac{\ell_{\text{eff}}}{\ell} \right)^{3/2} = \zeta_{\text{exc}} + \zeta_{\text{elec}},$$

where

$$\zeta_{\text{exc}} = \frac{4}{3} \left(\frac{3}{2\pi} \right)^{3/2} v \sqrt{N},$$

$$\zeta_{\text{elec}} = \frac{4}{3} \sqrt{\frac{6}{\pi}} \alpha^2 z_p^2 (\ell_B/\ell)(\ell_{\text{eff}}/\ell) N^{3/2} \Theta(a),$$

with

$$\Theta(a) = \frac{\sqrt{\pi}}{2a^{5/2}} \left[(a^2 - 4a + 6) e^a \, erfc \left(\sqrt{a} \right) - 6 - 2a + \frac{12}{\sqrt{\pi}} \sqrt{a} \right]. \tag{A2.34}$$

Here $a = \kappa^2 \ell \ell_{\text{eff}} / 6$. The above result is the desired derivation of Eqs. (5.6.33) and (5.6.34).

When the electrostatic interaction is absent, that is $\zeta_{\text{elec}} = 0$, Eq. (A2.34) reduces to Eq. (2.5.7), where $\ell_{\text{eff}}/\ell = \langle R^2 \rangle / N \ell^2$. For the electrostatic part, the low-salt and high-salt limits follow from Eq. (A2.34) as

$$\left(\frac{\ell_{\text{eff}}}{\ell} \right)^{5/2} - \left(\frac{\ell_{\text{eff}}}{\ell} \right)^{3/2} = \begin{cases} \frac{4}{3} \left(\frac{3}{2\pi} \right)^{3/2} \left(\frac{4\pi \alpha^2 z_p^2 \ell_B}{\kappa^2 \ell^3} \right) \sqrt{N}, & \kappa \to \infty \\ \frac{8}{15\sqrt{\pi}} \frac{\alpha^2 z_p^2 \ell_B \ell_{\text{eff}}}{\ell^2} N^{3/2}, & \kappa \to 0. \end{cases} \tag{A2.35}$$

These are the results quoted in Eqs. (5.6.9) and (5.6.17).

A2.2   Free Energy of a Polyelectrolyte Chain

Writing the free energy of a chain as a function of ℓ_{eff}, $F(\ell_{\text{eff}}) = -TS(\ell_{\text{eff}})$, where $S(\ell_{\text{eff}})$ is given by Eq. (A2.26), $\partial F(\ell_{\text{eff}}) / \partial \ell_{\text{eff}}$ is given by $k_B T$ times the negative of the left-hand side of Eq. (A2.32). The free energy of the chain is therefore obtained from

$$\frac{F(\ell_{\text{eff}})}{k_B T} = \int_\ell^{\ell_{\text{eff}}} d\ell_{\text{eff}} \frac{\partial F(\ell_{\text{eff}})}{\partial \ell_{\text{eff}}}, \tag{A2.36}$$

and using Eq. (A2.32). The result of this integration is Eq. (5.6.28). Using $\ell_{\text{eff}} = R^2/N\ell$, Eqs. (A2.36) and (5.6.28) reduce to Eqs. (5.6.7) and (5.6.16), respectively, for the high-salt and salt-free limits.

Appendix 3

A3.1 Dipole Interactions

The two-body pseudopotentials for charge–dipole and dipole–dipole interactions are derived below using the Debye–Hückel potential between charges.

A3.1.1 Charge–Dipole Interaction Energy

Consider a dipole $\mathbf{p}_1$ made of two charges $+q$ and $-q$ separated by a distance ℓ, as shown in Fig. 3.16a. The magnitude of the dipole is $q\ell$. Let there be a charge $Q = \alpha z_p e$ at the location $\mathbf{r}$ from the center of the dipole. The angle between $\mathbf{p}_1$ and $\mathbf{r}$ is θ, so that $\mathbf{p}_1 \cdot \mathbf{r} = p_1 r \cos\theta$, where $p_1 = q\ell$ and r are the magnitudes of the corresponding vectors. The electric potential $\psi(\mathbf{r})$ due to the dipole at the location of the charge Q in a solution of dielectric constant ϵ is

$$\psi(\mathbf{r}) = \frac{q}{4\pi\epsilon_0\epsilon}\left[\frac{e^{-\kappa r_+}}{r_+} - \frac{e^{-\kappa r_-}}{r_-}\right], \tag{A3.1}$$

where r_+ and r_- are, respectively, the distances from the $+q$ and $-q$ charges to the location $\mathbf{r}$. Using elementary geometry,

$$r_\pm = \left[\left(r \mp \frac{\ell}{2}\cos\theta\right)^2 + \frac{\ell^2}{4}\sin^2(\theta)\right]^{1/2}. \tag{A3.2}$$

For $r \gg \ell$, the dipole approaches the limit of a point-like dipole, and in this limit, the above equation becomes

$$r_\pm = r\left(1 \mp \frac{\ell}{2r}\cos\theta + \cdots\right). \tag{A3.3}$$

Using Eq. (A3.3), Eq. (A3.1) becomes

$$\psi(\mathbf{r}) = \frac{q\ell}{4\pi\epsilon_0\epsilon}\frac{e^{-\kappa r}}{r^2}(1 + \kappa r)\cos\theta. \tag{A3.4}$$

Writing $q\ell r \cos\theta$ as $\mathbf{p}_1 \cdot \mathbf{r}$, we get

$$\psi(\mathbf{r}) = \frac{1}{4\pi\epsilon_0\epsilon}\frac{\mathbf{p}_1 \cdot \mathbf{r}}{r^3}e^{-\kappa r}(1 + \kappa r). \tag{A3.5}$$

Note that this equation reduces to Eq. (3.7.5) in the absence of electrostatic screening.

The charge-dipole interaction energy between a charge $Q = \alpha z_p e$ and dipole $\mathbf{p}_1$ is (see Eq. (3.1.13))

$$u_{cd}(\mathbf{r}) = \frac{\alpha z_p e}{4\pi\epsilon_0\epsilon} \frac{1}{r^3} (\mathbf{p}_1 \cdot \mathbf{r}) e^{-\kappa r} (1 + \kappa r). \tag{A3.6}$$

Since the angular average of u_{cd} over the dipole orientation is zero, we define an average $\bar{u}_{cd}$ through its Boltzmann weight as

$$e^{-\bar{u}_{cd}/k_B T} = \frac{\int d\Omega e^{-u_{cd}(\mathbf{r})/k_B T}}{\int d\Omega}, \tag{A3.7}$$

where Ω denotes the solid angle for the dipole orientation. Expanding $\exp[-u_{cd}(\mathbf{r})/k_B T]$ and performing the angular integrals, we get

$$e^{-\bar{u}_{cd}/k_B T} = \left[1 + \frac{\alpha^2 z_p^2 p_1^2 \ell_B^2}{6r^4} e^{-2\kappa r} (1 + \kappa r)^2 + \cdots\right], \tag{A3.8}$$

where ℓ_B is the Bjerrum length, and p_1 is in units of the electronic charge e. Therefore, within the leading order of the above expansion, $\bar{u}_{cd}$ is approximated as

$$\frac{\bar{u}_{cd}}{k_B T} = -\frac{1}{6} \frac{\alpha^2 z_p^2 p_1^2 \ell_B^2}{r^4} e^{-2\kappa r} (1 + \kappa r)^2. \tag{A3.9}$$

We define the pseudopotential v_{cd} for the charge–dipole interaction, analogous to the definition of the excluded volume parameter in Eq. (2.4.1), as

$$v_{cd} = \frac{1}{\ell^3} \int d\mathbf{r} \left[1 - e^{-\bar{u}_{cd}/k_B T}\right]. \tag{A3.10}$$

Performing the high temperature expansion and keeping only the leading term, we get from the above two equations

$$v_{cd} = -\frac{2\pi}{3} \frac{\alpha^2 z_p^2 p_1^2 \ell_B^2}{\ell^3} \int_\ell^\infty dr \frac{e^{-2\kappa r}}{r^2} (1 + \kappa r)^2, \tag{A3.11}$$

where the minimum distance between the charge and the dipole is taken as ℓ. After the integration, we get the charge-dipole pseudopotential as

$$v_{cd} = -\frac{\pi}{3} \frac{\alpha^2 z_p^2 p_1^2 \ell_B^2}{\ell^4} e^{-2\kappa\ell} (2 + \kappa\ell). \tag{A3.12}$$

This reduces to Eq. (5.11.10) for monovalent charges.

A3.1.2 Dipole–Dipole Interaction Energy

Consider the presence of a second dipole $\mathbf{p}_2$ at $\mathbf{r}$ instead of the charge addressed above. The electric field $\mathbf{E}(\mathbf{r})$ due to the dipole $\mathbf{p}_1$ at the location $\mathbf{r}$ of the second dipole $\mathbf{p}_2$ follows from $\mathbf{E}(\mathbf{r}) = -\nabla\psi(\mathbf{r})$ and Eq. (A3.5) as

$$\mathbf{E}(\mathbf{r}) = -\frac{1}{4\pi\epsilon_0\epsilon} \frac{e^{-\kappa r}}{r^3} \left[(1 + \kappa r) \mathbf{p}_1 - \left(3 + 3\kappa r + \kappa^2 r^2\right) \hat{r} (\mathbf{p}_1 \cdot \hat{r})\right], \tag{A3.13}$$

where $\hat{r}$ is the unit vector along $\mathbf{r}$. Note that this equation reduces to Eq. (3.7.6) for $\kappa = 0$. Using Eqs. (3.7.8) and (A3.13), the dipole–dipole interaction energy u_{dd}

between the two dipoles $\mathbf{p}_1$ and $\mathbf{p}_2$ separated by distance $\mathbf{r}$ becomes (within the Debye–Hückel theory)

$$u_{dd}(\mathbf{r}) = \frac{1}{4\pi\epsilon_0\epsilon}\frac{e^{-\kappa r}}{r^3}\left[(1+\kappa r)\,\mathbf{p}_1\cdot\mathbf{p}_2 - \left(3+3\kappa r+\kappa^2 r^2\right)(\hat{r}\cdot\mathbf{p}_1)(\hat{r}\cdot\mathbf{p}_2)\right].$$

(A3.14)

Note that this equation reduces to Eq. (3.7.15) for $\kappa = 0$.

The pseudopotential for the dipole–dipole interaction is obtained by following the same procedure as in the preceding section. Since the angular average of u_{dd} over the orientations of the two dipoles is zero, we define an average $\bar{u}_{dd}$ through the Boltzmann weight as

$$e^{-\bar{u}_{dd}/k_BT} = \frac{\int d\Omega_1 d\Omega_2 e^{-u_{dd}(\mathbf{r})/k_BT}}{\int d\Omega_1 d\Omega_2},$$

(A3.15)

where Ω_1 and Ω_2 denote the solid angles for the orientation of the two dipoles. Expanding $\exp[-\bar{u}_{dd}/k_BT]$ and performing the angular integrals, we get

$$e^{-\bar{u}_{dd}/k_BT} = 1 + \frac{1}{3}\left(\frac{\ell_B}{r^3}\right)^2 p_1^2 p_2^2 e^{-2\kappa r}\left[1+2\kappa r+\frac{5}{3}(\kappa r)^2+\frac{2}{3}(\kappa r)^3+\frac{1}{6}(\kappa r)^4\right]+\cdots,$$

(A3.16)

where the factor of the electronic charge in p_1 and p_2 is absorbed in the Bjerrum length. Therefore, within the leading order (which is quadratic in u_{dd}) in the above expansion, $\bar{u}_{dd}$ is approximated as

$$\frac{\bar{u}_{dd}}{k_BT} = -\frac{1}{3}\frac{\ell_B^2 p_1^2 p_2^2}{r^6}e^{-2\kappa r}\left[1+2\kappa r+\frac{5}{3}(\kappa r)^2+\frac{2}{3}(\kappa r)^3+\frac{1}{6}(\kappa r)^4\right].$$

(A3.17)

This is the result quoted in Eq. (1.3.8).

We define the pseudopotential v_{dd} for the dipole–dipole interaction, analogous to Eqs. (2.4.1) and (A3.10), as

$$v_{dd} = \frac{1}{\ell^3}\int d\mathbf{r}\left[1-e^{-\bar{u}_{dd}/k_BT}\right].$$

(A3.18)

Performing the high temperature expansion and keeping only the leading term, we get

$$v_{dd} = \frac{4\pi}{\ell^3}\int_\ell^\infty dr\, r^2\frac{\bar{u}_{dd}}{k_BT},$$

(A3.19)

where the minimum distance between the centers of the dipoles is taken as ℓ. Substituting Eq. (A3.17) and performing the integral, we get

$$v_{dd} = -\frac{\pi}{9}\frac{\ell_B^2 p_1^2 p_2^2}{\ell^6}e^{-2\kappa\ell}\left[4+8\kappa\ell+4(\kappa\ell)^2+(\kappa\ell)^3\right].$$

(A3.20)

This result is quoted in Eq. (5.11.11).

Appendix 4

A4.1   Donnan Equilibrium

Consider a solution of charged macromolecules and a monovalent salt in the "in" compartment, which is equilibrated with a solution of the same monovalent salt in the "out" compartment (Fig. 5.27b). The "in" and "out" compartments are separated by a semipermeable membrane which is permeable only to the solvent and the salt, but not to the macromolecules. Let each macromolecule in the "in" compartment be positively charged with a net charge $Q = \alpha z_p N$. The negatively charged counterion from the macromolecule is taken to be chemically identical to the anion of the electrolyte.

At equilibrium, let the number concentration of cations and anions in the "in" compartment be c_+^{in} and c_-^{in}, respectively. The corresponding concentrations in the "out" compartment are c_+^{out} and c_-^{out}, respectively. The condition of electroneutrality for the "in" and "out" compartments are

$$\frac{c}{N}Q + c_+^{in} = c_-^{in} \qquad \text{(in)} \qquad (A4.1)$$

$$c_+^{out} = c_-^{out} = c_s \qquad \text{(out)} \qquad (A4.2)$$

where c is the monomer number concentration ($c = nN/V$) and c_s is the monovalent electrolyte concentration. Note that c_s is not the initial electrolyte concentration in the "out" compartment before reaching the equilibrium distribution of ions in both the "in" and "out" compartments.

The Donnan criterion of equilibrium is the same as for any multi-component system, namely that there are no gradients in the chemical potentials of the various species in the system. For the present situation, the Donnan criterion is that the chemical potentials of the electrolyte in the "in" and "out" compartments are the same,

$$c_+^{in}c_-^{in} = c_+^{out}c_-^{out} = c_s^2, \qquad (A4.3)$$

where the activity of the electrolyte is approximated by the concentration under the ideal conditions assumed in this section. c_-^{in} follows from Eq. (A4.3) as

$$c_-^{in} = \frac{c_s^2}{c_+^{in}}. \qquad (A4.4)$$

Substituting this result in Eq. (A4.1), we get the quadratic equation for c_+^{in} as

$$\left(c_+^{in}\right)^2 + \frac{c}{N}Qc_+^{in} - c_s^2 = 0. \tag{A4.5}$$

The physically allowed solution of this equation is

$$c_+^{in} = \frac{1}{2}\left[-\frac{Qc}{N} + \sqrt{\left(\frac{Qc}{N}\right)^2 + 4c_s^2}\right]. \tag{A4.6}$$

Similarly c_-^{in} can be obtained by substituting $c_+^{in} = c_s^2/c_-^{in}$ in Eq. (A4.1) and solving the resulting quadratic equation as

$$c_-^{in} = \frac{1}{2}\left[\frac{Qc}{N} + \sqrt{\left(\frac{Qc}{N}\right)^2 + 4c_s^2}\right]. \tag{A4.7}$$

The difference in the osmotic pressure between the "in" and "out" compartments, called the Donnan pressure Π_D, follows from the van't Hoff law as

$$\frac{\Pi_D}{k_B T} = \left(\frac{c}{N} + c_+^{in} + c_-^{in}\right) - \left(c_+^{out} + c_-^{out}\right). \tag{A4.8}$$

Using Eqs. (A4.3), (A4.6) and (A4.7) in Eq. (A4.8), we get

$$\frac{\Pi_D}{k_B T} = \frac{c}{N} + \sqrt{\left(\frac{Qc}{N}\right)^2 + 4c_s^2} - 2c_s, \tag{A4.9}$$

which is the same equation as Eq. (5.10.12).

Appendix 5

A5.1 Flory–Huggins Theory

Consider a solution of n_1 solvent molecules and n_2 flexible polymer chains, each of N_2 segments. The goal here is to estimate the free energy of mixing ΔF_m associated with the formation of the solution as a final state from its pure components as the initial state. ΔF_m is given in terms of energy of mixing ΔE_m and entropy of mixing ΔS_m as

$$\Delta F_m = \Delta E_m - T\Delta S_m, \tag{A5.1}$$

where ΔE_m and ΔS_m are given by

$$\Delta E_m = E_f - E_i; \qquad \Delta S_m = S_f - S_i, \tag{A5.2}$$

where the subscripts i and f denote the initial and final states, respectively. We derive expressions for ΔE_m and ΔS_m by adopting the following lattice model. Assume that the volume of the solution is discretized into a lattice of uniform sized cells as in Fig. 6.3. Let the volume of a solvent molecule and that of a polymer segment be the same, equal to the volume of a cell in the lattice. Therefore, there are $n_2 N_2$ sites occupied by polymer segments and n_1 sites occupied by the solvent in the mixed state. Let us assume that the lattice is incompressible so that the total number of sites n_0 is equal to the sum of n_1 and $n_2 N_2$. Since the volumes of segment and solvent molecule are taken to be the same, the volume fractions of the solvent and polymer are given, respectively, by ϕ_1 and ϕ_2 as

$$\phi_1 = \frac{n_1}{n_1 + n_2 N_2}, \qquad \phi_2 = \frac{n_2 N_2}{n_1 + n_2 N_2}. \tag{A5.3}$$

In addition, we make the important assumption that the polymer segments and solvent molecules are **randomly mixed**. Let $N_2 = N$ in the following derivation of ΔF_m for polymer solutions.

A5.1.1 Energy of Mixing, ΔE_m

The energy of the final state is given as (see Fig. 6.3)

$$E_f = v_{12}\epsilon_{12} + v_{11}\epsilon_{11} + v_{22}\epsilon_{22}, \tag{A5.4}$$

where v_{ij} is the average number of $i-i$ contacts and ϵ_{ij} is the energy of contact between i and j. Here, i or $j=1$ denotes solvent molecule and i or $j =2$ denotes a polymer segment. The average number of solvent-polymer segment contacts v_{12} is obtained as follows. Let us randomly pick a site in the lattice which is occupied by a solvent molecule. The probability of finding a polymer segment in one of the neighboring sites of this particular site is ϕ_2. Therefore, the average number of polymer segments which are in the neighboring sites is $z\phi_2$, where z is the lattice coordination number (number of neighboring sites at a given site in the empty lattice). Since there are n_1 sites occupied by solvent molecules, the average number of solvent–polymer segment–interactions v_{12} is $zn_1\phi_2$. Similarly, the average number of solvent–solvent contacts and segment–segment contacts follow as

$$v_{12} = zn_1\phi_2, \qquad v_{11} = \frac{1}{2}zn_1\phi_1, \qquad v_{22} = \frac{1}{2}zn_2N\phi_2. \qquad (A5.5)$$

The factor of 2 avoids double counting the interactions between two species of the same type. Substituting Eq. (A5.5) into Eq. (A5.4), we get

$$E_f = z\left[n_1\phi_2\epsilon_{12} + \frac{1}{2}n_1\phi_1\epsilon_{11} + \frac{1}{2}n_2N\phi_2\epsilon_{22}\right]. \qquad (A5.6)$$

The energy of the pure solvent component in the initial state (where $\phi_1 = 1$) is

$$E_{i1} = \frac{1}{2}zn_1\epsilon_{11}. \qquad (A5.7)$$

Similarly, for the pure polymer component in the initial state ($\phi_2 = 1$),

$$E_{i2} = \frac{1}{2}zn_2N\epsilon_{22}. \qquad (A5.8)$$

Substituting Eqs. (A5.6)–(A5.8) into Eq. (A5.2), the energy of mixing is given by

$$\Delta E_m = z\left[\epsilon_{12} - \frac{1}{2}(\epsilon_{11} + \epsilon_{22})\right]\phi_1\phi_2n_0, \qquad (A5.9)$$

where the incompressibility condition, $\phi_1 + \phi_2 = 1$, and Eq. (A5.3) are used. Since the three interaction energy terms $\epsilon_{11}, \epsilon_{22}$, and ϵ_{12} appear only through the combination inside the square brackets of Eq. (A5.9), they can be grouped together as a single parameter χ by rewriting ΔE_m as

$$\Delta E_m = k_BT\chi\phi_1\phi_2n_0, \qquad (A5.10)$$

where k_BT is the Boltzmann constant times the absolute temperature and χ is known as the Flory–Huggins chi parameter defined as

$$\chi = \frac{z}{k_BT}\left[\epsilon_{12} - \frac{1}{2}(\epsilon_{11} + \epsilon_{22})\right]. \qquad (A5.11)$$

Thus, the chi parameter is a measure of the chemical mismatch between the components of the mixture.

A5.1.2   Entropy of Mixing, ΔS_m

The entropy S_f of the final state where n_1 solvent molecules and n_2 chains, each of N segments, are randomly distributed into a lattice of total number of sites $n_0 = n_1 + n_2N$ is given by

$$S_f = k_B \ln \Omega_f, \tag{A5.12}$$

where Ω_f is the total number of configurations of solvent molecules and polymer segments. Ω_f is obtained as follows. Let i chains be already in the lattice and consider the subsequent insertion of the $(i + 1)$th chain into the lattice. The number of ways of inserting the first segment of the new chain is $(n_0 - Ni)$, which are the number of unoccupied sites available for the insertion. After the insertion of the first segment of the $(i+1)$th chain, the number of available sites is $(n_0 - Ni - 1)$. Therefore the number of ways of inserting the second segment of the $(i+1)$th chain is $z(n_0 - Ni - 1)/n_0$, which is the lattice coordination number times the probability of available sites. Similarly, the number of ways of inserting the third segment of $(i+1)$th chain is $(z-1)(n_0 - Ni - 2)/n_0$, because there are now only $(n_0 - Ni - 2)$ available sites and one adjacent neighbor of the lattice must have been occupied by the second segment of the $(i + 1)$th chain. This procedure is repeated for the rest of the segments of the $(i + 1)$th chain. The above results based on an average insertion procedure lead to the number of ways v_{i+1} of inserting the $(i + 1)$th chain as

$$v_{i+1} = (n_0 - Ni) \left[z \frac{(n_0 - Ni - 1)}{n_0} \right] \left[(z - 1) \frac{(n_0 - Ni - 2)}{n_0} \right] \cdots$$
$$\left[(z - 1) \frac{(n_0 - Ni - N + 1)}{n_0} \right]. \tag{A5.13}$$

Let us further assume that $z \simeq z - 1$, comparable to the approximation made in writing the $(z - 1)$ factors in the various terms by ignoring any topological correlations arising from chain connectivity and excluded volume effects. In view of this, the above equation becomes

$$v_{i+1} \simeq \frac{(n_0 - Ni)!}{(n_0 - Ni - N)!} \left[\frac{(z - 1)}{n_0} \right]^{N-1}. \tag{A5.14}$$

Therefore, the total number of configurations in the mixed solution is

$$\Omega_f = \frac{1}{n_2!} \prod_{i=1}^{n_2} v_i, \tag{A5.15}$$

where $n_2!$ accounts for the indistinguishability of the chains and v_i is the number of ways of inserting ith chain into the lattice. Substituting Eq. (A5.14) for v_i,

$$v_i \simeq \frac{(n_0 - N(i - 1))!}{(n_0 - Ni)!} \left[\frac{(z - 1)}{n_0} \right]^{N-1}. \tag{A5.16}$$

Therefore Eq. (A5.15) gives Ω_f as

$$\Omega_f = \frac{1}{n_2!} \frac{n_0!}{(n_0 - N)!} \frac{(n_0 - N)!}{(n_0 - 2N)!} \cdots \frac{(n_0 - N(n_2 - 1))!}{(n_0 - n_2 N)!} \left[\frac{(z - 1)}{n_0} \right]^{n_2(N-1)}$$

$$= \frac{n_0!}{(n_0 - n_2 N)! n_2!} \left[\frac{(z - 1)}{n_0} \right]^{n_2(N-1)}. \tag{A5.17}$$

Using the Stirling approximation, $\ln n! \simeq n \ln n - n$, for the various factorials in the above equation, we get from Eqs. (A5.12) and (A5.17)

$$\frac{S_f}{k_B} = n_0 \ln n_0 - n_0 - (n_0 - n_2 N) \ln(n_0 - n_2 N) + (n_0 - n_2 N) - n_2$$

$$\ln n_2 + n_2 + n_2(N-1) \ln\left(\frac{(z-1)}{n_0}\right), \tag{A5.18}$$

which simplifies to

$$\frac{S_f}{k_B} = -n_1 \ln\left(\frac{n_1}{n_0}\right) - n_2 \ln\left(\frac{n_2}{n_1 + n_2 N}\right) + n_2(N-1) \ln\left(\frac{z-1}{e}\right). \tag{A5.19}$$

For the pure polymer component ($n_1 = 0$), the entropy in the initial state, S_{i2} follows from the above equation as

$$S_{i2} = n_2 \ln N + n_2(N-1) \ln\left(\frac{z-1}{e}\right). \tag{A5.20}$$

The entropy of pure solvent in the initial state, S_{i1}, is zero, since the number of arranging n_1 solvent molecules in n_1 sites is one,

$$S_{i1} = 0. \tag{A5.21}$$

Combining Eq. (A5.2) and Eqs. (A5.19)–(A5.21), the entropy of mixing simplifies to

$$\frac{\Delta S_m}{k_B} = -n_1 \ln \phi_1 - n_2 \ln \phi_2, \tag{A5.22}$$

where ϕ_1 and ϕ_2 are given in Eq. (A5.3), where N_2 is taken as N.

A5.1.3   Free Energy of Mixing, ΔF_m

Substituting Eq. (A5.10) and Eq. (A5.22), respectively, for ΔE_m and ΔS_m in Eq. (A5.1), ΔF_m is

$$\frac{\Delta F_m}{k_B T} = n_1 \ln \phi_1 + n_2 \ln \phi_2 + \chi n_0 \phi_1 \phi_2. \tag{A5.23}$$

This is known as the Flory–Huggins free energy of mixing for a polymer solution [Flory (1942), Huggins (1942a, b, c)].

A generalization of the above derivation for a mixture of n_1 polymer molecules of type 1 with uniform number of segments per chain as N_1 and n_2 polymer molecules of type 2 with N_2 segments per chain, yields the same equation as Eq. (A5.23) with only change in the volume fractions as

$$\phi_1 = \frac{n_1 N_1}{n_1 N_1 + n_2 N_2} \quad \text{and} \quad \phi_2 = \frac{n_2 N_2}{n_1 N_1 + n_2 N_2}. \tag{A5.24}$$

For mixtures of small molecules, $N_1 = 1$ and $N_2 = 1$, and in this case, the Flory–Huggins theory reduces to the Bragg–Williams theory [Bragg & Williams (1934, 1935)]. For polymer blends of two kinds of polymers, Eqs. (A5.23) and (A5.24) provide an expression for ΔF_m, from which all qualitative aspects of phase behavior of blends of uncharged polymers can be predicted.

Appendix 6

A6.1   Models of Polymer Dynamics

Consider a collection of n polyelectrolyte chains distributed uniformly in a solvent. Each chain is modeled as a coarse-grained chain of N segments of segment length ℓ. Let us represent these segments as $N+1$ beads, as illustrated in Fig. 7.13. Let $\mathbf{R}_i$ denote the position vector of the ith bead. The set of vectors $\{\mathbf{R}_i\}(0 \leq i \leq N)$ represents a particular conformation of the chain. The ith bead of a chain in this solution is subjected to forces arising from chain connectivity to its adjacent segments and van der Waals and the electrostatic interactions with all other segments of the same chain and other chains. In addition to these previously treated forces in equilibrium conditions, the mth bead experiences forces from its inertia, friction against the background fluid, and hydrodynamic interaction with all monomers in the system. The potential interaction V and hydrodynamic interaction $\mathbf{G}$ between a pair of beads are illustrated in Fig. 7.14a. Furthermore, the mth bead experiences a random force from incessant collisions with solvent molecules. The equation of motion for the ith bead is the Newton equation,

$$m_b \frac{\partial^2 \mathbf{R}_i(t)}{\partial t^2} = \mathbf{f}_{i,\text{connectivity}} + \mathbf{f}_{i,\text{interactions}} + \mathbf{f}_{i,\text{friction}} + \mathbf{f}_{i,\text{random}} + \mathbf{f}_{i,\text{ext}}, \qquad (A6.1)$$

where $\mathbf{R}_i(t)$ is the position vector of the ith bead of mass m_b at time t. The first four terms on the right-hand side of Eq. (A6.1) are, respectively, the aforementioned forces from chain connectivity, intersegment potential interactions, friction against the background fluid, and random collisions by solvent molecules, which all act on the ith bead. $\mathbf{f}_{i,\text{ext}}$ is an external force, such as an electrical force acting on the ith bead. The above equation in the presence of a random force, called Langevin equation, is an example of a stochastic equation. The dynamics given by the Langevin equation can be described only in terms of statistical averages, in contrast with the deterministic Newton's equations. For time scales relevant to typical situations associated with macromolecules, the inertia term on the left-hand side of Eq. (A6.1) can be ignored. Therefore, we get

$$-\mathbf{f}_{i,\text{connectivity}} - \mathbf{f}_{i,\text{interactions}} = \mathbf{f}_{i,\text{friction}} + \mathbf{f}_{i,\text{random}} + \mathbf{f}_{i,\text{ext}}. \qquad (A6.2)$$

The above equation of motion for the ith bead is coupled to the hydrodynamic properties of the solvent, which is treated as a continuum (Section 7.2). This coupling is treated using the "no-slip" boundary condition between the velocity of the ith bead of the αth chain, $\partial \mathbf{R}_{\alpha i}(t)/\partial t$ and the fluid velocity $\mathbf{v}(\mathbf{R}_{\alpha i}(t))$ at its location,

$$\frac{\partial \mathbf{R}_{\alpha i}(t)}{\partial t} = \mathbf{v}(\mathbf{R}_{\alpha i}(t)). \tag{A6.3}$$

In view of this boundary condition, the coupled equations for the i-th bead of the αth chain (Eq. (A6.2)) and the linearized Navier–Stokes equation (Eq. (7.2.2)) become

$$-\mathbf{f}_{i,\text{connectivity}} - \mathbf{f}_{i,\text{interactions}} = -\sigma_{\alpha i} + \mathbf{f}_{i,\text{random}} + \mathbf{f}_{i,\text{ext}}, \tag{A6.4}$$

$$-\eta_0 \nabla^2 \mathbf{v}(\mathbf{r}) + \nabla p(\mathbf{r}) = \mathbf{F}(\mathbf{r}) + \sum_{\alpha=1}^{n} \sum_{j=0}^{N} \delta(\mathbf{r} - \mathbf{R}_{\alpha i})\sigma_{\alpha i}, \tag{A6.5}$$

where $\sigma_{\alpha i}$ is the force from the bead on the fluid, with $-\sigma_{\alpha i}$ as the frictional force $\mathbf{f}_{i,\text{friction}}$ acting on the ith bead of the α-th chain. As in Eq. (7.2.2), $\mathbf{F}(\mathbf{r})$ is the external force acting on the fluid element at location $\mathbf{r}$. Determination of $\sigma_{\alpha i}$ from the boundary condition and the above coupled equations yields the equation of motion for the chain beads and the modified equation of motion for the solvent as follows:

Upon averaging over chain conformations, the effective equation of motion for the solution at large length scales (comparable to the size of the system) can be derived as [Edwards & Freed (1974), Freed & Edwards (1974)]

$$-\eta_0 \nabla^2 \langle \mathbf{v}(\mathbf{r}) \rangle + \nabla p(\mathbf{r}) + \int d\mathbf{r}' \Sigma(\mathbf{r} - \mathbf{r}') \cdot \langle \mathbf{v}(\mathbf{r}) \rangle = \mathbf{F}(\mathbf{r}), \tag{A6.6}$$

where the angular brackets denote conformational averages, and the third term on the left-hand side denotes the net effect on the solvent due to all polymer chains. Σ contains all information about interactions among all beads mediated by solvent and structure of the solute molecules. This term can be expressed as a series expansion in $\nabla \mathbf{v}(\mathbf{r})$ as in the development of the linearized Navier–Stokes equation. It can be shown [Muthukumar (1981)] that the leading term of the third term on the left-hand side of Eq. (A6.6) is proportional to $\nabla^2 \langle \mathbf{v}(\mathbf{r}) \rangle$ so that the change in viscosity of the solvent due to the presence of polymers is given by

$$\int d\mathbf{r}' \Sigma(\mathbf{r} - \mathbf{r}') \cdot \langle \mathbf{v}(\mathbf{r}) \rangle = (\eta - \eta_0) \nabla^2 \langle \mathbf{v}(\mathbf{r}) \rangle + O(\nabla \mathbf{v})^3. \tag{A6.7}$$

Thus Eq. (A6.4) is used to describe the dynamics of beads and polymer chains in solutions and Eq. (A6.6) is used to describe the hydrodynamic properties of the solution.

Accounting for all inter-chain and intra-chain interactions, the net contribution $\Sigma(\mathbf{r} - \mathbf{r}')$ in Eq. (A6.6) from all chains can be approximately derived using the multiple scattering formalism [Edwards & Freed (1974), Freed & Edwards (1974), Muthukumar & Edwards (1982b)]. Defining the Fourier transform $\hat{\Sigma}(\mathbf{k})$ of $\Sigma(\mathbf{r})$ in terms of the scattering wave vector $\mathbf{k}$,

$$\hat{\Sigma}(\mathbf{k}) = \int d\mathbf{r} e^{-i\mathbf{k} \cdot \mathbf{r}} \Sigma(\mathbf{r}), \tag{A6.8}$$

$\hat{\Sigma}(\mathbf{k})$ is given by the self-consistent equation

$$\hat{\Sigma}(\mathbf{k}) = 12\pi^2 c\ell \int_{2\pi/N\ell}^{\infty} dq \frac{S(\mathbf{k}, q)}{\int d\mathbf{j} \frac{1}{\eta_0 j^2 + \hat{\Sigma}(\mathbf{j})} S(\mathbf{j}, q)}. \tag{A6.9}$$

Here, $S(\mathbf{k}, q)$ is the structure factor of a chain in the solution accounting for chain connectivity, excluded volume, and electrostatic interactions as described in the double screening theory (Section 6.3.1). q is proportional to the Rouse mode variable p in the continuous notation. Note that the kth mode of $\hat{\Sigma}(\mathbf{k})$ is coupled to all other jth modes appearing through $\hat{\Sigma}(\mathbf{j})$ in the denominator of the right-hand side of the above equation. Therefore, $\hat{\Sigma}(\mathbf{k})$ needs to be calculated using a self-consistent procedure. Substitution of $\hat{\Sigma}(\mathbf{k})$ in Eq. (A6.6) gives a modified hydrodynamic interaction $\mathcal{G}(\mathbf{r} - \mathbf{r}')$ instead of the bare Oseen tensor $\mathbf{G}(\mathbf{r} - \mathbf{r}')$ given in Eq. (7.2.5). The result is

$$\langle \mathbf{v}(\mathbf{r}) \rangle = \int d\mathbf{r}' \mathcal{G}(\mathbf{r} - \mathbf{r}') \cdot \mathbf{F}(\mathbf{r}'), \tag{A6.10}$$

where the angular brackets denote averaging over chain conformations and angular average over the hydrodynamic interaction tensor.

Calculations based on the multiple scattering formalism using suitable physically relevant assumptions give $\mathcal{G}$, for length scales smaller than R_g but larger than the segment size ℓ, as

$$\mathcal{G}(\mathbf{r} - \mathbf{r}') = \frac{e^{-|\mathbf{r} - \mathbf{r}'|/\xi_h}}{6\pi\eta_0 |\mathbf{r} - \mathbf{r}'|}, \tag{A6.11}$$

where ξ_h is a measure of the range of distance over which the hydrodynamic interaction is significant. Thus, interpenetration of chains leads to **hydrodynamic screening** as described in Fig. 7.27.

At very high polymer concentrations, entanglement effects arising from the uncrossability of polymer strands dominate the dynamics of the system (Fig. 7.14b). This situation belongs to a different class of dynamics where topological constraints play a major role.

A6.2   Rouse Dynamics of a Gaussian Chain

Consider a Gaussian chain of N segments, each of Kuhn length ℓ. There are $N+1$ beads connecting these segments, as illustrated in Fig. 7.13. Let $\mathbf{R}_m$ denote the position vector of the m-th bead. The set of vectors $\{\mathbf{R}_m\}(0 \leq m \leq N)$ represents a particular conformation of the chain. Let us assume that there are no potential interactions or hydrodynamic interactions among the various segments. Ignoring inertia, the Langevin equation for the m-th bead is

$$\zeta_b \frac{\partial \mathbf{R}_m(t)}{\partial t} + \frac{\partial}{\partial \mathbf{R}_m} F(\{\mathbf{R}_m\}) = \mathbf{f}_m(t). \tag{A6.12}$$

The two terms on the left-hand side are due to friction against solvent and chain connectivity, respectively. The term on the right-hand side is a random force. Here, ζ_b is the bead friction coefficient and $F(\mathbf{R}_i)$ is the free energy of a chain conformation due to chain connectivity. As seen in Section 2.1, the probability to realize a conformation is given by

$$P \sim \exp\left(-\frac{3}{2\ell^2} \sum_{j=0}^{N} \left[\mathbf{R}_j(t) - \mathbf{R}_{j-1}(t)\right]^2\right), \tag{A6.13}$$

so that the free energy, $F = -k_B T \ln P$, is given as

$$F(\{\mathbf{R}_j\}) = \frac{3k_B T}{2\ell^2} \sum_{j=0}^{N} \left[\mathbf{R}_j(t) - \mathbf{R}_{j-1}(t)\right]^2. \tag{A6.14}$$

This yields

$$\frac{\partial F}{\partial \mathbf{R}_m} = -\frac{3k_B T}{\ell^2} (\mathbf{R}_{m+1} - 2\mathbf{R}_m + \mathbf{R}_{m-1}). \tag{A6.15}$$

Therefore, the Langevin equation, Eq. (A6.12), for the m-th bead becomes,

$$\zeta_b \frac{\partial \mathbf{R}_m(t)}{\partial t} - \frac{3k_B T}{\ell^2} (\mathbf{R}_{m+1} - 2\mathbf{R}_m + \mathbf{R}_{m-1}) = \mathbf{f}_m(t). \tag{A6.16}$$

This equation is known as the **Rouse equation**. Adopting the continuous representation of the chain, that is, m is a continuous variable,

$$\zeta_b \frac{\partial \mathbf{R}(m,t)}{\partial t} - \frac{3k_B T}{\ell^2} \frac{\partial^2 \mathbf{R}(m,t)}{\partial m^2} = \mathbf{f}(m,t). \tag{A6.17}$$

Note that the end effects are included in this equation by using the boundary conditions, $\partial \mathbf{R}(m,t)/\partial m \to 0$, for $m = 0$ and $m = N$.

Eq. (A6.17) is like a wave equation, and can be easily solved with Fourier transforms (normal modes of vibrations). Let us choose the Fourier series pair,

$$\mathbf{R}(m,t) = \sum_{p=-\infty}^{\infty} \hat{R}_p(t) \cos\left(\frac{\pi p m}{N}\right) \tag{A6.18}$$

$$\hat{R}_p(t) = \frac{1}{N} \int_0^N dm\, \mathbf{R}(m,t) \cos\left(\frac{\pi p m}{N}\right), \tag{A6.19}$$

Note that $\hat{R}_p = \hat{R}_{-p}$.

The normal modes of the polymer $\hat{R}_p(t)$ are called **Rouse modes**. As can be seen from Eq. (A6.19), $p = 0$ denotes the center of mass motion of the chain; $p = 1$ denotes an excitation over the whole chain; $p = 2$ denotes excitations over half of the distance along the string, etc., as illustrated in Fig. 7.15 for a polymer chain with its two ends fixed. When a chain relaxes from plucking, $p = 10$ represents the relaxation of one-tenth of the contour length of the chain. Expressing the Rouse equation, Eq. (A6.17), in terms of the Rouse modes,

$$\zeta_b \frac{\partial \hat{R}_p(t)}{\partial t} + \frac{3\pi^2 k_B T}{\ell^2} \left(\frac{p}{N}\right)^2 \hat{R}_p(t) = \hat{f}_p(t), \tag{A6.20}$$

where $\hat{f}_p(t)$ is defined analogous to Eq. (A6.19).

(1) Intersegment dynamical correlation function

The time correlation function for the positions of two monomers is given by

$$\left\langle [\mathbf{R}_m(t) - \mathbf{R}_n(0)]^2 \right\rangle = 6D_{cm}t + |m - n|\ell^2 + \frac{4N\ell^2}{\pi^2}$$

$$\sum_{p=1}^{\infty} \frac{1}{p^2} \cos\left(\frac{\pi pm}{N}\right) \cos\left(\frac{\pi pn}{N}\right) \left(1 - e^{-p^2 t/\tau_R}\right). \quad (A6.21)$$

This general expression yields Eqs. (7.5.12) and (7.5.14) for the center of mass of the chain as well as the mean square displacement of a labeled monomer given below.

(2) Viscoelasticity

Consider a shear flow with a shear rate $\dot{\gamma}(t)$ (Fig. 7.17), where the velocity components are $v_x = \dot{\gamma}y$, and $v_y = 0 = v_z$.

In general, the shear stress $\sigma_{xy}(t)$ is given by [Doi & Edwards (1986)]

$$\sigma_{xy}(t) = \int_{-\infty}^{t} dt'\, G(t - t')\, \dot{\gamma}(t'), \quad (A6.22)$$

where G is the shear relaxation modulus. We can calculate $\sigma_{xy}(t)$ from the Rouse equation in the presence of flow and then match the result with the right-hand side of Eq. (A6.22), resulting in

$$G(t) = \frac{cRT}{M} \sum_p \exp\left(-\frac{t}{\tau_p}\right). \quad (A6.23)$$

Here, c is the monomer molar concentration, R is the ideal gas constant, M is the molar mass of the chain, and τ_p is now given by

$$\tau_p = \frac{\zeta_b N^2 \ell^2}{6\pi^2 k_B T} \frac{1}{p^2}. \quad (A6.24)$$

Note that τ_p defined for $G(t)$ is half of the Rouse time defined in Eq. (7.5.10).

For oscillatory shear, $\dot{\gamma}(t) = \dot{\gamma}e^{i\omega t}$, the frequency dependent modulus becomes

$$G(\omega) = i\omega \int_0^{\infty} dt\, e^{-i\omega t}\, G(t). \quad (A6.25)$$

Substituting Eq. (A6.23) into Eq. (A6.25), we get the complex modulus as

$$G(\omega) = \frac{cRT}{M} \sum_p \frac{i\omega\tau_p}{\left(1 + i\omega\tau_p\right)}. \quad (A6.26)$$

Writing $G(\omega)$ in terms of its real and imaginary parts G' and G'',

$$G(\omega) = G'(\omega) + iG''(\omega), \quad (A6.27)$$

the **storage modulus** G' and the **loss modulus** G'' are given by

$$G' = \frac{cRT}{M} \sum_p \frac{\omega^2 \tau_p^2}{\left(1 + \omega^2 \tau_p^2\right)} \tag{A6.28}$$

$$G'' = \frac{cRT}{M} \sum_p \frac{\omega \tau_p}{\left(1 + \omega^2 \tau_p^2\right)}. \tag{A6.29}$$

The contribution of the chains to viscosity $\eta(\omega) = G(\omega)/i\omega$ follows from Eq. (A6.26) as

$$\eta(\omega) = \frac{cRT}{M} \sum_p \frac{\tau_p}{(1 + i\omega\tau_p)}. \tag{A6.30}$$

The intrinsic viscosity of a polymer solution, defined as $[\eta] = \lim\limits_{c \to 0} (\eta - \eta_0)/c\eta_0$, is

$$[\eta] = \frac{RT}{M\eta_0} \sum_p \frac{\tau_p}{(1 + i\omega\tau_p)}. \tag{A6.31}$$

In the zero-frequency limit, Eqs. (A6.24) and (A6.31) give

$$[\eta] = \frac{N_A}{M\eta_0} \frac{\zeta_b N^2 \ell^2}{36}. \tag{A6.32}$$

Therefore, according to the Rouse model, the change in viscosity due to the chains is given as

$$\frac{\eta - \eta_0}{\eta_0} \sim c \, \zeta_b \, N. \tag{A6.33}$$

In the high frequency limit, $\omega\tau_R > 1$, Eqs. (A6.24) and (A6.31) give the frequency dependence of intrinsic viscosity as

$$[\eta] \sim \frac{1}{\sqrt{\omega}}. \tag{A6.34}$$

Equivalently, the shear modulus at higher frequencies exploring the internal chain dynamics is given by

$$G \sim \sqrt{\omega}. \tag{A6.35}$$

A6.3   Zimm Dynamics of a Gaussian Chain

Considering a Gaussian chain, we will now include hydrodynamic interactions among all segments. Recall from Section 7.2 that the long-ranged Oseen tensor $\mathbf{G}(\mathbf{r} - \mathbf{r}')$ transmits a force at $\mathbf{r}'$ to a velocity at $\mathbf{r}$ in the solvent. Therefore, in addition to the force from chain connectivity acting on the m-th bead given by Eq. (7.5.5), there are multiple forces acting on the m-th bead which emanate from the connectivity forces at all other segments. For example, the force at $\mathbf{R}_m$ due to the connectivity force at $\mathbf{R}_j$ of the j-th bead is $\zeta_b \mathbf{G}(\mathbf{R}_m - \mathbf{R}_j) \left[-(3k_B T/\ell^2)\partial^2 \mathbf{R}_j/\partial j^2\right]$ (where G converts force into velocity and ζ_b converts the resultant velocity at the bead into force). This pairwise hydrodynamic interaction is then summed over all j beads except $j = m$. Therefore, inclusion of hydrodynamic interaction among all beads leads to the equation of motion of the m-th bead as [Kirkwood & Riseman (1948), Zimm (1956)]

$$\zeta_b \frac{\partial \mathbf{R}(m,t)}{\partial t} + \sum_j \left[\mathbf{1} \delta_{mj} + (1 - \delta_{mj}) \zeta_b \mathbf{G}(\mathbf{R}_m - \mathbf{R}_j) \right] \left[-\frac{3k_B T}{\ell^2} \frac{\partial^2 \mathbf{R}(j,t)}{\partial j^2} \right] = \mathbf{f}(m,t),$$

$$(A6.36)$$

where $\mathbf{f}(m,t)$ is the net random force acting on the m-th bead arising from both directly from the solvent medium and all hydrodynamic coupling between the m-th bead and all $j \neq m$ beads with their own random forces $\mathbf{f}(j,t)$. The above equation is rewritten as

$$\frac{\partial \mathbf{R}(m,t)}{\partial t} + \sum_j \mathbf{D}[\mathbf{R}(m) - \mathbf{R}(j)] \left[-\frac{3k_B T}{\ell^2} \frac{\partial^2 \mathbf{R}(j,t)}{\partial j^2} \right] = \frac{\mathbf{f}(m,t)}{\zeta_b}, \qquad (A6.37)$$

where

$$\mathbf{D}[\mathbf{R}(m) - \mathbf{R}(j)] = \frac{1}{\zeta_b} \left[\mathbf{1} \delta_{mj} + (1 - \delta_{mj}) \zeta_b \mathbf{G}(\mathbf{R}_m - \mathbf{R}_j) \right]. \qquad (A6.38)$$

Eq. (A6.37) is known as the Zimm equation, although it was derived earlier by Kirkwood and Riseman (1948). This equation is nonlinear and no exact solution can be derived. We make the **"preaveraging approximation,"** that is, we replace $\mathbf{D}[\mathbf{R}(m) - \mathbf{R}(j)]$ by its configurational average,

$$\mathbf{D}[\mathbf{R}(m) - \mathbf{R}(j)] \simeq \langle \mathbf{D}[\mathbf{R}(m) - \mathbf{R}(j)] \rangle = \frac{1}{\zeta_b} \delta_{mj} + \frac{1}{\sqrt{6}\pi^{3/2}\eta_0 \ell} \frac{1}{|m - j|^{1/2}}. \quad (A6.39)$$

Writing the Zimm equation using the Rouse modes,

$$\frac{\partial \hat{R}_p(t)}{\partial t} + D_p \frac{3\pi^2 k_B T}{\ell^2} \left(\frac{p}{N} \right)^2 \hat{R}_p(t) = \frac{\hat{f}_p(t)}{\zeta_b}, \qquad (A6.40)$$

where

$$D_p = \frac{1}{\zeta_b} + \frac{1}{\sqrt{3}\pi^3 \eta_0 \ell} \left(\frac{N}{p} \right)^{1/2}. \qquad (A6.41)$$

If hydrodynamic interactions are absent, we recover the Rouse result (the so-called "free draining" limit) from the first term on the right-hand side. If hydrodynamic interactions dominate, the second term on the right-hand side overwhelms the first term and leads to the Zimm results. This limit is referred to as the "non-free draining" limit.

The key consequences of the Zimm model are summarized in Section 7.5.2, and details of the calculations are available in Doi & Edwards (1986).

A6.4   Effect of Potential Interactions on Rouse-Zimm Dynamics

The force on the mth bead arising from the van der Waals and electrostatic interactions between the mth bead and all other beads (labeled j) is

$$\mathbf{f}_{m,\,\text{interactions}} = -\frac{\partial}{\partial \mathbf{R}_m} \sum_j V\left(\mathbf{R}_m - \mathbf{R}_j\right), \tag{A6.42}$$

where V denotes the net intersegment potential interaction energy. To address the role of ∇V in Rouse–Zimm dynamics, let us first return to Eq. (7.5.1). Accounting for chain connectivity and intersegment potential interactions, we get from Eqs. (7.5.1) and (A6.42),

$$\zeta_b \frac{\partial \mathbf{R}_m(t)}{\partial t} - \frac{3k_B T}{\ell^2} \frac{\partial^2 \mathbf{R}_m(t)}{\partial m^2} + \frac{\partial}{\partial \mathbf{R}_m} \sum_j V\left(\mathbf{R}_m - \mathbf{R}_j\right) = \mathbf{f}(m,t). \tag{A6.43}$$

Note that the third term on the left-hand side is nonlinear in the position vectors of the various beads. Upon Fourier transform of this equation in terms of the Rouse modes, the first two terms result in Eq. (7.5.8) for the p-th Rouse mode, whereas the third term results in a function where the p-th mode is coupled to all other Rouse modes [Muthukumar & Edwards (1982b)]. This kind of "mode-mode coupling" can be approximated by an effective p-th mode. This approximation is equivalent to the assumption of uniform swelling of a polymer chain due to the intersegment interactions (see Sections 5.6 and 9.2.1). In general, the mean square end-to-end distance $\langle R^2 \rangle$ of a chain is expressed in terms of an effective segment length ℓ_{eff} (see Eq. (5.6.32))

$$\langle R^2 \rangle = N \ell \ell_{\text{eff}}, \tag{A6.44}$$

where ℓ_{eff} depends on the excluded volume parameter v, Bjerrum length ℓ_B, inverse Debye length κ, and N. For example, ℓ_{eff} for dilute solutions in the high salt limit follows from Eq. (5.6.9) as

$$\left(\frac{\ell_{\text{eff}}}{\ell}\right)^{5/2} - \left(\frac{\ell_{\text{eff}}}{\ell}\right)^{3/2} = \frac{4}{3}\left(\frac{3}{2\pi}\right)^{3/2}\left[v + \frac{4\pi\alpha^2 z_p^2 \ell_B}{\kappa^2 \ell^3}\right]\sqrt{N}, \tag{A6.45}$$

where α is the degree of ionization of the polyelectrolyte chain and z_p is the number of ionizable repeat units in a Kuhn segment. The equation is strictly valid only for the $p = 1$ Rouse mode. With the above mentioned approximation of uniform expansion of the chain, Eq. (A6.45) is generalized to all modes as [Muthukumar & Edwards (1982b)]

$$\left(\frac{\ell_{\text{eff}}}{\ell}\right)^{5/2} - \left(\frac{\ell_{\text{eff}}}{\ell}\right)^{3/2} = \frac{\sqrt{6}}{\pi}\left[v + \frac{4\pi\alpha^2 z_p^2 \ell_B}{\kappa^2 \ell^3}\right]\left(\frac{\pi p}{N}\right)^{-1/2}. \tag{A6.46}$$

Therefore, the dependence of ℓ_{eff} on the Rouse modes for strong intersegment interactions is given as

$$\ell_{\text{eff}} \sim \left(\frac{\pi p}{N}\right)^{-1/5}. \tag{A6.47}$$

In general, for any polymeric fractal with size exponent ν,

$$\ell_{\text{eff}} \sim \left(\frac{\pi p}{N}\right)^{1-2\nu}.$$ (A6.48)

The approximate mode-mode decoupling described above is equivalent to absorbing the third term on the left-hand side of Eq. (A6.43) into the second term by replacing one factor of ℓ by ℓ_{eff}. Performing this approximation, the generalized Rouse and Zimm equations are obtained from Eqs. (A6.20) and (A6.40) as follows.

A6.4.1   Generalized Rouse Dynamics

Including the intersegment potential interaction in terms of the effective segment length ℓ_{eff} as described above, Eq. (A6.20) gives

$$\zeta_b \frac{\partial \hat{R}_p(t)}{\partial t} + \frac{3\pi^2 k_B T}{\ell \ell_{\text{eff}}} \left(\frac{p}{N}\right)^2 \hat{R}_p(t) = \hat{f}_p(t). \qquad \text{(Rouse)}$$ (A6.49)

The relaxation time of the p-th mode of a Rouse chain with intersegment potential interactions follows from a dimensional analysis of Eq. (A6.49) as

$$\tau_p = \frac{\zeta_b \ell \ell_{\text{eff}}}{3\pi^2 k_B T} \left(\frac{N}{p}\right)^2 \sim \frac{\zeta_b}{T} \left(\frac{N}{p}\right)^{2\nu+1}.$$ (A6.50)

Therefore, the longest relaxation time (Rouse time) is

$$\tau_R \sim \frac{\zeta_b}{T} N^{2\nu+1}.$$ (A6.51)

It can be shown that the mean square displacement of a labeled monomer m during time intervals shorter than τ_R is given as

$$\left\langle [\mathbf{R}_m(t) - \mathbf{R}_m(0)]^2 \right\rangle \sim t^{2\nu/(2\nu+1)}.$$ (A6.52)

The corresponding relaxation rate Γ_k in the dynamic structure factor for $k R_g \gg 1$ is given by

$$\Gamma_k \sim k^{(2\nu+1)/\nu}.$$ (A6.53)

The results given in Eqs. (A6.51)-(A6.53) for $\nu \neq 1/2$ are of limited use, since it is difficult to realize experimental conditions where the potential interactions are not screened while the hydrodynamic interaction is screened.

A6.4.2   Generalized Zimm Dynamics

Using the same mode-mode decoupling approximation as above, the Zimm equation for a polymer chain with intersegment potential interactions is obtained from Eq. (A6.40) as

$$\frac{\partial \hat{R}_p(t)}{\partial t} + D_p \frac{3\pi^2 k_B T}{\ell \ell_{\text{eff}}} \left(\frac{p}{N}\right)^2 \hat{R}_p(t) = \frac{\hat{f}_p(t)}{\zeta_b}. \qquad \text{(Zimm)}$$ (A6.54)

Note that the preaveraged Oseen tensor $\langle G[\mathbf{R}_m - \mathbf{R}_j]\rangle$ defined in Eq. (A6.39) depends on the size exponent ν as

$$\langle G[\mathbf{R}_m - \mathbf{R}_j]\rangle \sim \frac{1}{|m-j|^\nu},\qquad(A6.55)$$

where the uniform chain expansion approximation is used. Expressing $\langle G[\mathbf{R}_m - \mathbf{R}_j]\rangle$ in terms of the Rouse modes, its dependence on the p-th mode is

$$G_p \sim \left(\frac{\pi p}{N}\right)^{\nu-1}.\qquad(A6.56)$$

Therefore, when hydrodynamic interactions dominate over monomer friction, D_p in Eq. (A6.41) is modified as

$$D_p \sim \frac{1}{\eta_0}\left(\frac{\pi p}{N}\right)^{\nu-1}.\qquad(A6.57)$$

The main features of the generalized Zimm dynamics that are of experimental relevance follow from Eqs. (A6.40), (A6.54), and (A6.57) as summarized in Section 7.5.2. In particular, the relaxation time of the p-th Rouse mode is obtained from a dimensional analysis of Eq. (A6.54) as

$$\tau_p = \frac{\ell\ell_{\text{eff}}}{3\pi^2 k_B T}\frac{1}{D_p}\left(\frac{N}{p}\right)^2.\qquad(A6.58)$$

Using Eq. (A6.46) for ℓ_{eff} and Eq. (A6.57) for D_p in the limit of dominance by hydrodynamic interactions between the segments, Eq. (A6.58) gives

$$\tau_p \sim \frac{\eta_0}{T}\left(\frac{N}{p}\right)^{3\nu}.\qquad(A6.59)$$

The longest relaxation time τ_Z (Zimm time) scales with N and consequently on R_g as

$$\tau_Z \sim \frac{\eta_0}{T}N^{3\nu} \sim \frac{\eta_0}{T}R_g^3.\qquad(A6.60)$$

A6.5   Dilute Polyelectrolyte Solutions

Consider a dilute solution of polyelectrolyte chains containing added salt. Here, we provide derivations of the results for the translational friction coefficient, electrophoretic mobility, and the diffusion coefficient of an isolated chain, and the modification of the viscosity of the solvent due to the presence of the polyelectrolyte chains.

A6.5.1   Translational Friction Coefficient

Due to friction against the background solvent, let $-\sigma_i$ be the frictional force exerted on the i-th bead as described in Eq. (A6.4) and Fig. 7.14a. Therefore, σ_i is the force exerted by the bead on the fluid at the location of the bead, as given in Eq. (A6.5). σ_i is then evaluated using the "no-slip" boundary condition Eq. (A6.3). We get the net average frictional force on the chain by summing $-\sigma_i$ over all values of the bead variable i and averaging over all conformations of the chain. This net frictional force is $-f_t\dot{\mathbf{R}}^0$, where f_t is the translational friction coefficient of the chain and $\dot{\mathbf{R}}^0$ is the net drift velocity of the center of mass of the chain,

$$-\sum_i \langle \sigma_i \rangle = -f_t \dot{\mathbf{R}}^0, \tag{A6.61}$$

where the angular brackets denote the averaging over chain conformations.

The velocity $\mathbf{v}(\mathbf{r})$ of the fluid at location $\mathbf{r}$ for the present situation follows from Eq. (A6.5) as

$$-\eta_0 \nabla^2 \mathbf{v}(\mathbf{r}) + \nabla p(\mathbf{r}) = \mathbf{F}_0 + \mathbf{F}_{\text{ext}}(\mathbf{r}) + \sum_{j=0}^N \delta(\mathbf{r} - \mathbf{R}_j)\sigma_j. \tag{A6.62}$$

Here, $\mathbf{F}_0(\mathbf{r})$ is the force responsible for any solvent flow with local velocity $\mathbf{v}^0(\mathbf{r})$ in the absence of polymer chains and salt. $\mathbf{F}_{\text{ext}}(\mathbf{r})$ denotes the external force arising from the charge distribution of the ions in the solution. The last term on the right-hand side of the equation is due to the frictional force exerted at $\mathbf{r}$ by all beads of the chain. Using the Oseen tensor $G(\mathbf{r} - \mathbf{r}')$ described in Section 7.2, the above equation becomes

$$\mathbf{v}(\mathbf{r}) = \mathbf{v}^0(\mathbf{r}) + \int d\mathbf{r}' G(\mathbf{r} - \mathbf{r}') \cdot \mathbf{F}_{\text{ext}}(\mathbf{r}') + \sum_{j=0}^N G(\mathbf{r} - \mathbf{R}_j) \cdot \sigma_j. \tag{A6.63}$$

The velocity of the fluid at the location of the i-th bead is therefore given as

$$\mathbf{v}(\mathbf{R}_i) = \mathbf{v}_i^0 + \sum_{j=0}^N G(\mathbf{R}_i - \mathbf{R}_j) \cdot \sigma_j + \int d\mathbf{r}' G(\mathbf{R}_i - \mathbf{r}') \cdot \mathbf{F}_{\text{ext}}(\mathbf{r}'), \tag{A6.64}$$

where $\mathbf{v}_i^0$ is $\mathbf{v}^0(\mathbf{R}_i)$. Since the Oseen tensor is applicable only to intersegment hydrodynamic interactions (and not for $i = j$), and the velocity of the i-th bead due to a force σ_i is σ_i/ζ_b (where ζ_b is the bead friction coefficient), the $j = i$ term in the summation of Eq. (A6.64) is separated out. Therefore, Eq. (A6.64) becomes

$$\mathbf{v}(\mathbf{R}_i) = \mathbf{v}_i^0 + \frac{1}{\zeta_b}\sigma_i + \sum_{j \neq i} G(\mathbf{R}_i - \mathbf{R}_j) \cdot \sigma_j + \int d\mathbf{r}' G(\mathbf{R}_i - \mathbf{r}') \cdot \mathbf{F}_{\text{ext}}(\mathbf{r}'). \tag{A6.65}$$

According to the "no-slip" boundary condition (Eq. (A6.3)), the instantaneous velocity of the i-th bead is equal to the velocity of the fluid at its location,

$$\dot{\mathbf{R}}_i = \mathbf{v}(\mathbf{R}_i). \tag{A6.66}$$

Using Eq. (A6.66) in Eq. (A6.65), we get

$$\sigma_i = \zeta_b \left(\dot{\mathbf{R}}_i - \mathbf{v}_i^0 \right) - \zeta_b \sum_{j \neq i} G(\mathbf{R}_i - \mathbf{R}_j) \cdot \sigma_j - \zeta_b \int d\mathbf{r}' G(\mathbf{R}_i - \mathbf{r}') \cdot \mathbf{F}_{\text{ext}}(\mathbf{r}'). \tag{A6.67}$$

Note that the force exerted by the i-th bead on the fluid is coupled to forces from all other beads of the chain. For the time being, let us assume that $\mathbf{F}_{\text{ext}} = 0$ and $\mathbf{v}_i^0 = 0$. Averaging over chain conformations, the above equation becomes

$$\langle \sigma_i \rangle = \zeta_b \langle \dot{\mathbf{R}}_i \rangle - \zeta_b \sum_{j \neq i} \left\langle G(\mathbf{R}_i - \mathbf{R}_j) \cdot \sigma_j \right\rangle. \tag{A6.68}$$

Since this equation relating the various one-dimensional bead indices involves $G(\mathbf{R}_i - \mathbf{R}_j)$, which depends on the three-dimensional vector $\mathbf{R}_i - \mathbf{R}_j$, an exact solution is not possible. To avoid this difficulty, we employ the "preaveraging approximation" (Section 7.5.2) and take the conformational average of the product $G \cdot \sigma$ on the right-hand side of Eq. (A6.68) as the product of averages of G and σ, so that we get

$$\langle \sigma_i \rangle = \zeta_b \langle \dot{\mathbf{R}}_i \rangle - \zeta_b \sum_{j \neq i} \langle G(\mathbf{R}_i - \mathbf{R}_j) \rangle \langle \sigma_j \rangle, \tag{A6.69}$$

where $\langle G(\mathbf{R}_i - \mathbf{R}_j) \rangle$ follows from Eq. (7.2.6) as

$$\langle G(\mathbf{R}_i - \mathbf{R}_j) \rangle = \frac{1}{6\pi\eta_0} \left\langle \frac{1}{|\mathbf{R}_i - \mathbf{R}_j|} \right\rangle. \tag{A6.70}$$

Upon averaging over the chain conformations, the right-hand side is a function of only $\left| \mathbf{R}_i - \mathbf{R}_j \right|$ enabling the solution of $\langle \sigma_i \rangle$ from Eq. (A6.69), which is carried out using Fourier transforms. Substituting the resultant solution for $\langle \sigma_i \rangle$ in Eq. (A6.61) and assuming that the average velocity of each bead of the chain is the same as that of the center of mass of the chain, the translational friction coefficient is given by [Yamakawa (1972), Muthukumar (1997), Muthukumar (2005)]

$$f_t = \frac{\zeta_b N}{\left(1 + \frac{8}{3\sqrt{\pi}} \frac{\zeta_b N}{6\pi\eta_0 R_g} \right)}. \tag{A6.71}$$

The second term in the denominator arises from the intersegment hydrodynamic interactions, and the numerator denotes the friction coefficient of the chain as a sum of the frictional coefficients from the individual beads.

A6.5.2   Electrophoretic Mobility

In the presence of an externally applied constant electric field $\mathbf{E}_0$, a polyelectrolyte chain undergoes electrophoretic drift, as a balance between the electric and frictional forces acting on it. Let the polyelectrolyte chain have a total charge $Q = N\alpha z_p e$ (where the charge sign is absorbed in z_p). The electric force on the chain is $Q\mathbf{E}_0$ and the net frictional force on the chain is $-\sum_i \langle \sigma_i \rangle$ as shown in the preceding section. In the steady state where the chain does not accelerate, the net force acting on the chain $Q\mathbf{E}_0 - \sum_i \langle \sigma_i \rangle$ is zero so that

$$\sum_i \langle \sigma_i \rangle = Q\mathbf{E}_0. \tag{A6.72}$$

Using Eq. (A6.61), the Einstein electrophoretic mobility follows from the above equation as

$$\mu_{\text{Einstein}} = \frac{Q}{f_t}. \tag{A6.73}$$

Since $f_t \sim R_g \sim N^\nu$ and $Q \sim N$, we get Eq. (7.4.14) as

$$\mu_{\text{Einstein}} \sim N^{1-\nu}, \tag{A6.74}$$

which is unphysical because this equation predicts an increasing electrophoretic mobility with an increase in N for $\nu < 1$. As discussed in Section 7.4, the polyelectrolyte chain does not move under the electric field as an isolated entity, but its motion is coupled to that of its ion cloud. Each polyelectrolyte chain is surrounded by $|Nz_p/z_c|$ counterions, each with a charge $z_c e$. In addition, there are dissociated ions from the added salt. Let $\rho_j(\mathbf{r})$ be the charge density at $\mathbf{r}$ due to the counterion cloud surrounding the j-th bead of the chain, as illustrated in Fig. 7.24. In the presence of the electric field $\mathbf{E}_0$, the charge density $\rho(\mathbf{r})$ arising from counterion clouds of all beads exerts an electric force on the fluid element at $\mathbf{r}$ according to

$$\mathbf{F}_{ext}(\mathbf{r}) = \rho(\mathbf{r})\mathbf{E}_0, \tag{A6.75}$$

where

$$\rho(\mathbf{r}) = \sum_{j=0}^{N} \rho_j(\mathbf{r}). \tag{A6.76}$$

Substituting Eq. (A6.75) for $\mathbf{F}_{ext}(\mathbf{r})$ in Eq. (A6.65), we get

$$\langle \sigma_i \rangle = \zeta_b \langle \dot{\mathbf{R}}_i \rangle - \zeta_b \sum_{j \neq i} \langle G(\mathbf{R}_i - \mathbf{R}_j) \rangle \langle \sigma_j \rangle - \zeta_b \int d\mathbf{r} \left\langle G(\mathbf{R}_i - \mathbf{r}) \sum_{j=0}^{N} \rho_j(\mathbf{r}) \right\rangle \cdot \mathbf{E}_0, \tag{A6.77}$$

where the preaveraging approximation is implemented and the solvent velocity in the absence of polymer chains and salt is taken as zero ($\mathbf{v}_i^0 = 0$).

Using the Poisson equation for $\rho_j(\mathbf{r})$ in terms of the electric potential $\psi_j(\mathbf{r})$ around the j-th bead, and the Debye–Hückel theory for $\psi_j(\mathbf{r})$, σ_i in Eq. (A6.77) is calculated. Substitution of the calculated result in Eq. (A6.72) yields the electrophoretic mobility as [Muthukumar (1997, 2005)]

$$\mu = Q \left(\frac{1}{f_t} + \hat{A} \right), \tag{A6.78}$$

where

$$\hat{A} = \frac{1}{N^2} \sum_{i=0}^{N} \sum_{j \neq i} \frac{1}{6\pi\eta_0} \left\langle \frac{e^{-\kappa|\mathbf{R}_i - \mathbf{R}_j|}}{|\mathbf{R}_i - \mathbf{R}_j|} - \frac{1}{|\mathbf{R}_i - \mathbf{R}_j|} \right\rangle. \tag{A6.79}$$

Whereas the first term inside the brackets on the right-hand side of Eq. (A6.78) gives the Einstein electrophoretic mobility, the second term gives the contribution from the counterion clouds around the charged segments of the polymer. Note that $\hat{A}$ is negative due to the motion of the counterion cloud in the opposite direction to that of the polymer, as illustrated in Fig. 7.25a. Furthermore, $\hat{A}$ is zero in the limit of $\kappa \to 0$.

Performing the conformational average in Eq. (A6.79) and combining with Eq. (A6.78) we get [Muthukumar (1997, 2005)]

$$\mu = \frac{2}{3} \frac{Q}{\eta_0 N} \int \frac{d\mathbf{k}}{(2\pi)^3} \frac{S(\mathbf{k})}{(k^2 + \kappa^2)}, \tag{A6.80}$$

where $S(\mathbf{k})$ is the static structure factor for a chain (Section 2.9), with the limits

$$S(k) = \begin{cases} N\left(1 - \frac{k^2 R_g^2}{3} + \cdots\right) & kR_g \ll 1 \\ \sim \frac{1}{(k\ell)^{1/\nu}}, & kR_g \gg 1. \end{cases} \tag{A6.81}$$

Although the integration in Eq. (A6.80) does not provide a closed-form formula, it can be performed analytically in the limits of no added salt ($\kappa \to 0$) and high salt ($\kappa R_g > 1$). The corresponding results for the electrophoretic mobility are

$$\mu \sim Q \times \begin{cases} \frac{1}{\eta_0 R_g}, & \text{low salt} \\ \frac{1}{\eta_0 R_g^{1/\nu} \kappa^{(1-\nu)/\nu}}, & \text{high salt.} \end{cases} \tag{A6.82}$$

A6.5.3 Coupled Diffusion Coefficient

For dilute salt-free polyelectrolyte solutions, the derivation of Eq. (7.6.17) from Eqs. (7.6.14)–(7.6.16) is as follows. Combining Eqs. (7.6.15) and (7.6.16), we get

$$\frac{\partial c_i(\mathbf{r},t)}{\partial t} = D_i \nabla^2 c_i(\mathbf{r},t) - \nabla \cdot [c_i(\mathbf{r},t)\mu_i \mathbf{E}_{\text{loc}}(\mathbf{r},t)]. \tag{A6.83}$$

From the Poisson equation (Eqs. (3.1.6) and (3.2.12))

$$\nabla \cdot \mathbf{E}_{\text{loc}}(\mathbf{r},t) = \frac{\rho(\mathbf{r})}{\epsilon_0 \epsilon}, \tag{A6.84}$$

where $\rho(\mathbf{r})$ is the local charge density

$$\rho(\mathbf{r}) = z_1 e c_1(\mathbf{r},t) + z_2 e c_2(\mathbf{r},t). \tag{A6.85}$$

Combining Eqs. (7.6.14), (A6.84), and (A6.85), we get

$$\nabla \cdot \mathbf{E}_{\text{loc}}(\mathbf{r},t) = \frac{1}{\epsilon_0 \epsilon}\left[\sum_{j=1}^{2} z_j e c_j^0 + \sum_{j=1}^{2} z_j e \delta c_j(\mathbf{r},t)\right]. \tag{A6.86}$$

The first term inside the square brackets is zero due to the electroneutrality condition given in Eq. (7.6.13). Therefore, we get

$$\nabla \cdot \mathbf{E}_{\text{loc}}(\mathbf{r},t) = \frac{e}{\epsilon_0 \epsilon}[z_1 \delta c_1(\mathbf{r},t) + z_2 \delta c_2(\mathbf{r},t)]. \tag{A6.87}$$

Since the diffusion coefficient is experimentally determined by measuring the pair correlation function of fluctuations in local concentration, it is sufficient to consider Eqs. (A6.83) and (A6.87) only to first order in $\delta c_i(\mathbf{r},t)$. Thus, the term $c_i(\mathbf{r},t)\mathbf{E}_{\text{loc}}(\mathbf{r},t)$ becomes $c_i^0 \mathbf{E}_{\text{loc}}(\mathbf{r},t)$ to first order in the concentration fluctuations. Using Eqs. (A6.83) and (A6.87), the time dependence of $\delta c_i(\mathbf{r},t)$, to first order in fluctuations, is

$$\frac{\partial}{\partial t}\delta c_i(\mathbf{r},t) = D_i \nabla^2 \delta c_i(\mathbf{r},t) - c_i^0 \frac{\mu_i e}{\epsilon_0 \epsilon}\sum_{j=1}^{2} z_j \delta c_j(\mathbf{r},t). \tag{A6.88}$$

Thus the fluctuations in the local concentrations of the polyelectrolyte and the counterions, $\delta c_1(\mathbf{r},t)$ and $\delta c_2(\mathbf{r},t)$, are coupled. This completes the derivation of Eq. (7.6.17).

Taking the Fourier transform of $\delta c_i(\mathbf{r},t)$ as

$$\delta \hat{c}_i(\mathbf{k},t) = \int d\mathbf{r}\, \delta c_i(\mathbf{r},t) e^{i\mathbf{k}\cdot\mathbf{r}}, \tag{A6.89}$$

where $\mathbf{k}$ is the scattering wave vector, Eq. (A6.88) becomes

$$\frac{\partial}{\partial t} \delta \hat{c}_i(\mathbf{k},t) = -D_i k^2 \delta \hat{c}_i(\mathbf{k},t) - c_i^0 \frac{\mu_i e}{\epsilon_0 \epsilon} \sum_{j=1}^{2} z_j \delta \hat{c}_j(\mathbf{k},t). \tag{A6.90}$$

The electrophoretic mobility of the polyelectrolyte chain is given by Eq. (7.6.9) and that of the counterion is given by the Einstein relation so that

$$\mu_1 = \frac{z_1 e D_1 \gamma_1}{k_B T}, \qquad \mu_2 = \frac{z_2 e D_2}{k_B T}. \tag{A6.91}$$

Recall that the factor γ_1 arises from the contribution from the counterion cloud of the polymer and generally depends on the concentration of added salt. Using Eq. (A6.91) and defining

$$\kappa_i^2 = \frac{(z_i e)^2 c_i^0}{\epsilon_0 \epsilon k_B T}, \qquad (i = 1,2) \tag{A6.92}$$

the coupled equations for the polyelectrolyte chain ($i = 1$) and the counterion ($i = 2$) follow from Eq. (A6.90) as

$$\frac{\partial}{\partial t} \delta \hat{c}_1(\mathbf{k},t) = -D_1 \left(k^2 + \gamma_1 \kappa_1^2\right) \delta \hat{c}_1(\mathbf{k},t) + D_1 \gamma_1 \kappa_1^2 \left|\frac{z_2}{z_1}\right| \delta \hat{c}_2(\mathbf{k},t) \tag{A6.93}$$

$$\frac{\partial}{\partial t} \delta \hat{c}_2(\mathbf{k},t) = -D_2 \left(k^2 + \kappa_2^2\right) \delta \hat{c}_2(\mathbf{k},t) + D_2 \kappa_2^2 \left|\frac{z_1}{z_2}\right| \delta \hat{c}_1(\mathbf{k},t). \tag{A6.94}$$

Due to the dynamical coupling between the fluctuations $\delta \hat{c}_1$ and $\delta \hat{c}_2$, two new modes with decay rates Γ_1 and Γ_2 emerge which replace the two individual uncoupled diffusion modes (with decay rates $D_1 k^2$ and $D_2 k^2$).

The coupled rates are obtained as the roots of the characteristic determinant of the matrix connecting $\partial \hat{c}_i(\mathbf{k},t)/\partial t$ and $\delta \hat{c}_i(\mathbf{k},t)$, given by

$$\frac{\partial}{\partial t} \begin{pmatrix} \delta \hat{c}_1 \\ \delta \hat{c}_2 \end{pmatrix} = \begin{pmatrix} -D_1 \left(k^2 + \gamma_1 \kappa_1^2\right) & D_1 \gamma_1 \kappa_1^2 \left|\frac{z_2}{z_1}\right| \\ D_2 \kappa_2^2 \left|\frac{z_1}{z_2}\right| & -D_2 \left(k^2 + \kappa_2^2\right) \end{pmatrix} \begin{pmatrix} \delta \hat{c}_1 \\ \delta \hat{c}_2 \end{pmatrix}. \tag{A6.95}$$

The normal modes are obtained by diagonalizing the rate matrix. The normal mode decay rates are given by the roots of the characteristic determinant,

$$\begin{vmatrix} -D_1(k^2 + \gamma_1 \kappa_1^2) - s & D_1 \gamma_1 \kappa_1^2 \left|\frac{z_2}{z_1}\right| \\ D_2 \kappa_2^2 \left|\frac{z_1}{z_2}\right| & -D_2(k^2 + \kappa_2^2) - s \end{vmatrix} = 0,$$

which is quadratic in s,

$$s^2 + s \left[D_1 \left(k^2 + \gamma_1 \kappa_1^2\right) + D_2 \left(k^2 + \kappa_2^2\right)\right] + D_1 D_2 k^2 \left(k^2 + \gamma_1 \kappa_1^2 + \kappa_2^2\right) = 0. \tag{A6.96}$$

In the limit of $k^2 \ll \kappa_1^2, \kappa_2^2$, the two roots are

$$s_- = -\left(D_1 \gamma_1 \kappa_1^2 + D_2 \kappa_2^2\right), \qquad s_+ = -\frac{D_1 D_2 \left(\gamma_1 \kappa_1^2 + \kappa_2^2\right)}{\left(D_1 \gamma_1 \kappa_1^2 + D_2 \kappa_2^2\right)} k^2. \tag{A6.97}$$

Therefore, the fluctuation $\delta\hat{c}_1(\mathbf{k},t)$ decays as a superposition of exponentials of these two decay rates,

$$\delta\hat{c}_1(\mathbf{k},t) = \left(A_1 e^{s_- t} + A_2 e^{s_+ t}\right)\delta\hat{c}_1(\mathbf{k},0). \tag{A6.98}$$

The amplitudes are obtained as follows. Defining the Laplace transform of $\delta\hat{c}_i(\mathbf{k},t)$ for $i = 1,2$ as

$$\delta\tilde{c}_i(\mathbf{k},s) = \int_0^\infty ds\, e^{-st}\delta\hat{c}_1(\mathbf{k},t), \tag{A6.99}$$

and noting that

$$\int_0^\infty dt\, e^{-st}\frac{\partial}{\partial t}\delta\hat{c}_i(\mathbf{k},t) = s\delta\tilde{c}_i(\mathbf{k},s) - \delta\hat{c}_i(\mathbf{k},t=0), \tag{A6.100}$$

the coupled equations become

$$\left[s + D_1(k^2 + \gamma_1\kappa_1^2)\right]\delta\tilde{c}_1(\mathbf{k},s) = \delta\hat{c}_1(\mathbf{k},0) + D_1\gamma_1\kappa_1^2\left|\frac{z_2}{z_1}\right|\delta\tilde{c}_2(\mathbf{k},s)$$

$$\left[s + D_2(k^2 + \kappa_2^2)\right]\delta\tilde{c}_2(\mathbf{k},s) = \delta\hat{c}_2(\mathbf{k},0) + D_2\kappa_2^2\left|\frac{z_1}{z_2}\right|\delta\tilde{c}_1(\mathbf{k},s). \tag{A6.101}$$

Eliminating $\delta\tilde{c}_2(\mathbf{k},s)$ from Eq. (A6.101) in terms of $\delta\tilde{c}_1(\mathbf{k},s)$ and $\delta\hat{c}_2(\mathbf{k},0)$, Eq. (A6.93) yields

$$\delta\tilde{c}_1(\mathbf{k},s) = \frac{1}{\Delta(s)}\left\{\left[s + D_2\left(k^2 + \kappa_2^2\right)\right]\delta\hat{c}_1(\mathbf{k},0) + D_1\gamma_1\kappa_1^2\left|\frac{z_2}{z_1}\right|\delta\hat{c}_2(\mathbf{k},0)\right\}, \tag{A6.102}$$

where $\Delta(s)$ is the determinant in Eq. (A6.96), and as seen above

$$\Delta(s) = (s - s_-)(s - s_+). \tag{A6.103}$$

Multiplying Eq. (A6.102) by $\delta\hat{c}_1(-\mathbf{k},0)$ and taking the thermal average,

$$\langle\delta\tilde{c}_1(\mathbf{k},s)\delta\hat{c}_1(-\mathbf{k},0)\rangle = \frac{(s + D_2\kappa_2^2)}{(s - s_-)(s - s_+)}\langle\delta\hat{c}_1(\mathbf{k},0)\delta\hat{c}_1(-\mathbf{k},0)\rangle, \tag{A6.104}$$

where the cross-correlation between δc_1 and δc_2 is ignored and the limit of $k^2 < \kappa_2^2$ is assumed. Performing the inverse Laplace transform of the above equation, we get Eqs. (7.6.18) and (7.6.19).

In the case of a polyelectrolyte chain and its counterions, the small ions are expected to relax must faster than the larger polyelectrolyte chain. In view of this, the left-hand side of Eq. (A6.94) can be ignored so that

$$\delta\hat{c}_2(\mathbf{k},t) = \frac{\kappa_2^2\left|\frac{z_1}{z_2}\right|}{\left(k^2 + \kappa_2^2\right)}\delta\hat{c}_1(\mathbf{k},t). \tag{A6.105}$$

Substituting this result in Eq. (A6.93),

$$\frac{\partial}{\partial t}\delta\hat{c}_1(\mathbf{k},t) = -D_1 k^2\delta\hat{c}_1(\mathbf{k},t) - D_1\gamma_1\kappa_1^2\left[1 - \frac{\kappa_2^2}{k^2 + \kappa_2^2}\right]\delta\hat{c}_1(\mathbf{k},t). \tag{A6.106}$$

This simplifies to

$$\frac{\partial}{\partial t}\delta\hat{c}_1(\mathbf{k},t) = -D_{\text{fast}}k^2\delta\hat{c}_1(\mathbf{k},t),\qquad(A6.107)$$

where D_{fast} is given by [Muthukumar (1997), (2016b)]

$$D_{\text{fast}} = D_1\left[1 + \frac{\gamma_1\kappa_1^2}{k^2+\kappa_2^2}\right],\qquad(A6.108)$$

which is k-dependent. However, for $k^2 < \kappa_2^2$, D_{fast} becomes

$$D_{\text{fast}} = D_1\left[1 + \frac{\gamma_1\kappa_1^2}{\kappa_2^2}\right],\qquad(A6.109)$$

which is Eq. (7.6.25).

A6.6   Semidilute Polyelectrolyte Solutions

A6.6.1   Scaling Argument for Tracer Diffusion Coefficient

As noted in Eq. (7.7.1), since the dimensionless concentration variable is $c/c^\star$, D_t can be expressed generally as

$$D_t = \frac{k_BT}{6\pi\eta_0 R_h}f_{D_t}\left(\frac{c}{c^\star}\right),\qquad(A6.110)$$

where f_{D_t} is some unknown function. Since $R_h \sim N^\nu$ and $c^\star \sim N^{1-3\nu}$, and since D_t follows the Rouse law $D_t \sim 1/N$ for $c > c^\star$, due to hydrodynamic screening, $f_{D_t}(c/c^\star)$ must be such that

$$D_t \sim N^{-\nu}f_{D_t}\left(cN^{3\nu-1}\right) \sim N^{-1},\qquad c > c^\star.\qquad(A6.111)$$

This condition is satisfied only if $f_{D_t}\left(cN^{3\nu-1}\right)$ is the power law $\left(cN^{3\nu-1}\right)^{(\nu-1)/(3\nu-1)}$. Thus we obtain

$$D_t \sim \frac{1}{N}c^{-(1-\nu)/(3\nu-1)},\qquad c > c^\star.\qquad(A6.112)$$

This scaling result is the same as Eq. (7.7.10) for the salt-free condition ($\nu = 1$) and high salt condition ($\nu = 3/5$).

A6.6.2   Scaling Arguments for Cooperative Diffusion Coefficient

For polymer concentrations above the overlap concentration, chains interpenetrate as sketched in Fig. 7.27b. The diffusion of chain segments that constitute the fluctuations in the local concentration is independent of N. Therefore, we expect D_c to be independent of N for $c > c^\star$. D_c can in general be written as (see Eq. (7.7.2))

$$D_c = \frac{k_BT}{6\pi\eta_0 R_h}f_{D_c}\left(\frac{c}{c^\star}\right),\qquad(A6.113)$$

where f_{D_c} is another unknown function. However this function must be such that $D_c \sim N^0$ for $c > c^\star$. Since $R_h \sim N^\nu$ and $c^\star \sim N^{1-3\nu}$, we get

$$D_c \sim N^{-\nu} f_{D_c} \left(cN^{3\nu-1}\right) \sim N^0. \qquad (c > c^\star) \qquad (A6.114)$$

This proportionality is equivalent to the statement that $f_{D_c} \left(cN^{3\nu-1}\right) \sim \left(cN^{3\nu-1}\right)^{\nu/(3\nu-1)}$. Therefore, the concentration dependence of D_c follows as

$$D_c \sim c^{\frac{\nu}{3\nu-1}}. \qquad (c > c^\star). \qquad (A6.115)$$

This scaling relation is consistent with Eqs. (7.7.23) and (7.7.25), for salt-free limit ($\nu = 1$) and the high-salt limit ($\nu = 3/5$), respectively.

A6.6.3   Coupled Diffusion and Ordinary–Extraordinary Transition

The coupled equations given in Eq. (7.7.31) are solved using the same method presented in Appendix 6.5.3. Using the Poisson equation,

$$\nabla \cdot \mathbf{E}_{\text{loc}}(\mathbf{r},t) = \frac{1}{\epsilon_0\epsilon}\rho(\mathbf{r},t), \qquad (A6.116)$$

where the local charge density $\rho(\mathbf{r})$ is

$$\rho(\mathbf{r},t) = z_1 e c_1(\mathbf{r},t) + z_2 e c_2(\mathbf{r},t), \qquad (A6.117)$$

and the electroneutrality condition (Eq. (7.7.28)), Eq. (7.7.31) becomes to first order in $\delta c_1(\mathbf{r},t)$ and $\delta c_2(\mathbf{r},t)$

$$\frac{\partial}{\partial t}\delta c_i(\mathbf{r},t) = D_i\nabla^2\delta c_i(\mathbf{r},t) - c_i^0\frac{\mu_i e}{\epsilon\epsilon}\sum_{j=1}^{2}z_j\delta c_j(\mathbf{r},t). \qquad (A6.118)$$

Taking the Fourier transform as in Appendix 6.5.3, we get

$$\frac{\partial}{\partial t}\delta\hat{c}_i(\mathbf{k},t) = -D_i k^2\delta\hat{c}_i(\mathbf{k},t) - c_i^0\frac{\mu_i e}{\epsilon_0\epsilon}\sum_{j=1}^{2}z_j\delta\hat{c}_j(\mathbf{k},t). \qquad (A6.119)$$

Using Eqs. (7.7.32) and (7.7.33) for the electrophoretic mobility of segments and counterions, respectively, and Eq. (7.7.35), we obtain the coupled equations for fluctuations in local concentrations of the polymer and counterion as

$$\frac{\partial}{\partial t}\delta\hat{c}_1(\mathbf{k},t) = -D_1\left(k^2 + \gamma_1\kappa_1^2\right)\delta\hat{c}_1(\mathbf{k},t) + D_1\gamma_1\kappa_1^2\left|\frac{z_2}{z_1}\right|\delta\hat{c}_2(\mathbf{k},t) \qquad (A6.120)$$

$$\frac{\partial}{\partial t}\delta\hat{c}_2(\mathbf{k},t) = -D_2\left(k^2 + \kappa_2^2\right)\delta\hat{c}_2(\mathbf{k},t) + D_2\kappa_2^2\left|\frac{z_1}{z_2}\right|\delta\hat{c}_1(\mathbf{k},t). \qquad (A6.121)$$

These equations are identical to Eqs. (A6.93) and (A6.94) derived for dilute solutions, except that D_1 and γ_1 have different expressions in semidilute solutions. The solution of the above two coupled equations gives two coupled relaxation modes, with the same form as Eq. (7.6.18), combined with Eqs. (7.7.32), (7.7.33), and (7.7.35). The consequences are described in Eqs. (7.7.36)–(7.7.42).

Breakdown of Coupled Dynamics with Salt

The coupled equation for the fluctuations in the monomer concentration, counterion concentration, and the two kinds of salt ions is given by Eq. (7.7.43) as

$$\frac{\partial}{\partial t}\delta\hat{c}_i(\mathbf{k},t) = -D_i k^2 \delta\hat{c}_i(\mathbf{k},t) - c_{i,0}\frac{\mu_i e}{\epsilon_0\epsilon}\sum_{j=1}^{4} z_j \delta\hat{c}_j(\mathbf{k},t). \tag{A6.122}$$

With the approximation $\partial\delta c_i(\mathbf{k},t)\partial t = 0$ for $i = 2,3$, and 4, the above equation simplifies to

$$\frac{\partial}{\partial t}\delta\hat{c}_1(\mathbf{k},t) = -D_1 k^2 \delta\hat{c}_1(\mathbf{k},t) - \frac{c_1^0 \mu_1 z_1 e}{\epsilon_0\epsilon}\frac{k^2}{(k^2+\kappa^2)}\delta\hat{c}_1(\mathbf{k},t), \tag{A6.123}$$

where

$$\kappa^2 = \frac{e^2}{\epsilon_0\epsilon k_B T}\left(z_2^2 c_{2,0} + z_3^2 c_{3,0} + z_4^2 c_{4,0}\right). \tag{A6.124}$$

Using $D_1 = D_c, \mu_1 = z_1 e D_{seg}/k_B T, z_1 = \alpha z_p$, and $c_{1,0} = c$, the above equation simplifies to

$$\frac{\partial}{\partial t}\delta\hat{c}_1(\mathbf{k},t) = -\left[D_c + \frac{c(\alpha z_p e)^2}{\epsilon_0\epsilon k_B T}\frac{D_{seg}}{k^2+\kappa^2}\right]k^2\delta\hat{c}_1(\mathbf{k},t). \tag{A6.125}$$

In DLS experiments, k^2 is typically smaller than κ^2 so that the k-dependence inside the square brackets can be ignored. For monovalent salt ions and $z_p = 1$, the above equation becomes

$$\frac{\partial}{\partial t}\delta\hat{c}_1(\mathbf{k},t) = -\left[D_c + \frac{c\alpha^2 D_{seg}}{(\alpha c + 2c_s)}\right]k^2\delta\hat{c}_1(\mathbf{k},t), \tag{A6.126}$$

where we have taken c_3^0 and c_4^0 as the number concentration c_s of added salt, and $c_2^0 = \alpha c$. The consequences of the above equation are discussed in Eqs. (7.7.45)–(7.7.48).

A6.7 Polymer Translocation

A6.7.1 Polymer Capture through a Barrier

The time-evolution of the center of mass position $x(t)$ of the macromolecule is generally given by the Langevin equation,

$$\zeta\frac{dx(t)}{dt} = -\frac{\partial}{\partial x}F(x) + \sqrt{\zeta k_B T}\Gamma(t), \tag{A6.127}$$

where ζ is the translational friction coefficient of the macromolecule if there were no free energy barriers, and $\Gamma(t)$ is the random force acting on the molecule due to the host medium. Let the average random force $\langle\Gamma(t)\rangle$ be zero and the two-point correlation function satisfy the fluctuation-dissipation theorem, $\langle\Gamma(t)\Gamma(t')\rangle = 2\delta(t-t')$. The diffusion coefficient D of the center of mass of the macromolecule in the absence of barriers and external fields is given by the Einstein law $D = k_B T / \zeta$. The first term on the right-hand side of the above equation gives the force on the molecule arising from the free energy landscape and the second term is due to thermal noise leading to diffusion. Using the standard procedures in treating stochastic processes [Risken (1989), Muthukumar (2011)], the probability $P(x,t)$ of finding the center of mass position of the molecule at location x and time t follows from the above Langevin equation as the continuity equation

$$\frac{\partial P(x,t)}{\partial t} = -\frac{\partial}{\partial x} J(x,t), \tag{A6.128}$$

where the flux $J(x,t)$ is given by the Fokker–Planck equation

$$J(x,t) = -D \left[\left(\frac{\partial}{\partial x} \left(\frac{F(x)}{k_B T} \right) \right) P(x,t) + \frac{\partial}{\partial x} P(x,t) \right]. \tag{A6.129}$$

Rewriting this equation for $J(x,t)$, we get

$$J(x,t) = -D e^{-F(x)/k_B T} \frac{\partial}{\partial x} \left[e^{F(x)/k_B T} P(x,t) \right]. \tag{A6.130}$$

In the steady state, the flux J, which is the velocity v of the macromolecule in the present situation, is a constant. Integrating the above equation between $x = 0$ and $x = L$, with $J = v =$ constant, we get

$$v = D \frac{\left[e^{F(x=0)/k_B T} P(x = 0, t) - e^{F(x=L)/k_B T} P(x = 2R_g, t) \right]}{\int_0^L dx e^{F(x)/k_B T}}. \tag{A6.131}$$

Since $P(x = 0, t) = 1$ and $P(x = L, t) = 0$ at early times, and since $F(x = 0) = 0$, we get

$$v = \frac{D}{\int_0^L dx e^{F(x)/k_B T}}. \tag{A6.132}$$

Substituting Eq. (7.11.1) for $F(x)$ in the above equation we get the velocity of the macromolecule in the presence of the barrier and electric field as given in Eq. (7.11.2).

Appendix 7

A7.1   Free Energy of Corona in Spherical Micelles

As sketched in Fig. 8.19, one end of the charged block is localized at the interface and m chains extend radially outward. As a result, the correlation length for monomer concentration ξ depends on the radial distance r from the center of the micelle. Generalizing the scaling laws developed in Section 6.2.3 to local variations in polymer concentration, the r-dependence of ξ is given as

$$\xi(r) \sim \ell v_{\text{eff}}^{-1/4} \phi(r)^{-3/4}, \tag{A7.1}$$

where the monomer volume fraction $\phi(r)$ is the dimensionless local polymer concentration.

Since we are ignoring the various numerical details in evaluating the role of the key experimental variables (N_A, N_B, v, T, polymer concentration, and salt concentration), we present here only the scaling arguments based on the blob picture for ξ and free energy [Halperin (1987)]. As depicted in Fig. 8.19, each polyelectrolyte chain can be imagined to be a succession of blobs, each of size $\xi(r)$, which increases in size as r increases. This construct enables the calculation of the corona thickness L and F_{corona} (which is $k_B T$ times the number of blobs in the corona (see Eq. (6.2.51)). From the spherical geometry, the surface area at r is proportional to the cross-sectional area $\xi^2(r)$ of a blob times the number of chains, so that

$$r^2 \sim m \xi^2(r). \tag{A7.2}$$

The volume of each chain ($N_A \ell^3$) is the total volume of the corona divided by the number of chains,

$$N_A \ell^3 \sim \frac{1}{m} \int_{R_0}^{R} dr \; r^2 \phi(r), \tag{A7.3}$$

where R_0 and R are the radial limits for the corona. Substituting Eq. (A7.1) into Eq. (A7.3), we get

$$N_A \sim \frac{1}{m \ell^3} v_{\text{eff}}^{-1/3} \int_{R_0}^{R} dr \; r^2 \left(\frac{\xi(r)}{\ell} \right)^{-4/3}. \tag{A7.4}$$

Using Eq. (A7.2), we obtain

$$N_A \sim \frac{1}{m\ell^3} v_{\text{eff}}^{-1/3} \int_{R_0}^{R} dr\, r^2 \left(\frac{r}{\ell\sqrt{m}}\right)^{-4/3}, \tag{A7.5}$$

which leads to

$$N_A \sim \frac{1}{m^{1/3} v_{\text{eff}}^{1/3}} \left[\left(\frac{R}{\ell}\right)^{5/3} - \left(\frac{R_0}{\ell}\right)^{5/3}\right]. \tag{A7.6}$$

Inverting this equation, we get

$$\left(\frac{R}{\ell}\right)^{5/3} - \left(\frac{R_0}{\ell}\right)^{5/3} \sim v_{\text{eff}}^{1/3} N_A m^{1/3}. \tag{A7.7}$$

Therefore, the corona thickness $L = R - R_0$ follows as

$$R = R_0 \left[1 + \frac{v_{\text{eff}}^{1/3} N_A m^{1/3}}{(R_0/\ell)^{5/3}}\right]^{3/5}, \tag{A7.8}$$

where the numerical prefactor in front of the second term inside the square brackets is left out. For small values of R_0 in comparison with R, we get the corona thickness L from Eqs. (A7.7) or (A7.8) as

$$L \sim R \sim \ell v_{\text{eff}}^{1/5} N_A^{3/5} m^{1/5}. \tag{A7.9}$$

In general, the free energy of a semidilute solution is the number of blobs times the thermal energy $k_B T$ as discussed in Section 6.2.2.2. Therefore, the free energy of the corona is $k_B T$ times the number of blobs so that

$$\frac{F_{\text{corona}}}{k_B T} \sim \int_{R_0}^{R} dr\, r^2 \frac{1}{\xi(r)^3}. \tag{A7.10}$$

Substituting Eq. (A7.2) into Eq. (A7.10), we get

$$\frac{F_{\text{corona}}}{k_B T} \sim m^{3/2} \ln\left(\frac{R}{R_0}\right). \tag{A7.11}$$

Using Eqs. (8.6.10) and (A7.9) for R_0 and R, we obtain the free energy of the corona per chain as

$$\frac{F_{\text{corona}}}{m k_B T} \sim \sqrt{m} \ln\left(\frac{y}{m^{2/15}}\right), \tag{A7.12}$$

where

$$y = v_{\text{eff}}^{1/5} \tau^{1/3} N_A^{3/5} N_B^{-1/3}. \tag{A7.13}$$

Appendix 8

A8.1 Kinetics of Fibril Elongation

Consider the model portrayed in Fig. 8.32 where monomers (modeled as spherical particles of diameter R and net charge ze) diffuse and drift toward the end of a growing fibril. The end of the fibril is also taken to be spherical with diameter R and charge ze. The diffusing particles approach a distance R from the center of the molecule at the growth end; they are absorbed one at a time. The concentration of the monomers in the bulk solution is c_0 and the solution contains salt so that the inverse Debye length is κ. The continuity equation for the local concentration $c(\mathbf{r},t)$ of the unaggregated molecules at location $\mathbf{r}$ and time t is

$$\frac{\partial c(\mathbf{r},t)}{\partial t} = -\nabla \cdot \mathbf{J}(\mathbf{r},t), \tag{A8.1}$$

where the flux is from both diffusion and drift,

$$\mathbf{J}(\mathbf{r},t) = -D\nabla c(\mathbf{r},t) + c(\mathbf{r},t)\mathbf{v}(\mathbf{r},t), \tag{A8.2}$$

with

$$\mathbf{v}(\mathbf{r},t) = \frac{D}{k_B T}(-\nabla u(\mathbf{r})). \tag{A8.3}$$

Ignoring excluded volume (hydrophobic) interactions, and using the extended Debye–Hückel theory, $u(\mathbf{r})$ is given as (Eq. (3.8.30))

$$\frac{u(\mathbf{r})}{k_B T} = \frac{z^2 \ell_B}{(1 + \kappa R)} \frac{e^{-\kappa(r-R)}}{r}, \tag{A8.4}$$

where r is the magnitude of $\mathbf{r}$. Defining the left-hand side as $\tilde{u}$ we get

$$\tilde{u}(r) = \frac{u(\mathbf{r})}{k_B T} = \gamma z^2 \ell_B \frac{e^{-\kappa r}}{r}; \qquad \gamma = \frac{e^{\kappa R}}{(1 + \kappa R)}. \tag{A8.5}$$

Assuming that the molecules can diffuse throughout the whole space, radial symmetry can be invoked so that Eqs. (A8.1)–(A8.3) become

$$\frac{\partial c(r,t)}{\partial t} = \frac{1}{r^2} \frac{\partial}{\partial r} \left\{ r^2 \left[D \frac{\partial c(r,t)}{\partial r} + D c(r,t) \frac{\partial \tilde{u}(r)}{\partial r} \right] \right\}. \tag{A8.6}$$

In the steady state $\partial c(\mathbf{r},t)/\partial t = 0$, so that we get from the above equation,

$$r^2 \left[D\frac{\partial c(r)}{\partial r} + Dc(r)\frac{\partial \tilde{u}(r)}{\partial r} \right] = \text{constant} = \theta. \tag{A8.7}$$

Therefore,

$$\frac{\partial c(r)}{\partial r} + c(r)\frac{\partial \tilde{u}(r)}{\partial r} = \frac{\theta}{Dr^2}. \tag{A8.8}$$

Using the integrating factor $\exp(\tilde{u}(r))$, the above equation becomes

$$\frac{\partial}{\partial r}\left[e^{\tilde{u}(r)}c(r) \right] = \frac{\theta}{Dr^2}e^{\tilde{u}(r)}. \tag{A8.9}$$

Integration of this equation from $r = R$ gives

$$e^{\tilde{u}(r)}c(r) - e^{\tilde{u}(R)}c(R) = \frac{\theta}{D}\int_R^r \frac{dr'}{(r')^2}e^{\tilde{u}(r')}. \tag{A8.10}$$

According to the absorption boundary condition at $r = R$, we take $c(R) = 0$ so that the concentration profile in front of the fibril follows as

$$c(r) = \frac{\theta}{D}e^{-\tilde{u}(r)}\int_R^r \frac{dr'}{(r')^2}e^{\tilde{u}(r')}. \tag{A8.11}$$

Since $c(r)$ is the bulk concentration c_0 and $\tilde{u}(r) = 0$ at distances far from the fibril front ($r \to \infty$), we get from the above equation

$$c_0 = \frac{\theta}{D}\int_R^\infty \frac{dr'}{(r')^2}e^{\tilde{u}(r')}. \tag{A8.12}$$

Substituting the value of the constant θ from Eq. (A8.12) into Eq. (A8.11), we get

$$c(r) = c_0 e^{-\tilde{u}(r)}\frac{\int_R^r \frac{dr'}{(r')^2}e^{\tilde{u}(r')}}{\int_R^\infty \frac{dr'}{(r')^2}e^{\tilde{u}(r')}}. \tag{A8.13}$$

The flux of the molecules into the sphere at the fibril growth end per unit area is

$$J_R = -D\frac{\partial c(r)}{\partial r}\bigg|_{r=R}. \tag{A8.14}$$

Substituting Eq. (A8.13) into the above equation, we obtain

$$J_R = -\frac{Dc_0}{R^2}\left[\int_R^\infty \frac{dr'}{(r')^2}e^{\tilde{u}(r')} \right]^{-1}. \tag{A8.15}$$

Using Eq. (A8.5), the integral in the above equation becomes

$$\int_R^\infty \frac{dr'}{(r')^2}e^{\tilde{u}(r')} = \int_R^\infty \frac{dr'}{(r')^2}\exp\left[\frac{\gamma z^2 \ell_B e^{-\kappa r'}}{r'} \right]. \tag{A8.16}$$

This integral can be only done numerically. However, for very low ionic strengths such that $\kappa r' \ll 1$, the exponential in the argument of the exponential can be expanded as $\exp(-\kappa r') = 1 - \kappa r' + \cdots$ so that we get from the above equation (to the leading order in κ)

$$\int_R^\infty \frac{dr'}{(r')^2} e^{\tilde{u}(r')} = \int_R^\infty \frac{dr'}{(r')^2} \exp\left[\frac{\gamma z^2 \ell_B}{r} - \gamma z^2 \ell_B \kappa + \cdots\right]. \tag{A8.17}$$

Performing the integral yields

$$\int_R^\infty \frac{dr'}{(r')^2} e^{\tilde{u}(r')} = e^{-\gamma z^2 \ell_B \kappa} \frac{1}{\gamma z^2 \ell_B} \left(e^{\gamma z^2 \ell_B / R} - 1\right). \tag{A8.18}$$

Substituting this result in Eq. (A8.15), we get

$$J_R = -\frac{Dc_0}{R^2} e^{\gamma z^2 \ell_B \kappa} \left(\gamma z^2 \ell_B\right) \left(e^{\gamma z^2 \ell_B / R} - 1\right)^{-1}. \tag{A8.19}$$

The total flux J_{total} into the spherical fibril end monomer of capture radius R is $4\pi R^2$ times J_R so that

$$J_{\text{total}} = -4\pi D R c_0 e^{\gamma z^2 \ell_B \kappa} \left(\frac{\gamma z^2 \ell_B}{R}\right) \left(e^{\gamma z^2 \ell_B / R} - 1\right)^{-1}. \tag{A8.20}$$

The total rate of elongation I_{total} is the negative of the above total flux, as given in Eq. (8.8.13).

Appendix 9

A9.1   Charged Brushes in High-Salt Limit

In the limit of high salt concentrations inside a charged brush, expressions for the free energy and thickness of the brush are derived below based on scaling arguments. As described in Section 9.3.2 and Fig. 9.6b, the bristle length is $N\ell$ and the brush is treated as a semidilute solution of blobs of size ξ. Let g be the number of segments per blob. Therefore, there are N/g blobs per chain. The grafting density and polymer concentration are given in Eq. (9.3.12).

Recall that in semidilute solutions with good solvents, the chain statistics is self-avoiding walk statistics with the Flory exponent $\nu = 3/5$ for distances shorter than the correlation length ξ, and becomes Gaussian chain statistics for distances longer than ξ. Therefore, within each blob containing g segments, we get from Eq. (5.6.5)

$$\xi \sim \ell v_{\text{eff}}^{1/5} g^{3/5}, \qquad (A9.1)$$

where v_{eff} is given by Eq. (8.6.6) for monovalent salt as

$$v_{\text{eff}} = v + \frac{\alpha^2 z_p^2}{2c_s \ell^3}, \qquad (A9.2)$$

and v is the excluded volume parameter $(1 - 2\chi)$. The local polymer volume fraction ϕ is

$$\phi = \frac{g\ell^3}{\xi^3} \sim v_{\text{eff}}^{-1/3} \left(\frac{\xi}{\ell}\right)^{-4/3}. \qquad (A9.3)$$

Inverting this equation, we get the concentration dependence of the correlation length ξ as

$$\xi \sim \ell v_{\text{eff}}^{-1/4} \phi^{-3/4}. \qquad (A9.4)$$

This result is the same as Eqs. (6.2.45) and (A7.1) derived previously in the discussion of semidilute solutions and corona of star-like micelles.

The brush height L is given by the product of the number of blobs per chain and the linear blob size as

$$L = \left(\frac{N}{g}\right)\xi. \qquad (A9.5)$$

Combining Eqs. (9.3.12), (A9.1), and (A9.5), we get

$$L \sim \ell v_{\text{eff}}^{1/3} \sigma_g^{1/3} N. \tag{A9.6}$$

The free energy F of the brush is the number of blobs times the thermal energy $k_B T$ (see Section 6.2.2.2), so that

$$\frac{F}{k_B T} = \frac{N}{g}. \tag{A9.7}$$

Using Eqs. (9.3.12), (A9.1), and (A9.7), we get

$$\frac{F}{k_B T} \sim N v_{\text{eff}}^{1/3} \sigma_g^{5/6}. \tag{A9.8}$$

Appendix 10

A10.1   Genome Packaging in Viruses

As described in Section 9.4, consider a zeroth-order model for icosahedral ssRNA viruses. The capsid is assumed to be a spherical polyelectrolyte brush with n_h bristles (capsid protein tails), each with N_h segments of charge ez_h, protruding into the interior of the capsid. The total positive charge on the bristle is uniformly distributed along the backbone. The ssRNA genome, that complexes with the protein tails, is taken as a flexible polyelectrolyte chain of N segments, each of segment length ℓ and charge ez_p ($z_p < 0$). The excluded volume and electrostatic interactions between the protein tails and the RNA are present.

The free energy of the present model system is derived by generalizing Eq. (6.3.1) to the presence of the negatively charged guest genome chain with the constraint that one end of the bristles is permanently anchored at the internal surface of the capsid wall. For the sake of simplicity, assume that the capsid surface is planar around the anchoring location of a protein tail. This assumption enables the description of the probability distribution function for a tail as a one-dimensional problem in the x-coordinate normal to the surface as in Section 9.3. However, the effective potential acting on every segment of the tail emerges from all species including small ions and the genome in the 3-dimensional brush. By repeating the field theoretic method described in Section 6.3, the probability distribution function $G_h[x, x_0; N_h; \Phi]$ for each cationic bristle follows as

$$
G_h[x, x_0; N_h; \Phi] = \int_{x_0}^{x N_h} \mathcal{D}[x(s)]
$$
$$
\exp\left[-\frac{3}{2\ell_h^2} \int_0^{N_h} ds \left(\frac{\partial x(s)}{\partial s}\right)^2 - \int_0^{N_h} ds z_h \Phi[x(s)]\right], \quad (A10.1)
$$

where the ends of the chain are at x_0 and x. $\Phi[x(s)]$ is the net potential at position $x(s)$, generated by all interactions in the system including short-ranged and long-ranged interactions and independent of whether the genome is present or not. We shall self-consistently determine Φ as shown below.

Analogous to Eq. (A10.1), the distribution function $G[\mathbf{r}, \mathbf{r}_0; N; \Phi]$ of the genome existing in three-dimensional space of the brush is

$$G[\mathbf{r},\mathbf{r}_0; N; \Phi] = \int_{\mathbf{r}_0}^{\mathbf{r}} \mathcal{D}[\mathbf{r}(s)] \exp\left[-\frac{3}{2\ell^2} \int_0^N ds \left(\frac{\partial \mathbf{r}(s)}{\partial s}\right)^2 - \int_0^N ds z_p \Phi[\mathbf{r}(s)]\right].$$

$$(A10.2)$$

The rest of the calculation has two major steps. The first is the determination of the potential Φ in Eqs. (A10.1) and (A10.2). The second step is to use the derived potential Φ in Eq. (A10.1), where the sign of z_p is opposite to that of z_h, and determine the critical condition for complexation between the genome and protein brush. We shall adopt the quantum mechanical analogy outlined in Section 9.2 in obtaining the optimal condition for complexation.

The situation of Eq. (A10.1) is equivalent to a classical system with the Lagrangian $\mathcal{L}$,

$$\mathcal{L} = T - V, \tag{A10.3}$$

where

$$T = -\frac{3}{2\ell_h^2}\left(\frac{\partial x}{\partial s}\right)^2, \qquad V(x) = z_h \Phi[x(s)]. \tag{A10.4}$$

The equation of conservation of energy for the classical system is [Young & Freedman (2000)]

$$T + V = \text{constant}. \tag{A10.5}$$

At the free end of the bristle, $x(s = N_h) = x_{N_h}$, the force is zero so that

$$\left(\frac{dx_{N_h}}{ds}\right) = 0. \tag{A10.6}$$

This boundary condition is used to determine the constant term in Eq. (A10.5) as $V(x_{N_h})$. Therefore, Eqs. (A10.4) and (A10.5) give

$$V(x) - V(x_{N_h}) = \frac{3}{2\ell_h^2}\left(\frac{\partial x}{\partial s}\right)^2, \tag{A10.7}$$

where $V(x)$ is still unknown.

The potential $\Phi(x) = V(x)/z_h$ is determined by using the constraint that the number of segments in a bristle is a fixed number N_h,

$$N_h = \int_0^{x_{N_h}} dx \frac{\partial s}{\partial x}. \tag{A10.8}$$

In other words, upon substitution of Eq. (A10.7) for the integrand in Eq. (A10.8), the right-hand side integral must be N_h. Substituting Eq. (A10.7) in Eq. (A10.8) gives

$$N_h = \sqrt{\frac{3}{2}}\frac{1}{\ell_h} \int_0^{x_{N_h}} \frac{dx}{\sqrt{V(x) - V(x_{N_h})}}. \tag{A10.9}$$

The self-consistent result for the x-dependence of $V(x)$ is obtained by insisting that the right-hand side of this equation is exactly N_h. The only unique solution is [Belyi & Muthukumar (2006), Muthukumar (2012b)]

$$V(x) = \text{constant} - \frac{3\pi^2}{8\ell_h^2 N_h^2} x^2. \tag{A10.10}$$

The parabolic profile for $V(x)$ is universal independent of whether the system is charged or uncharged and whether the genome is present or absent. A convenient choice of the constant term in Eq. (A10.10) is made by letting the potential to be zero at the brush height H so that

$$V(x) = \frac{3\pi^2}{8\ell_h^2 N_h^2} \left(H^2 - x^2 \right). \tag{A10.11}$$

Hence the potential $\Phi(x)$ follows from Eqs. (A10.4) and (A10.11) as

$$\Phi(x) = \left(\frac{H}{H_0} \right)^2 - \left(\frac{x}{H_0} \right)^2, \tag{A10.12}$$

where

$$H_0^2 = \frac{8 z_h \ell_h^2 N_h^2}{3\pi^2}. \tag{A10.13}$$

The above derivation completes the first step of determining the potential Φ.

In the second step, we can use the derived potential in Eq. (A10.2) for the genome. Since the potential depends only on x (and not the three-dimensional coordinate), $G[\mathbf{r}, \mathbf{r}_0; N; \Phi]$ can be written as the product of the two-dimensional Gaussian chain distribution function and the one-dimensional distribution function $G_1(x, x_0; N; \Phi)$ (see Eq. (9.2.17)), where the latter is given in the form of time-dependent Schrödinger equation

$$\left(\frac{\partial}{\partial N} - \frac{\ell^2}{6} \frac{\partial^2}{\partial x^2} + z_p \Phi \right) G_1(x, x_0; N; \Phi) = \delta(x - x_0)\delta(N). \tag{A10.14}$$

Since the genome is negatively charged, $z_p = -|z_p|$, the potential term in the above equation is negative. Expressing $G_1(x, x_0; N; \Phi)$ as a bilinear expansion,

$$G_1(x, x_0; N; \Phi) = \sum_{m=0}^{\infty} \psi_m(x)\psi_m(x_0)e^{-\lambda_m N}, \tag{A10.15}$$

we get

$$\left(-\frac{\ell^2}{6} \frac{\partial^2}{\partial x^2} + z_p \Phi \right) \psi_m(x) = \lambda_m \psi_m(x). \tag{A10.16}$$

Substituting Eq. (A10.12) for Φ in Eq. (A10.16), we get

$$\frac{\partial^2 \psi_m(x)}{\partial x^2} - \omega^2 x^2 \psi_m(x) = -\frac{6}{\ell^2} \left(\lambda_m + \frac{|z_p|H^2}{H_0^2} \right) \psi_m(x), \tag{A10.17}$$

where

$$\omega^2 = \frac{6}{\ell^2} \frac{|z_p|}{H_0^2}.$$ (A10.18)

The above equation for $\psi_m(x)$ is the same as that for a quantum harmonic oscillator and its solution is well-taught in textbooks on quantum mechanics. Since the protein tail exists only for $x > 0$, and since it does not extend too far from the capsid wall, we use the boundary conditions, $\psi_m(x = 0) = 0$ and $\psi_m(x \rightarrow \infty) = 0$, to solve Eq. (A10.17). Furthermore, as we have seen in Section 9.2, the ground state dominance approximation is adequate if the genome length $N\ell$ is large. The ground state solutions for the harmonic oscillator of Eq. (A10.17) with the above mentioned boundary conditions are

$$\psi_0(x) = \frac{\sqrt{2}}{\pi^{1/4}} \omega^{3/4} x \exp\left[-\frac{\omega x^2}{2}\right],$$ (A10.19)

$$\lambda_0 = \frac{\ell^2 \omega}{2} - \frac{|z_p|H^2}{H_0^2}.$$ (A10.20)

Using the GSD approximation, the density profile of the genome inside the capsid is given by Eqs. (9.2.7) and (A10.19) as

$$\rho(x) = N\psi_0^2(x) = \frac{2N}{\sqrt{\pi}} \omega^{3/2} x^2 \exp\left[-\omega x^2\right].$$ (A10.21)

Therefore, the dependence of the density profile of the genome on the distance from the capsid wall follows as

$$\rho(x) \sim x^2 e^{-(x/x_{\max})^2},$$ (A10.22)

where the location of the maximum genome density $x_{\max}$ is given by

$$x_{\max} = \frac{1}{\sqrt{\omega}},$$ (A10.23)

which depends on N_h, z_h, and z_p through Eqs. (A10.13) and (A10.18).

The relation between the packaged genome length and the net charge on the protein tails is obtained from the Poisson equation, Eq. (3.8.2). In the brush region, the Poisson equation yields

$$-S\epsilon_0\epsilon k_B T \int_0^H dx \frac{d^2\Phi(x)}{dx^2} = n_h z_h N_h + z_p N + S \sum_{\gamma=+,-} \int_0^H dx z_\gamma \rho_\gamma(x),$$ (A10.24)

where S is the total surface area and the integrals correspond to the one-dimensional region of a single bristle inside the brush. The right-hand side of Eq. (A10.24) gives the total charge in the brush region. The index γ denotes the cations and anions of the electrolyte solution inside the capsid, and $\rho_\gamma(x)$ is the number concentration of the γ-th type electrolyte ion at the location x inside the brush. Performing the integrals in the above equation yields [Muthukumar (2012b)]

$$z_p N + n_h z_h N_h = O\left(\frac{H}{H_0}\right). \tag{A10.25}$$

Since the brush height is generally smaller than the extended bristle length $N_h \ell_h$ before complexation, the right-hand side can be ignored. With this approximation, the relation between the packaged genome length and the total charge on the protein tails emerges from the above equation as an effective electroneutrality equation,

$$\left|z_p\right| N \simeq n_h z_h N_h. \tag{A10.26}$$

Therefore, the optimal condition for genome packaging is that the net negative charge of the packaged genome balances the net positive charge on all protein tails inside the virus, as discussed in Section 9.4.

References

Abramowitz, M. and Stegun, I. A. (1965). *Handbook of Mathematical Functions*. New York: Dover.

Acheson, N. H. (2007). *Fundamentals of Molecular Virology*. New York: John Wiley & Sons.

Adhikari, S., Leaf, M. A. and Muthukumar, M. (2018). Polyelectrolyte complex coacervation by electrostatic dipolar interactions. *J. Chem. Phys.*, **149**, 163308.

Adhikari, S., Prabhu, V. M. and Muthukumar, M. (2019). Lower critical solution temperature behavior in polyelectrolyte complex coacervates. *Macromolecules*, **52**, 6998–7004.

Alberts, B., Johnson, A., Lewis, J., Morgan, D., Raff, M., Roberts, K. and Walter, P. (2015). *Molecular Biology of the Cell*, 6th edn. New York: Garland Science.

Alexander, S. (1977). Adsorption of chain molecules with a polar head a scaling description. *J. Phys. France*, **38**, 983–987.

Alfrey, T., Berg, P. W. and Morawetz, H. (1951). The counterion distribution in solutions of rod-shaped polyelectrolytes. *J. Polym. Sci.* **7**, 543–547.

Alfrey, T., Fuoss, R. M., Morawetz, H. and Pinner, H. (1952). Amphoteric polyelectrolytes. II. Copolymers of methacrylic acid and diethylaminoethyl methacrylate. *J. Am. Chem. Soc.*, **74**, 438–441.

Antipova, O. and Orgel, J. P. R. O. (2010). In situ D-periodic molecular structure of type II collagen. *J. Biol. Chem.*, **285**, 7087–7096.

Baker, T. S., Olson, N. H. and Fuller. S. D. (1999). Adding the third dimension to virus cycles: Three-dimensional reconstruction of icosahedral viruses from cryo-electron micrographs. *Microbiol. Mol. Biol. Rev.*, **63**, 862–922.

Baltimore, D. (1970). Viral RNA-dependent DNA polymerase: RNA-dependent DNA polymerase in virions of RNA tumour viruses. *Naturè*, **226**, 1209–1211.

Banani, S. F., Lee, H. O., Hyman, A. A. and Rosen, M. K. (2017). Biomolecular condensates: organizers of cellular biochemistry. *Nat. Rev. Mol. Cell Biol.* **18**, 285–298.

Banavar, J.R., Hong, T.X. and Maritan, A. (2005). Proteins and polymers. *J. Chem. Phys.***122**, 234910.

Barthel, J. M. G., Krienke, H. and Kunz, W. (1998). *Physical Chemistry of Electrolyte Solutions*. New York: Springer.

Beer, M., Schmidt, M. and Muthukumar, M. (1997). The electrostatic expansion of linear polyelectrolytes: Effects of gegenions, co-ions, and hydrophobicity. *Macromolecules*, **30**, 8375–8385.

Belyi, V. A. and Muthukumar, M. (2006). Electrostatic origin of the genome packing in viruses. *Proc. Natl. Acad. Sci. USA.*, **103**, 17174–17178.

Bemporad, F., Calloni, G., Campioni, S., Plakoutsi, G., Taddei, N. and Chiti, F. (2006). Sequence and structural determinants of amyloid fibril formation. *Acc. Chem. Res.*, **39**, 620–627.

Berne, B. J. and Pecora, R. (1976). *Dynamic Light Scattering*. New York: John Wiley & Sons.

Berry, R. S., Rice, S. A. and Ross, J. (2000). *Physical Chemistry*. Oxford: Oxford University Press.

Besteman, K., Eijk, K.V. and Lemay, S.G. (2007). Charge inversion accompanies DNA condensation by multivalent ions. *Nat. Phys.*, **3**, 641–644.

Bluhm, T. L. and Whitmore, M. D. (1985). Styrene/butadiene block copolymer micelles in heptane. *Can. J. Chem.* **63**, 249.

Bockris, J. O. and Reddy, A. K. N. (1970). *Modern Electrochemistry 1*. New York: Plenum Press.

Bordi, F., Cametti, C. and Colby, R. H. (2004). Dielectric spectroscopy and conductivity of polyelectrolyte solutions. *J. Phys.: Condens. Matter*, **16**, R1423–R1463.

Borgia, A., Borgia, M. B., Bugge, K., Kissling, V. M., Heidarsson, P. O., Fernandes, C. B., Sottini, A., Soranno, A., Buholzer, K. J., Nettels, D., Kragelund, B. B., Best, R. B. and Schuler, B. (2018). Extreme disorder in an ultrahigh-affinity protein.*Nature*, **555**, 61–66.

Borisov, O. V. and Zhulina, E. B. (2002). Effect of salt on self-assembly in charged block copolymer micelles. *Macromolecules*, **35**, 4472–4480.

Boyd, R. H. and Phillips, P. J. (1993). *The Science of Polymer Molecules*. Cambridge: Cambridge University Press.

Bragg, W. L. and Williams, E. J. (1934). The effect of thermal agitation on atomic arrangement in alloys. *Proc. Roy. Soc. London,* **145A**, 699–730.

Bragg, W. L. and Williams, E. J. (1935). The effect of thermal agitation on atomic arrangement in alloys. II. *Proc. Roy. Soc. London,* **151A**, 540–566.

Brilliantov, N. V., Kuznetsov, D. V. and Klein, R. (1998). Chain collapse and counterion condensation in dilute polyelectrolyte solutions. *Phys. Rev. Lett.*, **81**, 1433–1436.

Brinkman, H. C. and Hermans, J. J. (1949). The effect of non-homogeneity of molecular weight on the scattering of light by high polymer solutions. *J. Chem. Phys.*, **17**, 574–576.

Brinkers, S., Dietrich, H. R. C., de Groote, F. H., Young, I. T. and Rieger, B. (2009). The persistence length of double stranded DNA determined using dark field tethered particle motion. *J. Chem. Phys.*, **130**, 215105.

Brown, W., 1993. *Dynamic Light Scattering*. Oxford: Clarendon Press.

Budkov, Y. A., Kolesnikov, A. L., Georgi, N., Nogovitsyn, E. A. and Kiselov, M. G. (2015). A new equation of state of a flexible-chain polyelectrolyte solution: Phase equilibria and osmotic pressure in the salt-free case, *J. Chem. Phys.*, **142**, 174901.

Buell, A. K., Blundell, J. R., Dobson, C. M., Welland, M. E., Terentjev, E. M. and Knowles, T. P. J. (2010). Frequency factors in a landscape model of filamentous protein aggregation. *Phys. Rev. Lett.*, **104**, 228101.

Buell, A. K., Hung, P., Salvatella, X., Welland, M. E., Dobson, C. M. and Knowles, T. P. J. (2013). Electrostatic effects in filamentous protein aggregation. *Biophys. Journal*, **104**, 1116–1126.

Buell, A. K., Galvagnion, C., Gasper, R., Sparr, E., Vendruscoio, M., Knowles, T. P.J. and Dobson, C. M. (2014). Solution conditions determine the relative importance of nucleation and growth processes in α-synuclein aggregation. *Proc. Natl. Acad. Sci.* (USA), **111**, 7671–7676.

Bungenberg de Jong, H. G. and Kruyt, H. R. (1929). Coacerrvation (Partial miscibility in colloid systems). *Proc. K. Ned. Akad. Wet.*, **32**, 849–855.

Cadena-Nava, R. D., Comas-Garcia, M., Garmann, R. F., Rao, A. L. N., Knobler, C. M. and Gelbart, W. M. (2012). Self-assembly of viral capsid protein and RNA molecules of different sizes: Requirement for a specific high protein/RNA mass ratio,. *Virol.*, **86**, 3318–3326.

Calladine, C. R., Collis, C. M., Drew, H. R. and Mott, M. R. (1991). A study of electrophoretic mobility of DNA in agarose an polyacrylamide gels. *J. Mol. Biol.*, **221**, 981–1005.

Candau, S., Bastide, J. and Delsanti, M., (1982). Structural, Elastic, and Dynamic Properties of Swollen Polymer Networks. *Adv. Polym. Sci.*, **44**, 27–71.

Carri, G. A. and Muthukumar, M. (1999). Attractive interactions and phase transitions in solutions of similarly charged rod-like polyelectrolytes. *J. Chem. Phys.* **111**, 1765–1777.

Chandrasekhar, S. (1943). Stochastic processes in physics and astronomy. *Rev. Mod. Phys.*, **15**, 1–89.

Chandrasekhar, S. (1992). *Liquid Crystals*. Cambridge: Cambridge University Press.

Chen, K., Jou, I., Ermann, N., Muthukumar, M., Keyser, U. F. and Bell, N. A. W. (2021). Dynamics of driven polymer transport through a nanopore. *Nat. Phys.*, **17**, 1043–1049.

Chen, K. and Muthukumar, M. (2021). Entropic barrier of topologically immobilized DNA in hydrogels. *Proc. Natl. Acad. Sci. USA*, **118**, https://doi.org/10.1038/s41567-021-01268-2.

Cherstvy, A. G. and Winkler, R. G. (2011). Polyelectrolyte adsorption onto oppositely charged interfaces: Unified approach for plane, cylinder, and sphere. *Phys. Chem. Chem. Phys.*, **13**, 11686–11693.

Choi, J.-M., Holehouse, A. S. and Pappu, R. V. (2020a). Physical principles underlying the complex biology of intracellular phase transitions. *Annu. Rev. Biophys.*, **49**, 107–133.

Choi, J.-M., Hyman, A. A. and Pappu, R. V. (2020b). Generalized models for bond percolation transitions of associative polymers. *Phys. Rev. E.* **102**, 042403.

Chremos, A. and Dougla, J. F. (2017). Communication: Counter-ion solvation and anomalous low-angle scattering in salt-free polyelectrolyte solutions. *J. Chem. Phys.*, **147**, 241103.

Chu, B. (1991). *Laser Light Scattering*. Boston: Academic Press.

Cohen, J., Priel, Z. and Rabin, Y. (1988). Viscosity of dilute polyelectrolyte solutions. *J. Chem. Phys.*, **88**, 7111.

Colby, R. H., Boris, D. C., Krause, W. E. and Tan, J. S. (1997). Polyelectrolyte conductivity. *J. Polym. Sci. Part B: Polym. Phys.*, **35** 2951–2960.

Conway, B. E., Bockris, J. O' M. and Ammart, I. A. (1951). The dielectric constant of the solution in the diffuse and Helmholtz double layers at a charged interface in aqueous solution. *Trans. Faraday Soc.*, **47**, 756–766.

Conway, J. F. Steven, A. C. (1999). Methods for reconstructing density maps of "single" particles from cryoelectron micrographs to subnanometer resolution. *J. Struct. Biol.*, **128**, 106–118.

Cotton, J. P., Decker, D., Benoit, H., Farnoux, B., Higgins, J., Jannink, G., Ober, R., Picot, C., and des Cloizeaux, J. (1974). Conformation of polymer chain in the bulk. *Macromolecules*, **7**, 863–872.

Creighton, T. E. (1993). *Proteins*. New York: W. H. Freeman and Company.

Crick, F. (1970). Central dogma of molecular biology. *Nature*, **227**, 561–563.

Dan, N. and Tirrell, M. (1993). Self-assembly of block copolymers with a strongly charged and a hydrophobic block in a selective, polar solvent. Micelles and adsorbed layers. *Macromolecules*, **26**, 4310–4315.

Daoud, M., Cotton, J. P., Farnoux, B., Jannink, G., Sarma, G., Benoit, H., Duplessix, R., Picot, C. and de Gennes, P. G. (1975). Solutions of flexible polymers. Neutron experiments and interpretation. *Macromolecules*, **8**, 804–818.

Dar, F. and Pappu, R. V. (2020). Phase separation: Restricting the sizes of condensates. *Elife*, **9**, e59663.

Das, R. K. and Pappu, R. V. (2013). Conformations of intrinsically disordered proteins are influenced by linear sequence distributions of oppositely charged residues. *Proc. Natl. Acad. Sci. USA*, **110**, 13392–13397.

Das, R. K., Ruff, K. M. and Pappu, R. V. (2015). Relating sequence encoded information to form and function of intrinsically disordered proteins. *Current Opinion in Structural Biology*, **32**, 102–112.

Das, S. and Muthukumar, M. (2022). Microstructural organization in α-synuclein solutions. *Macromolecules*, **55**, 4228–4236.

Daune, M. (1999). *Molecular Biophysics*. Oxford: Oxford University Press.

Dautzenberg, H., Jaeger, W., J. Kotz, B. P., Seidel, C. and Stscherbina, D. (1994). *Polyelectrolytes*. New York: Hanser Publishers.

Debenedetti, P. G. (1996) *Metastable Liquids*. Princeton: Princeton University Press.

Debye, P. (1925). Marx' Handbuch der Radiologie. *Akademische Verlagsgesellschaft, Liepzig*, **6**, 618, 680.

Debye, P. (1928). *Polar Molecules*. USA: The Chemical Catalogue Company.

de Gennes, P. G. (1968). Statistics of branching and hairpin helices for the dAT copolymer. *Biopolymers* **6**, 715–729.

de Gennes, P. G. (1971). Reptation of a polymer chain in the presence of fixed obstacles. *J. Chem. Phys.*, **55**, 572–579.

de Gennes, P. G. (1979). *Scaling Concepts in Polymer Physics*. Ithaca: Cornell University Press.

de Gennes, P. G., Pincus, P., Velasco, R. M. and Brochard, F. (1976). Remarks on polyelectrolyte conformation. *J. Phys. France*, **37**, 1461–1473.

de Gennes, P. G. & Prost, J. (1993). *The Physics of Liquid Crystals*. Oxford: Clarendon Press.

de la Cruz, M. O., Belloni, L., Delsanti, M., Dalbiez, J. P., Spalla, O. and Drifford, M. (1995). Precipitation of highly charged polyelectrolyte solutions in the presence of multivalent salts. *J. Chem. Phys.*, **103**, 5781–5791.

des Closeaux, J. and Jannink, G. (1990). *Polymers in Solution*. Oxford: Clarendon Press.

Dobrynin, A. V., Colby, R. H. and Rubinstein, M. (1995). Scaling theory of polyelectrolyte solutions. *Macromolecules*, **28**, 1859–1871.

Dobrynin, A. V., Colby, R. H. and Rubinstein, M. (2004). Polyampholytes. *J. Polym. Sci. Part B: Polym. Phys.*, **42**, 3513–3538.

Dobrynin, A. V. and Rubinstein, M. (2005). Theory of polyelectrolytes in solutions and at surfaces. *Prog. Polym. Sci.*, **30**, 1049–1118.

Doi, M. and Edwards, S. F. (1986). *The Theory of Polymer Dynamics*. Oxford: Clarendon Press.

Dou, S. and Colby, R. H. (2006). Charge density effects in salt-free polyelectrolyte solution rheology. *Journal of Polymer Science: Part B: Polymer Physics*, **44**, 2001–2013.

Drifford, M. and Dalbiez, J-P. (1984). Light scattering by dilute solutions of salt-free polyelectrolytes, *J. Phys. Chem.*, **88**, 5368–5375.

Drifford, M. and Dalbiez, J. P. (1985). Effect of salt on sodium polystyrene sulfonate measured by light scattering. *Biopolymers*, **24**, 1501–1514.

Dusek, K. and Patterson, D., (1968). Transition in swollen polymer networks induced by intramolecular condensation, *J. Polym. Sci: Part A-2*, **6**, 1209–1216.

Dyson, H. J. and Wright, P. E. (2005). Intrinsically unstructured proteins and their functions. *Nat. Rev. Mol. Cell Biol.*, **6**, 197–208.

Edwards, S. F. (1966). The theory of polymer solutions at intermediate concentration, *Proc. Phys. Soc.*, **88**, 265–280.

Edwards, S. F. (1967). The statistical mechanics of polymerized material. *Proc. Phys. Soc. (London)*, **92**, 9–16.

Edwards, S. F. and Freed, K. F. (1974).Theory of the dynamical viscosity of polymer solutions. *J. Chem. Phys.,* 61, 1189–1202.

Ehrlich, G. and Doty, P. (1954). Macro-ions. III. The solution behavior of a polymeric ampholyte. *J. Am. Chem. Soc.*, **76**, 3764–3777.

Einstein, A. (1956). *Investigations on the Theory of Brownian Movement*, New York: Dover.

Eisenberg, H. and Mohan, G. R. (1959). Aqueous solutions of polyvinylsulfonic acid: Phase separation and specific interaction with ions, viscosity, conductance and potentiometry. *J. Phys. Chem.* **63**, 671–680.

Eisenberg, H. and Casassa, E. F. (1960). Aqueous solutions of salts of poly(vinylsulfonic acid). *J. Polym. Sci.* **47**, 29–44.

Eisenberg, D., Nelson, R., Sawaya, M. R., Balbirnie, M., Sambashivan, S., Ivanova, M. I., Madsen, A. O. and Riekel, C. (2006). The structural biology of protein aggregation diseases: Fundamental questions and some answers. *Acc. Chem. Res.*, **39**, 568–575.

Ellison, W. J., Lamkaouchi, K. and Moreau, J. -M. (1996). Water: A dielectric reference. *J Mol. Liq.*, **68**, 171–279.

England, J. L. and Haran, G. (2010). To fold or expand—a charged question. *Proc. Natl. Acad. Sci. USA*, **107**, 14519–14520.

Essafi, W., Lafuma, F. and Williams, C. E. (1995). Effect of solvent quality on the behavior of highly charged polyelectrolytes. *J. Phys.II France*, **5**, 1269–1275.

Evans, D. F. and Wennerström. (1999). *The Colloidal Domain*. New York: Wiley-VCH.

Everaers, R., Grosberg, A. Y., Rubinstein, M. and Rosa, A. (2017). Flory theory of randomly branched polymers. *Soft Matter*, **13**, 1223–1234.

Faxén, H. (1922). Der widerstand gegen bewegung einer starren kugel in einer zaehen. *Ann. Phys.*, **373**, 89–119.

Ferry, J. D. (1936). Statistical evaluation of sieve constants in ultrafiltration. *J. Gen. Physiol.*, **20**, 95–104.

Fink, A. L. (2006). The aggregation and fibrillization of α-synuclein. *Acc. Chem. Res.*, **39**, 628–634.

Finkelstein, A. V. and Ptitsyn, O. B. (2002). *Protein Physics*. Amsterdam: Academic Press.

Firman, T. and Ghosh, K. (2018). Sequence charge decoration dictates coil-globule transition in intrinsically disordered proteins. *J. Chem. Phys.* **148**, 123305.

Fisher, M. E. and Levin, Y. (1993). Criticality in ionic fluids: Debye-Hückel theory, Bjerrum, and beyond. *Phys. Rev. Lett.* **71**, 3826–3829.

Fitzkee, N. Z. and Rose, G. D., (2004). Reassessing random-coil statistics in unfolded proteins, *Proc. Natl. Acad. Sci. USA*, **101**, 12497–12502.

Fleer, G. J., Cohen Stuart, M. A., Scheutjens, J. M. H. M., Cosgrove, T. and Vincent, B. (1993). *Polymers at Interfaces*, London: Chapma & Hall.

Flory, P. J. (1942). Thermodynamics of high polymer solutions. *J. Chem. Phys.*, **10**, 51- 61.

Flory, P. J. (1949). The configuration of real polymer chains.*J. Chem. Phys.,* **17**, 303–310.

Flory, P. J. (1953a). *Principles of Polymer Chemistry*. Ithaca: Cornell University Press.

Flory, P. J. (1953b). Molecular configuration of polyelectrolytes. *J. Chem. Phys.* **21**, 162–163.

Flory, P. J. (1956). Phase equilibria in solutions of rod-like particles. *Proc. R. Soc.* **A234**, 73–89.

Flory, P. J. (1969). *Statistical Mechanics of Chain Molecules*. New York: John Wiley & Sons.

Flory, P. J. and Rehner, J. (1943). Statistical mechanics of crosslinked polymer networks ii. swelling, *J. Chem. Phys.*, **11**, 521.

Flory, P. J. and Krigbaum, W. R. (1950). Statistical mechanics of dilute polymer solutions. II. *J. Chem. Phys.*, **18**, 1086–1094.

Förster, S., Schmidt, M. and Antonietti, M. (1990). Static and dynamic light scattering by aqueous polyelectrolyte solutions: Effect of molecular weight, charge density and added salt. *Polymer* **31**, 781–792.

Förster, S. and Schmidt, M. (1995). Polyelectrolytes in solution. *Adv. Polym. Sci.* **120**, 51–133.

Fowler, R. H. (1966). *Statistical Mechanics*. Cambridge: Cambridge University Press.

Fraden, S., Maret, G., Casper, D. L. D. and Meyer, R. B. (1989). Isotropic-nematic phase transition and angular correlations in isotropic suspensions of tobacco mosaic virus. *Phys. Rev. Lett.*, **63**, 2068–2071.

Fraden, S., Maret, G. and Casper, D. L. D. (1993). Angular correlations and the isotropic-nematic phase transition in suspensions of tobacco mosaic virus. *Phys. Rev. E*, **48**, 2816–2837.

Franks, F. (2000). *Water a Matrix of Life* (Royal Society of Chemistry, UK).

Fredrickson, G. H. (2006). The Equilibrium Theory of Inhomogeneous Polymers. Oxford: Clarendon Press.

Freed, K. F. (1972). Functional integrals and polymer statistics. *Adv. Chem. Phys.* **22**, 1–128.

Freed, K. F. (1987). *Renormalization Group Theory of Macromolecules*. New York: John Wiley & Sons.

Freed, K. F. and Edwards, S. F. (1974). Polymer viscosity in concentrated solutions. J. *Chem. Phys.*, **61**, 3626–3633.

Frolich, H. (1958). *Theory of Dielectrics*. Oxford: Clarendon Press.

Fujita, H. (1990). *Polymer Solutions*. Amsterdam: Elsevier.

Fuoss, R. M., Katchalsky, A. and Lifson, S. (1951). The potential of an infinite rod-like molecule and the distribution of the counterions. *Proc. Natl. Acad. Sic. USA*, **37**, 579–589.

Ghosh, K., Carri, G. A. and Muthukumar, M. (2001). Configurational properties of a single semiflexible polyelectrolyte. *J. Chem. Phys.*, **115**, 4367–4375.

Gitlin, L., Carbeck, J. D. and Whitesides, G. M. (2006). Why are proteins charged? Networks of charge-charge interactions in proteins measured by charge ladders and capillary electrophoresis. *Angew. Chem. Int. Ed.*, **45**, 3022–3060.

Gong, H., Hocky, G. and Freed, K. F. (2008). Influence of nonlinear electrostatics on transfer energies between liquid phases: Charge burial is far less expensive than Born model. *Proc. Natl. Acad. Sci. USA*, **105**, 11146–11151.

Griffiths, D. J. (1999). *Introduction to Electrodynamics*. Upper Saddle River: Prentice Hall.

Grosberg, A. Y. and Khokhlov, A. R. (1994). *Statistical Physics of Macromolecules*. New York: AIP Press.

Grosberg, A. Yu., Nguyen, T. T., and Shklovskii, B. I., (2002). Colloquium: The physics of charge inversion in chemical and biological systems. *Rev. Mod. Phys.*, 74, 329–345.

Gross, R. J. and Osterle, J. F. (1968). Membrane transport characteristics of ultrafine capillaries. *J. Chem. Phys.* 49, 228–234.

Gunton, J. D., San Miguel, M. and Sahni, P. S. (1983). The dynamics of first-order phase transitions. *Phase Transitions*, **8**, 267–482.

Gutin, A. M. and Shakhnovich, E. I. (1994). Effect of a net charge on the conformation of polyampholytes. *Phys. Rev. E*, **50**, R3322–R3325.

Halperin, A. (1987). Polymeric micelles: A star model. *Macromolecules*, **20**, 2943–2946.

Hamada, F., Kinugasa, S., Hayashi, H. and Nakajima, A. (1985). Small-angle x-ray scattering from semidilute polymer solutions. I. Polystyrene in toluene. *Macromolecules*, **18**, 2290–2294.

Han, C. C., and Akcasu, A. Z., (2011). *Scattering and Dynamics of Polymers*. Singapore: John Wiley & Sons.

Hänggi, P., Talkner, P. and Borkovec, M. (1990). Reaction-rate theory: Fifty years after Kramers. *Rev. Mod. Phys.*, **62**, 251–341.

Harmon, T. S., Holehouse, A. S., Rosen, M. K. and Pappu, R. V. (2017). Intrinsically disordered linkers determine the interplay between phase separation and gelation in multivalent proteins. *Elife*, **6**, e30294.

Hendrix, R. W. (1999). Evolution: The long evolutionary reach of viruses. *Curr. Bio.*, **9**, R914–R917.

Henry, D. C. (1931). Cataphoresis of suspended particles. Part I. The equation of cataphoresis. *Proc. R. Soc. London, Ser. A*, **133**, 106–129.

Hiemenz, P. C. and Rajagopalan, R. (1997). *Principles of Colloid and Surface Chemistry*. New York: Marcel Dekker, Inc.

Hiemenz, P. C. and Lodge, T. P. (2007). *Polymer Chemistry*. Boca Raton: CRC Press.

Higgins, J. S. and Benoit, H. (1994). *Polymers and Neutron Scattering*. Oxford: Clarendon Press.

Higgs, P. G. and Joanny, J-F. (1991). Theory of polyampholyte solutions. *J. Chem. Phys.*, **94**, 1543–1554.

Hill, T. L. (1986). *An Introduction to Statistical Thermodynamics*. New York: Dover.

Hirotsu, S., Hirokawa, Y. and Tanaka, T. (1987). Volume-phase transitions of ionized n-isopropylacrylamide gels, *J. Chem. Phys.*, 87, 1392–1395.

Hoagland, D. (2003). Polyelectrolytes. *Encycl. Polym. Sci. Technol.*, **7**, 439–504.

Hoagland, D., Arvanitidou, E. and Welch, C. (1999). Capillary electrophoresis measurements of the free solution mobility for several model polyelectrolyte systems. *Macromolecules*, **32**, 6180–6190.

Hofmann, H., Soranno, A., Borgia, A., Gast, K., Nettels, D. and Schuler, B. (2012). Polymer scaling laws of unfolded and intrinsically disordered proteins quantified with single-molecule spectroscopy. *Proc. Natl. Acad. Sci. USA*, **109**, 16155–16160.

Holm, C., Joanny, J. F., Netz, R. R., Reineker, P., Seidel, C., Vilgis, T. A. and Winkler, R. G. (2004). Polyelectrolyte theory. *Adv. Polym. Sci.* **166**, 67–111.

Holmes, D. F. and Kadler, K. E. (2006). The 10+4 microfibril structure of thin cartilage fibrils. *Proc. Natl. Acad. Sci. (USA)*, **103**, 17249–17254.

Horkay, F., Tasaki, I., and Basser, P. (2001). Effect of monovalent-divalent cation exchange on the swelling of polyacrylate hydrogels in physiological salt solutions, *Biomacromolecules*, **2**, 195–199.

Horkay, F., Nishi, K., and Shibayama, M. (2017). Decisive test of the ideal behavior of tetra-PEG gels, *J. Chem. Phys.*, 146, 164905.

Hu, Y., Zandi, R., Anavitarte, A., Knobler, C. M. and Gelbart, W. M. (2008). Packaging of a polymer by a viral capsid: The interplay between polymer length and capsid size, *Biophys. J.*, **94**, 1428–1436.

Hua, J., Mitra, M. K., and Muthukumar, M. (2012). Theory of volume transition in polyelectrolyte gels with charge regularization, *J. Chem. Phys.*, 136, 134901.

Huber, K. (1993). Calcium-induced shrinking of polyacrylate chains in aqueous solution. *J. Phys. Chem.*, **97**, 9825–9830.

Hückel, E. (1924). Die kataphoresese der kugel. *Phys. Z.*, **25**, 204–210.

Huggins, M. L. (1942a). Thermodynamic properties of solutions of long-chain compounds. *Ann. N. Y. Acad. Sci.*, **43**, 1–32.

Huggins, M. L. (1942b). Some properties of solutions of long-chain compounds. *J. Phys. Chem.* **46**, 151–158.

Huggins, M. L. (1942c). Theory of solutions of high polymers. *J. Am. Chem. Soc.,* **64**, 1712–1719.

Huihui, J., Firman, T. and Ghosh, K. (2018). Modulating charge patterning and ionic strength as a strategy to induce conformational changes in intrinsically disordered proteins. *J. Chem. Phys.* **149**, 085101.

Huihui, J. and Ghosh, K. (2020). An analytical theory to describe sequence-specific inter-residue distance profiles for polyampholytes and intrinsically disordered proteins. *J. Chem. Phys.* **152**, 161102.

Hulmes, D. J. S., Wess, T. J., Prockop, D. J. and Fratzl, P. (1995). Radial packing, order, and disorder in collagen fibril. *Biophys. J.*, **68**, 1661–1670.

Ikeda, Y., Beer, M., Schmidt, M. and Huber, K. (1998). Ca2+ and Cu2+ induced conformational changes of sodium polymethacrylate in dilute aqueous solution, *Macromolecules*, **31**, 728–733.

Innes-Gold, S. N., Jacobson, D. R., Pincus, P. A., Stevens, M. J. and Saleh, O. A. (2021). Flexible, charged biopolymers in monovalent and mixed-valence salt: Regimes of anomalous electrostatic stiffening and of salt insensitivity. *Phys. Rev, E*, **104**, 014504.

Isrealachvilli, J. N. (2011). *Intermolecular and Surface Forces* (Academic Press, London).

Izumrudov, V. A., Kargov, S. I., Zhiryakova, M. V., Zezin, A. B. and Kabanov, V. A. (1995). Competitive reactions in solutions of DNA and water-soluble interpolyelectrolyte complexes. *Biopolymers*, **35**, 523–531.

Jackson, J. D. (1999). *Classical Electrodynamics*. New York: John Wiley & Sons.

Jeon, B. J. and Muthukumar, M. (2014). Polymer capture by α-hemolysin pore upon salt concentration gradient. *J. Chem. Phys.* **140**, 015101.

Jia, D. and Muthukumar, M. (2018). Topologically frustrated dynamics of crowded charged macromolecules in charged gels. *Nat. Commun.*, **9**, 2248.

Jia, D. and Muthukumar, M. (2019). Effect of salt on the ordinary-extraordinary transition in solutions of charged macromolecules. *J. Am. Chem. Soc.*, **141**, 5886–5896.

Jia, D. and Muthukumar, M., (2020). Interplay between microscopic and macroscopic properties of charged hydrogels. *Macromolecules*, 53, 90–101.

Jia, D. and Muthukumar, M. (2021a). Electrostatically driven topological freezing of polymer diffusion at intermediate confinements. *Phys. Rev. Lett.*, **126**, 057802.

Jia, D. and Muthukumar, M. (2021b). Theory of charged gels: Swelling, elasticity, and dynamics. *Gels*, **7**, 49.

Joanny, J. F. and Leibler, L. (1990). Weakly charged polyelectrolytes in a poor solvent. *J. Phys. France*, **51**, 545–557.

Johner, C., Kramer, H., Batzill, S., Graf, C., Hagenbüchle, M., Martin, C. and Weber, R. (1994). Static light scattering and electric birefringence experiments on saltfree solutions of poly(styrenesulfonate), *J. Phys. II France*, **4**, 1571–1584.

Joosten, J. G. H., McCarthy, J. L., and Pusey, P. N., (1991). Dynamic and static light scattering by aqueous polyacrylamide gels. *Macromolecules*, **24**, 6690–6699. 1991, 24, 6690.

Kaji, K., Urakawa, H., Kanaya, T. and Kitamaru, R. (1988). Phase diagram of polyelectrolyte solutions, *J. Phys. France*, **49**, 993–1000.

Kanai, S. and Muthukumar, M. (2007). Phase separation kinetics of polyelectrolyte solutions. *J. Chem. Phys.* **127**, 244908.

Kasianowicz, J. J., Brandin, E., Branton, D. and Deamer, D. W. (1996). Characterization of individual polynucleotide molecules using a membrane channel. *Proc. Natl. Acad. Sci. U. S. A.* **93**, 13770–13773 (1996).

Katchalsky, A., Künzle, O. and Kuhn, W. (1950). Behavior of polyvalent polymeric ions in solution. *J. Polym. Sci.* **5**, 283–300.

Katchalsky, A., Lifson, S., and Eisenberg, H., (1951). Equation of swelling for polyelectrolyte gels, *J. Polymer Sci.*, 7, 571–574.

Katchalsky, A. and Miller, R. (1954). Polyampholytes. *J. Polym. Sci.*, **13**, 57–68.

Katchalsky, A., Shavit, N. and Eisenberg, H. (1954). Dissociation of weak polymeric acids and bases. *J. Polym. Sci.* **13**, 69–84.

Kato, T., Miyaso, K., Noda, I., Fujimoto, T. and Nagasawa, M. (1970). Thermodynamic and hydrodynamic properties of linear polymer solutions. i. Light scattering of monodisperse poly(?-methylstyrene), *Macromolecules*, **3**, 777–786.

Kato, M., Han, T. W., Xie, S., Shi, K., Du, X., Wu, L. C., Mirzaei, H., Goldsmith E. J., Longgood, J, Pei, J., Grishin, N. V., Frantz, D. E., Schneider, J. W., Chen, S., Li, L., Sawaya, M. R., Eisenberg, D., Tycko, R and McKnight, S. L. (2012). Cell-free formation of RNA granules: Low complexity sequence domains form dynamic fibers within hydrogels. *Cell*, **149**, 753–767.

Khokhlov, A. R. (1980). On the collapse of weakly charged polyelectrolytes. *J. Phys. A*, **13**, 979–987.

Khokhlov, A. R. and Kramarenko, E. Y. (1994). Polyelectrolyte/ionomer behavior in polymer gel collapse. *Macromol. Theory Simul.*, **3**, 45–59.

Kim, H., Chang, T., Yohanan, J. M., Wang, L. and Yu, H. (1986). Polymer diffusion in linear matrices: Polystyrene in toluene. *Macromolecules*, **19**, 2737–2744.

Kirkwood, J. and Riseman, J. (1948). The intrinsic viscosities and diffusion constants of flexible macromolecules in solution. *J. Chem. Phys.*, **16**, 565–573.

Kirkwood, J. G., and Goldberg, J. (1950). Light scattering arising from composition fluctuations in multi-component systems. *J. Chem. Phys.*, **18**, 54–57.

Kirste, R. G., Kruse, W. A. and K. Ibel. (1975). Determination of the conformation of polymers in the amorphous solid state and in concentrated solution by neutron diffraction. *Polymer*, **16**, 120–124.

Kittel, C. (1996). *Introduction to Solid State Physics*. New York: John Wiley & Sons, Inc.

Kizilay, E., Kayitmazer, A. B. and Dubin, P. L. (2011). Complexation and coacervation of polyelectrolytes with oppositely charged colloids, *Adv. Colloid Interface Sci.*, **167**, 24–37.

Klein J. (1996). Shear, friction, and lubrication forces between polymer-bearing surfaces. *Annu. Rev. Mater. Sci.*, **26**, 581–612.

Kleman, M. and Lavrentovich, O. D. (2003). *Soft Matter Physics*. New York: Springer-Verlag.

Klooster, N. T. M., Van der Touw, F. and Mandel, M. (1984). Solvent effects in polyelectrolyte solutions. 2. Osmotic, elastic light scattering, and conductometric measurements on (partially) neutralized poly(acrylic acid) in methanol. *Macromolecules*, **17**, 2078–2086.

Knipe, D. M. and Howley, P. M. (2001). *Fundamental Virology*, 4th edn. Philadelphia: Lippincott Williams & Wilkins.

Kohn, J. E., Millett, I. S., Jacob, J., Zagrovic, B., Dillon, T. M., Cingel, N., Dothager, R. S., Seifert, S., Thiyagarajan, P., Sosnick, T. R., Hasan, M. Z., Pande, V. S., Ruczinski, I., Doniach, S. and Plaxco, K. W. (2004). Random-coil behavior and the dimensions of chemically unfolded proteins, *Proc. Natl. Acad. Sci. USA*, **101**, 12491–12496.

Kozak, D., Kristan, J. Dolar, D. (1971). Osmotic coefficient of polyelectrolyte solutions. *Z. Phys. Chem. Neue Folge*, **76**, 85–92.

Kremer, F. and Schönhals, A. (2003). *Broadband Dielectric Spectroscopy*. Berlin: Springer.

Kudaibergenov, S., Jaeger, W. and Laschewsky, A. (2006). Polymeric betaines: Synthesis, characterization, and application. *Adv. Polym. Sci.*, **201**, 157–224.

Kudlay, A., Ermoshkin, A. V. and de la Cruz, M. O. (2004). Complexation of Oppositely Charged Polyelectrolytes: Effect of Ion Pair Formation, *Macromolecules*, **37**, 9231–9241.

Kuhn, W., Hargitay, B., Katchalsky, A. and Eisenberg, H. (1950). Reversible dilation and contraction by changing the state of ionization of high-polymer acid networks. *Nature* **165**, 514–516.

Kumar, R. and Muthukumar, M. (2007). Microphase separation in polyelectrolyte diblock copolymer melt: Weak segregation limit. *J. Chem. Phys.* **126**, 214902.

Kumar, R. and Fredrickson, G. H. (2009). Theory of polyzwitterion conformations. *J. Chem. Phys.*, **131**, 104901.

Kundagrami, A. and Muthukumar, M. (2008). Theory of competitive counterion adsorption on flexible polyelectrolytes: Divalent salts. *J. Chem. Phys.*, **128**, 244901.

Kundagrami, A. and Muthukumar, M. (2010). Effective charge and coil-globule transition of a polyelectrolyte chain. *Macromolecules*, **43**, 2574–2581.

Lagueci, A., Ulrich, S., Labille, J., Fatin-Rouge, N., Stoll, S. and Buffle, J. (2006). Size and pH effect on electrical and conformational behavior of poly(acrylic acid): Simulation and experiment. *Eur. Polym. J.*, **42**, 1135–1144.

Lamm, G. and Pack, G. R. (1997). Calculation of dielectric constants near polyelectrolytes in solution, *J. Phys. Chem. B*, **101**, 959–965.

Landau, L. D. and Lifshitz, E. M. (1980). *Statistical Physics*. Oxford: Pergamon Press.

Landau, L. D. and Lifshitz, E. M., (1959). *Fluid Mechanics*, Oxford: Pergamon Press.

Landau, L. D. and Lifshitz, E. M., (1986). *Theory of Elasticity*, Oxford: Pergamon Press.

Laschewsky, A. (2014). Structures and synthesis of zwitterionic polymers. *Polymers*, **6**, 1544–1601.

Lee, C.-L. and Muthukumar, M. (2009). Phase behavior of polyelectrolyte solutions with salt. *J. Chem. Phys.* **130**, 024904.

Leibler, L. (1980). Theory of microphase separation in block copolymers. 1980. *Macromolecules*, **13**, 1602–1617.

Li, Y. and Tanaka, T. (1990). Kinetics of swelling and shrinking of gels, *J. Chem. Phys.*, **92**, 1365–1371.

Liaw, D-J. and Huang, C-C. (1997). Dilute solution properties of poly(3-dimethyl acryloyloxyethyl ammonium propiolactone). *Polymer*, **38**, 6355–6362.

Lin, S.-C., Lee, W. I. and Schurr, J. M. (1978). Brownian motion of highly charged poly(L-lysine). Effects of salt and polyion concentration. *Biopolymers* **17**, 1041–1064.

Lin, Y-H., Brady, J. P., Chan, H. S. and Ghosh, K. (2020). A unified analytical theory of heteropolymers for sequence-specific phase behaviors of polyelectrolytes and polyampholytes. *J. Chem. Phys.* **152**, 045102.

Liu, S. and Muthukumar, M. (2002). Langevin dynamics simulation of counterion distribution around isolated flexible polyelectrolyte chains. *J. Chem. Phys.*, **116**, 9975–9982.

Liu, S., Ghosh, K. and Muthukumar, M. (2003). Polyelectrolyte solutions with added salt: A simulation study. *J. Chem. Phys.*, **119**, 1813–1823.

Lodge, T. P. and Rotstein, N. A. (1991). Tracer diffusion of linear and star polymers in entangled solutions and gels. *J. Non-Cryst. Solids*, **131–133**, 671–675.

Loh, P., Deen, R., Vollmer, D., Fischer, K., Schmidt, M., Kundagrami, A. and Muthukumar, M. (2008). Collapse of linear polyelectrolyte chains in a poor solvent: When does a collapsing polyelectrolyte collect its counterions? *Macromolecules*, **41**, 9352–9358.

Lopez, C. G. (2019a). Entanglement properties of polyelectrolytes in salt-free and excess-salt solutions. *ACS Macro Lett.*, **8**, 979–983.

Lopez, C. G. (2019b). Scaling and entanglement properties of neutral and sulfonated polystyrene. *Macromolecules*, **52**, 9409–9415.

Lowe, A. B. and McCormick, C. L. (2002). Synthesis and solution properties of zwitterionic polymers. *Chem. Rev.*, **102**, 4177–4189.

Luo, S., Jiang, X., Zou, L., Wang, F., Yang, J., Chen, Y. and Zhao, J. (2013). Resolving the difference in electric potential within charged macromolecule. *Macromolecules*, **46**, 3132–3136.

Mahalik, J. P., Hildebrandt, B. and Muthukumar, M. (2013). Langevin dynamics simulation of DNA ejection from a phage. *J. Biol. Phys.*, **39**, 229–245.

Maier, W. and Saupe, A. Z. (1958). Eine einfache molekulare Theorie des nematischen kristallinflüssigen Zustandes. *Z. Natureforsch.*, **A13**, 564–566.

Malmberg, C. G. and Maryott, A. A. (1956). Dielectric constant of water from 0° to 100°. *J. Res. Nat. Bur. Stand.* **56**, 1–8.

Mandel, M. (1988). Polyelectrolytes. *Encyclopedia of Polymer Science and Technology*, 2nd Ed., **11**, 739–829.

Mandel, M. (2000). The dielectric increments of aqueous polyelectrolyte solutions: A scaling approach. *Biophys. Chem.*, **85**, 125–139.

Mandel, M., Leyte, J. C. and Stadhouder, M. G. (1967). The conformational transition of poly(methacrylic acid) in solution. *J. Phys. Chem.* **71**, 603–612.

Manning, G. S. (1969). Limiting laws and counterion condensation in polyelectrolyte solutions. I: Colligative properties. II: Self-diffusion of the small ions. *J. Chem. Phys.* **51**, 924–938.

Manning, G. S. (1975). Limiting law for the conductance of the rod model of a salt-free polyelectrolyte solution. *J. Phys. Chem.*, **79**, 262–265.

Manning, G. S. (1978). The molecular theory of polyelectrolyte solutions with application to the electrostatic properties of polynucleotides. *Quarter. Rev. Biophys.* bf 11, 179–246.

Manning, G. S. (1981). Limiting laws and counterion condensation in polyelectrolyte solutions. 7. Electrophoretic mobility and conductance. *J. Phys. Chem.*, bf 85, 1506–1515.

Mao, A. H., Crick, S. L., Vitalis, A., Chicoine, C. L. and Pappu, R. V. (2010). Net charge per residue modulates conformational ensembles of intrinsically disordered proteins. *Proc. Natl. Acad. Sci. USA*, **107**, 8183–8188.

Marcus, Y. (2015). *Ions in Solution and their Solvation*. New York: Wiley.

Mark, J. E. and Erman, B., (1988). *Rubberlike Elasticity A Molecular Primer*, New York: John Wiley & Sons

Marko, J. F. and Rabin, Y. (1992). Microphase separation of charged diblock copolymers: Melts and solutions. *Macromolecules*, **25**, 1503–1509.

Matsumoto, T., Nishioka, N. and Fujita, H. (1972). Excluded-volume effects in dilute polymer solutions. iv. Polyisobutylene, *J. Polym. Sci. A-2*, **10**, 23–42.

Maxwell, J. C. (1875). On the dynamical evidence of the molecular constitution of bodies. *Nature*, **11**, 357–359.

McQuarrie, D. A. (1976). *Statistical Mechanics*. New York: Harper & Row.

Mehler, A. and Eichele, G. (1984). Electrostatic effects in water-accessible regions of proteins. *Biochemistry* **23**, 3887–3891.

Meller, A. (2003). Dynamics of polynucleotide transport through nanometre-scalepores. *J. Phys. Condes. Matter*, **15**, R581–R607 (2003).

Miklavic, S. J. and Marčelja, S. (1988). Interaction of surface carrying grafted polyelectrolytes. *J. Phys. Chem.*, **92**, 6718–6722.

Milner, S. T., Witten, T. A. and Cates, M. E. (1988). Theory of the grafted polymer brush. *Macromolecules*, **21**, 2610–2619.

Misra, S., Varanasi, S. and Varanasi, P. P. (1989). A polyelectrolyte brush theory. *Macromolecules*, **22**, 4173–4179.

Miyaki, Y., Einaga, Y., Hirosye, T. and Fujita, H. (1977). Solution properties of poly(d-β-hydroxybutyrate). 2. Light scattering and viscosity in trifluoroethanol and behavior of highly expanded polymer coils, *Macromolecules*, **10**, 1356–1364.

Miyaki, Y., Einaga, Y. Fujita, H. (1978). Excluded-volume effects in dilute polymer solutions. 7. Very high molecular weight polystyrene in benzene and cyclohexane, *Macromolecules*, **11**, 1180–1186.

Molliex, A., Temirov, J., Lee, J., Coughlin, M., Kanagaraj, A. P., Kim, H. J., Mittag, T. and Taylor, J. P. (2015). Phase separation by low complexity domains promotes stress granule assembly and drives pathological fibrillization. *Cell*, **163**, 123–133.

Montiel-Garcia, D., Santoyo-Rivera, N., Ho, P., Carrillo-Tripp, M., Brooks III, C. L., Johnson, J. E. Reddy, V. S. (2021). VIPERdb v3. 0: a structure-based data analytics platform for viral capsids. *Nucleic Acid Research*, **49**, D809–D816.

Morozova, S., Hu, G., Emrick, T. and Muthukumar, M. (2016). Influence of dipole orientation on solution properties of polyzwitterions. *ACS Macro Lett.*, **5**, 118–122.

Morozova, S. and Muthukumar, M. (2017). Elasticity of swelling equilibrium of ultrasoft polyelectrolyte gels: Comparisons of theory and experiments, *Macromolecules*, **50**, 2456–2466.

Morozova, S. Muthukumar, M. (2018). Electrostatic effects in collagen fibril formation. *J. Chem. Phys.*, **149**, 163333.

Müller-Späth, S., Soranno, A., Hirschfeld, V., Hofmann, H., Rüegger, S., Reymond, L., Nettels, D. and Schuler, B. (2010). Charge interactions can dominate the dimensions of intrinsically disordered proteins. *Proc. Natl. Acad. Sci. USA*, **107**, 14609–14614.

Muthukumar, M. (1981). Viscosity of polymer solution. *J. Phys. A: Math. Gen.*, **14**, 2129–2148.

Muthukumar, M. (1984). Collapse transition of a stiff chain. *J. Chem. Phys.* **81**, 6172–6276.

Muthukumar, M. (1987). Adsorption of a polyelectrolyte chain to a charged surface. *J. Chem. Phys.*, **86**, 7230–7235.

Muthukumar, M., (1989). Polyelectrolyte gels: Replica theory, *Springer Proc. Phys.*, **42**, 28–34.

Muthukumar, M. (1991). Entropic barrier model for polymer diffusion in concentrated polymer solutions and random media. *J. Non-Cryst Solids*, **131–133**, 654–666.

Muthukumar, M. (1996a). Double screening in polyelectrolyte solutions: Limiting laws and crossover formulas. *J. Chem. Phys.*, **105**, 5183–5199.

Muthukumar, M. (1996b). Localized structures of polymers with long-range interactions. *J. Chem. Phys.*, **104**, 691–700.

Muthukumar, M. (1997). Dynamics of polyelectrolyte solutions. *J. Chem. Phys.*, **107**, 2619–2635.

Muthukumar, M. (1999). Polymer translocation through a hole. *J. Chem. Phys.*, **111**, 10371–10374.

Muthukumar, M. (2001). Theory of viscoelastic properties of polyelectrolyte solutions. *Polymer*, **42**, 5921–5923.

Muthukumar, M. (2002). Phase diagram of polyelectrolyte solutions: Weak polymer effect. *Macromolecules*, **35**, 9142–9145.

Muthukumar, M. (2004). Theory of counterion condensation flexible polyelectrolytes: Adsorption mechanism. *J. Chem. Phys.* **120**, 9343–9350.

Muthukumar, M. (2005). Polyelectrolyte dynamics. *Adv. Chem. Phys.*, **131**, 1–60.

Muthukumar, M. (2010). Theory of capture rate in polymer translocation, *J. Chem. Phys.* **132**, 195101.

Muthukumar, M. (2011).*Polymer Translocation*. USA: CRC Press, Boca Raton.

Muthukumar, M. (2012a). Counterion adsorption theory of dilute polyelectrolyte solution: Apparent molecular weight, second virial coefficient, and intermolecular structure factor. *J. Chem. Phys.* **137**, 034902.

Muthukumar, M. (2012b). Polymers under confinement. *Adv. Chem. Phys.* **149**, 129–196.

Muthukumar, M. (2014). Communication: Charge, diffusion, and mobility of proteins through nanopores. *J. Chem. Phys.*, **141**, 081104.

Muthukumar, M. (2016a). Electrostatic correlations in polyelectrolyte solutions, *Polym. Sci. - A*, **58**, 852–863.

Muthukumar, M. (2016b). Ordinary-extraordinary transition in dynamics of solutions of charged macromolecules, *Proc. Natl. Acad. Sci. (USA)*, **113**, 12627–12632.

Muthukumar, M. (2017). A Perspective on polyelectrolyte solutions. *Macromolecules*, **50**, 9528–9560.

Muthukumar, M. (2019). Collective dynamics of semidilute polyelectrolyte solutions with salt. *J. Polym. Sci. Part B: Polym. Phys.*, **57**, 1263–1269.

Muthukumar, M. (2021). Theory of ionic conductivity with morphological control in polymers. *ACS Macro Lett.*, **10**, 958–964.

Muthukumar, M. and Baumgärtner, A. (1989a). Effects of entropic barriers on polymer dynamics. *Macromolecules*, **22**, 1937–1941.

Muthukumar, M. and Baumgärtner, A. (1989b). Diffusion of a polymer chain in random media. *Macromolecules*, **22**, 1941–1946.

Muthukumar, M. and Edwards, S. F. (1982a). Extrapolation formulas for polymer solution properties, *J. Chem. Phys.*, **76**, 2720–2730.

Muthukumar, M. and Edwards, S. F. (1982b). Screening concepts in polymer solution dynamics. *Polymer*, **23**, 345–348.

Muthukumar, M., Hua, J. and Kundagrami, A. (2010). Charge regularization in phase separating polyelectrolyte solutions. *J. Chem. Phys.*, **132**, 08490.

Muthukumar, M. and Nickel, B. G. (1984). Perturbation theory for a polymer chain with excluded volume interaction. *J. Chem. Phys.* **80**, 5839–5850.

Muthukumar, M. and Nickel, B. G. (1987). Expansion of a polymer chain with excluded volume interaction. *J. Chem. Phys.* **86**, 460–476.

Muthukumar, M., Plesa, C. and Dekker, C. (2015). Single-molecule sensing with nanopores. *Physics Today* **68**, 32.

Naji, A., Kanduc, M., Forsman, J., Podgornik, R. Coulomb fluids – Weak coupling, strong coupling, in between and beyond. *J. Chem. Phys.* 2013, **139**, 150901.

Nelson, D. L. and Cox, M. M. (2005). *Lehninger Principles of Biochemistry*. New York: W. H. Freeman and Company.

Nguyen, T. T., Rouzina, I. and Shklovskii, B. I. (2000). Reentrant condensation of DNA induced by multivalent counterions. *J. Chem. Phys.*, **112**, 2562–2568.

Nierlich, M., Williams, C. E., Boue, F., Cotton, J. P., Daoud, M., Farnoux, B., Jannink, G., Picot, C., Moan, M., Wolff, C., Rinaudo, M. and de Gennes, P. G. (1979). Small angle neutron scattering by semi-dilute solutions of polyelectrolyte, *J. Physique*, **40**, 701–704.

Nishida, K., Kaji, K. and Kanaya, T. (2001). High concentration crossovers of polyelectrolyte solutions, *J. Chem. Phys.*, **114**, 8671–8677.

Nishida, K., Kaji, K., Kanaya, T. and Shibano, T. (2002). Added salt effect on the intermolecular correlation in flexible polyelectrolyte solutions: Small-angle scattering study. *Macromolecules*, **35**, 4084–4089.

Niu, A., Liaw, D-J., Sang, H-C. and Wu, C. (2000). Light scattering study of a zwitterionic polycarboxybetaine in aqueous solution. *Macromolecules*, **33**, 3492–3494.

Noda, I., Kato, N., Kitano, T. and Nagasawa, M. (1981). Thermodynamic properties of moderately concentrated solutions of linear polymers. *Macromolecules*, **14**, 668–676.

Norisuye, T., Miyata, Q. T., and Shibayama, M., (2004). Dynamic inhomogeneities in polymer gels investigated by dynamic light scattering, *Macromolecules*, **37**, 2944–2953.

Nott, T. J., Petsalaki, E., Farber, P., Jervis, D., Fussner, E., Plochowietz, A., Craggs, T. D., Bazett-Jones, D. P., Pawson, T., Forman-Kay, J. D. and Baldwin, A. (2015). Phase transition of a disordered nuage protein generates environmentally responsive membraneless organelles. *Molecular Cell*, **57**, 936–947.

Odijk, T. (1977). Polyelectrolytes near the rod limit. *J. Polym. Sci., Polym. Phys.*, **15**, 477–483.

Ohmine, I. and Tanaka, T. (1982). Salt effects on the phase transition of ionic gels, *J. Chem. Phys.*, **77**, 5725–5729.

Olvera de la Cruz, M., Belloni, M. L., Delsanti, M., Dalbiez, J. P., Spalla, O. and Drifford, M. (1995). Precipitation of highly charged polyelectrolyte solutions in the presence of multivalent salts. *J. Chem. Phys.*, **103**, 5781.

Onsager, L. (1949). The effects of shape on the interaction of colloidal particles. *Ann. N. Y. Acad. Sci.*, **51**, 627–659.

Onuki, A., (2002). *Phase Transition Dynamics*, Cambridge: Cambridge University Press.

Oosawa, F. (1957). A simple theory of thermodynamic properties of polyelectrolyte solutions. *J. Polym, Sci.* **23**, 421–430.

Orkoulas, G., Kumar, S. K. and Panagiotopoulos, A. Z. (2003). Monte-Carlo study of coulombic criticality in polyelectrolytes. *Phys. Rev. Lett.* **90**, 048303.

Ou, Z. and Muthukumar, M. (2005). Langevin dynamics of semiflexible polyelectrolytes: Rod-toroid-globule-coil structures and counterion distribution. *J. Chem. Phys.*, **123**, 074905.

Ou, Z. and Muthukumar, M. (2006). Entropy and enthalpy of polyelectrolyte complexation: Langevin dynamics simulations. *J. Chem. Phys.*, **124**, 154902.

Pak, C. W., Kosno, M., Holehouse, A. S., Padrick, S. B., Mittal, A., Ali, R., Yunus, A. A., Liu, D. R., Pappu, R. V. and Rosen, M. K. (2016). Sequence determinants of intracellular phase separation by complex coacervation of a disordered protein. *Mol. Cell*, **63**, 72–85.

Palmenberg, A. C. and Sgro, J-Y. (1997). Topological organization of picornaviral genomes: Statistical prediction of RNA structural signals. *Semin. Virol.*, **8**, 231–241.

Peng, B. and Muthukumar, M. (2015). Modeling competitive substitution in a polyelectrolyte complex. *J. Chem. Phys.*, **143**, 243133.

Pincus, P. (1991). Colloid stabilization with grafted polyelectrolytes. *Macromolecules*, **24**, 2912–2919.

Prabhu, V. M., Muthukumar, M., Wignall, G. D. and Melnichenko, Y. B. (2001). Dimensions of polyelectrolyte chains and concentration fluctuations in semidilute solutions of sodium-poly(styrene sulfonate) as measured by small-angle neutron scattering, *Polymer*, **42**, 8935–8946.

Prabhu, V. M., Muthukumar, M., Wignall, G. D. and Melnichenko, Y. B. (2003). Polyelectrolyte chain dimensions and concentration fluctuations near phase boundaries, *J. Chem. Phys.*, **119**, 4085–4098.

Prabhu, V. M., Amis, E. J., Bossov, D. P. and Rossov, N. (2004). Counterion associative behavior with flexible polyelectrolytes. *J. Chem. Phys.*, **121**, 4424–4429.

Probstein, R. F. (1989). *Physicochemical Hydrodynamics*, Boston: Butterworths.

Pum, D. and Sleytr, U. B. (2014). Reassembly of S-layer proteins. *Nanotechnology*, **25**, 312001.

Rabin, Y. and Panyukov, S., (1997). Scattering profiles of charged gels: Frozen inhomogeneities, thermal fluctuations, and microphone separation, *Macromolecules*, 30, 301–312.

Raman, B., Chatani, E., Kihara, M., Ban, T., Sakai, M., Hasegawa, K., Naiki, H., Rao, C. M. and Goto, Y. (2005). Critical balance of electrostatic and hydrophobic Interactions Is required for β2-microglobulin amyloid fibril growth and stability. *Biochemistry*, **44**, 1288–1299.

Ray, S., Singh, N., Kumar, R., Patel, K., Pandey, S., Datta, D., Mahato, J., Panigrahi, R., Navalkar, A., Mehra, S., Gadhe, L., Chatterjee, D., Sawner, A. S., Maiti, S., Bhatia, S., Gerez, J. A., Chowdhury, A., Kumar, A., Padinhateeri, R., Riek, R., Krishnamorthy, G. and Maji, S. K. 2020. α-Synuclein aggregation nucleates through liquid-liquid phase separation. (2020). *Nat. Chem.*, **12** , 705–716.

Renkin, E. M. (1954). Filtration, diffusion, and molecular sieving through porous cellulose membranes. *J. Gen. Physiol.*, **38**, 225–243.

Rice, S. A. and Nagasawa, M. (1961). *Polyelectrolyte Solutions*, New York: Academic Press.

Richards, R. W., Maconnachie, A. and Allen, G. (1978). Temperature dependence of chain dimensions. *Polymer*, **19**, 266–270.

Risken, H. (1989). *The Fokker-Planck Equation,*.

Robertson, H. S. (1993). *Statistical Thermophysics*, Englewood Cliffs: Prentice Hall.

Robinson, R. A. and Stokes, R. H., (1959). *Electrolyte Solutions*, New York: Dover.

Rotstein, N. A. and Lodge, T. P. (1992). Tracer diffusion of linear polystyrenes in poly(vinyl methyl ether) gels. *Macromolecules*, **25**, 1316–1325.

Rouse, P. E. (1953). A theory of linear viscoelastic properties of dilute solutions of coiling polymers. *J. Chem. Phys.* **21**, 1272–1280.

Rousseau, J., Drouin, G. and Slater, G. W. (1997). Entropic trapping of DNA during gel electrophoresis: Effect of field intensity and gel concentration. *Phys. Rev. Lett.*, **79**, 1945.

Routh, A., Domitrovic, T. and Johnson, J. E. (2012). Host RNAs, including transposons, are encapsidated by a eukaryotic single-stranded RNA virus. *Proc. Natl. Acad. Sic. (USA)*, **109**, 1907–1912.

Rowghanian, P. and Grosberg, A. Y. (2012). Propagation of tension along a polymer chain. *Phys. Rev. E*, **86**, 011803.

Rubinstein, M. and Colby, R. H. (2003). *Polymer Physics*, Oxford: Oxford University Press.

Sabbagh, I. and Delsanti, M. (2000). Solubility of highly charged anionic polyelectrolytes in presence of multivalent cations: Specific interaction effect. *Eur. Phys. J. E*, **1**, 75–86.

Saha, S., Fischer, K., Muthukumar, M. and Schmidt, M. (2013). Apparent molar mass of a polyelectrolyte in an organic solvent in the low ionic strength limit as revealed by light scattering. *Macromolecules*, **46**, 8296–8303.

Sakaue, T. (2007). Nonequilibrium dynamics of polymer translocation and straightening. *Phys. Rev. E*, **76**, 021803.

Salehi, A. and Larson, R. G. (2016). A molecular thermodynamic model of complexation in mixtures of oppositely charged polyelectrolytes with explicit account of charge association/dissociation. *Macromolecules*, **49**, 9706–9719.

Samanta, H. S., Chakraborty, D. and Thirumalai, D. (2018). Charge fluctuation effects on the shape of flexible polyampholytes with applications to intrinsically disordered proteins. *J. Chem. Phys.* **149**, 163323.

Sasaki, S. and Schipper, F. J. M., (2001). Coupled diffusion of segments and counterions in polyelectrolyte gels and solutions, *J. Chem. Phys.*, **115**, 4349–4354.

Sawle, L. and Ghosh, K. (2015). A theoretical method to compute sequence dependent configurational properties in charged polymers and proteins. *J. Chem. Phys.*, **143**, 085101.

Schiessel, H. (1999). Counterion condensation on flexible polyelectrolytes: Dependence on ionic strength and chain concentration. *Macromolecules*, **32**, 5673–5680.

Schmitz, K. S. (1993). *Macroions in Solution and Colloidal Suspensions*, New York: VCH Publishers.

Schosseler, F., Ilmain, F., and Candau, S. J., 1991. Structure and properties of partially neutralized poly(acrylic acid) gels, *Macromolecules*, **24**, 225–234.

Schuler, B., Soranno, A., Hofmann, H. and Nettels, D. (2016). Single-molecule FRET spectroscopy and the polymer physics of unfolded and intrinsically disordered proteins. *Annu. Rev. Biophys.*, **45**, 207–231.

Schuler, B., Borgia, A., Borgia, M. B., Heidarsson, P. O., Holmstrom, E. D., Nettels, D. and Sottini, A. (2020). Binding without folding - the biomolecular function of disordered polyelectrolyte complexes. *Curr. Opin. Struct. Biol.*, **60**, 66–76.

Schwarzenbach, G. (1936). Thermodynamik kinetik electrochemie eigenschaftslehre. *Z. Phys. Chem.* **A 176**, 133–153.

Sedlak, M. and Amis, E. J. (1992a). Dynamics of moderately concentrated salt-free polyelectrolyte solutions: Molecular weight dependence. *J. Chem. Phys.*, **96**, 817–825.

Sedlak, M. and Amis, E. J. (1992b). Concentration and molecular weight regime diagram of salt-free polyelectrolyte solutions as studied by light scattering. *J. Chem. Phys.*, **96**, 826–834.

Severin, M. (1993). Thermal maximum in the size of short polyelectrolyte chains. A Monte Carlo study. *J. Chem. Phys.*, **99**, 628–633.

Shew, C-Y. and Yethiraj, A. (1999). Monte Carlo simulations and self-consistent integral equation theory for polyelectrolyte solutions, *J. Chem. Phys.*, **110**, 5437–5443.

Shew, C-Y. and Yethiraj, A. (2000). Self-consistent integral equation theory for semiflexible chain polyelectrolyte solutions, *J. Chem. Phys.*, **113**, 8841–8847.

Shibayama, M. and Tanka, T. (1993). Volume phase transition and related phenomena of polymer gels, *Adv. Poly. Sci.*, **109**, 1–62.

Shibayama, M., Ikkai, F., Shiwa, Y., and Rabin, Y., (1997). Effect of degree of cross-linking on spatial inhomogeneity in charged gels. I. Theoretical predictions and light scattering study, *J. Chem. Phys.*, **107**, 5227–5235.

Shojaei, H. R. and Muthukumar, M. (2017). Adsorption and encapsulation of flexible polyelectrolytes in charged spherical vesicles. *J. Chem. Phys.*, **146**, 244901.

Shoulders, M. D. and Raines, R. T. (2009). Collagen structure and stability. *Ann. Rev. Biochemistry*, **78**, 929–958.

Shultz, A. R. and Flory, P. J. (1952). Phase equilibria in polymer-solvent systems. *J. Amer. Chem. Soc.* **74**, 4760–4767.

Siber, A. and Podgornik, R. (2008). Nonspecific interactions in spontaneous assembly of empty versus functional single-stranded RNA viruses, *Phys. Rev. E*, **78**, 05191.

Siber, A., Bozic, A. L. and Podgornik, R. (2012). Energies and pressures in viruses: Contribution of nonspecific electrostatic interactions, *Phys. Chem. Chem. Phys.*, **14**, 3746–3765.

Skolnick, J. and Fixman, M. (1977). Electrostatic persistence length of a wormlike polyelectrolyte. *Macromolecules*, **10**, 944–948.

Slagowski, E., Tsai, B. and McIntyre, D. (1976.) The dimensions of polystyrene near and below the theta temperature. *Macromolecules*, **9**, 687–688.

Slater, G. W. and Wu, S. Y. (1995). Reptation, entropic trapping, percolation, and rouse dynamics of polymers in "random" environments. *Phys. Rev. Lett.*, **75**, 164.

Spruijt, E., Westphal, A. H., Borst, J. W., Cohen Stuart, M. A. and van der Gucht, J. (2010). Binodal compositions of polyelectrolyte complexes. *Macromolecules*, **43**, 6476–6484.

Srivastava, D. and Muthukumar, M. (1996). Sequence dependence of conformations of polyampholytes. *Macromolecules*, **29**, 2324–2326.

Srivastava, S. and Tirrell, M. V. (2016). Polyelectrolyte complexation. *Adv. Chem. Phys.*, **161**, 499–543.

Stell, G., Wu, K. C., and Larsen, B. (1976). Critical point in a fluid of charged hard spheres. *Phys. Rev. Lett.* **37**, 1369–1372.

Stellwagen, E., Lu, Y. and Stellwagen, N. (2003). Unified description of electrophoresis and diffusion for DNA and other polyions. *Biochemistry*, **42**, 11745–11750.

Stevens, M. J., and Kremer, K. (1995). The nature of flexible linear polyelectrolytes in salt free solution: A molecular dynamics study. *J. Chem. Phys.*, **103**, 1669–1690.

Stockley, P. G., Twarock, R., Bakker, S. E., Barker, A. M., Borodavka, A., Dykeman, E., Ford, R. J., Pearson, A. M., Phillips, S. E. V., Ranson, N. A. and Tuma, R. (2013). Packaging signals in single-stranded RNA viruses: Nature's alternative to a purely electrostatic assembly mechanism. *J. Biol. Phys.*, **39**, 277–287.

Stockmayer, W. H. (1950). Light scattering in multi-component systems. *J. Chem. Phys.*, **18**, 58–61.

Störkle, D., Duschner, S., Heimann, N., Maskos, M. and Schmidt, M. (2007). Complex formation of DNA with oppositely charged polyelectrolytes of different chain topology: Cylindrical brushes and dendrimers. *Macromolecules* **40**, 7998–8006.

Strauss, U. P., Helfgott, C. and Pink, H. (1967). Interactions of polyelectrolytes with simple electrolytes. II. Donnan equilibria obtained with DNA in solutions of 1–1 electrolytes. *J. Phys. Chem.*, **71**, 2550–2556.

Stroobants, A., Lekkerkerker, H. N. W. and Odijk, Th. (1986). Effect of electrostatic interaction on the liquid crystal phase transition in solutions of rodlike polyelectrolytes. *Macromolecules*, **19**, 2232–2238.

Strzelecka, T. E. and Rill, R. L. (1990). Phase transitions of concentrated DNA solutions in low concentrations of 1:1 supporting electrolyte. *Biopolymers*, **30**, 57–71.

Sung, W. and Park, P. J. (1996). Polymer translocation through a pore in a membrane. *Phys. Rev. Lett.*, **77**, 783–788.

Tanaka, T, Hocker, L. O., and Benedek, G. B., (1973). Spectrum of light scattered from a viscoelastic gel, *J. Chem. Phys.*, 59, 5151–5159.

Tanaka, T. (1978). Collapse of gels and the critical point, *Phys. Rev. Lett.*, 40, 820–823.

Tanaka, T., Fillmore, D., Sun, S-T., Nishio, I., Swislow, G., and Shah, A., (1980). Phase transitions in ionic gels, *Phys. Rev. Lett.*, 45, 1636–1639.

Tanaka, F. (2011). *Polymer Physics*. Cambridge: Cambridge University Press.

Tanford, C. (1980). *The Hydrophobic Effect*. New York: Wiley-Interscience.

Temin, H. M. and Mizutani, S. (1970). Viral RNA-dependent DNA polymerase: RNA-dependent DNA polymerase in virions of Rous Sarcoma Virus. *Nature*, **226**, 1211–1213.

Tihova, M., Dryden, K. A., Le, T. L., Harvey, S. C., Johnson, J. E., Yeager, M. and Schneemann, A. (2004). Nodavirus coat protein imposes dodecahedral RNA structure independent of nucleotide sequence and length. *J. Virol.*, **78**, 2897–2905.

Ting, C. L., Wu, J. and Wang, Z-G. (2011). Thermodynamic basis for the genome to capsid charge relationship in viral encapsidation, *Proc. Natl. Acad. Sci. (USA)*, **108**, 16986–16991.

Tohyama, K. and Miller, W. G. (1981). Network structure in gels of rod-like polypeptides. *Nature*, **289**, 813–814.

Tokita, M. Tanaka, T. (1991). Friction coefficient of polymer network of gels. *J. Chem. Phys.*, **95**, 4613–4619

Tompa, P. (2010). *Structure and Function of Intrinsically Disordered Proteins*. Boca Raton: CRC Press.

Treloar, L. R. G. (1958). *The Physics of Rubber Elasticity*, Oxford: Clarendon Press.

Uversky, V. N. (2002). Natively unfolded proteins: A point where biology waits for physics. *Protein Sci.*, **11**, 739–756.

Uversky, V. N., Oldfield, C. J. and Dunker, A. K. (2008). Intrinsically disordered proteins in human diseases: Introducing the D2 concept. *Annu. Rev. Biophys.*, **37**, 215–246.

van der Lee, R., Buljan, M., Lang, B., Robert J. Weatheritt, R. J., Gary W. Daughdrill, G. W., Dunker, A. K., Fuxreiter, M., Gough, J., Gsponer, J., David T. Jones, D. T., Kim, P. M., Kriwacki, R. W., Oldfield, C. J., Pappu, R. V., Tompa, P., Uversky, V. N., Wright, P. E. and Babu, M. M. (2014). Classification of intrinsically disordered regions and proteins. *Chem. Rev.*, **114**, 6589–6631.

Vink, H. (1982). Electrolytic conductivity of polyelectrolyte solutions. *Makromol. Chem.*, **183**, 2273–2283.

Viovy, J.-L. (2000). Electrophoresis of DNA and other polyelectrolytes: Physical mechanisms. *Rev. Mod. Phys.*, **72**, 813.

Volk, N., Vollmer, D., Schmidt, M., Oppermann, W., and Huber, K. (2004). Polyelectrolyte theory, *Adv. Polym. Sci.*, **166**, 29–65.

von Goeler, F. and Muthukumar, M. (1994). Adsorption of polyelectrolytes onto curved surfaces. *J. Chem. Phys.*, **100**, 7796–7803.

Wallace, D. G. (1990). The relative contribution of electrostatic interactions to stabilization of collagen fibrils. *Biopolymers*, **29**, 1015–1026.

Wandrey, C. (1999). Concentration regimes in polyelectrolyte solutions. *Langmuir*, **15**, 4069–4075.

Wandrey, C., Hunkeler, D., Wendler, U. and Jaeger, W. (2000). Counterion activity of highly charged strong polyelectrolytes. *Macromolecules*, **33**, 7136–7143.

Wang, J. and Muthukumar, M. (2011). Encapsulation of a polyelectrolyte chain by an oppositely charged spherical surface. *J. Chem. Phys.*, **135**, 194901.

Wang, Q. and Schlenoff, J. B. (2014). The polyelectrolyte complex/coacervate continuum. *Macromolecules*, **47**, 3108–3116.

Wanunu, M. (2012). Nanopores: A journey towards DNA sequencing. *Phys. Life. Rev.* **9**, 125–158.

Wanunu, M., Morrison, W., Rabin, Y., Grosberg, A. Y. and Meller, A. (2010). Electrostatic focusing of unlabelled DNA into nanoscale pores using a salt gradient. *Nat. Nanotech.* **5**, 160–165.

Webb, T. J. (1926). The free energy of hydration of ions and the electrostriction of the solvent. *J. Amer. Chem. Soc.*, **48**, 2589–2603.

Wetzel, R. (2006). Kinetics and thermodynamics of amyloid fibril assembly. *Acc. Chem. Res.*, **39**, 671–679.

Wiegel, F. W. (1977). Adsorption of a macromolecule to a charged surface. *J. Phys. A: Math. Gen.*, **10**, 299–303.

Wiegel, F. W. (1986). *Introduction to Path-Integral Methods in Physics and Polymer Science.* Singapore: World Scientific.

Winkler, R. G., Gold, M. and Reineker, P. (1998). Collapse of polyelectrolyte macromolecules by counterion condensation and ion pair formation: A molecular dynamics simulation study. *Phys. Rev. Lett.*, **80**, 3731–3734.

Winkler, R. G. and Cherstvy, A. G. (2014). Strong and weak polyelectrolyte adsorption onto oppositely charged curved surfaces. *Adv. Polym. Sci.*, **255**, 1–56.

Wissenburg, P., Odijk, T., Cirket, P. and Mandel, M. (1995). Multimolecular aggregation of mononucleosomal DNA in concentrated isotropic solutions. *Macromolecules* **28**, 2315–2328.

Wittmer, J., Johner, A. and Joanny, J. F. (1995). Precipitation of polyelectrolytes in the presence of multivalent salts. *J. Phys. II France*, **5**, 635–654.

Yamakawa, H. (1971). *Modern Theory of Polymer Solutions.* New York: Harper & Row.

Yethiraj, A. (1998). Theory for chain conformations and static structure of dilute and semidilute polyelectrolyte solutions, *J. Chem. Phys.*, **108**, 1184–1192.

Yethiraj, A. (2009). Liquid state theory of polyelectrolyte solutions, *J. Phys. Chem. B*, **113**, 1539–1551.

Young, H. D. and Freedman, R. A. (2000). *University Physics*, 10th edn. San Francisco: Addison-Wesley.

Zandi, R., Reguera, D., Rudnick, J. and Gelbart, W. M. (2003). What drives the translocation of stiff chains? *Proc. Natl. Acad. Sci. USA*, **100**, 8649–8653.

Zhang, J. and Muthukumar, M. (2009). Simulations of nucleation and elongation of amyloid fibrils. *J. Chem. Phys.*, **130**, 035102.

Zhou, K., Li, J., Lu, Y., Zhang, G., Xie, Z. and Wu, C. (2009). Re-examination of dynamics of polyeletrolytes in salt-free dilute solutions by designing and using a novel neutral-charged-neutral reversible polymer. *Macromolecules*, **42**, 7146–7154.

Zimm, B. H. (1956). Dynamics of polymer molecules in dilute solution: Viscoelasticity, flow birefringence and dielectric loss. *J. Chem. Phys.* **24**, 269–278.

Zimm, B. H. (1988). Size fluctuations can explain anomalous mobility in field-inversion electrophoresis of DNA. *Phys. Rev. Lett.* **61**, 2965–2968.

Zimm, B. H. (1991). "Lakes-straits" model of field-inversion gel electrophoresis of DNA. *J. Chem. Phys.* **94**, 2187–2206.

Zimm, B. H. (1996). A gel as an array of channels. *Electrophoresis*, **17**, 996–1002.

Zimm, B. H. and Stockmayer, W. H. (1949). The dimensions of chain molecules containing branches and rings. *J. Chem. Phys.* **17**, 1301–1314.

Zlotnick, A., Cheng, N., Stahl, S. J., Conway, J. F., Steven, A. C. and Wingfield, P. T. (1997). Localization of the C terminus of the assembly domain of hepatitis B virus capsid protein: Implications for morphogenesis and organization of encapsidated RNA. *Proc. Natl. Acad. Sci. USA*, **94**, 9556–9561.

Index